LATEX Edition

Solutions of The Examples

In

Higher Algebra

By

Henry Sinclair Hall
Formerly Scholar of Christ's College, Cambridge
Master of The Military And Engineering Side, Clifton College

Samuel Ratcliffe Knight
Formerly Scholar Of Trinity College, Cambridge
Late Assistant-Master At Marlborough College

Edited By

Neeru Singh
Founder, Ancient Science Publishers
Co-Founder, Ancient Kriya Yoga Mission

Chandra Shekhar Kumar
Founder, Ancient Kriya Yoga Mission
Co-Founder, Ancient Science Publishers
Integrated M. Sc. in Physics, IIT Kanpur, India

Ancient Science Publishers

TEX is a trademark of the American Mathematical Society.

METAFONT is a trademark of Addison-Wesley.

Care has been taken in the preparation of this book, but makes no ex-pressed or implied warranty of any kind and assumes no responsibility for errors or omissions. No liability is assumed for incidental or con-sequential damages in connection with or arising out of the use of the information contained herein.

For comments, suggestions and or feedback, send mail to :
ancientsciencepublishers@gmail.com

First Printing, 2015 (small font)
Second Printing, 2019 (medium font)

ISBN-13: 978-1508763505
ISBN-10: 150876350X

Preface

This work forms a Key or Companion to the Higher Algebra, and contains full solutions of nearly all the Examples. In many cases more than one solution is given, while throughout the book frequent reference is made to the text and illustrative Examples in the Algebra. The work has been undertaken at the request of many teachers who have introduced the Algebra into their classes, and for such readers it is mainly intended; but it is hoped that, if judiciously used, the solutions may also be found serviceable by that large and increasing class of students who read Mathematics without the assistance of a teacher.

H. S. HALL & S. R. KNIGHT

LaTeX Edition, Second Printing

In this edition and printing, typesetting of the entire manu-script is done in a medium size font [9 pt : 'DejaVu Serif'] (honoring readers' suggestions) using the LaTeX 2_ε document processing system originally developed by Leslie Lamport, based on TeX typesetting system created by Donald Knuth. The typesetting software used the XǝLaTeX distribution.

We are grateful for this opportunity to put the materials into a consistent format, and to correct errors in the original publication that have come to our attention. Most of the hard work of preparing this edition was accomplished by Neeru Singh, who expertly keyboarded and edited the text of the original manuscript. She helped us put hundreds of pages of typographically difficult material into a consistent digital format.

The process of compiling this book has given us an incentive to improve the layout, to double-check almost all of the mathematical rendering, to correct all known errors, to improve the original illustrations by redrawing them with Till Tantau's marvelous TikZ. Thus the book now appears in a form that we hope will remain useful for at least another generation.

September, 2019.   Ancient Science Publishers.

List of Chapters

EXAMPLES I : Ratio

§8. Let $r = \dfrac{a}{b} = \dfrac{b}{c} = \dfrac{c}{d}$; then $c = dr, b = cr = dr^2, a = br = dr^3$; and by

substituting for a, b, c in terms of d, we have $\dfrac{a^5 + b^2c^2 + a^3c^2}{b^4c + d^4 + b^2cd^2} = r^6 = \dfrac{a^2}{d^2}$.

§9. Let $k = \dfrac{x}{q+r-p} = \dfrac{y}{r+p-q} = \dfrac{z}{p+q-r}$;

then $(q-r)x + (r-p)y + (p-q)z = k\{(q-r)(q+r-p) + \ldots + \ldots\} = 0$.

§10. $\dfrac{y}{x-z} = \dfrac{y+x}{z} = \dfrac{x}{y}$.

Each ratio $= \dfrac{\text{sum of numerators}}{\text{sum of denominators}} = \dfrac{2(x+y)}{x+y}$.

Thus each ratio is equal to 2 unless $x + y = 0$. In the first case

$$\dfrac{x+y}{z} = \dfrac{x}{y} = 2; \text{ whence } x : y : z = 4 : 2 : 3.$$

In the second case, $y = -x$, and $\dfrac{y}{x-z} = \dfrac{x}{y}$; whence $x : y : z = 1 : -1 : 0$.

§11.

$$\text{Each ratio} = \dfrac{\text{sum of numerators}}{\text{sum of denominators}} = \dfrac{2(x+y+z)}{(p+q)(a+b+c)} \qquad (1)$$

Multiply the numerator and denominator of each ratio by a, b, c respectively and add, then each of the given ratios

$$= \dfrac{(b+c)x + (c+a)y + (a+b)z}{(p+q)(bc+ca+ab)} \qquad (2)$$

From equations (1) and (2), the result follows.

§12. See Example 2, Art. 12.

§13. $\dfrac{2y+2z-x}{a} = \dfrac{2z+2x-y}{b} = \dfrac{2x+2y-z}{c}$. Multiply the numerator and denominator of each of the given ratios by $-1, 2, 2$ and add; then each ratio

$$= \dfrac{-(2y+2z-x) + 2(2z+2x-y) + 2(2x+2y-z)}{-a+2b+2c} = \dfrac{5x}{2b+2c-a}.$$

Similarly each of the given ratios is equal to $\dfrac{5y}{2c+2a-b}$ and to $\dfrac{5z}{2a+2b-c}$.

§14. Multiplying out and transposing,

$$b^2z^2 + c^2y^2 - 2bcyz + c^2x^2 + a^2z^2 - 2cazx + a^2x^2 + b^2y^2 - 2abxy = 0;$$

That is $(bz - cy)^2 + (cx - az)^2 + (ay - bx)^2 = 0$.

$$\therefore \ bz - cy = 0, \ cx - az = 0, \ ay - bx = 0.$$

§15. Dividing throughout by lmn,

$$\dfrac{my+nz-lx}{mn} = \dfrac{nz+lx-my}{nl} = \dfrac{lx+my-nz}{lm}$$
$$= \dfrac{(nz+lx-my) + (lx+my-nz)}{nl+lm} = \dfrac{2lx}{nl+lm} = \dfrac{x}{m+n}.$$

Thus we have

$$\dfrac{x}{m+n} = \dfrac{y}{n+l} = \dfrac{z}{l+m} = \dfrac{y+z-x}{(n+l)+(l+m)-(m+n)} = \dfrac{y+z-x}{2l}.$$

Hence the result.

§16. From $ax+cy+bz = 0$, $cx+by+az = 0$, we have by cross multiplication,

$$\dfrac{x}{ac-b^2} = \dfrac{y}{bc-a^2} = \dfrac{z}{ab-c^2} = k, \text{ say.}$$

Substituting in the third equation $bx + ay + cz = 0$, we have

$$b(ac - b^2) + a(bc - a^2) + c(ab - c^2) = 0.$$

§17. From the first two equations we have by cross multiplication,
$$\frac{x}{hf - bg} = \frac{y}{gh - af} = \frac{z}{ab - h^2}.$$
Substituting in the third equation, we get
$$g(hf - bg) + f(gh - af) + c(ab - h^2) = 0.$$

§18. From the first and second equations,
$$\frac{x}{ac + b} = \frac{y}{bc + a} = \frac{z}{1 - c^2} \tag{3}$$
From the first and third equations,
$$\frac{x}{ab + c} = \frac{y}{1 - b^2} = \frac{z}{bc + a} \tag{4}$$
From (3) and (4)
$$\frac{y}{bc + a} \times \frac{y}{1 - b^2} = \frac{z}{1 - c^2} \times \frac{z}{bc + a}, \text{ or}$$
$$\frac{y^2}{1 - b^2} = \frac{z^2}{1 - c^2}.$$

§19. From the first two equations,
$$\frac{x}{ab + a} = \frac{y}{ab + b} = \frac{z}{1 - ab}.$$
Substitute in the third equation,
$$\therefore \; c(ab + a) + c(ab + b) = 1 - ab.$$

§22. From the first and second equations we have by cross multiplication,
$$\frac{yz}{18} = \frac{zx}{14} = \frac{xy}{84}, \text{ that is } \frac{x}{7} = \frac{y}{9}, \text{ and } z = \frac{y}{6}.$$

§23. From the first and second equations by cross multiplication,
$$\frac{x^2}{45} = \frac{y^2}{80} = \frac{z^2}{5}; \text{ whence } x = \pm 3z; \; y = \pm 4z.$$

§24. From the given equations by cross multiplication we obtain the ratios of $l : m : n$. The value of l is proportional to
$$\frac{1}{(\sqrt{b} - \sqrt{c})(\sqrt{c} + \sqrt{a})} - \frac{1}{(\sqrt{b} + \sqrt{c})(\sqrt{c} - \sqrt{a})},$$
that is, proportional to $\dfrac{c - \sqrt{ab}}{(b - c)(c - a)}$,

so that we may put $l = \dfrac{c - \sqrt{ab}}{(b - c)(c - a)} k.$

Hence $\dfrac{l}{(a - b)(c - \sqrt{ab})} = \dfrac{k}{(b - c)(c - a)(a - b)}.$

By symmetry we obtain the required result.

§25. From the first two equations,
$$\frac{x}{a(b^2 - c^2)} = \frac{y}{b(c^2 - a^2)} = \frac{z}{c(a^2 - b^2)} = k \text{ say;}$$
substituting in the third equation, we find that $k^2 = 1$. See Example 3, Art. 16.

§26. From the first two equations,
$$\frac{x}{bc(b - c)} = \frac{y}{ca(c - a)} = \frac{z}{ab(a - b)} = k \text{ say;}$$
and from the third equation, we find that $k = 1$.

§27. From the first two equations,
$$\frac{x}{ab + a} = \frac{y}{ab + b} = \frac{z}{1 - ab} \tag{5}$$
From the second and third equations,
$$\frac{x}{1 - bc} = \frac{y}{bc + b} = \frac{z}{bc + c} \tag{6}$$

From (5) and (6) $\dfrac{x}{a(b+1)} \times \dfrac{x}{1-bc} = \dfrac{z}{1-ab} \times \dfrac{z}{c(b+1)}$;

$$\therefore \dfrac{x^2}{a(1-bc)} = \dfrac{z^2}{c(1-ab)}.$$

§28. From the first and second equations,

$$\dfrac{x}{hf-bg} = \dfrac{y}{gh-af} = \dfrac{z}{ab-h^2} \tag{7}$$

From the second and third equations,

$$\dfrac{x}{bc-f^2} = \dfrac{y}{fg-ch} = \dfrac{z}{hf-bg} \tag{8}$$

From the first and third equations,

$$\dfrac{x}{fg-ch} = \dfrac{y}{ca-g^2} = \dfrac{z}{gh-af} \tag{9}$$

From (8) and (9) $\dfrac{x}{bc-f^2} \times \dfrac{x}{fg-ch} = \dfrac{y}{fg-ch} \times \dfrac{y}{ca-g^2}$;

$$\therefore \dfrac{x^2}{bc-f^2} = \dfrac{y^2}{ca-g^2}.$$

From (7), (8) and (9)

$$\dfrac{x}{bc-f^2} \cdot \dfrac{y}{ca-g^2} \cdot \dfrac{z}{ab-h^2} = \dfrac{y}{fg-ch} \cdot \dfrac{z}{gh-af} \cdot \dfrac{x}{hf-bg};$$

by equating the denominators the second result follows.

EXAMPLES II : Proportion

Examples $4, 5, 6, 7$ may all be solved in a similar manner; thus take Example 6, and put $\dfrac{a}{b} = \dfrac{c}{d} = k$, so that $a = bk$, $c = dk$; then

$$\frac{a-c}{b-d} = \frac{bk-dk}{b-d} = k = \frac{k\sqrt{b^2+d^2}}{\sqrt{b^2+d^2}} = \frac{\sqrt{k^2b^2+k^2d^2}}{\sqrt{b^2+d^2}} = \frac{\sqrt{a^2+c^2}}{\sqrt{b^2+d^2}}.$$

Examples $8, 9, 10$ may all solved in a similar manner.

Put $\dfrac{a}{b} = \dfrac{b}{c} = \dfrac{c}{d} = k$, so that $c = dk$, $b = dk^2$, $a = dk^3$, then in Examples 9,

$$\frac{2a+3d}{3a-4d} = \frac{2dk^3+3d}{3dk^3-4d} = \frac{2k^3+3}{3k^3-4} = \frac{2b^3k^3+3b^3}{3b^3k^3-4b^3} = \frac{2a^3+3b^3}{3a^3-4b^3}.$$

§11. Put $\dfrac{a}{b} = \dfrac{b}{c} = k$; then $a = bk$, $c = \dfrac{b}{k}$, and

$$\frac{a^2-b^2+c^2}{a^{-2}-b^{-2}+c^{-2}} = \frac{b^2\left(k^2-1+\dfrac{1}{k^2}\right)}{\dfrac{1}{b^2}\left(\dfrac{1}{k^2}-1+k^2\right)} = b^4.$$

§13. Componendo and dividendo, $\dfrac{2x^3-3x^2}{x+1} = \dfrac{3x^3-x^2}{5x-13}$,

whence $x = 0$, or

$$\frac{2x-3}{x+1} = \frac{3x-1}{5x-13}.$$

§15. Componendo and dividendo, $\dfrac{mx-a}{nx+b} = \dfrac{mx+a}{nx+c}$; by clearing of fractions we obtain a simple equation.

§16. We have

$$a(a-b-c+d) = a^2-ab-ac+ad = a^2-ab-ac+bc = (a-b)(a-c).$$

$$\therefore a-b-c+d = \frac{(a-b)(a-c)}{a}.$$

§18. The work done by $x-1$ men in $x+1$ days is proportional to $(x-1)(x+1)$; hence $\dfrac{(x-1)(x+1)}{(x+2)(x-1)} = \dfrac{9}{10}$.

§19. Denote the proportionals by $x, y, 19-y, 21-x$. Then

$$x(21-x) = y(19-y) \tag{10}$$

and

$$x^2+y^2+(19-y)^2+(21-x)^2 = 442 \tag{11}$$

From (10) : $x^2-y^2-21x+19y = 0$.
From (11) : $x^2+y^2-21x-19y+180 = 0$.
Add : $x^2-21x+90 = 0$,
$x = 6$ or 15.
Subtract : $y^2-19y+90 = 0$,
$y = 9$ or 10.

§20. Let the quantities taken from A and B be x and y gallons respectively. Then

$$\frac{2}{9}x + \frac{1}{6}y = 2, \quad \frac{7}{9}x + \frac{5}{6}y = 9.$$

§21. Suppose the cask contains x gallons; after the first drawing there are $x-9$ gallons of wine and 9 gallons of water. At the second drawing $\dfrac{x-9}{x} \times 9$ gallons of wine are taken, and therefore the quantity of wine left is $(x-9) - \dfrac{9(x-9)}{x} = \dfrac{(x-9)^2}{x}$. Hence the quantity of water in the cask is $x - \dfrac{(x-9)^2}{x}$.

$$\therefore \frac{(x-9)^2}{x} : x - \frac{(x-9)^2}{x} = 16 : 9.$$
$$\therefore (x-9)^2 : 18x - 81 = 16 : 9,$$
$$\text{or } (x-9)^2 = 16(2x-9).$$

§22. Denote the quantities by a, ar, ar^2, ar^3.

$\dfrac{\text{Difference between first and last}}{\text{Difference between the other two}} = \dfrac{ar^3 \sim a}{ar^3 \sim r}$

$= \dfrac{r^2 + r + 1}{r} = \dfrac{3r + (1-r)^2}{r} = 3 + \dfrac{(1-r)^2}{r};$

and this is greater than 3.

§23. Let T and C denote the town and country populations ;

the increase in the town population is $\dfrac{18}{100}T$;

the increase in the country population is $\dfrac{4}{100}C$;

the increase in the total population is $\dfrac{15 \cdot 9}{100}(T + C)$;

$$\therefore 18T + 4c = 15 \cdot 9(T + C).$$

§24. Let $5x$ and x denote the amounts of tea and coffee respectively.

On the first supposition, the increase of tea is $\dfrac{a}{100} \times 5x$; the increase of

coffee is $\dfrac{b}{100} \times x$; and the total increase is $\dfrac{7c}{100} \times 6x$.

$$\therefore 5a + b = 42c \tag{12}$$

On the second supposition, we have

$$\therefore 5b + a = 18c \tag{13}$$

From (12) and (13), $\dfrac{5a + b}{5b + a} = \dfrac{7}{3}.$

§25. Suppose that in 100 parts of bronze there are x parts of copper and $100 - x$ of zinc; also suppose that in the fused mass there are $100a$ parts of brass, and $100b$ parts of bronze. $100a$ parts of brass contain ax parts of copper and $a(100 - x)$ parts of zinc. Also $100b$ parts of bronze contain $80b$ parts of copper, $4b$ parts of zinc, and $16b$ parts of tin. Hence in the fused mass there are $ax + 80b$ parts of copper; $a(100 - x) + 4b$ parts of zinc, and $16b$ parts of tin.

$$\therefore \frac{ax + 80b}{74} = \frac{a(100 - x) + 4b}{16} = \frac{16b}{10}.$$

$\therefore 10(ax + 80b) = 74 \times 16b$; that is $10ax = 384b$.

Also $10\{a(100 - x) + 4b\} = 16 \times 16b$; that is $10a(100 - x) = 216b$.

$$\therefore \frac{10ax}{10a(100 - x)} = \frac{384b}{216b}, \text{ or } \frac{x}{100 - x} = \frac{16}{9}.$$

§26. Let x be the rate of rowing in still water, y the rate of the stream, and a the length of the course.

Then the times taken to row the course against the stream, in still water, and with the stream are $\dfrac{a}{x - y}, \dfrac{a}{x}, \dfrac{a}{x + y}$ minutes respectively.

Thus

$$\frac{a}{x - y} = 84 \tag{14}$$

$$\frac{a}{x} - \frac{a}{x + y} = 9 \tag{15}$$

From (14), $a = 84(x - y)$,

$$\therefore \frac{84(x - y)}{x} - \frac{84(x - y)}{x + y} = 9,$$

$$\therefore \; 28y(x - y) = 3x(x + y),$$
$$\text{or, } 3x^2 - 25xy + 28y^2 = 0,$$
$$x = 7y, \text{ or } 3x = 4y.$$

If $x = 7y$, then $a = 84 \times 6y$, and time down stream $= \dfrac{a}{8y} = 68$ minutes.

Similarly in the other case.

EXAMPLES III : Variation

§7. $P = \dfrac{mQ}{R}$, where m is constant; hence $\dfrac{2}{3} = m \times \dfrac{3}{7} \times \dfrac{14}{9}$;

thus $m = 1$, and $Q = PR = \sqrt{48} \times \sqrt{75} = 60$.

§9. Here $y = mx + \dfrac{n}{x}$, where m and n are constants; whence $6 = 4m + \dfrac{n}{4}$.

and $\dfrac{10}{3} = 3m + \dfrac{n}{3}$. From these equations we find $m = 2$, $n = -8$.

§10. Here $y = mx + \dfrac{n}{x^2}$, so that $19 = 2m + \dfrac{n}{4}$, and $19 = 3m + \dfrac{m}{9}$; whence

$m = 5$, $n = 36$.

§11. $A = \dfrac{m\sqrt{B}}{C^3}$, so that $3 = \dfrac{m\sqrt{256}}{8}$, and therefore $m = \dfrac{3}{2}$; hence

$$\sqrt{B} = \dfrac{2}{3} AC^3 = \dfrac{2}{3} \times 24 \times \dfrac{1}{8} = 2.$$

§12. Here $x + y = m\left(z + \dfrac{1}{z}\right)$, and $x - y = n\left(z - \dfrac{1}{z}\right)$.

From the numerical data,

$$4 = m \times 2\dfrac{1}{2}, \text{ and } 2 = n \times 1\dfrac{1}{2};$$

$$\text{thus } x + y = \dfrac{8}{5}\left(z + \dfrac{1}{z}\right) \text{ and } x - y = \dfrac{4}{3}\left(z - \dfrac{1}{z}\right).$$

$$\text{By addition, } 2x = \dfrac{44}{15}z + \dfrac{4}{15z}.$$

§14. Here $y = m + nx + px^2$.

From the numerical data,

$$0 = m + n + p; \quad 1 = m + 2n + 4p; \quad 4 = m + 3n + 9p;$$

whence $m = 1$, $n = -2$, $p = 1$; and

$$y = 1 - 2x + x^2 = (x - 1)^2.$$

§15. Let s denote the distance in feet, t the time in seconds : then $s \propto t^2$,

so that $s = mt^2$. Now $400\dfrac{1}{2} = m \times 5^2$, hence $m = 16 \cdot 1$.

$$\text{In 10 seconds, } s = 16 \cdot 1 \times 10^2 = 1610.$$
$$\text{In 9 seconds, } s = 16 \cdot 1 \times 9^2 = 1304 \cdot 1.$$

The difference gives the distance fallen through in the 10th second.

§16. Let r denote the radius in feet, V the volume in cubic feet; then

$V \propto r^3$, so that $V = mr^3$.

Hence $179\dfrac{2}{3} = m \times \left(\dfrac{7}{2}\right)^3$;

when $r = \dfrac{7}{4}$, $V = m \times \left(\dfrac{7}{4}\right)^3 = \dfrac{1}{8} \times m \times \left(\dfrac{7}{2}\right)^3 = \dfrac{1}{8} \times 179\dfrac{2}{3} = 22\dfrac{11}{24}$.

§17. Let w denote the weight of the disc, r the radius and t the thickness; then w varies jointly as r^2 and t; hence $w = mtr^2$. If w', r', t' denote corresponding quantities for a second disc, $w' = mt'(r')^2$.

Hence $\dfrac{w}{w'} = \dfrac{tr^2}{t'(r')^2}$.

If $\dfrac{t}{t'} = \dfrac{9}{8}$ and $\dfrac{w}{w'} = 2$, we have $2 = \dfrac{9r^2}{8(r')^2}$, that is $3r = 4r'$.

§18. Suppose that the regatta lasted a days and that the days in question were the $(x-1)^{th}$, x^{th}, $(x+1)^{th}$. Then the number of races on the xth day varies as the product $x(a - x - 1)$. Similarly the number of races on the $(x-1)^{th}$ and $(x+1)^{th}$ days are proportional to $(x-1)(a - x - 2)$ and $(x+1)(a - x)$.

Hence

$$(x-1)(a - \overline{x-2}) = 6k \qquad (16)$$

$$x(a - \overline{x-1}) = 5k \qquad (17)$$

$$(x+1)(a-x) = 3k \qquad (18)$$

Subtracting (17) from (16),

$$2x - 2 - a = k \qquad (19)$$

Subtracting (18) from (17),

$$2x - a = 2k \qquad (20)$$

Subtracting (19) from (20), $2 = k$.
Hence from (20),

$$2x - a = 4; \text{that is } a = 2x - 4 \qquad (21)$$

Also from (17), $x(a - x + 1) = 10$,
and substituting from (21), $x(x - 3) = 10$.
Thus $x = 5$ and $a = 6$.

§19. Let $\pounds p$ be the cost of workmanship;
w carats the weight of the ring;
$\pounds x$ the cost of a diamond of one carat;
$\pounds y$ the value of a carat of gold.
Thus $a = p + (w - 3)y + 9x$,
$b = p + (w - 4)y + 16x$,
$c = p + (w - 5)y + 25x$,
$\therefore a + c - 2b = 2x$, whence $x = \dfrac{a+c}{2} - b$.

§20. Let $\pounds P$ denote the value of the pension, Y the number of years;
then by the question $P \propto \sqrt{Y}$, that is

$$P = m\sqrt{Y} \qquad (22)$$
$$\text{Also } P + 50 = m\sqrt{Y + 9} \qquad (23)$$

$$\text{and } \frac{\sqrt{Y + 4\frac{1}{4}}}{\sqrt{Y}} = \frac{9}{8} \qquad (24)$$

From this last equation $Y = 16$, and therefore from (22) and (23),
$P = 4m$, $P + 50 = 5m$.

§21. Let F denote the force of attraction, T the time of revolution; then

$$F \propto \frac{M}{D^2}, \text{ and } T^2 \propto \frac{D}{F}.$$

Thus $D \propto FT^2$; that is $D \propto \dfrac{M}{D^2}T^2$, or $MT^2 \propto D^3$; $\therefore MT^2 = kD^3$.

Thus $m_1 t_1^2 = kd_1^3$, and $m_2 t_2^2 = kd_2^3$; that is $\dfrac{m_1 t_1^2}{m_2 t_2^2} = \dfrac{d_1^3}{d_2^3}$.

Using the numerical data, $\dfrac{d_1}{d_2} = \dfrac{35}{31}$,

$$\frac{m_1}{m_2} = 343, \text{ and } t_2 = 27.32 \text{ days.}$$

$$\therefore \frac{343 t_1^2}{(27.32)^2} = \frac{35 \times 35 \times 35}{31 \times 31 \times 31},$$

$$\therefore t_1^2 = (27.32)^2 \times \frac{5 \times 5 \times 5}{31 \times 31 \times 31};$$

$$\therefore t_1 = \frac{27.32 \times 5}{31} \times \sqrt{\frac{5}{31}} = \frac{13.66}{31 \times 2.49} = 1.77 \text{ days.}$$

§22. Let x be the rate of the train in miles per hour,
q the quantity of fuel used per hour, estimated in tons;
then $q = kx^2$;

but $2 = k \times (16)^2$;

$\therefore q = \dfrac{2}{256}x^2$;

$\therefore$ the cost of the fuel per hour is $£\dfrac{1}{2} \times \dfrac{2}{256}x^2 = £\dfrac{x^2}{256}$,

$\therefore$ cost of fuel per mile is $£\dfrac{1}{x} \times \dfrac{x^2}{256} = £\dfrac{x}{256}$.

Also cost for journey of one mile, due to "other expenses," is

$$£\dfrac{11\frac{1}{4}}{20} \times \dfrac{1}{x} = £\dfrac{9}{16x};$$

$\therefore$ cost of journey per mile is $£\left(\dfrac{x}{256} + \dfrac{9}{16x}\right)$,

and this has to be as small as possible.

Now this expression $= \left(\dfrac{\sqrt{x}}{16} - \dfrac{3}{4\sqrt{x}}\right)^2 + \dfrac{3}{32}$, and therefore is least when

$\dfrac{\sqrt{x}}{16} - \dfrac{3}{4\sqrt{x}} = 0$; that is, $x = 12$.

Hence the least cost of the journey per mile is $£\dfrac{3}{32}$, and the cost for

100 miles is $£\dfrac{300}{32} = £9.\ 7s.\ 6d.$

EXAMPLES IV : Arithmetical Progression

EXAMPLES IV a

§18. $s = \dfrac{n}{2}(a + l)$; thus $155 = \dfrac{n}{2}(2 + 29)$, and $n = 10$.

Again $l = a + (n - 1)d$, that is, $29 = 2 + 9d$.

§20. Here $18 = a + 2d$, $30 = a + 6d$, so that $a = 12$, $d = 3$.

§21. Denote the numbers by $a - d$, a, $a + d$,

then $3a = 27$, that is, $a = 9$.

Hence $(9 - d) \times 9 \times (9 + d) = 504$.

§22. The middle number is clearly 4, so that the three numbers are $4 - d$, 4, $4 + d$.

Thus $(4 - d)^3 + 4^3 + (4 + d)^3 = 408$.

§23. Put $n = 1$; then the first term $= 5$;

put $n = 15$; then the last term $= 61$.

$$\text{Sum} = \frac{15}{2}(\text{first term} + \text{last term}) = \frac{15}{2} \times 66 = 495.$$

Example 24 may be solved in the same way.

§25. Put $n = 1$; then the first term $= \dfrac{1}{a} + b$;

put $n = p$; then the last term $= \dfrac{p}{a} + b$;

$$\therefore \quad sum = \frac{p}{2}(\text{first term} + \text{last term}) = \frac{p}{2}\left(\frac{p + 1}{a} + 2b\right).$$

§26. The series $= 2a - \dfrac{1}{a}$, $4a - \dfrac{3}{a}$, $6a - \dfrac{5}{a}$, $\ldots$

$$S = (2a + 4a + 6a + \ldots \text{ to } n \text{ terms}) - \left(\frac{1}{a} + \frac{3}{a} + \frac{5}{a} + \ldots \text{ to } n \text{ terms}\right).$$

EXAMPLES IV b

§3. Here $a + 2d = 4a$, and $a + 5d = 17$; hence $a = 2$, $d = 3$.

§4. Here $a + d = \dfrac{31}{4}$, $a + 30d = \dfrac{1}{2}$, $a + (n - 1)d = -\dfrac{13}{2}$; so that

$$d = -\frac{1}{4}, \quad a = 8, \quad n = 59.$$

§6. Denote the installments by a, $a + d$, $a + 2d$, $\ldots$;

then sum of 40 terms $= 3600$;

and sum of 30 terms $= \dfrac{2}{3}$ of $3600 = 2400$.

$\therefore$ $20(2a + 39d) = 3600$, and $15(2a + 29d) = 2400$;

$\therefore$ $2a + 39d = 180$, $2a + 29d = 160$.

§7. Denote the numbers by a and l, and the number of means by $2m$.

Then $a + l = \dfrac{13}{6}$, and the sum of the means $= 2m \times \dfrac{a + l}{2} = m(a + l)$. But

this sum $= 2m + 1$;

$$\therefore \quad 2m + 1 = m(a + l) = \frac{13}{6}m, \text{ whence } m = 6;$$

and the number of means is 12.

§9. The series is $\dfrac{1 - \sqrt{x}}{1 - x}$, $\dfrac{x}{1 - x}$, $\dfrac{1 + \sqrt{x}}{1 - x}$, and is therefore an A.P. whose

first term is $\dfrac{1 - \sqrt{x}}{1 - x}$ and difference $\dfrac{\sqrt{x}}{1 - x}$.

$$S = \frac{n}{2}\left\{\frac{2(1 - \sqrt{x})}{1 - x} + \frac{(n - 1)\sqrt{x}}{1 - x}\right\} = \frac{n}{2(1 - x)}\{2 + (n - 3)\sqrt{x}\}.$$

§10. We have $\dfrac{7}{2}\{2a+6d\}=49$, that is $a+3d=7$.

Similarly $\dfrac{17}{2}\{2a+16d\}=289$, that is $a+8d=17$.

Thus $a=1$, $d=2$.

§11. Let x be the first term, y the common difference; then
$$a=x+(p-1)y,\ b=x+(q-1)y,\ c=x+(r-1)y;$$
$$\therefore\ (q-r)a+(r-p)b+(p-q)c=0,$$
since the coefficients of x and y will both be found to vanish.

§12. Here $\dfrac{p}{2}\{2a+(p-1)d\}=q$;

that is, $2a+(p-1)d=\dfrac{2q}{p}$.

Similarly $2a+(q-1)d=\dfrac{2p}{q}$.

Whence $d=-2\left(\dfrac{1}{p}+\dfrac{1}{q}\right)$, $a=\dfrac{p}{q}+\dfrac{q}{p}-\dfrac{1}{p}-\dfrac{1}{q}+1$.

$$\therefore\ s=\frac{p+q}{2}\left\{\frac{2p}{q}+\frac{2q}{p}-\frac{2}{p}-\frac{2}{q}+2-2(p+q-1)\left(\frac{1}{p}+\frac{1}{q}\right)\right\}$$
$$=\frac{p+q}{2}(-2)=-(p+q).$$

§13. Assume for the integers $a-3d,\ a-d,a+d,a+3d$; the sum of these is $4a$; thus $4a=24$ and $a=6$.
$$\therefore\ (6-3d)(6-d)(6+d)(6+3d)=945,$$
$$\text{that is, } 9(2-d)(2+d)(6-d)(6+d)=945.$$

§14. Assume for the integers $a-3d,\ a-d,a+d,a+3d$; thus from the first part of the question $a=5$; and from the second
$$\frac{(5-3d)(5+3d)}{(5-d)(5+d)}=\frac{2}{3};\text{ whence } d=1.$$

§15. Here $a+(p-1)d=q$, and $a+(q-1)d=p$;
whence $d=-1$, $a=p+q-1$.
Thus the mth term $=p+q-1+(m-1)(-1)=p+q-m$.

§17. Putting $n=r$, the sum of r terms is $2r+3r^2$; putting $n=r-1$, the sum of $(r-1)$ terms is $2(r-1)+3(r-1)^2$. The difference gives the rth term.

§18. We have $\dfrac{m(2a+\overline{m-1}\cdot d)}{n(2a+\overline{n-1}\cdot d)}=\dfrac{m^2}{n^2}$;

that is, $n(2a+\overline{m-1}\cdot d)=m(2a+\overline{n-1}\cdot d)$; whence $2a=d$.

Thus $\dfrac{m^{th}\text{ term}}{n^{th}\text{ term}}=\dfrac{a+(m-1)d}{a+(n-1)d}=\dfrac{1+2(m-1)}{1+2(n-1)}=\dfrac{2m-1}{2n-1}$.

§19. Let m be the middle term, d the common difference, and $2p+1$ the number of terms; then the pairs of terms equidistant from the middle term are
$$m-d,\ m+d;\ m-2d,\ m+2d;\ m-3d,\ m+3d;\ \ldots\ m-(p-1)d,\ m+(p-1)d.$$
Thus the result follows at once.

§20. See the solution of Example 17 above.

§21. Let the number of terms be $2n$.
Denote the series by
$$a,\ a+d,\ a+2d,\ a+3d,\ \ldots\ a+(2n-1)d.$$
Then we have the equations :

$$\frac{n}{2}\{2a+(n-1)2d\}=24 \tag{25}$$

$$\frac{n}{2}\{2(a+d)+(n-1)2d\}=30 \tag{26}$$

$$(2n - 1)d = 10\frac{1}{2} \tag{27}$$

. From (25) and (26),

$$nd = 6 \tag{28}$$

From (27) and (28), $n = 4$, and the number of terms is 8.

§22. In each set the middle term is 5 [Art. 46, Ex. 1].

Denote the first set of numbers by $5 - d$, 5, $5 + d$; then the second set will be denoted by $5 - (d - 1)$, 5, $5 + (d - 1)$; hence

$$\frac{(5 - d)(5 + d)}{(6 - d)(4 + d)} = \frac{7}{8};$$

whence $d = 2$ or -16.

The latter value Is rejected.

§23. In the first case the common difference is $\dfrac{2y - x}{n + 1}$ and the rth mean,

being the $(r + 1)^{th}$ term, is $x + \dfrac{r(2y - x)}{n + 1}$.

In the second case the rth mean is $2x + \dfrac{r(y - 2x)}{n + 1}$.

$$\therefore \ x + \frac{r(2y - x)}{n + 1} = 2x + \frac{r(y - 2x)}{n + 1};$$
$$\therefore \ (n + 1)x + r(2y - x) = 2(n + 1)x + r(y - 2x),$$
$$\therefore \ ry = (n + 1 - r)x.$$

§24. Here $\dfrac{p}{2}\{2a + (p - 1)d\} = \dfrac{q}{2}\{2a + (q - 1)d\}$,

$$\therefore \ (2a - d)p + p^2 d = (2a - d)q + q^2 d;$$
$$(2a - d)(p - q) + (p^2 - q^2)d = 0,$$
$$2a - d + (p + q)d = 0, \text{ or}$$
$$2a + (p + q - 1)d = 0.$$

$$\therefore \ \frac{p + q}{2}\{2a + (p + q - 1)d\} = 0;$$

that is, the sum of $p + q$ terms is zero.

EXAMPLES V : Geometrical Progression

EXAMPLES V a

§20. $\dfrac{a(r^6 - 1)}{r - 1} = \dfrac{9a(r^3 - 1)}{r - 1}$; $\therefore r^3 + 1 = 9$; $r = 2$.

§21. $ar^4 = 81$, $ar = 24$; $\therefore r = \dfrac{3}{2}$ and $a = 16$.

§22, 23. Use the formula $s = \dfrac{rl - a}{r - 1}$.

§24, 25. The solutions of these two questions are very similar. In Ex. 25, assume for the three numbers $\dfrac{a}{r}$, a, ar; then $\dfrac{a}{r} \times a \times ar = 216$; whence $a = 6$, and the numbers are $\dfrac{6}{r}$, 6, $6r$

Again, $\left(\dfrac{6}{r} \times 6\right) + (6 \times 6r) + \left(\dfrac{6}{r} \times 6r\right) = 156$;

that is, $\dfrac{3}{r} + 3r = 10$, whence $r = 3$ or $\dfrac{1}{3}$.

§26.

$$S_p = 1 + r^p + r^{2p} + \ldots = \frac{1}{1 - r^p}; \quad \therefore S_{2p} = \frac{1}{1 - r^{2p}}.$$

$$s_p = 1 - r^p + r^{2p} + \ldots = \frac{1}{1 + r^p}.$$

$$\therefore S_p + s_p = \frac{1}{1 - r^p} + \frac{1}{1 + r^p} = \frac{2}{1 - r^{2p}} = 2S_{2p}.$$

§27. Let f denote the first term, x the common ratio; then

$$a = fx^{p-1}, \ b = fx^{q-1}, \ c = fx^{r-1}.$$

$$\therefore a^{q-r}b^{r-p}c^{p-q} = f^{q-r+r-p+p-q}x^{(p-1)(q-r)+(q-1)(r-p)+(r-1)(p-q)} = f^0 x^0 = 1.$$

§28. Here $\dfrac{a}{1 - r} = 4$, and $\dfrac{a^3}{1 - r^3} = 192$.

From the first equation $a = 4(1 - r)$; hence

$$\frac{64(1 - r)^3}{1 - r^3} = 192, \text{ or } (1 - r)^2 = 3(1 + r + r^2),$$

$$\text{that is, } 2r^2 - 5r + 2 = 0, \text{ whence } r = 2 \text{ or } \frac{1}{2}.$$

The first of these values is inadmissible in an infinite geometrical progression; the other value gives $a = 2$.

EXAMPLES V b

§1.

$$
\begin{aligned}
S &= 1 + 2a + 3a^2 + \ldots + & na^{n-1}, \\
aS &= \quad\; a + 2a^2 + \ldots + (n-1)a^{n-1} + na^n; \\
\therefore\; S(1-a) &= 1 + \; a \; + \; a^2 \; + \ldots + & a^{n-1} - na^n \\
&= \frac{1-a^n}{1-a} - na^n.
\end{aligned}
$$

§2.

$$
S = 1 + \frac{3}{4} + \frac{7}{16} + \frac{15}{64} + \frac{31}{256} + \ldots
$$

$$
\therefore \quad \frac{1}{4}S = \quad \frac{1}{4} + \frac{3}{16} + \frac{7}{64} + \frac{15}{256} + \ldots
$$

By subtraction, $\dfrac{3}{4}S = 1 + \dfrac{2}{4} + \dfrac{4}{16} + \dfrac{8}{64} + \dfrac{16}{256} + \ldots$

$$
= 1 + \; \frac{1}{2} + \frac{1}{4} + \frac{1}{8} + \frac{1}{16} + \ldots \; = 2.
$$

§3.

$S = 1 + 3x + 5x^2 + 7x^3 + 9x^4 + \ldots$
$\therefore\; xS = x + 3x^2 + 5x^3 + 7x^4 + \ldots$

By subtraction,
$(1-x)S = 1 + 2x + 2x^2 + 2x^3 + 2x^4 + \ldots$
$= 1 + \dfrac{2x}{1-x} = \dfrac{1+x}{1-x}.$

§4.

$$
S = 1 + \frac{2}{2} + \frac{3}{2^2} + \frac{4}{2^3} + \ldots + \frac{n}{2^{n-1}};
$$

$$
\therefore \quad \frac{1}{2}S = \quad \frac{1}{2} + \frac{2}{2^2} + \frac{3}{2^3} + \ldots + \frac{n-1}{2^{n-1}} + \frac{n}{2^n}.
$$

By subtraction, $\dfrac{1}{2}S = 1 + \dfrac{1}{2} + \dfrac{1}{2^2} + \dfrac{1}{2^3} + \ldots + \dfrac{1}{2^{n-1}} - \dfrac{n}{2^n}.$

$$
= \frac{1 - \dfrac{1}{2^n}}{1 - \dfrac{1}{2}} - \frac{n}{2^n} = 2 - \frac{2}{2^n} - \frac{n}{2^n}.
$$

§5.

$$
S = 1 + \frac{3}{2} + \frac{5}{4} + \frac{7}{8} + \ldots
$$

$$
\therefore \quad \frac{1}{2}S = \quad \frac{1}{2} + \frac{3}{4} + \frac{5}{8} + \ldots
$$

By subtraction, $\dfrac{1}{2}S = 1 + 1 + \dfrac{1}{2} + \dfrac{1}{4} + \ldots = 1 + 2 = 3.$

§6.

$S = 1 + 3x + 6x^2 + 10x^3 + \ldots$
$\therefore\; xS = \quad x + 3x^2 + 6x^3 + \ldots$

By subtraction,

$(1-x)S = 1 + 2x + 3x^2 + 4x^3 + \dots$

$= \dfrac{1}{(1-x)^2}$, Ex. 1, Art. 60.

§7. Let p and q be the common ratios of the two progressions; then

$$b = ap^2, \text{ and } b = aq^4; \text{ hence } p = q^2.$$
$$\therefore \ ap^n = aq^{2n};$$

that is, $(n+1)^{th}$ term of first series $= (2n+1)^{th}$ term of second series.

§8. The sums are $\dfrac{a(r^{2n}-1)}{r-1}$ and $\dfrac{b(r^{2n}-1)}{r^2-1}$ respectively; and since these

are equal $\dfrac{a}{r-1} = \dfrac{b}{r^2-1}; \ \therefore \ b = a(r+1) = a + ar.$

§9.

$$S = 1 + (1+b)r + (1+b+b^2)r^2 + (1+b+b^2+b^3)r^3 + \dots$$

$$\therefore \ rS = r + (1+b)r^2 + (1+b+b^2)r^3 + \dots$$

By subtraction, $(1-r)S = 1 + br + b^2r^2 + b^3r^3 + \dots = \dfrac{1}{1-br}$.

§10. We have

$$a + ar + ar^2 = 70 \tag{29}$$

$$4a + 4ar^2 = 10ar \tag{30}$$

From (30), $r = 2$ or $\dfrac{1}{2}$.

§11. We shall first shew that the sum of an infinite G.P. commencing at any term, say the $(n+1)^{th}$, is equal to the preceding term multiplied by $\dfrac{r}{1-r}$; for

$$ar^n + ar^{n+1} + ar^{n+2} + \dots = \dfrac{ar^n}{1-r} = ar^{n-1} \times \dfrac{r}{1-r}.$$

In this particular example, the value of $\dfrac{r}{1-r}$ is $\dfrac{1}{3}$, so that $r = \dfrac{1}{4}$. Again $a + ar = 5$, hence $a = 4$.

§12. $S = (x + x^2 + x^3 + \dots) + (a + 2a + 3a + \dots)$; the first series is in G.P., the second in A.P.

§13. $S = (x^2 + x^4 + x^6 + \dots) + (xy + x^2y^2 + x^3y^3 + \dots)$; here both series are in G.P.

§14. $S = (a - 3a + 5a + \dots) + \left(\dfrac{1}{3} - \dfrac{1}{6} + \dfrac{1}{12} - \dots\right)$; the first series is in A.P., the second in G.P.

§15. The series may be expressed as the sum of two infinite series in G.P.

§16. The series may be expressed as the difference of two infinite series in G.P.

§17. Here $\dfrac{d}{c} = \dfrac{c}{b} = \dfrac{b}{a}$; hence $b^2 = ac$, $c^2 = bd$, $ad = bc$. Thus

$(b-c)^2 + (c-a)^2 + (d-b)^2 = b^2 - 2bc + c^2 + c^2 - 2ca + a^2 + d^2 - 2bd + b^2$
$= a^2 - 2bc + d^2 = a^2 - 2ad + d^2 = (a-d)^2.$

§18. Here $\dfrac{a+b}{2} = 2\sqrt{ab}$; so that $(a+b)^2 = 16ab$, or $a^2 - 14ab + b^2 = 0$;

that is, $\left(\dfrac{a}{b}\right)^2 - 14\left(\dfrac{a}{b}\right) + 1 = 0.$

Hence $\dfrac{a}{b} = 7 \pm 4\sqrt{3} = \dfrac{2+\sqrt{3}}{2-\sqrt{3}}, \textbf{ or } \dfrac{2-\sqrt{3}}{2+\sqrt{3}}.$

§19. Giving to r the values $1, 2, 3, \ldots n$, we have

$$S = 3 \cdot 2 + \quad 5 \cdot 2^2 + 7 \cdot 2^3 + \ldots + (2n+1)2^n;$$
$$\therefore \quad 2S = \quad 3 \cdot 2^2 + 5 \cdot 2^3 + \ldots + (2n-1)2^n$$
$$+ (2n+1)2^{n+1}.$$

Subtracting the upper line from the lower,

$$S = (2n+1)2^{n+1} - 3 \cdot 2 - (2 \cdot 2^2 + 2 \cdot 2^3 + \ldots + 2 \cdot 2^n)$$

$$= (2n+1)2^{n+1} - 6 - \frac{8(2^{n-1} - 1)}{2 - 1}$$

$$= (2n+1)2^{n+1} - 6 - 2 \cdot 2^{n+1} + 8 = n \cdot 2^{n+2} - 2^{n+1} + 2.$$

§20. The series is $1 + a + ac + a^2c + a^2c^2 + a^3c^2 + \ldots$ to $2n$ terns

$$= (1 + ac + a^2c^2 + \ldots \text{ to n terms})$$
$$+ a(1 + ac + a^2c^2 + \ldots \text{ to n terms})$$
$$= (1 + a)(1 + ac + a^2c^2 + \ldots \text{ to n terms})$$
$$= \frac{(1 + a)(a^n c^n - 1)}{ac - 1}.$$

§21. We have $S_n = \dfrac{a(r^n - 1)}{r - 1}$, and by putting in succession $n = 1, 3, 5, \ldots$ we obtain the values of $S_1, S_3, S_5, \ldots$ Thus the required sum

$$= \frac{a}{r - 1} \left\{ (r - 1) + (r^3 - 1) + (r^5 - 1) + \ldots \text{ to n terms} \right\}$$

$$= \frac{a}{r - 1} \left\{ r + r^3 + r^5 + \ldots \text{ to n terms } - n \right\}$$

$$= \frac{a}{r - 1} \left\{ \frac{r(r^{2n} - 1)}{r^2 - 1} - n \right\}.$$

§22. We have $S_1 = \dfrac{1}{1 - \dfrac{1}{2}} = 2;$ $\quad S_2 = \dfrac{2}{1 - \dfrac{1}{3}} = 3;$

$S_3 = \dfrac{3}{1 - \dfrac{1}{4}} = 4, \&c.;$ $\quad S_p = \dfrac{p}{1 - \dfrac{1}{p+1}} = p + 1.$

$$\therefore \quad sum = 2 + 3 + 4 + \ldots \text{ to p terms } = \frac{p}{2}\{4 + (p - 1)\} = \frac{p}{2}(p + 3).$$

§23. We have $1 + r + r^2 + r^3 + \ldots + r^{2m} = \dfrac{1 - r^{2m+1}}{1 - r}.$

Now $(1 - r^m)^2$ is positive; that is, $1 - 2r^m + r^{2m} > 0$, or $1 + r^{2m} > 2r^m$; and generally $r^p(1 + r^{m-p})^2 > 0$, that is $r^p - 2r^m + r^{2m-p} > 0$, that is $r^p + r^{2m-p} > 2r^m$.

Now $1 + r + r^2 + r^3 + r^m + \ldots + r^{2m}$
$= (1 + r^{2m}) + (r + r^{2m-1}) + (r^2 + r^{2m-2}) + \ldots + r^m,$
and is therefore greater than $2r^m + 2r^m + \ldots + r^m$, that is greater than $(2m + 1)r^m.$

$$\therefore \quad (2m + 1)r^m < \frac{1 - r^{m+1}}{1 - r}, \text{ that is } (2m + 1)r^m(1 - r) < 1 - r^{2m+1}.$$

Multiply both side by r^{m+1}, thus

$$(2m + 1)r^{2m+1}(1 - r) < r^{m+1}(1 - r^{2m+1}).$$

Put $2m + 1 = n$, then $nr^n(1 - r) < r^{\frac{n+1}{2}}(1 - r^n).$

Making n indefinitely great $r^{\frac{n+1}{2}}$ is indefinitely small, and therefore nr^n is indefinitely small.

EXAMPLES VI : Harmonical Progression

EXAMPLES VI a

§4. Here $\sqrt{ab} = 12$, $\dfrac{2ab}{a+b} = 9\dfrac{3}{5}$,

$\therefore a+b = 30$, $\sqrt{ab} = 12$, which give 6 and 24 for the two numbers.

§5. Here $\dfrac{2ab}{a+b} : \sqrt{ab} = 12 : 13$,

$$\therefore \frac{2\sqrt{ab}}{a+b} = \frac{12}{13};$$

whence $\qquad 6a - 13\sqrt{ab} + 6b = 0$,

or $\qquad (3\sqrt{a} - 2\sqrt{b})(2\sqrt{a} - 3\sqrt{b}) = 0$;

$$\therefore \sqrt{a} : \sqrt{b} = 3 : 2, \text{ or } 2 : 3,$$

that is, the two quantities are as 4 to 9.

§6. We have $\dfrac{a}{c} = \dfrac{a-b}{b-c}$;

$$\therefore \frac{a}{a+c} = \frac{a-b}{(a-b)+(b-c)} = \frac{a-b}{a-c};$$

$$\therefore a : a-b = a+c : a-c.$$

§7. Let a, d be the 1st term and common diff. of the corresponding A.P., then we have

$$\frac{1}{n} = a + (m-1)d, \quad \frac{1}{m} = a + (n-1)d;$$

whence $d = \dfrac{1}{nm}$, and $a = \dfrac{1}{nm}$;

$$\therefore (m+n)^{th} \text{ term of A.P. } = \frac{1}{nm} + \frac{m+n-1}{nm}$$

$$= \frac{m+n}{nm};$$

that is, the $(m+n)^{th}$ term of the H.P. $= \dfrac{mn}{m+n}$.

§8. We have

$$\left. \begin{array}{l} \dfrac{1}{a} = a + (p-1)\beta \\[2mm] \dfrac{1}{b} = a + (q-1)\beta \\[2mm] \dfrac{1}{c} = a + (r-1)\beta \end{array} \right\} \text{ where } \alpha \text{ and } \beta \text{ are } 1^{st} \text{ terms and common diff. of the A.P.}$$

Multiply these equations by $q-r$, $r-p$, $p-q$ respectively, and add the results.

§9. We have $b = \dfrac{2ac}{a+c}$;

$$\therefore \frac{1}{b-a} + \frac{1}{b-c} = \frac{1}{\dfrac{2ac}{a+c} - a} + \frac{1}{\dfrac{2ac}{a+c} - c} = \frac{a+c}{ac-a^2} + \frac{a+c}{ac-c^2}$$

$$= \frac{a+c}{c-a}\left\{ \frac{1}{a} - \frac{1}{c} \right\} = \frac{a+c}{ac} = \frac{1}{a} + \frac{1}{c}.$$

§10. Here

$$S = 3\Sigma n^2 - \Sigma n$$

$$= \frac{n(n+1)(2n+1)}{2} - \frac{n(n+1)}{2}$$

$$= n^2(n+1).$$

§11. Here

$$S = \Sigma n^3 + \frac{3}{2}\Sigma n = \left\{\frac{n(n+1)}{2}\right\}^2 + \frac{3n(n+1)}{4}$$

$$= \frac{n(n+1)}{4}\{n(n+1)+3\} = \frac{1}{4}n(n+1)(n^2+n+3).$$

§12. Here

$$S = \Sigma n^2 + 2\Sigma n = \frac{n(n+1)(2n+1)}{6} + n(n+1)$$

$$= \frac{1}{6}n(n+1)(2n+7).$$

§13. The nth term $= 2n^3 + 3n^2$;

$$\therefore\ S = 2\Sigma n^3 + 3\Sigma n^2 = \frac{n^2(n+1)^2}{2} + \frac{n(n+1)(2n+1)}{2}$$

$$= \frac{1}{2}n(n+1)\{n(n+1)+2n+1\}$$

$$= \frac{1}{2}n(n+1)(n^2+3n+1).$$

§14. Here

$$\therefore\ S = \Sigma 3^n - \Sigma 2^n$$

$$= \frac{3(3^n-1)}{3-1} - \frac{2(2^n-1)}{2-1}$$

$$= \frac{3^{n+1}-3}{2} - (2^{n+1}-2)$$

$$= \frac{1}{2}\left(3^{n+1}+1\right) - 2^{n+1}.$$

§15. The nth term $= 3\cdot 4^n + 6n^2 - 4n^3$.

$$\therefore\ S = 3\Sigma 4^n + 6\Sigma n^2 - 4\Sigma n^3$$

$$= 3\cdot 4\frac{(4^n-1)}{4-1} + n(n+1)(2n+1) - n^2(n+1)^2$$

$$= 4^{n+1} - 4 - n(n+1)(n^2-n-1).$$

§16. We have

$$\frac{a+md}{a+nd} = \frac{a+nd}{a+rd}$$

$$\therefore\ a^2 + (m+r)ad + mrd^2 = a^2 + 2nad + n^2a^2;$$

$$\therefore\ \frac{d}{a} = \frac{m+r-2n}{n^2-mr}.$$

Now $$\frac{2}{n} = \frac{1}{m} + \frac{1}{r};$$

$$\therefore\ 2mr = nm + nr;$$

$$\therefore\ \frac{d}{a} = \frac{2(m+r-2n)}{2n^2-(nm+nr)} = -\frac{2}{n}.$$

§17. We have $a + (l-1)d,\ a + (m-1)d,\ a + (n-1)d$ in H.P.

$$\therefore\ \frac{a+(l-1)d}{a+(n-1)d} = \frac{(l-m)d}{(m-n)d} = \frac{l-m}{m-n}$$

$$= \frac{l}{m},\ \text{since}\ \frac{l}{m} = \frac{m}{n};$$

$$\therefore\ a(l-m) = d\{m(l-1) - l(n-1)\};$$

$$\therefore\ \frac{a}{d} = \frac{l(m-n)+(l-m)}{l-m}$$

$$= m+1,\ \text{for}\ \frac{l(m-n)}{l-m} = m.$$

§18. Putting $n = 1, 2, 3, \ldots$ successively, we get

$$1^{st} \text{ term} = a + b + c = s_1 \text{ suppose};$$
$$\text{sum of 2 terms} = a + 2b + 4c = s_2 \text{ suppose};$$
$$\text{sum of 3 terms} = a + 3b + 9c = s_3 \text{ suppose};$$
$$\therefore s_2 - s_1 = b + 3c = 2^{nd} \text{ term},$$
$$\text{and} \quad s_3 - s_2 = b + 5c = 3^{rd} \text{ term};$$

∴ the first three terms are $a + b + c$, $b + 3c$, $b + 5c$;

∴ after the first term the series is an A.P. whose common diff. is $2c$. Also the nth term $= b + (2n - 1)c$.

§19. The nth term $= 4n^3 - 6n^2 + 4n - 1$,

∴ $S = 4\Sigma n^3 - 6\Sigma n^2 + 4\Sigma n - n = n^4$, after reduction.

§20. Let x, y be the two quantities, then

$$y - A_2 = A_1 - x, \text{ or } x + y = A_1 + A_2 \tag{31}$$

$$\frac{y}{G_2} = \frac{g_1}{x}, \text{ or } xy = G_1 G_2 \tag{32}$$

$$\frac{1}{y} - \frac{1}{H_2} = \frac{1}{H_1} - \frac{1}{x}, \text{ or } \frac{x+y}{xy} = \frac{H_1 + H_2}{H_1 H_2} \tag{33}$$

Divide (31) by (32) and equate to (33).

§21. We have $p = \dfrac{na + b}{n + 1}, \qquad q = \dfrac{ab(n + 1)}{a + nb}$;

by eliminating b we get the equation

$$na^2 - a\{(n + 1)p + (n - 1)q\} + npq = 0.$$

For real roots we must have

$$\{(n + 1)p + (n - 1)q\}^2 - 4n^2 pq \text{ positive};$$
$$\text{that is, } (n + 1)^2 p^2 - 2pq(n^2 + 1) + (n - 1)^2 q^2,$$
$$\text{or } \{(n + 1)^2 p - (n - 1)^2 q\}(p - q) \text{ must be positive:}$$

∴ q cannot lies between p and $\left(\dfrac{n+1}{n-1}\right)^2 p$.

§22.

$$S = \Sigma \left(a + \overline{n - 1} \cdot d\right)^3$$
$$= na^3 + \frac{3a^2 d(n - 1)n}{2} + \frac{3ad^2(n - 1)n(2n - 1)}{6} + \frac{d^3(n - 1)^2 n^2}{4}$$
$$= \frac{n}{4}\left\{4a^3 + 6a^2 d(n - 1) + 2ad^2(n - 1)(2n - 1) + d^3 n(n - 1)^2\right\}$$
$$= \frac{n}{4}\left(2a + \overline{n - 1} \cdot d\right)\left\{2a^2 + 2(n - 1)ad + n(n - 1)d^2\right\}$$
$$= \frac{n}{2}\left(2a + \overline{n - 1} \cdot d\right)\left\{a^2 + (n - 1)ad + \frac{n(n - 1)}{2}d^2\right\},$$

which proves the proposition, since $\dfrac{n(n - 1)}{2}$ is an integer.

EXAMPLES. VI b

§4. Place on the given pile a triangular pile having 13 shot in each side of the base; then

$$\text{No. of shot in the complete pile} = \frac{25 \cdot 26 \cdot 27}{6},$$

$$\text{No. of shot in the added pile} = \frac{13 \cdot 14 \cdot 15}{6},$$

$$\therefore \text{ required number} = 3 \times \frac{13}{6}\{50 \times 9 - 14 \times 5\}$$
$$= 2470.$$

§5. The required number is $\dfrac{40\cdot 41\cdot 81}{6} - \dfrac{13\cdot 14\cdot 27}{6}$, which reduces to 21321.

§6. We have to find m from the equation
$$\frac{34\cdot 35(3m-33)}{6} = 23495,$$
$17\cdot 35(m-11) = 23495$, whence $m = 52$.

§7. The no. of shots in a complete pile which has 33 in a side of the base is $\dfrac{33\times 34\times 67}{6}$, or $11\times 17\times 67$, that is 12529.

In a pile which has 12 shot in each side of the base there are $\dfrac{12\times 13\times 25}{6}$, or 650 shot;

∴ the required number $= 12529 - 650 = 11879$.

§8. Since there are 15 courses, and the pile is complete, $n = 15$, and $m = 20$;

∴ by the formula of Art. 73, the number is 1840.

§9. Add a rectangular pile having 10 and 17 shot in the sides of its base, then the no. of shot in this pile is $\dfrac{10\times 11\times 42}{6}$, or 770. Also there are 20 courses, so that the base of the complete pile has 30 and 37 shot in its sides ;

∴ no. of shot in the complete pile $= \dfrac{30\times 31\times 82}{6} = 12710$;

∴ no. in the incomplete pile $= 11940$.

§10. By formula of Art.73 the required number is $\dfrac{5\times 6\times 38}{6}$, or 190.

§11. Let n be the no. of layers, then, by Arts. 71 and 72, we have
$$\frac{n(n+1)(n+2)}{6} - \frac{n(n+1)(2n+1)}{12} = 150,$$
and we have to find the value of $\dfrac{n(n+1)}{2}$.

Now $\dfrac{n(n+1)}{2}\left\{\dfrac{n+2}{3} - \dfrac{2n+1}{6}\right\} = 150$, whence $\dfrac{n(n+1)}{2} = 300$.

§12. Let n be the number of shot in a side of the base, then we have
$$n^2 - (n-15)^2 = 1005, \text{ whence } n = 41.$$
We have now to find the number of shot in an incomplete square pile of 16 courses when there are 41 shot in a side of the base. This is
$$\frac{41\cdot 42\cdot 83}{6} - \frac{25\cdot 26\cdot 51}{6}.$$
which reduces to 18296.

§13. We have to shew that
$$\frac{n(n+1)(2n+1)}{6} = \frac{1}{4}\cdot\frac{2n(2n+1)(2n+2)}{6}.$$

§14. We have $\dfrac{n(n+1)(n+2)}{2n(2n+1)(4n+1)} = \dfrac{13}{175}$;

whence $11n^2 - 123n - 108 = 0$,
or $(11n+9)(n-12) = 0$; whence $n = 12$.

Thus the number of shot in triangular pile $= \dfrac{12\cdot 13\cdot 14}{6} = 364$;

and the number of shot in square pile $= \dfrac{24\cdot 25\cdot 49}{6} = 4900$.

§15. The no. of shot in the pile $= \dfrac{51 \times 20}{10\frac{1}{2}} \times \dfrac{112}{16} = 680,$

$$\therefore \frac{n(n+1)(n+2)}{6} = 680,$$

$$n(n+1)(n+2) = 6 \times 17 \times 40 = 15 \times 16 \times 17;$$

whence $n = 15, \therefore \dfrac{n(n+1)}{2} = 120.$

§16. The number of shot in square pile $= \dfrac{n(n+1)(2n+1)}{6}.$

The number of shot in triangular pile $= \dfrac{n(n+1)(n+2)}{6}.$

The difference $= \dfrac{n(n+1)}{6}(2n+1-n-2) = \dfrac{(n-1)n(n+1)}{6}$, and this is the number of shot in a triangular pile which has $n-1$ shot in a side of the base.

EXAMPLES VII : Scales of Notation

EXAMPLES VII a

§1.
```
 23241
  4032
300421
──────
333244
```

§2.
```
303478
150732
264305
──────
728626
```

§3.
```
3673124
1732765
───────
1740137
```

§4.
```
3te756
2e46t2
──────
 e7074
```

§5.
```
   1131315
    235143
4)452132
  ──────
  112022
```

§6.
```
  6431
    35
─────
 45115
 25623
──────
334345
```

§7.
```
 4685
 3483
─────
15276
42154
21072
15276
────────
17832126
```

§8.
```
36)102432(1625
   36
   ───
   334
   321
   ───
   133
   105
   ───
    252
    252
```

§9.
```
       11022201
        121012
1201)10201112(2012
     10102
     ─────
      2211
      1201
      ─────
      10102
      10102
```

§10.
```
          3̇0̇0i1i̇4342
   114|14        |
  1232|1101
      |1021
      |────
      |3014
      |3014
```

§11.
```
   tttt
   tttt
  ─────
  9ttt1
  9ttt1
  9ttt1
  9ttt1
  ────────
  ttt90001
```

Or thus : $tttt = 10000 - 1,$
and $(10000 - 1)^2 = 100000000 + 1 - 20000$
$= ttt90001.$

§12.
```
1,2541,3102,1
 231 |2541|
1|231 | 231|
 |231 |
 ──── ────
```

§13.
```
6541)14332216(1456
     6541
     ────
     44612
     36124
     ─────
     54551
     45665
     ─────
     55536
     55536
```

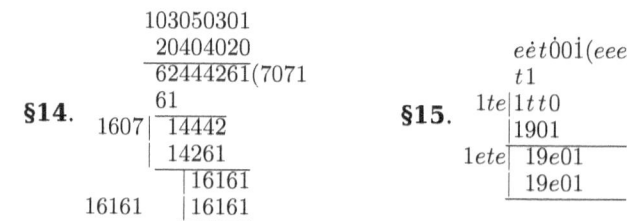

§14.

```
        103050301
         20404020
         62444261(7071
         61
  1607| 14442
      | 14261
        |16161
  16161 |16161
```

§15.

```
         eėtȮ0ì(eee
          t1
    1te|1tt0
       |1901
  1ete| 19e01
     | 19e01
```

§16.

```
   2|3102|31141|10;
    |242 |3102
   2| 242|  121
    | 242|
```

∴ the G. C. M. = 121.

In the scale of six we have

23 = 3 × 5,	24 = 4 × 4,
30 = 3 × 3 × 2,	32 = 4 × 5,
40 = 2 × 3 × 4,	41 = 5 × 5,
43 = 3 × 3 × 3,	50 = 3 × 5 × 2;

∴ the L. C. M. $= 3^3 \times 5^2 \times 4^2 = 122000.$

EXAMPLES VII b

§1.
```
7)4954
7)707...5
7)101...0
7)14...3
   2...0
```

§2.
```
5)624
5)124...4
5)24...4
   4...4
```

§3.
```
2)206
2)103...0
2)51...1
2)25...1
2)12...1
2)6...0
2)3...0
  1...1
```

§4.
```
3)1458
3)486...0
3)162...0
3)54...0
3)18...0
3)6...0
  2...0
```

§5.
```
9)5381
9)597...8
3)66...3
  7...3
```

§6.
```
5)212231
5)13233...2
5)1203...0
5)103...4
    3...4
```

§7.
```
t)398e
t)46t...7
t)55...8
  6...5
```

§8.
```
e)6t12
e)756...8
e)81...7
  8...9
```

§9.
```
9)213014
9)1300 1...1
9)100 0...1
9)40...0
  2...6
```

§10.
```
8)23861
8)2663...4
8)307...1
8)34...2
  3...7
```

§11.
```
5)400803
5)71872...2
5)13885...4
5)2534...3
5)460...4
5)83...3
5)16...0
   3...0
```

§12.
```
T)20665152
T)1151414...3
T)50500...e
T)2655...t
T)151...0
T)10...1
   0...7
```

§13.

```
      ttteee
       T
      130
       T
     1570
       T
    18851
       T
   226223
       T
  2714687
```

or

```
t)tttteee
t)111124...7
t)13862...8
t)16t2...6
t)1t7...4
t)23...1
   2...7
```

§14. $\dfrac{3 \times 7}{10} = 2\dfrac{1}{10}$; $\dfrac{1 \times 7}{10} = 0\dfrac{7}{10}$; $\dfrac{7 \times 7}{10} = 4\dfrac{9}{10}$; $\dfrac{9 \times 7}{10} = 6\dfrac{3}{10}$;

after this the figures recur.

§15. 17 in scale ten = 15 in scale twelve.

```
·15625
    T
1·875
    T
t·5
  T
6
```

Or thus: $.15625 = \dfrac{5}{32}$; $\dfrac{5 \times 12}{32} = 1\dfrac{7}{8}$; $\dfrac{7 \times 12}{8} = t + \dfrac{1}{2}$; $\dfrac{1 \times 12}{2} = 6.$

§16.

```
9)200
   2...0
```

```
  ·211
    9
 7·1
   9
   3
```

§17.

```
8)71
8)t...5
  1...2
```

```
  ·03
   8
0·2
  8
1·4
  8
2·8
  8
5·4
```

after this the

figures recur.

§18. The septenary numbers 1552 and 2626 are equal to the denary numbers 625 and 1000 respectively; and $\dfrac{625}{1000} = \dfrac{5}{8}$.

§19. $\dot{4} = .44444\ldots = \dfrac{4}{7} + \dfrac{4}{7^2} + \dfrac{4}{7^3} + \ldots = \dfrac{4}{7} \div \left(1 - \dfrac{1}{7}\right) = \dfrac{4}{6} = \dfrac{2}{3}.$

$.\dot{4}\dot{2} = .42424242\ldots = \left(\dfrac{4}{7} + \dfrac{2}{7^2}\right) + \left(\dfrac{4}{7^3} + \dfrac{2}{7^4}\right) + \left(\dfrac{4}{7^5} + \dfrac{2}{7^6}\right) + \ldots$

$= \left(\dfrac{4}{7} + \dfrac{2}{7^2}\right) \div \left(1 - \dfrac{1}{7^2}\right) = \dfrac{30}{48} = \dfrac{5}{8}.$

§20. If r be the radix of the scale, then

$$182 = 2r^2 + 2r + 2; \text{ that is } r^2 + r - 90 = 0, \text{ or } r = 9.$$

§21. Let r denote the radix of the scale, then

$$\dfrac{25}{128} = \dfrac{3}{r^2} + \dfrac{2}{r^4}; \text{ that is } 25r^4 - 384r^2 - 256 = 0,$$

or $(25r^2 + 16)(r^2 - 16) = 0$; thus $r = 4$.

§22.

Here $\qquad\qquad 5r^2 + 5r + 4 = (2r + 4)^2;$

that is, $\qquad\qquad r^2 - 11r - 12 = 0, \text{ or } r = 12.$

§23. The second number appears the greater, and therefore its radix is less than ten; also the radix must be greater than 7; thus the radix is either 8 or 9; and by trial we find that it is 8.

§24.

Here
$$(4r^2 + 7r + 9) + (9r^2 + 7) = 2(6r^2 + 9r + 8);$$

that is,
$$r^2 - 11r = 0, \text{ or } r = 11.$$

§25.

Here
$$\left(\frac{1}{r} + \frac{6}{r^2}\right)\left(\frac{2}{r} + \frac{8}{r^2}\right) = \left(\frac{2}{r}\right)^2,$$

that is
$$2 - \frac{20}{r} - \frac{48}{r^2} = 0,$$

or
$$r^2 - 10r - 24 = 0, \text{ and } r = 12.$$

§26. The second number appears the smaller; hence the radix must be greater than 6; also it must be greater than 8; hence it must be one of the numbers $9, 10, \ldots$; by trial we find that it is 10.

§27. $r^2 + 4r + 8 + \dfrac{8}{r} + \dfrac{4}{r^2}$ is the square of $r + 2 + \dfrac{2}{r}$.

§28. $r^6 + 2r^5 + 3r^4 + 4r^3 + 3r^2 + 2r + 1$ is the square of $r^3 + r^3 + r + 1$.

§29. $1 + \dfrac{3}{r} + \dfrac{3}{r^2} + \dfrac{1}{r^3}$ is the cube of $1 + \dfrac{1}{r}$.

§30. One tone= 2240 lbs., and we have to express 2240 in the binary scale.

```
2)2240
2)1120...0
 2)560...0
 2)280...0
 2)140...0
  2)70...0
  2)35...0
  2)17...1
   2)8...1
   2)4...0
   2)2...0
    1...0
```

Thus, $2240 = 2^{11} + 2^7 + 2^6$.

§31. We proceed as in the last Example, and express 10000 in the scale of three.

```
3)10000
3)3333...1
3)1111...0
 3)370...1
 3)123...1
  3)41...0
  3)14...-1
   3)5...-1
   3)2...-1
    1...-1
```

In dividing 41 by 3 we have a quotient 13 and remainder 2; since, however, only one weight of each kind is to be used we put 14 as the quotient and -1 as the remainder, the negative sign indicating that the corresponding weight 3^5 is to be placed in the opposite scale to those indicated by the positive remainders. Thus, weights 3^9, 3^3, 3^2, 1 must be placed in one scale and 3^8, 3^7, 3^6, 3^5 in the other scale.

§32. This follows from the fact that
$$r^5 + 3r^5 + 6r^4 + 7r^3 + 6r^2 + 3r + 1 = (r^2 + r + 1)^3.$$

§33. Let the number be denoted by
$$a \cdot 10^n + b \cdot 10^{n-1} + c \cdot 10^{n-2} + \ldots + p \cdot 10^3 + q \cdot 10^2 + r \cdot 10 + s;$$
now 10^3, 10^4, $10^5, \ldots$ are all divisible by 8; hence the number is divisible by 8 if $q \cdot 10^2 + r \cdot 10 + s$ is divisible by 8.

§34. Since $r = s - 1$, the number $rrrr$ in the scale of s is equal to $10000 - 1$, and the square of this is $100000000 + 1 - 20000$; hence we have the result, since $s - 2 = q$, and $s - 1 = r$.

§35. Let s denote the sum of the digits; then $\dfrac{N - S}{r - 1}$ and $\dfrac{N' - S}{r - 1}$ are both integers.   [Art. 88.]

Hence $\dfrac{N - N'}{r - 1}$ is also an integer.

§36. Let $2n$ denote the number of digits; then the number may be represented by
$$ar^{2n-1} + br^{2n-2} + cr^{2n-3} + \ldots + cr^2 + br + a.$$
This expression may be written
$$a\left(r^{2n-1} + 1\right) + br\left(r^{2n-3} + 1\right) + cr^2\left(r^{2n-5} + 1\right) + \ldots,$$
and is therefore divisible by $r + 1$.

§37. It follows from Art. 82 that $\dfrac{N - S_1}{9}$ is an integer; hence $\dfrac{N - S_1}{3}$ is also an integer.

Similarly $\dfrac{3N - 3S_2}{9}$, or $\dfrac{N - S_2}{3}$ is an integer.

Hence $\dfrac{S_1 - S_2}{3}$ is an integer.

§38. The number will be denoted by $abcabc$; thus the number
$$= a \cdot 10^5 + b \cdot 10^4 + c \cdot 10^3 + a \cdot 10^2 + b \cdot 10 + c$$
$$= a(10^3 + 1)10^2 + b(10^3 + 1)10 + c(10^3 + 1)$$
$$= (10^3 + 1)(a \cdot 10^2 + b \cdot 10 + c).$$
Thus the number is divisible by 1001, that is by $7 \times 11 \times 13$.

This is a particular case of Example 40.

§39. Let N be the number, S the sum of its digits, and r the radix; then $N - S = I(r - 1)$, where I is an integer. But $r - 1$ is even; hence $N - S$ is even, and therefore N and S are either both even or both odd.

§40. Denote ten by t, and let the number be
$$p_1 t^{n-1} + p_2 t^{n-2} + \ldots + p_{n-1} t + p_n;$$
on repeating the n digits $p_1, p_2, p_3, \ldots p_n$ the new number will be
$$p_1 t^{2n-1} + p_2 t^{2n-2} + \ldots + p_{n-1} t^{n+1} + p_n t^n$$
$$+ p_1 t^{n-1} + p_2 t^{n-2} + \ldots + p_{n-1} t + p_n$$
$$= \left(p_1 t^{n-1} + p_2 t^{n-2} + \ldots + p_{n-1} t + p_n\right) t^n$$
$$+ \left(p_1 t^n + p_2 t^{n-2} + \ldots + p_{n-1} t + p_n\right)$$
$$= \left(p_1 t^{n-1} + p_2 t^{n-2} + \ldots + p_{n-1} t + p_n\right)(t^n + 1).$$
Thus the number is divisible by the original number and also by $t^n + 1$.

Also, since n is odd, $t^n + 1$ is divisible by $t + 1$, that is by eleven, and it can easily be seen that the quotient is
$$9090 \ldots 9091;$$

$$\therefore 100001 = 11 \times 9091; \ 10000001 = 11 \times 909091.$$

EXAMPLES VIII : Surds and Imaginary Quantities

EXAMPLES VIII a

§1.

$$\frac{1}{1+\sqrt{2}-\sqrt{3}} = \frac{1+\sqrt{2}+\sqrt{3}}{(1+\sqrt{2})^2-(\sqrt{3})^2}$$

$$= \frac{1+\sqrt{2}+\sqrt{3}}{2\sqrt{2}} = \frac{\sqrt{2}+2+\sqrt{6}}{4}.$$

§2. $\dfrac{\sqrt{2}}{\sqrt{2}+\sqrt{3}-\sqrt{5}} = \dfrac{\sqrt{2}(\sqrt{2}+\sqrt{3}+\sqrt{5})}{(\sqrt{2}+\sqrt{3})^2-(\sqrt{5})^2} = \dfrac{\sqrt{2}+\sqrt{3}+\sqrt{5}}{2\sqrt{3}}$

$$= \frac{\sqrt{6}+\sqrt{3}+\sqrt{15}}{6}.$$

§3.

$$\frac{1}{\sqrt{a}+\sqrt{b}+\sqrt{a+b}} = \frac{\sqrt{a}+\sqrt{b}-\sqrt{a+b}}{(\sqrt{a}+\sqrt{b})^2-(a+b)} = \frac{\sqrt{a}+\sqrt{b}-\sqrt{a+b}}{2\sqrt{ab}}$$

$$= \frac{a\sqrt{b}+b\sqrt{a}-\sqrt{ab(a+b)}}{2ab}.$$

§4.

$$\frac{2\sqrt{a+1}}{\sqrt{a-1}+\sqrt{a+1}-\sqrt{2a}} = \frac{2\sqrt{a+1}(\sqrt{a-1}+\sqrt{a+1}+\sqrt{2a})}{(\sqrt{a-1}+\sqrt{a+1})^2-(\sqrt{2a})^2}$$

$$= \frac{\sqrt{a-1}+\sqrt{a+1}+\sqrt{2a}}{\sqrt{a-1}}$$

$$= \frac{a-1+\sqrt{a^2-1}+\sqrt{2a(a-1)}}{a-1}.$$

§5.

$$\text{The expression} = \frac{(\sqrt{10}+\sqrt{5}-\sqrt{3})(\sqrt{10}+\sqrt{5}+\sqrt{3})}{(\sqrt{10}+\sqrt{3}-\sqrt{5})(\sqrt{10}+\sqrt{3}+\sqrt{5})}$$

$$= \frac{(15+10\sqrt{2})-3}{(13+2\sqrt{30})-5} = \frac{(6+5\sqrt{2})(\sqrt{30}-4)}{30-16}.$$

§6.

$$\text{The expression} = \frac{5+\sqrt{15}+\sqrt{10}+\sqrt{6}}{\sqrt{2}+\sqrt{3}+\sqrt{5}}$$

$$= \sqrt{5}+\frac{\sqrt{6}}{\sqrt{2}+\sqrt{3}+\sqrt{5}}$$

$$= \sqrt{5}+\frac{\sqrt{6}(\sqrt{2}+\sqrt{3}-\sqrt{5})}{2\sqrt{6}}.$$

§13.

$$\text{The expression} = \frac{(3^{\frac{1}{3}}-1)(3^{\frac{2}{3}}-3^{\frac{1}{3}}+1)}{3-1}$$

$$= \frac{3-2\cdot 3^{\frac{2}{3}}+2\cdot 3^{\frac{1}{3}}-1}{2}.$$

§14.

$$\text{The expression} = \frac{3^{\frac{1}{3}}-2^{\frac{1}{2}}}{3^{\frac{1}{3}}+2^{\frac{1}{2}}}$$

$$= \frac{(3^{\frac{1}{3}}-2^{\frac{1}{2}})(3^{\frac{5}{3}}-3^{\frac{4}{3}}\cdot 2^{\frac{1}{2}}+3^{\frac{3}{3}}\cdot 2^{\frac{2}{2}}-3^{\frac{2}{3}}\cdot 2^{\frac{3}{2}}+3^{\frac{1}{3}}\cdot 2^{\frac{4}{2}}-2^{\frac{5}{2}})}{3^2-2^3};$$

the denominator is unity, and the numerator gives the result.

§15. The denominator is $3^{\frac{1}{3}} + 2^{\frac{1}{2}}$, hence as in the preceding example, the expression

$$= \frac{2^{\frac{1}{2}} \cdot 3^{\frac{1}{3}} (3^{\frac{5}{3}} - 3^{\frac{4}{3}} \cdot 2^{\frac{1}{2}} + 3^{\frac{3}{3}} \cdot 2^{\frac{2}{2}} - 3^{\frac{2}{3}} \cdot 2^{\frac{3}{2}} + 3^{\frac{1}{3}} \cdot 2^{\frac{4}{2}} - 2^{\frac{5}{2}})}{3^2 - 2^3};$$

§16.

The expression

$$= \frac{3^{\frac{1}{3}}}{3^{\frac{1}{2}} + 3^{\frac{1}{3}}} = \frac{1}{3^{\frac{1}{6}} + 1}$$
$$= \frac{3^{\frac{5}{6}} - 3^{\frac{4}{6}} + 3^{\frac{3}{6}} - 3^{\frac{2}{6}} + 3^{\frac{1}{6}} - 1}{3 - 1}.$$

§17.

The expression

$$= \frac{2^{\frac{3}{2}} + 2^{\frac{2}{3}}}{2^{\frac{3}{2}} - 2^{\frac{2}{3}}} = \frac{2^{\frac{5}{6}} + 1}{2^{\frac{5}{6}} - 1}$$
$$= \frac{(2^{\frac{5}{6}} + 1)(2^{\frac{25}{6}} + 2^{\frac{20}{6}} + 2^{\frac{15}{6}} + 2^{\frac{10}{6}} + 2^{\frac{5}{6}} + 1)}{2^5 - 1}$$
$$= \frac{1}{31} \left(2^5 + 2 \cdot 2^{\frac{25}{6}} + 2 \cdot 2^{\frac{20}{6}} + 2 \cdot 2^{\frac{15}{6}} + 2 \cdot 2^{\frac{10}{6}} + 2 \cdot 2^{\frac{5}{6}} + 1 \right).$$

§18.

The expression $= \dfrac{3^{\frac{1}{2}}}{3 - 3^{\frac{1}{3}}} = \dfrac{3^{\frac{1}{6}}}{3^{\frac{2}{3}} - 1}$

$$= \frac{3^{\frac{1}{6}}(3^{\frac{4}{3}} + 3^{\frac{2}{3}} + 1)}{3^2 - 1} = \frac{1}{8} \left(3^{\frac{3}{2}} + 3^{\frac{5}{6}} + 3^{\frac{1}{6}} \right).$$

Examples 19 to 24 are solved by the method of Art. 87; the results however may generally be written down by inspection; thus in Ex.19 the quantities $20, 28, 35$ under the radicals are the products of the numbers $4, 5, 7$ taken two at a time; and the sum of these numbers is 16;

$$\therefore \ 16 - 2\sqrt{20} - 2\sqrt{28} + 2\sqrt{35} = (\sqrt{5} + \sqrt{7} - \sqrt{4})^2;$$

the two quantities $\sqrt{5}, \sqrt{7}$ having the same sign, because of the term $+2\sqrt{35}$.

§21. $6 + \sqrt{12} - \sqrt{24} - \sqrt{8} = 6 + 2\sqrt{3} - 2\sqrt{6} - 2\sqrt{2} = (\sqrt{3} + 1 - \sqrt{2})^2.$

§22. $5 - \sqrt{10} - \sqrt{15} + \sqrt{6} = \dfrac{1}{2} \left(10 - 2\sqrt{10} - 2\sqrt{15} + 2\sqrt{6} \right) = \dfrac{1}{2}(\sqrt{3} + \sqrt{2} - \sqrt{5})^2.$

§23. $a + 3b + 4 + 4\sqrt{a} - 4\sqrt{3b} - 2\sqrt{3ab} = (\sqrt{a} - \sqrt{3b} + 2)^2.$

§24. $21 + 3\sqrt{8} - 6\sqrt{3} - 6\sqrt{7} - \sqrt{24} - \sqrt{56} + 2\sqrt{21}$
$= 21 + 2\sqrt{18} - 2\sqrt{27} - 2\sqrt{63} - 2\sqrt{6} - 2\sqrt{14} + 2\sqrt{21} = (\sqrt{9} + \sqrt{2} - \sqrt{7} - \sqrt{3})^2;$
the numbers $9, 2, 7, 3$ are seen by inspection, and the signs before the radicals easily assigned by trial.

§25. Proceeding as in Art.89, we shall find $x^2 - y = \sqrt[3]{100 - 108} = -2$; and $x^3 + 3xy = 10$, whence $x = 1, \ y = 3$.

§26. Here $x^2 - y = \sqrt[3]{38^2 - 289 \times 5} = \sqrt[3]{1444 - 1445} = -1$; and $x^3 + 3xy = 38$; whence $x = 2, \ y = 5$.

§27. Here $x^2 - y = \sqrt[3]{9801 - 4900 \times 2} = \sqrt[3]{1} = 1$; and $x^3 + 3xy = 99$, whence $x = 3, \ y = 8$.

§28.

Here $\quad 38\sqrt{14} - 100\sqrt{2} = -2\sqrt{2}(50 - 19\sqrt{7})$;

and $\quad x^2 - y = \sqrt[3]{2500 - 361 \times 7} = \sqrt[3]{-27} = -3$;

also $x^3 + 3xy = 50$, whence $x = 2, \ y = 7$;

thus the cube root $= -\sqrt{2}(2 - \sqrt{7}) = \sqrt{14} - 2\sqrt{2}.$

§29.

$$\text{We have } 54\sqrt{3} + 41\sqrt{5} = 3\sqrt{3}\left(18 + \frac{41}{3}\sqrt{\frac{5}{3}}\right);$$

$$\text{here} \quad x^2 - y = \sqrt[3]{324 - \frac{1681}{9} \times \frac{5}{3}} = \sqrt[3]{\frac{343}{27}} = \frac{7}{3};$$

also $x^3 + 3xy = 18$, whence $x = 2, y = \dfrac{5}{3}$;

thus the cube root $= \sqrt{3}\left(2 + \sqrt{\dfrac{5}{3}}\right) = 2\sqrt{3} + \sqrt{5}.$

§30. We have $135\sqrt{3} - 87\sqrt{6} = 3\sqrt{3}(45 - 29\sqrt{2}).$

Here $x^2 - y = \sqrt[3]{2025 - 841 \times 2} = \sqrt[3]{343} = 7$;
and $x^3 + 3xy = 45$, whence $x = 3, \ y = 2$;
thus the cube root $= \sqrt{3}(3 - \sqrt{2}) = 3\sqrt{3} - \sqrt{6}.$

Examples 31 to 34 may be solved by inspection; thus

§31.

$$a + x + \sqrt{2ax + x^2} = (a + x) + 2\sqrt{\frac{x}{2}\left(a + \frac{x}{2}\right)}$$

$$= \left(\sqrt{\frac{x}{2}} + \sqrt{a + \frac{x}{2}}\right)^2.$$

§32.

$$2a - \sqrt{3a^2 - 2ab - b^2} = \frac{1}{2}\left(4a - 2\sqrt{(3a + b)(a - b)}\right)$$

$$= \frac{1}{2}\left(\sqrt{3a + b} - \sqrt{a - b}\right)^2.$$

§33.

$$1 + a^2 + \sqrt{1 + a^2 + a^4} = \frac{1}{2}\left(2 + 2a^2 + 2\sqrt{(1 + a + a^2)(1 - a + a^2)}\right)$$

$$= \frac{1}{2}\left(\sqrt{1 + a + a^2} + \sqrt{1 - a + a^2}\right)^2.$$

§34.

$$1 + (1 - a^2)^{-\frac{1}{2}} = 1 + \frac{1}{\sqrt{1 - a^2}} = \frac{1}{2\sqrt{1 - a^2}}\left(2 + 2\sqrt{1 - a^2}\right)$$

$$= \frac{1}{2\sqrt{1 - a^2}}\left(\sqrt{1 + a} + \sqrt{1 - a}\right)^2.$$

§35. Here $a = 2 + \sqrt{3}, \ b = 2 - \sqrt{3}$; thus $a + b = 4, \ a - b = 2\sqrt{3}, \ ab = 1$.
$$7a^2 + 11ab - 7b^2 = 7(a + b)(a - b) + 11ab = 56\sqrt{3} + 11.$$

§36. Here $x = 5 - 2\sqrt{6}, \ y = 5 + 2\sqrt{6}$; thus $x + y = 10, \ xy = 1$.
$$\therefore \ 3x^2 - 5xy + 3y^2 = 3(x + y)^2 - 11xy = 300 - 11 = 289.$$

§37.

$$\text{The expression} \quad = \frac{\sqrt{52 - 30\sqrt{3}}}{10 - \sqrt{76 + 10\sqrt{3}}} = \frac{3\sqrt{3} - 5}{10 - (5\sqrt{3} + 1)}$$

$$= \frac{3\sqrt{3} - 5}{9 - 5\sqrt{3}} = \frac{1}{\sqrt{3}}.$$

§38. Dividing numerator and denominator by $\sqrt{3}$, the expression under the radical

$$= \frac{2\sqrt{3} + 2}{11\sqrt{3} - 19} = \frac{2(\sqrt{3} + 1)(11\sqrt{3} + 19)}{2} = 52 + 30\sqrt{3} = (3\sqrt{3} + 5)^2.$$

§39. The expression $= (5 - \sqrt{3}) - \dfrac{1}{2 + \sqrt{3}} = (5 - \sqrt{3}) - (2 - \sqrt{3}).$

§40. The cube root of $26 + 15\sqrt{3}$ is $2 + \sqrt{3}$.
Hence the expression

$$= (2+\sqrt{3})^2 - \left(\frac{1}{2+\sqrt{3}}\right)^2$$
$$= (2+\sqrt{3})^2 - (2-\sqrt{3})^2$$
$$= 4 \times 2\sqrt{3} = 8\sqrt{3}.$$

§41. Multiply each numerator and denominator by $\sqrt{2}$; thus the expression

$$= \frac{20}{6 - \sqrt{6 + 2\sqrt{5}}} - \frac{2\sqrt{5}+6}{4 + \sqrt{6 - 2\sqrt{5}}}$$
$$= \frac{20}{5 - \sqrt{5}} - \frac{2(3+\sqrt{5})}{3+\sqrt{5}} = (5+\sqrt{5}) - 2.$$

§42. From the formula
$$a^3 + b^3 + c^3 - 3abc = (a+b+c)(a^2+b^2+c^2 - bc - ca - ab);$$
we have
$$x^3 + 2 - 1 + 3x\sqrt[3]{2} = x^3 + \left(\sqrt[3]{2}\right)^3 + (-1)^3 - 3x\left(\sqrt[3]{2}\right)(-1)$$
$$= \left(x + \sqrt[3]{2} - 1\right)\left(x^2 + \sqrt[3]{4} + 1 - x\sqrt[3]{2} + x + \sqrt[3]{2}\right).$$

§43. As in Art. 89 we have $x^3 + 3xy = 9ab^2$.
Again
$$(9ab^2)^2 - (b^2 + 24a^2)^2(b^2 - 3a^2)$$
$$= 1728a^6 - 432a^4b^2 + 36a^2b^4 - b^6 = (12a^2 - b^2)^3;$$
$$\therefore\ x^2 - y = 12a^2 - b^2;\text{ and thus } 4x^3 - 3x(12a^2 - b^2) - 9ab^2 = 0;$$

or
$$4x(x^2 - 9a^2) + 3b^2(x - 3a) = 0;\text{ whence } x = 3a,\ y = b^2 - 3a^2.$$

§44.
$$4x^2 - 4 = \left(\sqrt{a} + \frac{1}{\sqrt{a}}\right)^2 - 4 = \left(\sqrt{a} - \frac{1}{\sqrt{a}}\right)^2;$$
$$\therefore\ 2\sqrt{x^2 - 1} = \sqrt{a} - \frac{1}{\sqrt{a}}.$$

Thus the expression

$$= \frac{\sqrt{a} - \dfrac{1}{\sqrt{a}}}{\left(\sqrt{a} + \dfrac{1}{\sqrt{a}}\right) - \left(\sqrt{a} - \dfrac{1}{\sqrt{a}}\right)} = \frac{a-1}{2}.$$

EXAMPLES VIII b

§4. The product $= (x+\omega)(x+\omega^2) = x^2 + (\omega + \omega^2)x + \omega^3 = x^2 - x + 1.$

§5. We have $\dfrac{1}{3 - \sqrt{-2}} = \dfrac{3 + \sqrt{-2}}{9 - (-2)} = \dfrac{3 + \sqrt{-2}}{11}.$

§6. The expression $= \dfrac{3\sqrt{2} + 2\sqrt{5}}{3\sqrt{2} - 2\sqrt{5}} = \dfrac{(3\sqrt{2} + 2\sqrt{5})^2}{18 - 20}.$

§7. The expression
$$= \frac{(3+2i)(2+5i) + (3-2i)(2-5i)}{4 - (-25)}$$
$$= \frac{2(6 + 10i^2)}{29} = -\frac{8}{29}.$$

§8. The expression $= \dfrac{(a+ix)^2 - (a - ix)^2}{a^2 - i^2x^2} = \dfrac{4iax}{a^2 + x^2}.$

§9. The expression

$$= \frac{(x+i)^3 - (x-i)^3}{x^2 - i^2} = \frac{2(3ix^2 + i^3)}{x^2 + 1} = \frac{2i(3x^2 - 1)}{x^2 + 1}.$$

§10. The expression $= \dfrac{2(3ia^2 + i^3)}{4ia} = \dfrac{3a^2 + i^2}{2a} = \dfrac{3a^2 - 1}{2a}.$

§11.

$$\left(-\sqrt{-1}\right)^{4n+3} = (-1)^{4n+3} \times \left(\sqrt{-1}\right)^{4n+3}$$
$$= (-1) \times \left(\sqrt{-1}\right)^3 = (-1) \times \left(-\sqrt{-1}\right) = \sqrt{-1}.$$

§12.

The square $\quad = (9 + 40i) + (9 - 40i) + 2\sqrt{81 - 1600i^2}$
$$= 18 + 2\sqrt{1681} = 100.$$

Examples 13 to 18 may be solved by the method of Art. 105, or by inspection as follows.

§13. $-5 + 12\sqrt{-1} = -5 + 2\sqrt{-36} = -9 + 4 + 2\sqrt{-9 \times 4} = \left(\sqrt{-9} + 2\right)^2.$

§14.

$$-11 - 60\sqrt{-1} = -11 - 2\sqrt{-900}$$
$$= -36 + 25 - 2\sqrt{-36 \times 25} = \left(5 - \sqrt{-36}\right)^2.$$

§15.

$$-47 + 8\sqrt{-3} = -47 + 2\sqrt{-48}$$
$$= (-48 + 1 + 2\sqrt{-48}) = (1 + \sqrt{-48})^2.$$

§16. $-8\sqrt{-1} = 0 - 2\sqrt{-16} = 4 - 4 - 2\sqrt{-4 \times 4} = (2 - \sqrt{-4})^2.$

§17. $a^2 - 1 + 2a\sqrt{-1} = (a + \sqrt{-1})^2.$

§18.

$$4ab - 2(a^2 - b^2)\sqrt{-1} = (a+b)^2 - (a-b)^2 - 2(a^2 - b^2)\sqrt{-1}$$
$$= \left\{(a+b) - (a-b)\sqrt{-1}\right\}^2.$$

§19. We have $\dfrac{3 + 5i}{2 - 3i} = \dfrac{(3+5i)(2+3i)}{4 - 9i^2} = \dfrac{-9 + 19i}{13}.$

§20. $\dfrac{\sqrt{3} - i\sqrt{2}}{2\sqrt{3} - i\sqrt{2}} = \dfrac{(\sqrt{3} - i\sqrt{2})(2\sqrt{3} + i\sqrt{2})}{12 - 2i^2} = \dfrac{8 - i\sqrt{6}}{14}.$

§21. We have $\dfrac{1+i}{1-i} = \dfrac{(1+i)(1+i)}{1 - i^2} = \dfrac{1 + i^2 + 2i}{2} = i.$

§22. $\dfrac{(1+i)^2}{3-i} = \dfrac{1 + i^2 + 2i}{3-i} = \dfrac{2i}{3-i} = \dfrac{2i(3+i)}{9 - i^2} = \dfrac{6i + 2i^2}{10} = \dfrac{3i - 1}{5}.$

§23. The expression $= \dfrac{(a + ib)^3 - (a - ib)^3}{(a + ib)(a - ib)} = \dfrac{2ib(3a^2 - b^2)}{a^2 + b^2}.$

§24. We have $1 + \omega^2 = -\omega$; thus $(1 + \omega^2)^4 = (-\omega)^4 = \omega^4 = \omega.$

§25. We have $1 - \omega + \omega^2 = (1 + \omega + \omega^2) - 2\omega = 0 - 2\omega = -2\omega.$

Similarly $\qquad\qquad 1 + \omega - \omega^2 = -2\omega^2.$

The product is $\qquad\qquad 4\omega^3 = 4.$

§26. Since $1 - \omega^4 = 1 - \omega$ and $1 - \omega^5 = 1 - \omega^2,$
the expression

$$= (1 - \omega)^2(1 - \omega^2)^2$$
$$= (1 - 2\omega + \omega^2)(1 - 2\omega^2 + \omega^4) = (-3\omega)(-3\omega^2) = 9.$$

§27. $2 + 5\omega + 2\omega^2 = 2(1 + \omega + \omega^2) + 3\omega = 3\omega,$ and $(3\omega)^6 = 729\omega^6 = 729.$

The solution of the second part is similar.

§28. The factors are equal to $1 - \omega + \omega^2$ and $1 - \omega^2 + \omega$ alternately, and the product of each pair is 2^2. (Ex. 25.)

§29.
$$x^3 + y^3 + z^3 - 3xyz = (x + y + z)(x^2 + y^2 + z^2 - yz - zx - xy)$$
$$= (x + y + z)(x + \omega y + \omega^2 z)(x + \omega^2 y + \omega z).$$
Ex. 3, Art. 110.

§30. $yz = (\omega a + \omega^2 b)(\omega^2 a + \omega b) = \omega^3 a^2 + \omega^3 b^2 + (\omega^2 + \omega^4)ab = a^2 - ab + b^2,$
$$y + z = (\omega + \omega^2)a + (\omega + \omega^2)b = -a - b.$$

Hence

(1) $xyz = (a + b)(a^2 - ab + b^2) = a^3 + b^3.$

(2)
$$x^2 + y^2 + z^2 = x^2 + (y + z)^2 - 2yz$$
$$= (a + b)^2 + (a + b)^2 - 2(a^2 - ab + b^2) = 6ab.$$

(3)
$$x^3 + y^3 + z^3 = x^3 + (y + z)(y^2 + z^2 - yz)$$
$$= x^3 + (y + z)\left\{(y + z)^2 - 3yz\right\}$$
$$= (a + b)^3 - (a + b)\left\{(a + b)^2 - 3(a^2 - ab + b^2)\right\}$$
$$= 3(a^3 + b^3).$$

EXAMPLES IX : The Theory of Quadratic Equations

EXAMPLES IX a

§13. If the roots of $Ax^2 + Bx + C = 0$ are real, then $B^2 - 4AC$ is positive.
Now in (1), $4a^2 - 4(a^2 - b^2 - c^2) = 4b^2 + 4c^2$, a positive quantity.
Again in (2), $16(a - b)^2 - 4(a - b + c)(a - b - c)$
$= 16(a - b)^2 - 4(a - b)^2 + 4c^2 = 12(a - b)^2 + 4c^2$, a positive quantity.
§14. Applying the test for equal roots to the equation
$$x^2 - 2mx + 8m - 15 = 0,$$
we have $\qquad m^2 = 8m - 15$; that is $(m - 5)(m - 3) = 0$.
§15. If the roots are equal $(1 + 3m)^2 = 7(3 + 2m)$ or $9m^2 - 8m - 20 = 0$.
that is $\qquad\qquad (9m + 10)(m - 2) = 0$.
§16. On reduction we have $(m + 1)x^2 - bx(m + 1) = ax(m - 1) - c(m - 1)$,
$$\therefore (m + 1)x^2 - \{b(m + 1) + a(m - 1)\}x + c(m - 1) = 0.$$
The required condition is obtained by equating to zero the coefficient
of x.
§17. If the roots of $Ax^2 + Bx + C = 0$ are rational, $B^2 - 4AC$ must be a
perfect square.
In (1),
$$4c^2 - 4(c + a - b)(c - a + b)$$
$$= 4c^2 - 4c^2 + 4(a - b)^2 = 4(a - b)^2, \text{ a perfect square.}$$
In (2),
$$(3a^2 + b^2)^2 c^2 - 4abc^2(-6a^2 - ab + 2b^2)$$
$$= c^2(9a^4 + 24a^3b + 10a^2b^2 - 8ab^3 + b^4)$$
$$= c^2(3a^2 + 4ab - b^2)^2 = \text{ a perfect square.}$$
In Examples 18 to 20 we have
$$\alpha + \beta = -\frac{b}{a}, \ \alpha\beta = \frac{c}{a}; \text{ whence } \alpha^2 + \beta^2 = \frac{b^2 - 2ac}{a^2}.$$
§18. $\qquad \dfrac{1}{\alpha^2} + \dfrac{1}{\beta^2} = \dfrac{\alpha^2 + \beta^2}{\alpha^2\beta^2} = \dfrac{b^2 - 2ac}{a^2} \div \dfrac{c^2}{a^2} = \dfrac{b^2 - 2ac}{c^2}.$
§19.
$$\alpha^4\beta^7 + \alpha^7\beta^4 = \alpha^4\beta^4(\alpha^3 + \beta^3) = \alpha^4\beta^4(\alpha + \beta)(\alpha^2 + \beta^2 - \alpha\beta)$$
$$= \frac{c^4}{a^4}\left(-\frac{b}{a}\right)\frac{b^2 - 3ac}{a^2} = \frac{bc^4(3ac - b^2)}{a^7}.$$
§20.
$$\left(\frac{\alpha}{\beta} - \frac{\beta}{\alpha}\right)^2 = \frac{(\alpha^2 - \beta^2)^2}{\alpha^2\beta^2}$$
$$= \frac{(\alpha + \beta)^2(\alpha - \beta)^2}{\alpha^2\beta^2} = \frac{(\alpha + \beta)^2\{(\alpha + \beta)^2 - 4\alpha\beta\}}{\alpha^2\beta^2}$$
$$= \frac{b^2}{a^2}\left(\frac{b^2 - 4ac}{a^2}\right) \div \frac{c^2}{a^2} = \frac{b^2(b^2 - 4ac)}{a^2c^2}.$$
§21. Form the quadratic equation whose roots are $1 \pm 2i$. This equation
is $x^2 - 2x + 5 = 0$. Therefore $x^2 - 2x + 5$ is a quadratic expression which
vanishes for each of the values $1 + 2i$, $1 - 2i$.
Now
$$x^3 + x^2 - x + 22 = x(x^2 - 2x + 5) + 3(x^2 - 2x + 5) + 7$$
$$= x \times 0 + 3 \times 0 + 7 = 7.$$
§22. The equation whose roots are $3 \pm i$ is $x^2 - 6x + 10 = 0$.

Now
$$x^3 - 3x^2 - 8x + 15 = x(x^2 - 6x + 10) + 3(x^2 - 6x + 10) - 15 = -15.$$
§23. The equation whose roots are $a(1 \mp \sqrt{-3})$ is $x^2 - 2ax + 4a^2 = 0$.
Now
$$x^3 - ax^2 + 2a^2x + 4a^3 = x(x^2 - 2ax + 4a^2) + a(x^2 - 2ax + 4a^2) = 0.$$
§24. Here $\qquad \alpha + \beta = -p, \ \alpha\beta = q.$
$$\text{Sum of roots } = (\alpha + \beta)^2 + (\alpha - \beta)^2 = 2(\alpha^2 + \beta^2) = 2(p^2 - 2q).$$
$$\text{Product of roots } = (\alpha + \beta)^2(\alpha - \beta)^2 = p^2(p^2 - 4q).$$
§25. In the equation $x^2 - (a+b)x + ab - h^2 = 0$, the condition for real roots is that $(a + b)^2 - 4(ab - h^2)$ should be positive; that is, $(a - b)^2 + 4h^2$ must be positive, which is clearly the case.
§26. From the equation $ax^2 + bx + c = 0$, we have $ax^2 + bx = -c$, that is,
$$ax + b = -\frac{c}{x} = -cx^{-1}; \text{ whence } (ax + b)^{-2} = (-cx^{-1})^{-2} = \frac{x^2}{c^2}.$$

In (1),
$$(ax_1 + b)^{-2} + (ax_2 + b)^{-2} = \frac{x_1^2 + x_2^2}{c^2}$$
$$= \frac{(x_1 + x_2)^2 - 2x_1x_2}{c_2} = \frac{1}{c^2}\left(\frac{b^2}{a^2} - \frac{2c}{a}\right).$$

In (2),
$$(ax_1 + b)^{-3} + (ax_2 + b)^{-3} = -\frac{x_1^3 + x_2^3}{c^3}$$
$$= -\frac{1}{c^3}(x_1 + x_2)(x_1^2 + x_2^2 - x_1x_2)$$
$$= -\frac{1}{c^3}\left(-\frac{b}{a}\right)\left(\frac{b^2}{a^2} - \frac{3c}{a}\right).$$
§27. Denote the roots by α and $n\alpha$; then
$$\alpha + n\alpha = -\frac{b}{a}, \text{ and } \alpha \times n\alpha = \frac{c}{a}.$$
Eliminating α, we have $\qquad \dfrac{nb^2}{a^2(1 + n)^2} = \dfrac{c}{a}.$
§28. Here $\alpha^2 + \beta^2 = \dfrac{b^2 - 2ac}{a^2},$
and $\alpha^{-2} + \beta^{-2} = \dfrac{\alpha^2 + \beta^2}{\alpha^2\beta^2} = \dfrac{b^2 - 2ac}{c^2};$
$$\therefore \text{ sum of roots } = \frac{(b^2 - 2ac)(a^2 + c^2)}{a^2c^2}, \text{ and}$$
$$\text{product } = \frac{(b^2 - 2ac)^2}{a^2c^2}.$$
§29. Here $\qquad \alpha + \beta = -(m + n), \qquad \alpha\beta = \dfrac{1}{2}(m^2 + n^2).$
$$\therefore (\alpha + \beta)^2 = (m + n)^2; \text{ and}$$
$$(\alpha - \beta)^2 = (m + n)^2 - 2(m^2 + n^2) = -(m - n)^2.$$
Thus we have to form the equation whose roots are $(m+n)^2, -(m-n)^2$; the sum of roots $= 4mn$, and product $= -(m + n)^2(m - n)^2$.

EXAMPLES IX b

§1. In the equation $2a^2x^2 + 2ancx + (n^2 - 2)c^2 = 0$, the condition for real roots is that $a^2n^2c^2 - 2a^2(n^2 - 2)c^2$ should be positive, that is, $4 - n^2$ should be positive. therefore n must lie between -2 and $+2$.

§2. Put $\dfrac{x}{x^2 - 5x + 9} = y$; then $yx^2 - (5y+1)x + 9y = 0$. If x is real, $(5y+1)^2 - 36y^2$ must be positive; $\therefore (1+11y)(1-y)$ must be positive; that is, y must lie between 1 and $-\dfrac{1}{11}$.

§3. Put $\dfrac{x^2 - x + 1}{x^2 + x + 1} = y$; then $(y-1)x^2 + (y+1)x + y - 1 = 0$. If x is real, $(y+1)^2 - 4(y-1)^2$ must be positive; $\therefore (y-3)(1-3y)$ must be positive.

§4. Put $\dfrac{x^2 + 34x - 71}{x^2 + 2x - 7} = y$; then $x^2(y-1) + 2(y-17)x - 7y + 71 = 0$. If x is real, $(y-17)^2 + (y-1)(7y-71)$ must be positive; $\therefore 8(y^2 - 14y + 45)$ must be positive; $\therefore 8(y-5)(y-9)$ must be positive.

§5. Sum of roots $= \dfrac{\sqrt{a}}{\sqrt{a} + \sqrt{a-b}} + \dfrac{\sqrt{a}}{\sqrt{a} - \sqrt{a-b}} = \dfrac{2a}{a - (a-b)} = \dfrac{2a}{b}$.

Product of roots $= \dfrac{\sqrt{a}}{\sqrt{a} + \sqrt{a-b}} \times \dfrac{\sqrt{a}}{\sqrt{a} - \sqrt{a-b}} = \dfrac{a}{b}$.

Hence the equation is $\qquad x^2 - \dfrac{2a}{b}x + \dfrac{a}{b} = 0$.

§6. (1) $\alpha^2(\alpha^2\beta^{-1} - \beta) + \beta^2(\beta^2\alpha^{-1} - \alpha)$

$= \dfrac{\alpha^2}{\beta}(\alpha^2 - \beta^2) + \dfrac{\beta^2}{\alpha}(\beta^2 - \alpha^2) = \dfrac{(\alpha^3 - \beta^3)(\alpha^2 - \beta^2)}{\alpha\beta}$

$= \dfrac{(\alpha + \beta)(\alpha - \beta)^2(\alpha^2 + \alpha\beta + \beta^2)}{\alpha\beta} = \dfrac{p}{q}(p^2 - 4q)(p^2 - q)$.

(2) From $x^2 - px + q = 0$, we have $x - p = \dfrac{q}{x}$; hence

$$(x - p)^{-4} = (qx^{-1})^{-4} = \dfrac{x^4}{q^4}.$$

Substituting α and β for x successively,

$$(\alpha - p)^{-4} + (\beta - p)^{-4} = \dfrac{\alpha^4 + \beta^4}{q^4}$$

$$= \dfrac{(\alpha^2 + \beta^2)^2 - 2\alpha^2\beta^2}{q^4}$$

$$= \dfrac{1}{q^4}\left\{ (p^2 - 2q)^2 - 2q^2 \right\}.$$

§7. Denote the roots by $p\alpha$ and $q\alpha$; then

$$p\alpha + q\alpha = -\dfrac{n}{l}; \ p\alpha \times q\times = \dfrac{n}{l}.$$

From the second equation $\alpha = \dfrac{1}{\sqrt{pq}} \cdot \sqrt{\dfrac{n}{l}}$.

Substituting in the first equation $\dfrac{p+q}{\sqrt{pq}} \cdot \sqrt{\dfrac{n}{l}} + \dfrac{n}{l} = 0$.

Dividing by $\sqrt{\dfrac{n}{l}}$ we have the required result.

§8. Put $\dfrac{(x+m)^2 - 4mn}{2x - 2n} = y$;

then $x^2 + 2(m - y)x + m^2 - 4mn + 2ny = 0$.
If x is real, $(m - y)^2 - m^2 + 4mn - 2ny$ must be positive;
$\therefore y^2 - (2m + 2n)y + 4mn$ must be positive;
that is, $(y - 2m)(y - 2n)$ must be positive.

§9. In the first equation we have $\alpha + \beta = -\dfrac{2b}{a}$, $\alpha\beta = \dfrac{c}{a}$;

$$\therefore (\alpha - \beta)^2 = \dfrac{4(b^2 - ac)}{a^2}.$$

Again, from the second equation we have
$$\{(\alpha + \delta) - (\beta + \delta)\}^2 = \frac{4(B^2 - AC)}{A^2};$$
$$\text{that is, } (\alpha - \beta)^2 = \frac{4(B^2 - AC)}{A^2};$$
whence the result follows.

§10. Put $\dfrac{px^2 + 3x - 4}{p + 3x - 4x^2} = y;$

then $(p + 4y)x^2 + 3x(1 - y) - (4 + py) = 0.$

If x is real, $9(1 - y)^2 + 4(p + 4y)(4 + py)$ must be positive;
$$\therefore (9 + 16p)y^2 + 2(2p^2 + 23)y + (9 + 16p) \text{ must be positive;}$$
$$\therefore (2p^2 + 23)^2 - (9 + 16p)^2 \text{ must be negative or zero,}$$
$$\text{and } 9 + 16p \text{ must be positive.}$$

Thus
$$4(p^2 + 8p + 16)(p^2 - 8p + 7) \text{ must be negative or zero;}$$
that is,
$$4(p + 4)^2(p - 1)(p - 7) \text{ must be negative or zero.}$$

§11. Put $\dfrac{x + 2}{2x^2 + 3x + 6} = y;$ then $2yx^2 + (3y - 1)x + 6y - 2 = 0.$

If x is real, $(3y - 1)^2 - 8y(6y - 2)$ must be positive;
$$\therefore (1 + 13y)(1 - 3y) \text{ must be positive.}$$

Hence y must lie between $\dfrac{1}{3}$ and $-\dfrac{1}{13}$, and its greatest value is $\dfrac{1}{3}$.

§12. Put $\dfrac{x^2 - bc}{2x - b - c} = y;$ then $x^2 - 2yx + by + cy - bc = 0.$ If x is real,

$y^2 - by - cy + bc$ must be positive; $\quad \therefore (y - b)(y - c)$ must be positive.

§13. In order that the roots of $ax^2 + 2bx + c = 0$ may be possible and different we must have $b^2 - ac$ positve.

The second equation may be written
$$(a^2 - ac + 2b^2)x^2 + 2b(a + c)x + c^2 - ac + 2b^2 = 0;$$
and the condition for roots possible and different is that
$$b^2(a + c)^2 - (a^2 - ac + 2b^2)(c^2 - ac + 2b^2)$$
should be positive. This expression reduces to $(ac - b^2)\{4b^2 + (a - c)^2\}$, so that its sign is contrary to that of $b^2 - ac$. Hence the required result follows at once.

§14. Denote the given expression by y; multiply up and re-arrange, then
$$(ad - bcy)x^2 - (ac + bd)(1 - y)x + (bc - ady) = 0.$$
If x is real, we must have
$$(ac + bd)^2(1 - y)^2 - 4(ad - bcy)(bc - ady) \text{ positive;}$$
$$\therefore \{(ac + bd)^2 - 4abcd\}(y^2 + 1) - 2y\{(ac + bd)^2 - 2(a^2d^2 + b^2c^2)\},$$
$$\text{or } (ac - bd)^2 y^2 - 2y\{(ac - bd)^2 - 2(ad - bc)^2\} + (ac - bd)^2$$
must be positive for all values of y.

This will be the case provided
$$(ac - bd)^4 > \{(ac - bd)^2 - 2(ad - bc)^2\}^2,$$
$$\therefore (ac - bd)^4 > (ac - bd)^4 - 4(ac - bd)^2(ad - bc)^2 + 4(ad - bc)^4,$$
$$\therefore (ac - bd)^2 > (ad - bc)^2;$$
$$\therefore (ac - bd - ad + bc)(ac - bd + ad - bc) \text{ is a positive quantity;}$$
$$\therefore (a + b)(c - d)(a - b)(c + d), \text{ or } (a^2 - b^2)(c^2 - d^2) \text{ must be +ve.}$$
Hence $a^2 - b^2$ and $c^2 - d^2$ must have the same sign.

EXAMPLES IX c

Questions 1 and 2 may be solved by application of the formula of Art. 127.

§1. Here $\quad m - 1 + 3 = 0$, whence $m = -2$.

Or thus : the given equation may be written $2x(y + 1) + y^2 + my - 3 = 0$; hence $y + 1$ must be a factor of $y^2 + my - 3$; that is, $y = -1$ must satisfy the equation $y^2 + my - 3 = 0$.

§2. Here the condition gives $-12 - \dfrac{25}{2} + \dfrac{m^2}{2} = 0$, whence $m^2 = 49$.

§3. The condition that the roots of
$$Ax^2 - (B - C)xy - Ay^2 = 0$$
should be real is that $(B - C)^2 + 4A^2$ should be a positive quantity: this condition is clearly satisfied.

§4. Since the equations are satisfied by a common root, we must have
$$(x^2 + px + q) - (x^2 + p'x + q') = 0 \tag{34}$$
Also by eliminating the absolute term, we obtain
$$q'(x^2 + px + q) - q(x^2 + p'x + q') = 0 \tag{35}$$
From (34) we get $x = \dfrac{q - q'}{p' - p}$ and from (35) $x = \dfrac{pq' - p'q}{q - q'}$.

§5. When the condition is fulfilled, the equations
$$lx^2 + mxy + ny^2 = 0 \text{ and } l'x^2 + m'xy + n'y^2 = 0$$
must be satisfied by a common value of the ratio $x : y$.

From these equations we have by cross multiplication
$$\frac{x^2}{mn' - m'n} = \frac{xy}{nl' - n'l} = \frac{y^2}{lm' - l'm};$$
whence $\quad (nl' - n'l)^2 = (mn' - m'n)(lm' - l'm)$.

§6. Applying the condition of Art. 127, we have
$$6 - 4aP - 12 - 2a^2 - P^2 = 0.$$

§7. If $y - mx$ is a factor of $ax^2 + 2hxy + by^2$, then this last expression vanishes when $y = mx$; that is, $a + 2hm + bm^2 = 0$.
Similarly if $my + x$ is a factor of $a'x^2 + 2h'xy + b'y^2$, we must have
$$a'm^2 - 2h'm + b' = 0.$$
From these equations, we have by cross multiplication
$$\frac{m^2}{2(b'h + ah')} = \frac{m}{aa' - bb'} = \frac{1}{-2(bh' + a'h)};$$
whence $\quad (aa' - bb')^2 = -4(ah' + b'h)(a'h + bh')$.

§8. Here $\quad x^2 - x(3y + 2) + 2y^2 - 3y - 35 = 0$;
whence solving as a quadratic in x.
$$2x = 3y + 2 \pm \sqrt{(3y + 2)^2 - 4(2y^2 - 3y - 35)} = 3y + 2 \pm (y + 12).$$
Giving to y any real value, we find two real values for x : or giving to x any real value we find two real values for y.

§9. Solving the equation $9x^2 + 2x(y - 46) + y^2 - 20y + 244 = 0$ as a quadratic in x, we have
$$9x = -(y - 46) \pm \sqrt{(y - 46)^2 - 9(y^2 - 20y + 244)}$$
$$= -(y - 46) \pm \sqrt{-8(y^2 - 11y + 10)}$$
$$= -(y - 46) \pm \sqrt{-8(y - 1)(y - 10)}.$$
Now the quantity under the radical is only positive when y lies between 1 and 10; and unless y lies between these limits the value of x will

be imaginary.

Again $\qquad\qquad y^2 + 2y(x - 10) + 9x^2 - 92x + 244 = 0;$

whence $\qquad y = -(x - 10) \pm \sqrt{(x - 10)^2 - (9x^2 - 92x + 244)}$

$$= -(x - 10) \pm \sqrt{-8(x - 6)(x - 3)}.$$

Thus in order that y may be real x must lie between 6 and 3.

§10. We have $\qquad x^2(ay + a') + x(by + b') + cy + c' = 0;$

solving this equation as a quadratic in x,

$$2(ay + a')x = -(by + b') \pm \sqrt{(by + b')^2 - 4(ay + a')(cy + c')}.$$

Now in order that x may be a rational function of y the expression under the radical, namely $(b^2 - 4ac)y^2 + 2(bb' - 2ac' - 2a'c)y + b'^2 - 4a'c'$, must be the square os a linear function of y;

hence $\qquad (bb' - 2ac' - 2a'c)^2 = (b^2 - 4ac)(b'^2 - 4a'c').$

Simplifying we have

$$a^2c'^2 + a'^2c^2 - ac'bb' - a'cbb' + 2aa'cc' = 4aa'cc' - acb'^2 - a'c'b^2;$$

$$\therefore a^2c'^2 + a'^2c^2 - 2aa'cc' = ac'bb' + a'cbb' - acb'^2 - a'c'b^2;$$

$$\therefore (ac' - a'c)^2 = (ab' - a'b)(bc' - b'c).$$

EXAMPLES X a

§1. $(x^{-1} - 4)(x^{-1} + 2) = 0$; whence $\dfrac{1}{x} = 4$ or -2.

§2. $(x^{-2} - 9)(x^{-2} - 1) = 0$; whence $\dfrac{1}{x^2} = 9$ or 1.

§3. $(2x^{\frac{1}{2}} - 1)(x^{\frac{1}{2}} - 2) = 0$; whence $\sqrt{x} = \dfrac{1}{2}$ or 2.

§4. $(3x^{\frac{1}{2}} - 2)(2x^{\frac{1}{2}} - 1) = 0$; whence $\sqrt{x} = \dfrac{2}{3}$ or $\dfrac{1}{2}$.

§5. $(x^{\frac{1}{n}} - 3)(x^{\frac{1}{n}} - 2) = 0$.   **§6.** $(x^{\frac{1}{2n}} - 1)(x^{\frac{1}{2n}} - 2) = 0$.

§7. Putting $y = \sqrt{\dfrac{x}{3}}$, we have $7y + \dfrac{5}{y} = \dfrac{68}{3}$; whence $y = \dfrac{5}{21}$ or 3.

§8. Putting $y = \sqrt{\dfrac{x}{1-x}}$, we have $y + \dfrac{1}{y} = \dfrac{13}{6}$; whence $y = \dfrac{3}{2}$ or $\dfrac{2}{3}$.

§9. $(3x^{\frac{1}{2}} - 1)(2x^{\frac{1}{2}} + 5) = 0$; whence $\sqrt{x} = \dfrac{1}{3}$ or $-\dfrac{5}{2}$.

The value $x = \dfrac{25}{4}$ satisfied a modified form of the given equation.

§10. $(8x^{\frac{3}{5}} + 1)(x^{\frac{3}{5}} + 1) = 0$; whence $x = \left(-\dfrac{1}{8}\right)^{\frac{5}{3}}$ or $(-1)^{\frac{5}{3}}$.

§11. $(3^x - 9)(3^x - 1) = 0$; whence $3^x = 9$ or 1.

§12. $(5 \cdot 5^x - 1)(5^x - 5) = 0$; whence $5^x = \dfrac{1}{5} = 5^{-1}$, and $5^x = 5$.

§13. $2^{2x+8} - 2 \cdot 2^{x+4} + 1 = 0$; that is $(2^{x+4} - 1)^2 = 0$; whence $x + 4 = 0$.

§14. $8 \cdot 2^{2x} - 65 \cdot 2^x + 8 = 0$; that is $(8 \cdot 2^x - 1)(2^x - 8) = 0$;

whence $\qquad\qquad 2^x = \dfrac{1}{8} = 2^{-3}$, and $2^x = 2^3$.

§15. $(\sqrt{2^x} - 1)^2 = 0$; whence $\sqrt{2^x} = 1$, and $2^x = 1$.

§16. Putting $y = \sqrt{2x}$, we have $\dfrac{3}{y} - \dfrac{y}{5} = \dfrac{59}{10}$; whence $y = \dfrac{1}{2}$ or -30.

§17. $(x - 7)(x + 5)(x - 3)(x + 1) = 1680$;

that is, $\qquad\qquad (x^2 - 2x - 35)(x^2 - 2x - 3) = 1680$;

this is a quadratic in $x^2 - 2x$ and gives $(x^2 - 2x - 63)(x^2 - 2x - 25) = 0$.

§18. $(x + 9)(x - 7)(x - 3)(x + 5) = 385$;

that is, $\qquad\qquad (x^2 + 2x - 63)(x^2 + 2x - 15) = 385$;

this is a quadratic in $x^2 + 2x$ and gives $(x^2 + 2x - 70)(x^2 + 2x - 8) = 0$.

§19. $x(2x - 3)(2x + 1)(x - 2) = 63$; that is $(2x^2 - 3x)(2x^2 - 3x - 2) = 63$; this is a quadratic in $2x^2 - 3x$ and gives $(2x^2 - 3x - 9)(2x^2 - 3x + 7) = 0$.

§20. $(2x - 7)(x + 3)(x - 3)(2x + 5) = 91$;

that is, $\qquad\qquad (2x^2 - x - 21)(2x^2 - x - 15) = 91$;

this is a quadratic in $2x^2 - x$ and gives $(2x^2 - x - 8)(2x^2 - x - 28) = 0$.

§21. Put $y^2 = x^2 + 6x$; then $y^2 + 2y - 24 = 0$; thus $y = 4$ or -6; and

$$x^2 + 6x = 16 \text{ or } 36.$$

N.B. In this and the following examples, the solution obtained by taking the negative value of y satisfies a modified form of the given equation.

§22. Put $y^2 = 3x^2 - 4x - 6$; then $y^2 + y - 12 = 0$; thus $y = 3$ or -4;

and therefore $\qquad\qquad 3x^2 - 4x - 6 = 9$ or 16.

§23. Put $y^2 = 3x^2 - 16x + 21$; then $y^2 + 3y - 28 = 0$;

thus $y = 4$ or -7; and $3x^2 - 16x + 21 = 16$ or 49.

§24. Put $y^2 = 3x^2 - 7x + 2$; then $y^2 - 9y - 10 = 0$; thus $y = 10$ or -1; and $3x^2 - 7x + 2 = 100$ or 1.

§25. Put $y^2 = 2x^2 - 5x + 3$; then $y^2 - 6y + 5 = 0$; thus $y = 1$ or 5; and $2x^2 - 5x + 3 = 1$ or 25.

§26. Put $y^2 = 3x^2 - 8x + 1$; then $2y^2 - y - 66 = 0$; thus $y = 6$ or $-\dfrac{11}{2}$; and $3x^2 - 8x + 1 = 36$ or $\dfrac{121}{4}$.

§27. Dividing by $\sqrt{x-3}$, we have $\sqrt{x-3} = 0$, and $\sqrt{4x+5} - \sqrt{x} = \sqrt{x+3}$; then see Art. 131.

§28. Dividing by $\sqrt{2x-1}$, we have $\sqrt{2x-1} = 0$, and $\sqrt{x-4}+3 = \sqrt{x+11}$.

§29. Dividing by $\sqrt{x-1}$; we have $\sqrt{x-1} = 0$,

and $$\sqrt{2x+7} + \sqrt{3(x-6)} = \sqrt{7x+1}.$$

§30. Dividing by $\sqrt{a+3x}$, we have $\sqrt{a+3x} = 0$,

and $$\sqrt{a-x} - \sqrt{a-2x} = \sqrt{2a-3x}.$$

Examples 31 to 34 may be solved as in Art. 132.

§31. Use the identity $(2x^2 + 5x - 2) - (2x^2 + 5x - 9) = 7$.

§32. Use the identity $(3x^2 - 2x + 9) - (3x^2 - 2x - 4) = 13$.

§33. Use the identity $(2x^2 - 7x + 1) - (2x^2 - 9x + 4) = 2x - 3$.

§34. Use the identity $(3x^2 - 7x - 4) - (2x^2 - 7x + 21) = x^2 - 25$.

Examples $35 - 37$ are reciprocal equations and may be solved by the method of Art. 133.

Example 38 may be solved by Art. 134.

§39. We have componendo et dividendo $\dfrac{x}{\sqrt{12a - x}} = \sqrt{a}$.

§40. Divide numerator and denominator by $\sqrt{a+2x}$;

then $$\frac{\sqrt{a+2x} + \sqrt{a-2x}}{\sqrt{a+2x} - \sqrt{a-2x}} = \frac{5x}{a};$$

thus $$\frac{\sqrt{a+2x}}{\sqrt{a-2x}} = \frac{5x+a}{5x-a}; \text{ or } \frac{a+2x}{a-2x} = \frac{(5x+a)^2}{(5x-a)^2}.$$

Whence componendo et dividendo $\dfrac{2x}{a} = \dfrac{10ax}{25x^2 + a^2}$.

§41. The simplified form of the left side is $4x\sqrt{x^2 - 1}$.

thus $$4x\sqrt{x^2 - 1} = 8x\sqrt{x^2 - 3x + 2};$$

dividing by $x\sqrt{x-1}$, we have

$$x\sqrt{x-1} = 0, \text{ and } \sqrt{x+1} = 2\sqrt{x-2}.$$

§42. $\dfrac{\sqrt{x-1}}{\sqrt{x^3 - x}} = \dfrac{\sqrt{x-1}}{\sqrt{x(x^2 - 1)}} = \dfrac{1}{\sqrt{x(x+1)}}$

thus $\sqrt{x(x+1)} + \dfrac{1}{\sqrt{x(x+1)}} = \dfrac{5}{2}$;

which is a quadratic in $\sqrt{x(x+1)}$.

§43. $$\frac{x^2 - x + 1}{x - 1} = x + \sqrt{\frac{6}{x}};$$

$\therefore$ by transposition and reduction $\dfrac{1}{x-1} = \sqrt{\dfrac{6}{x}}$;

hence $6x^2 - 13x + 6 = 0$, or $(2x - 3)(3x - 2) = 0$.

§44. $2^{x^2} = 8 \times 2^{2x} = 2^{2x+3}$; thus $x^2 = 2x + 3$.

§45. Divide by a^x; then $a^x(a^2 + 1) = (a^{2x} + 1)a$; thus $a \cdot a^{2x} - a^2 \cdot a^x - a^x + a = 0$; that is $(a \cdot a^x - 1)(a^x - a) = 0$; whence

$$a^x = \frac{1}{a} = a^{-1}, \text{ and } a^x = a.$$

§46. Clearing of fractions, $8(x-5)^{\frac{3}{2}} = (3x-7)^{\frac{3}{2}}$; taking the cute root of each side, $2\sqrt{x-5} = \sqrt{3x-7}$.

§47. The solution is similar to that of Ex. 46.

§48. Dividing each term by $(a^2 - x^2)^{\frac{1}{3}}$, or $(a+x)^{\frac{1}{3}} \cdot (a-x)^{\frac{1}{3}}$, we get

$$\left(\frac{a+x}{a-x}\right)^{\frac{1}{3}} + 4\left(\frac{a-x}{a+x}\right)^{\frac{1}{3}} = 5,$$

or $\qquad\qquad y + \dfrac{4}{y} = 5$, where $y = \left(\dfrac{a+x}{a-x}\right)^{\frac{1}{3}}$.

§49. We have identically $(x^2 + ax - 1) - (x^2 + bx - 1) = (a-b)x$; and by the question,

$$\sqrt{x^2 + ax - 1} - \sqrt{x^2 + bx - 1} = \sqrt{a} - \sqrt{b} :$$

hence by division,

$$\sqrt{x^2 + ax - 1} + \sqrt{x^2 + bx - 1} = (\sqrt{a} + \sqrt{b})x.$$

By addition,

$$2\sqrt{x^2 + ax - 1} = (\sqrt{a} + \sqrt{b})x + (\sqrt{a} - \sqrt{b}).$$

Squaring,

$$4(x^2 + ax - 1) = (\sqrt{a} + \sqrt{b})^2 x^2 + 2(a-b)x + (\sqrt{a} - \sqrt{b})^2;$$

$$\therefore \{(\sqrt{a} + \sqrt{b})^2 - 4\}x^2 - 2(a+b)x + \{(\sqrt{a} - \sqrt{b})^2 + 4\} = 0.$$

Now by inspection, the original equation is satisfied by $x = 1$; hence by the theory of quadratic equations

the other root is $\dfrac{(\sqrt{a} - \sqrt{b})^2 + 4}{(\sqrt{a} + \sqrt{b})^2 - 4}$.

§50. The simplified form of the left side is $2x^2 + 2(x^2 - 1)$;

thus $\qquad\qquad\qquad 2x^2 + 2(x^2 - 1) = 98.$

§51. This equation may be written $x^4 - 2x^3 + x^2 - x^2 + x = 380$; that is $x^2(x-1)^2 - x(x-1) = 380$; which is a quadratic in $x(x-1)$.

§52. This equation may be written $27x^3 + 1 + 21x + 7 = 0$, that is

$$(27x^3 + 1) + 7(3x + 1) = 0;$$

dividing by $3x + 1$, we have $3x + 1 = 0$, and $9x^2 - 3x + 1 + 7 = 0$.

EXAMPLES X b

§1. $y = \dfrac{20}{x}$; hence $3x - \dfrac{40}{x} = 7$.

§2. $y = 5x - 3$; hence $(5x-3)^2 - 6x^2 = 25$.

§3. $4x = 3y + 1$; hence $3y(3y + 1) + 13y^2 = 25$.

§4. By division $x^2 + xy + y^2 = 49$; combines this with $x^2 - xy + y^2 = 19$.

    Examples $5, 6, 7$ are solved by the method of Ex.1, Art. 136.

    Examples 8 to 12 : transpose if necessary; the equations will be found to be homogeneous, and may be solved by putting $y = mx$.

    Examples 13 to 15 may be solved by the method of Ex. 2, Art. 136.

§16. From (1), $y = \dfrac{4}{1-x}$; hence $\dfrac{4}{1-x} + \dfrac{4}{x} = 25$.

§17. From (2), $x + y = 3$; from (1), $2(x^3 + y^3) = 9xy$; by division

$$2(x^2 - xy + y^2) = 3xy, \text{ or } 2x^2 - 5xy + 2y^2 = 0;$$

$$\text{whence } (2x - y)(x - 2y) = 0.$$

§18. Put $\dfrac{x}{2} = u$, $\dfrac{y}{5} = v$; then $u + v = 5$, and $\dfrac{1}{u} + \dfrac{1}{v} = \dfrac{5}{6}$; whence we have $uv = 6$.

§19. Put $u = x^{\frac{1}{3}}$, $v = y^{\frac{1}{3}}$; then the equations become
$$u^3 + v^3 = 1072;\ u + v = 16.$$

§20. Put $u = x^{\frac{1}{2}}$, $v = y^{\frac{1}{2}}$; then the equations become
$$u^2 v + uv^2 = 20,\ \text{and } u^3 + v^3 = 65.$$

Multiply the first of these by 3 and add to the second; thus $(u+v)^3 = 125$; whence $u + v = 5$.

§21. Put $u = x^{\frac{1}{2}}$, $v = y^{\frac{1}{2}}$; then the equations become
$$u + v = 5,\ 6\left(\frac{1}{u} + \frac{1}{v}\right) = 5;\ \text{whence we find } uv = 6.$$

§22. Square the first equation; thus $2x + 2\sqrt{x^2 - y^2} = 16$; substituting from the second equation $2x + 6 = 16$,

§23. Square the second equation; thus $2x - 2\sqrt{x^2 - 1} = y = 2 - \sqrt{x^2 - 1}$ from the first equation; hence $\sqrt{x^2 - 1} = 2(x - 1)$.

§24. The first equation is a quadratic in $\sqrt{\dfrac{x}{y}}$, and gives $\sqrt{\dfrac{x}{y}} = 3$ or $\dfrac{1}{3}$; whence $x = 9y$, or $x = \dfrac{y}{9}$.

§25. The first equation is a quadratic in $\dfrac{\sqrt{x} + \sqrt{y}}{\sqrt{x} - \sqrt{y}}$, and gives

$\dfrac{\sqrt{x} + \sqrt{y}}{\sqrt{x} - \sqrt{y}} = 4$ or $\dfrac{1}{4}$; that is $\dfrac{\sqrt{x}}{\sqrt{y}} = \dfrac{5}{3}$ or $\dfrac{5}{-3}$; whence $\dfrac{x}{y} = \dfrac{25}{9}$.

§26. Multiply the second equation by 4 and add to the first; thus
$$(x^2 + 4xy + 4y^2) - 15(x + 2y) + 56 = 0.$$
This is a quadratic $x + 2y$, and gives $x + 2y = 7$ or 8. Combine each of these separately with $xy = 8$.

§27. The first equation is a quadratic in xy, and gives $xy = 25$ or 16. From the second equation $(x - y)(4x - y) = 0$.

§28. From the first equation, $(2x - 5y)^2 - (2x - 5y) - 6 = 0$. This is a quadratic in $2x - 5y$, and gives $2x - 5y = 3$ or -2. Combine with the second equation.

§29. From (1), $(3x - 2y)^2 + 11(3x - 2y) - 12 = 0$; whence $3x - 2y = 1$ or -12. Combine with the second equation.

§30. Divide (2) by (1); thus $(x^2 + y^2)(x + y) = 40xy$: divide this last equation by (1); thus $\dfrac{x^2 + y^2}{(x - y)^2} = \dfrac{40}{16} = \dfrac{5}{2}$; that is, $3x^2 - 10xy + 3y^2 = 0$; whence $(3x - y)(x - 3y) = 0$. Thus $x = 3y$ or $\dfrac{y}{3}$. Substitute in the first equation.

§31. By division $\dfrac{2x^2 - xy + y^2}{2x^2 + 4xy} = \dfrac{2y}{5y} = \dfrac{2}{5}$; that is, $6x^2 - 13xy + 5y^2 = 0$; whence $(2x - y)(3x - 5y) = 0$. Substitute $x = \dfrac{y}{2}$, and $x = \dfrac{5y}{2}$ successively in the second equation.

§32. From (1),
$$\dfrac{x^2 - xy + y^2}{x + y} + \dfrac{x^2 + xy + y^2}{x - y} = \dfrac{43x}{8};\ \text{that is } \dfrac{2x(x^2 + 2y^2)}{x^2 - y^2} = \dfrac{43x}{8};$$
whence $x = 0$, or $9x^2 = 25y^2$. Substitute $x = \pm\dfrac{5y}{3}$ in the second equation; $x = 0$ gives no solution.

33 and 34 are solved by the method of Ex. 4, Art. 136.

§33. Here $\dfrac{m(m^2 - 3m - 1)}{m^2 - 4m + 2} = \dfrac{24}{8} = 3$; thus $m^3 - 6m^2 + 11m - 6 = 0$; that is, $(m - 1)(m - 2)(m - 3) = 0$.

§34. Here $\dfrac{3 - 8m^2 + m^3}{m - 1} = -\dfrac{21}{1} = -21$; thus $m^2 - 8m^2 + 21m - 18 = 0$; that is, $(m - 2)(m - 3)(m - 3) = 0$.

35 and 36 are solved by the method of Ex.5, Art. 136.

§35. From (1), $x^4 - 9xy^3 - 4x^2y^2 = -108y^2 = -y^2(2x^2 + 9xy + y^2)$, by (2). Thus $x^4 - 2x^2y^2 + y^4 = 0$; that is $(x^2 - y^2)^2 = 0$; whence $x^2 - y^2 = 0$; that is $x = \pm y$.

§36. From (1), $(6x^4 + x^2y^2 - 2xy^3) - (4 \times 6x^2) + (4)^2 = 0$; substituting from (2), namely $4 = x^2 + xy - y^2$, we have

$$6x^4 + x^2y^2 - 2xy^3 - 6x^2(x^2 + xy - y^2) + (x^2 + xy - y^2)^2 = 0,$$

whence $x^4 - 4x^3y + 6x^2y^2 - 4xy^3 + y^4 = 0$; that is $(x - y)^4 = 0$, and $x = y$.

§37. From (1), $x^2 - y^2 = by - ax$; dividing by (2), $\dfrac{x - y}{x + y} = \dfrac{by - ax}{by + ax}$; whence $\dfrac{x}{y} = \dfrac{by}{ax}$; that is, $\dfrac{x}{\sqrt{b}} = \pm\dfrac{y}{\sqrt{a}} = k$, say. Substitute in either of the given equations.

§38. Square (1) and subtract (2); thus $2abxy = 4a^2x^2 - 2b^2y^2$; that is, $2a^2x^2 - abxy - b^2y^2 = 0$, or $(2ax + by)(ax - by) = 0$. Thus $x = -\dfrac{by}{2a}$, or $x = \dfrac{by}{a}$. Combine each of these with the first of the given equations.

§39. On equating the first expression to 0 and simplifying, we obtain $b^2x + a^2y = a^2b + ab^2$. Similarly from the second expression we find

$$xy - bx - ay + a^2 - ab + b^2 = 0.$$

Substituting for y from the first of these equations in the second, we obtain $b^2x^2 - 2ab^2x - a^3(a - 2b) = 0$, whence $(bx - a^2)\{bx + a(a - 2b)\} = 0$.

§40. Divide (1) by (2), and we get $\dfrac{b^3x^3}{a^3y^3} = \dfrac{10bx + 3ay}{10ay + 3bx}$; whence, by putting m for $\dfrac{bx}{ay}$, we obtain $m^3 = \dfrac{10m + 3}{10 + 3m}$;

$$3m^4 + 10m^3 = 10m + 3, \quad 3(m^4 - 1) + 10m(m^2 - 1) = 0;$$

whence $m^2 - 1 = 0$, or $3m^2 + 10m + 3 = 0$.

§41. From (1), we have $2ax^2 + (4a^2 - 1)xy - 2ay^2 = 0$;

$$\therefore (2ax - y)(x + 2ay) = 0; \text{ that is, } y = 2ax, \text{ or } y = -\dfrac{x}{2a}.$$

Substitute these values in the second equation.

EXAMPLES X c

§1. From (1) and (2) by cross multiplication,
$$\dfrac{x}{3} = \dfrac{y}{5} = \dfrac{z}{4}.$$

§2. From (1) and (2) by cross multiplication,
$$\dfrac{x}{5} = \dfrac{y}{-1} = \dfrac{z}{7}.$$

§3. From (2) and (3), $(x - y)^2 - z^2 = 12$; putting $u = x - y$, we have $u^2 - z^2 = 12$. Also from (1) $u - z = 2$. Whence $u = 4$, $z = 2$; thus $x - y = 4$. Combine with $xy = 5$.

§4. From (2) and (3), $(x - z)^2 - 4y^2 = -11$; putting $u = x - z$, this gives $4y^2 - u^2 = 11$; also from (1), $2y + u = 11$; whence $2y = 6$, and $u = 5$. Thus $x - z = 5$. Combine with $xz = 24$.

§5. From (1) and (2), $(x + y)^2 - 3z(x + y) - z^2 = 3$; putting $u = x + y$, this gives $u^2 - 3uz - z^2 = 3$. Also from (3), $u - z = 5$. These equations give $z = 2$

or $-\dfrac{11}{3}$, and $u = 7$ or $\dfrac{4}{3}$. Combine these results with the first equation

$$x^2 + y^2 - z^2 = 21.$$

§6. By addition of all three equations,

$$x^2 + y^2 + z^2 + 2xy + 2xz + 2yz = 36;$$

that is $(x + y + z)^2 = 36$, and $x + y + z = \pm 6$. Divide each of the given equations by this last result.

§7. The given equations may be written

$$x(x + 2y + 3z) = 50, \ y(x + 2y + 3z) = 10, \ z(x + 2y + 3z) = 10.$$

Thus $\dfrac{x}{50} = \dfrac{y}{10} = \dfrac{z}{10}$ or $\dfrac{x}{5} = \dfrac{y}{1} = \dfrac{z}{1} = k$, say.

Or, multiply the second equation by 2, the third equation by 3, and add to the first; thus $(x + 2y + 3z)^2 = 100$.

§8. Put $u = y - z$, $v = z + x$, $w = x - y$; then $uv = 22$, $vw = 33$, $wu = 6$; thus $u^2 v^2 w^2 = 22 \times 33 \times 6$; whence $uvw = \pm 66$, and $u = \pm 2$, $v = \pm 11$, $w = \pm 3$.

§9. By multiplication, $x^7 y^7 z^7 u^7 = 128 = 2^7$; thus $xyzu = 2$. Dividing each of the given equations by this last result, we have

$$xyz = 6, \ xyu = 4, \ xzu = \frac{1}{2}, \ yzu = \frac{2}{3}.$$

Now divide the equation $xyzu = 2$ by each of these four equations.

§10. Divide (1) by (2), thus $\dfrac{y}{z^2} = \dfrac{2}{9}$.

Multiply (1) by (2) and divide by (3), thus $\dfrac{z^2}{x} = 9$.

Substituting in (1), $z^{11} = 3 \times 9^5 = 3^{11}$; whence $z = 3$.

§11. These equations may be written

$$(x + 1)(y + 1) = 24, \ (x + 1)(z + 1) = 42, \ (y + 1)(z + 1) = 28.$$

Multiplying these together and taking the square root, we have

$$(x + 1)(y + 1)(z + 1) = \pm 168.$$

Divide this result by each of the three equations above.

§12. These equations may be written

$$(2x + 1)(y - 2) = 15, \ (y - 2)(3z + 1) = 50, \ (2x + 1)(3z + 1) = 30.$$

Whence $(2x + 1)(y - 2)(3z + 1) = \pm 150$.

Divide this result by each of the three equations above.

§13. From (1) and (2), $xz + yz + x + y = 15z$; that is $(x + y)(z + 1) = 15z$. Combining with (3), $(12 - z)(z + 1) = 15z$, whence $z = 2$ or -6.

Substitute these values of z successively in the equations

$$x + y = 12 - z \text{ and } xz + y = 7z.$$

§14. Subtract (2) from the square of (3), thus

$$yz + zx + xy = 0 \tag{α}$$

Subtract (1) from the product of (2) and (3),

thus $\qquad y^2 z + yz^2 + z^2 x + zx^2 + z^2 y + xy^2 = 0 \tag{β}$

Combining (1) and (α), we have

$$(x + y + z)(yz + zx + xy) = 0.$$

Subtracting (β) from this last result, we have $3xyz = 0$.

Hence one of the quantities x, y, or z must be zero. Let $x = 0$; substituting in (α), we have $yz = 0$; thus a second of the quantities must be zero. Hence from (3) the remaining quantity must be equal to a.

§15. From the first two equations, we have

$$x^2 + y^2 + z^2 + 2(yz + zx + xy) = 3a^2;$$

that is, $\qquad x + y + z = \pm a\sqrt{3}.$

From (3), $\qquad 3x - y + z = a\sqrt{3}.$

I. From $\qquad$ $x + y + z = a\sqrt{3}$ and $3x - y + z = a\sqrt{3}$,

we have $\qquad$ $y = x,\ z = a\sqrt{3} - 2x.$

    Substituting in the first equation, we find

$$3x^2 - 2\sqrt{3} \cdot ax + ax + a^2 = 0,\ \text{or}\ (x\sqrt{3} - a)^2 = 0;$$

that is, $\qquad\qquad\qquad\qquad x = \dfrac{a}{\sqrt{3}}.$

II. From $\qquad$ $x + y + z = -a\sqrt{3}$ and $3x - y + z = a\sqrt{3}$,

we have $\qquad$ $y = x - a\sqrt{3},\ z = -2x.$

    Substituting in the first equation, we find

$$3x^2 - \sqrt{3}ax + a^2 = 0,\ \text{whence}\ \frac{x}{a} = \frac{\sqrt{3} \pm \sqrt{-9}}{6}.$$

§16. From the first and second equations,

$$x^2 + y^2 + z^2 - 2yz - 2xz + 2xy = 9a^2,$$

that is, $\qquad\qquad\qquad x + y - z = \pm 3a.$

I. From $\qquad$ $x + y - z = 3a$ and $3x + y - 2z = 3a$,

we have $\qquad\qquad\qquad y = 3a + x,\ z = 2x.$

    Substituting in the first equation, we have

$$x^2 + ax - 2a^2 = 0,\ \text{whence}\ x = a\ \text{or}\ -2a.$$

II. From $\qquad$ $x + y - z = -3a$, and $3x + y - 2z = 3a$,

we have $\qquad\qquad\qquad y = x - 9a,\ z = 2x - 6a.$

    Substituting in the first equation, we have

$$x^2 - 7ax + 16a^2 = 0,\ \text{whence}\ \frac{x}{a} = \frac{7 \pm \sqrt{-15}}{2}.$$

EXAMPLES X d

§1. Divide by 3, then $x + 2y + \dfrac{2y}{3} = 34 + \dfrac{1}{3}$; thus $\dfrac{2y - 1}{3}$ = integer; multiply

by 2; thus $y + \dfrac{y - 2}{3}$ = integer; that is $\dfrac{y - 2}{3} = p$;

$$\text{hence}\ y = 3p + 2\ \text{and}\ x = 29 - 8p.$$

§2. Divide by 2, thus $2x + y + \dfrac{x}{2} = 26 + \dfrac{1}{2}$; therefore $\dfrac{x - 1}{2}$ = integer $= p$

say; hence $x = 2p + 1,\ y = 24 - 5p.$

§3. Divide by 7, then $x + 5y + \dfrac{5y}{7} = 21 + \dfrac{5}{7}$; thus $\dfrac{5y - 5}{7}$ = integer, and

therefore $\dfrac{y - 1}{7}$ = integer $= p$ say; thus $y = 7p + 1$, and $x = 20 - 12p.$

§4. Divide by 11, then $x + y + \dfrac{2x}{11} = 37 + \dfrac{7}{11}$; thus $\dfrac{2x - 7}{11}$ = integer; multiply

by 6, then $x - 3 + \dfrac{x - 9}{11}$ = integer; that is $\dfrac{x - 9}{11} = p$;

$$\text{hence}\ x = 9 + 11p\ \text{and}\ y = 27 - 13p.$$

§5. Divide by 23, then $x + y + \dfrac{2y}{23} = 39 + \dfrac{18}{23}$; thus $\dfrac{2y - 18}{23}$ = integer; and

therefore $\dfrac{y - 9}{23}$ = integer $= p$ say; thus $y = 9 + 23p,\ x = 30 - 25p.$

§6. Divide by 41, then $x + y + \dfrac{6y}{41} = 53 + \dfrac{18}{41}$; thus $\dfrac{6y - 18}{41}$ = integer, and

therefore $\dfrac{y - 3}{41}$ = integer $= p$ say ; thus $y = 3 + 41p,\ x = 50 - 47p.$

§7. Divide by 5, then $x - y - \dfrac{2y}{5} = \dfrac{3}{5}$; thus $\dfrac{2y+3}{5}$ = integer; multiply by 3, then $y + 1 + \dfrac{y+4}{5}$ = integer; thus $\dfrac{y+4}{5} = p$, or $y = 5p - 4$, $x = 7p - 5$.

§8. Divide by 6, then $x - 2y - \dfrac{y}{6} = \dfrac{1}{6}$, thus $\dfrac{y+1}{6}$ = integer $= p$;

hence $y = 6p - 1$ and $x = 13p - 2$.

§9. Divide by 8, then $x - 2y - \dfrac{5y}{8} = 4 + \dfrac{1}{8}$; thus $\dfrac{5y+1}{8}$ = integer; multiply by 5, then $3y + \dfrac{y+5}{8}$ = integer; thus $\dfrac{y+5}{8} = p$, or $y = 8p - 5$, $x = 21p - 9$.

§10. We have at once $\dfrac{x}{17} = \dfrac{y}{13} = p$ say; thus $x = 17p$, $y = 13p$.

§11. Divide by 19, then $y - x - \dfrac{4x}{19} = \dfrac{7}{19}$; thus $\dfrac{4x+7}{19}$ = integer; multiply by 5, then $x + 1 + \dfrac{x+16}{19}$ = integer; thus $\dfrac{x+16}{19} = p$;

hence $x = 19p - 16$ and $y = 23p - 19$.

§12. Divide by 30, then $2y + \dfrac{17y}{30} - x = 9 + \dfrac{25}{30}$;

thus $\dfrac{17y - 25}{30}$ = integer; multiply by 7,

then $4y - 5 - \dfrac{y+25}{30}$ = integer; that is $\dfrac{y+25}{30} = p$, or

$$y = 30p - 25, \quad x = 77p - 74.$$

§13. Let x be the number of horses, y the number of cows; then

$$37x + 23y = 752.$$

Divide by 23, then $x + y + \dfrac{14x}{23} = 32 + \dfrac{16}{23}$; thus $\dfrac{14x - 16}{24}$ = integer, and

therefore $\dfrac{7x - 8}{23}$ = integer. Multiply by 10, then $3x - 3 + \dfrac{x-11}{23}$ = integer;

thus $\dfrac{x-11}{23} = p$, and the general solution is $x = 23p + 11$, $y = 15 - 37p$.

§14. Let x denote the number of shillings, y the number of sixpences; then $2x + y = 200$; here x may have all values from 0 to 100, and therefore the number of ways is 101.

§15. A multiple of 8 may be denoted by $8x$, and a multiple of 5 by $5y$; thus the two numbers may be denoted by $8x$ and $5y$; then $8x + 5y = 81$. The general solution is $x = 5p + 2$, $y = 13 - 8p$.

§16. Let x be the number of guineas paid, y the number of half-crowns received; then reducing to sixpenny pieces, we have $42x - 5y = 21$; the general solution is $x = 5p + 3$, $y = 21p + 21$.

§17. Let x and y represent the quotients of the number by 39 and 56; then the number $= 39x + 16$; and the number also $= 56y + 27$; hence $39x + 16 = 56y + 27$, or $39x - 56y = 11$.

Divide by 39, then $x - y - \dfrac{17y}{39} = \dfrac{11}{39}$; thus $\dfrac{17y + 11}{39}$ = integer; multiply by 2, then $y + \dfrac{22 - 5y}{39}$ = integer; that is $\dfrac{22 - 5y}{39}$ = integer; multiply by 8,

then $4 - y + \dfrac{20 - y}{39}$ = integer; thus $\dfrac{y - 20}{39} = p$, or $y = 39p + 20$, $x = 56p + 29$.

§18. Let x be the number of florins paid, y the number of half-crowns received; then $4x - 5y = 53$; thus $x - y - \dfrac{y}{4} = 13 + \dfrac{1}{4}$; and therefore $\dfrac{y+1}{4} =$ an integer $= p$; whence the general solution is $y = 4p - 1$, $x = 5p + 12$.

§19. Let x denote the quotient of the part divided by 5, and y that of the part divided by 8; then the two parts may be represented by $5x + 2$ and

$8y + 3$. Thus $(5x + 2) + (8y + 3) = 136$, that is $5x + 8y = 131$. The general solution is $x = 23 - 8p$, and $y = 5p + 2$.

§20. Let x, y, z denote the number of rams, pigs and oxen respectively; then we have $x + y + z = 40$, and $4x + 2y + 17z = 301$. Whence $2x + 15z = 221$.

The general solution is $x = 15p + 13$, $z = 13 - 2p$; whence $y = 14 - 13p$.

§21. Let x, y, z denote the number of sovereigns, half-crowns, and shillings respectively; then we have $x + y + z = 27$, and $40x + 5y + 2z = 201$; whence $$38x + 3y = 147.$$

The general solution is $x = 3p$, $y = 49 - 38p$; whence $z = 35p - 22$.

EXAMPLES XI a

§5. We have $\quad 4n(n-1)(n-2) = 5(n-1)(n-2)(n-3);$
$$\therefore\ 4n = 5(n-3); \therefore\ n = 15.$$

§6. The number $= \lfloor 8$ without restriction; if t and e occupy specified places, we can arrange the remaining letters in $\lfloor 6$ ways.

§7. The number $= {}^6C_4 = 15$; if each such selection is arranged in all possible ways to form a number, we get $15 \times \lfloor 4 = 360$.

§8. Here $\qquad \dfrac{2n(2n-1)(2n-2)}{1\cdot2\cdot3} = \dfrac{44}{3} \times \dfrac{n(n-1)}{1\cdot2};$
whence $\qquad\qquad 2n-1 = 11$, or $n = 6$.

§10. We can now only change the order of 6 bells, therefore the no. of changes $= \lfloor 6 = 720$.

§11. The number of ways $= {}^{24}C_4 = 10626$. When the particular man is included we have to select 3 men out of the remaining 23; this can be done in $\dfrac{23\cdot22\cdot21}{1\cdot2\cdot3}$, or 1771 ways.

§12. Suppose the letters a, u fastened together; they count as one letter and we have six things to arrange. This can be done in 720 ways; but since a, u admit of two arrangements among themselves we must multiply this result by 2.

§13. The number $= {}^{25}C_5 \times {}^{10}C_3 = 6375600$.

§14. (1) There are 3 ways of choosing the capital, and then $\lfloor 5$ ways of arranging the other letters; therefore $3 \times \lfloor 5$, or 360 is the no. of arrangements.

(2) The no. of ways of placing the capitals at the beginning and end is 3×2; and the remaining letters can then be arranged in $\lfloor 4$ ways;
$$\therefore\ \text{no. of arrangements} = 6 \times 24 = 144.$$

§15. ${}^{50}C_{48} = {}^{50}C_4 = 230300$.

§16. We have $12 + 8 = n$ by Art. 145; $\therefore\ n = 20$, and ${}^{20}C_{17}$, ${}^{22}C_{20}$ may be easily found.

§17. Here we have 3 places in which two letters are to be placed; this gives rise to 3×2 or 6 ways. Then the four consonants can be arranged in $\lfloor 4$ ways; $\therefore$ required no. of ways $= 6 \times 24 = 144$.

§18. (1) $4 \times {}^8C_5 = 4 \times \dfrac{8\cdot7\cdot6}{1\cdot2\cdot3} = 224$.

(2) We must have 1 officer and 5 privates, or 2 officers and 4 privates, and 3 officers and 3 privates, or 4 officers and 2 privates;
$\therefore$ the required no. of ways is
$$4 \times {}^8C_5 + {}^4C_2 \times {}^8C_4 + {}^4C_3 \times {}^8C_3 + {}^8C_2,$$
which reduces to 896.

§19. The required number
$$= {}^{10}C_4 + {}^{10}C_5 + {}^{10}C_6 + {}^{10}C_7 + {}^{10}C_8 + {}^{10}C_9 + {}^{10}C_{10} = 848.$$

§20. We have $r + r + 2 = 18$; $\therefore\ r = 8$, and ${}^8C_5 = \dfrac{8\cdot7\cdot6}{1\cdot2\cdot3} = 56$.

§21. and **§22.** See Ex. 2, Art. 148.

§23. By Art. 147 the number of ways is $\dfrac{\lfloor 12}{(\lfloor 3)^4} = 369600$.

§24. As in Ex. 2, Art. 148, we have $3 \times {}^5C_3 \times {}^4C_2 \times \lfloor 5 = 21600$.

§25. By Art. 147, the number of ways is $\dfrac{\lfloor 45}{\lfloor 10 \ \lfloor 15 \ \lfloor 20}$.

§26. The Latin books can be chosen in $^7C_4$ ways; the English books can be chosen in 3 ways; and they admit of $\lfloor 4$ arrangements, since the English book keeps the middle place; ∴ the no. of ways $= \dfrac{7 \cdot 6 \cdot 5}{1 \cdot 2 \cdot 3} \times 3 \times \lfloor 4 = 2520$.

§27. There are 5 men who can row on either side; these can be subdivided into groups of 2 and 1 in $\dfrac{\lfloor 5}{\lfloor 2}$ ways. Each side can now be arranged in $\lfloor 4$ ways; therefore the required no. of ways is $\dfrac{\lfloor 5}{\lfloor 2} \times \lfloor 4 \times \lfloor 4$, or 34560.

§28. Suppose the vols. of the same work inseparable, then we have 4 works to be arranged (taken as a whole); since vols. of each work can be arranged in any order, we get $\lfloor 4 \times \lfloor 3 \times \lfloor 3 \times \lfloor 2 \times \lfloor 2$, or 3456.

§29. Suppose the best and worst papers fastened together, then the no. of ways in which they could come together is $2\lfloor 9$, since either may come before the other. We must subtract this no. from $\lfloor 10$, the whole no. of arrangements when there is no restriction. Thus we get $\lfloor 10 - 2\lfloor 9$.

§30. There are 8 men who can row, and of these two only on bow side. The remaining 6 may be allotted to the two sides in $\dfrac{\lfloor 6}{\lfloor 4 \ \lfloor 2}$ ways.

The coxswain can be chosen in 3 ways, and each side can be arranged in $\lfloor 4$ ways; thus we get $\quad 3 \times \dfrac{\lfloor 6}{\lfloor 4 \ \lfloor 2} \times \lfloor 4 \times \lfloor 4$, or 25920.

§31. If we write down all the positive signs there will be $p - 1$ places between them in which a negative sign may be placed. Also the row may begin and end with a negative sign. Therefore we have $p + 1$ places from which we have to choose n.

§32. Here $\dfrac{56 \cdot 55 \cdot 54 \ldots (51 - r)}{54 \cdot 53 \cdot 52 \ldots (52 - r)} = 30800;$

∴ $56 \cdot 55(51 - r) = 28 \times 11 \times 100;$ ∴ $51 - r = 10;$ $r = 41$.

§33. With all the flags the number of signals is $\lfloor 6$. With 5 flags the number of signals is $^6P_5$; with 4 flags the number of signals is $^6P_4$; and so on. Thus the number required is

$$720 + 720 + 360 + 120 + 30 + 6 \text{ or } 1956.$$

§34. Here $\dfrac{\lfloor 28}{\lfloor 2r \ \lfloor 28 - 2r} \times \dfrac{\lfloor 28 - 2r \ \lfloor 2r - 4}{\lfloor 24} = \dfrac{225}{11};$

that is, $\dfrac{28 \cdot 27 \cdot 26 \cdot 25}{2r(2r - 1)(2r - 2)(2r - 3)} = \dfrac{225}{11};$

∴ $2r(2r - 1)(2r - 2)(2r - 3) = 24024,$

$$(4r^2 - 6r)(4r^2 - 6r + 2) = 24024.$$

Put x for $4r^2 - 6r$; then $x(x + 2) = 24024$.

From this equation $x = 154$, or -146. Putting $4r^2 - 6r = 154$, we get $r = 7$. The other values are inadmissible.

EXAMPLES XI b

§1. See Art. 151.

§2. $\dfrac{\lfloor 17}{\lfloor 7\,\lfloor 6\,\lfloor 4} = 4084080.$

§3. $\dfrac{\lfloor 14}{(\lfloor 3)^2(\lfloor 2)^4} = 151351200.$

§4. If 0 could stand first the number would be $\dfrac{\lfloor 7}{\lfloor 2\,\lfloor 3}$. But there are $\dfrac{\lfloor 6}{\lfloor 2\,\lfloor 3}$ cases in which the number begins with 0; therefore deducting these we have $\dfrac{\lfloor 7 - \lfloor 6}{\lfloor 2\,\lfloor 3} = \dfrac{6 \times 720}{6 \times 2} = 360.$

§5. The consonants can be arranged in $\lfloor 4$ ways, and the vowels in $\dfrac{\lfloor 3}{\lfloor 2}$ ways; therefore the number of arrangements is $\lfloor 4 \times 3$, or 72.

§6. He can make each journey in 5 ways, and the three journeys in $5 \times 5 \times 5$, or 125 ways.

§7. The first place can be occupied in n ways, and then the second place can also be occupied in n ways; and so on, as in Art. 152.

§8. Each stall can be occupied in 3 ways and the twelve stalls in 3^{12} ways.

§9. Each thing may be given away in p ways; therefore the required number is p^n.

§10. The first thing may be given in two ways; so may the second; so may the third, and so on. Hence we have $2 \times 2 \times 2 \times 2 \times 2$, or 32 ways; but this includes two cases in which either person has all the five things. If we reject these the number of ways will be 30.

§11. We have to arrange 9 letters, three of which are a, two b, and four c; therefore the number of arrangements is $\dfrac{\lfloor 9}{\lfloor 3\,\lfloor 2\,\lfloor 4}$, or 1260.

§12. The first ring can be placed in fifteen different positions; so may the second; so may the third. Hence there are $15 \times 15 \times 15$ different positions possible, only one of which is the right one; therefore the number of unsuccessful attempts possible is 3374.

§13. We have to select three points for each triangle; thus the required number is $^{15}C_3$, or 455.

§14. The number $= \dfrac{\lfloor a + 2b + 3c + d}{\lfloor a\,(\lfloor b)^2(\lfloor c)^3}$, by Art. 151.

§15. Each number is to consist of not more than 4 figures; and we may suppose each number to be written with four figures, because if we have less than 4 we can insert ciphers to begin the number with. Thus 24 may be written 0024. Therefore every possible arrangement of 4 figures out of the given 8 will furnish one of the required numbers, except 0000. Thus by Art. 152 the required number is $8^4 - 1$, or 4095.

§16. The first Classical prize can be given in 20 ways,
...... second .... 19 ...
The first Mathematical .... 20...
...... second .... 19 ...
and the other two each in 20 ways;
∴ the number of ways $= 20^4 \times 19^2 = 57760000.$

§17. The first arm can be put in 4 district positions, so can the second; thus we can with these two form 4^2 signals. Then taking each arm in succession and combining the different positions each is capable of, we ultimately get 4^5. From this result we must subtract 1 for the case in which each arm is in the position of rest.

§18. (1) As we only have to consider the relative positions of the persons forming the ring, suppose one man to remain fixed; then we can permute the other 6 men about him in $\lfloor 6$ or 720 ways.

(2) Suppose one Englishman to remain fixed; then the others can take their appropriate places in $\lfloor 6$ ways; but corresponding to each arrangement of Englishmen, there are 7 places in which the Americans can sit;

$$\therefore \text{ required number of ways } = \lfloor 6 \times \lfloor 7 = 3628800.$$

§19. Each coin may be either taken or left, therefore, as in Art. 153, the number of ways $= 2^7 - 1 = 127$.

§20. By Art. 153 the number of ways of selecting one or more cocoanuts, one or more apples, one or more oranges respectively will be $2^3 - 1$, $2^4 - 1$, $2^2 - 1$; and any one of these selections may be associated with each of the others, giving $7 \times 15 \times 3$, or 315 selections in all.

§21. The number of different ways of dividing into n equal groups is

$$\frac{\lfloor mn}{(\lfloor m)^n \lfloor n}.$$

[See Art. 147, note.]

§22. (1) With one flag, the number of signals $= 4$;
.... two flags, ... $= {}^4P_2 = 12$;
.... three .... $= {}^4P_3 = 24$;
.... four .... $= {}^4P_4 = 24$;
$\therefore$ the whole number of signals is $4 + 12 + 24 + 24$, or 64.

(2) With 5 flags, the total number of signals is

$${}^5P_1 + {}^5P_2 + {}^5P_3 + {}^5P_4 + {}^5P_5, \text{ or } 325.$$

§23. There are 6 letters of four different sorts, namely s, s; e, e; r; i. In finding arrangements of three, these may be classified as follows:

(1) Two alike, one different.

(2) All three different.

(1) The selection can be made in 2×3 ways; for we have to select one of the two pairs $s, s; e, e$; and then one from the remaining three letters.

(2) The selection can be made in ${}^4C_3$, or 4 ways,

(1) gives rise to $6 \times \dfrac{\lfloor 3}{\lfloor 2}$ or 18,

(2) gives to $4 \times \lfloor 3$ or 24 ways;

$\therefore$ the whole number of arrangements is $24 + 18$, or 42.

§24. (1) If there were no three points in a straight line we should have ${}^pC_2$ lines; but since q points lie in a straight line we must subtract ${}^qC_2$ lines and add the one in which are the q points; thus we have

$$\frac{p(p-1)}{2} - \frac{q(q-1)}{2} + 1.$$

(2) If there were no three points in a straight line we should have ${}^pC_3$ triangles; from this we must subtract ${}^qC_3$ which is the number of triangles lost in consequence of q points coming into one straight line.

§25. Since three points are required to determine a plane, we have, by the method of (1) in the last question, $^{p}C_3 - {}^{q}C_3 + 1$,

or $\qquad \dfrac{p(p-1)(p-2)}{6} - \dfrac{q(q-1)(q-2)}{6} + 1.$

§26. In the case of each book we may take $0, 1, 2, 3, \ldots p$; that is, we may deal with each book in $p+1$ ways, and therefore with all the books in $(p+1)^n$ ways. But this includes the case where all the books are rejected and no selection is made;

$\qquad\qquad \therefore$ the required number $= (p+1)^n - 1$.

§27. Ten letters, namely, $e, e; s, s; x; p; r; i; o; n$. For groups of 4, the letters may be arranged as follows:
(1) Two alike, two others alike.
(2) Two alike, two different.
(3) All four different.

(1) gives rise to 1 selection,
(2) gives rise to $2 \times {}^{7}C_2$ or 42 selections,
(3) gives rise to $^{8}C_4$ or 70 selections;

$\qquad\qquad \therefore$ number of selections $= 1 + 42 + 70 = 113.$

The number of arrangements is $\dfrac{\lfloor\underline{4}}{\lfloor\underline{2}\,\lfloor\underline{2}} + 42 \times \dfrac{\lfloor\underline{4}}{\lfloor\underline{2}} + 70 \times \lfloor\underline{4}$,

or $6 + 504 + 1680$, that is, 2190.

§28. Eleven letters, namely, $a, a; i, i; n, n; e; x; a; t; u$.
For groups of 4 we may arrange these as follows:
(1) Two alike, two others alike.
(2) Two alike, two different.
(3) All four different.

(1) gives rise to $^{3}C_2$ selections,
(2) gives rise to $3 \times {}^{7}C_2$ selections,
(3) gives rise to $^{8}C_4$ selections;

$\qquad \therefore$ number of permutations

$$= 3 \times \frac{\lfloor\underline{4}}{\lfloor\underline{2}\,\lfloor\underline{2}} + 63 \times \frac{\lfloor\underline{4}}{\lfloor\underline{2}} + 70 \times \lfloor\underline{4} = 18 + 756 + 1680 = 2454.$$

§29. There are $\lfloor\underline{5}$ numbers altogether and if we consider any one of the digits, say 7, there are $\lfloor\underline{4}$ cases in which 7 occupies each of the five places. Thus the sum arising from the digit 7 alone is

$$\lfloor\underline{4}\ \{7 + 70 + 700 + 7000 + 70000\},$$

that is, $\qquad\qquad\qquad 7 \times \lfloor\underline{4} \times 11111.$

Proceeding in the same way with each of the other digits we get finally

$$(1 + 3 + 5 + 7 + 9) \times \lfloor\underline{4} \times 11111 \text{ or } 6666600.$$

§30. If 0 could stand in the first place we should have $\lfloor\underline{4} \times 20 \times 11111$ as in Example 29. The sum of all the numbers in which 0 would stand first is $\lfloor\underline{3} \times 20 \times 11111$. Hence by subtraction we obtain 519960.

§31. Of the p like things we may take $0, 1, 2, \ldots p$; that is we may dispose of them in $p+1$ ways. Similarly we may dispose of the q like things in $q+1$ ways. The r unlike things may each be disposed of in 2 ways and therefore the r things may be disposed of in 2^r ways. Hence, combining these results, and subtracting 1 for the case in which all the things are rejected and no selection made, we get the required result.

§32. $\dfrac{\lfloor\underline{2n}}{\lfloor\underline{r}\,\lfloor\underline{2n-r}} = $ the number of permutations of $2n$ letters r of which are a and $2n - r$ of which are b. But this also $= {}^{2n}C_r$, which by Art. 154 is greatest when $r = n$, in which case $2n - r = n$ also.

§33. Of the m letters a we can take $0, 1, 2, 3, \ldots m$, that is we can deal with these letters in $m + 1$ ways, each of which will give a different factor of a^m. Then the other n unlike letters may each be dealt with in two ways, either taken or left. Combining the results and subtracting 1 for the case in which none of the letters are taken we obtain the result $(m + 1)2^n - 1$.

EXAMPLES XIII : Binomial Theorem Positive Integral Index

EXAMPLES XIII a

§13. $^{13}C_3 x^{10} \times (-5)^3 = -\dfrac{13 \cdot 12 \cdot 11}{1 \cdot 2 \cdot 3} \times 125 x^{10} = -35750 x^{10}.$

§14. $^{12}C_9(-2x)^9 = -^{12}C_3 2^9 \cdot x^9 = -112640 x^9.$

§15. $^{13}C_{11}(2x)^2(-1)^{11} = -\dfrac{13 \cdot 12}{1 \cdot 2} \times 4x^2 = -312 x^2.$

§16. $^{30}C_{27}(5x)^3(8y)^{27} = \dfrac{\underline{|30}}{\underline{|3}\ \underline{|27}}(5x)^3(8y)^{27}.$

§17. $^{10}C_3 \left(\dfrac{a}{3}\right)^7 (9b)^3 = \dfrac{10 \cdot 9 \cdot 8}{1 \cdot 2 \cdot 3} \cdot \dfrac{3^6}{3^7} \cdot a^7 \cdot b^3 = 40a^7 b^3.$

§18. $^8C_4(2a)^4 \left(-\dfrac{b}{3}\right)^4 = \dfrac{8 \cdot 7 \cdot 6 \cdot 5}{1 \cdot 2 \cdot 3 \cdot 4} \cdot \dfrac{2^4}{3^4} a^4 b^4 = \dfrac{1120}{81} a^4 b^4.$

§19. $^9C_6 \left(\dfrac{4x}{5}\right)^3 \left(-\dfrac{5}{2x}\right)^8 = \dfrac{9 \cdot 8 \cdot 7}{1 \cdot 2 \cdot 3} \cdot \dfrac{5^3}{x^3} = \dfrac{10500}{x^3}.$

§20. $^8C_4 \left(\dfrac{x^{\frac{3}{2}}}{a^{\frac{1}{2}}}\right)^4 \cdot \left(-\dfrac{y^{\frac{5}{2}}}{b^{\frac{3}{2}}}\right)^4 = \dfrac{8 \cdot 7 \cdot 6 \cdot 5}{1 \cdot 2 \cdot 3 \cdot 4} \cdot \dfrac{x^6 y^{10}}{a^2 b^6} = \dfrac{70 x^6 y^{10}}{a^2 b^6}.$

§21. The terms of the two series are numerically the same, but in the first the terms are all positive, and in the second they are alternately positive and negative;

$$\therefore \text{ the value} = 2(x^4 + {^4C_2} \cdot 2x^2 + 4) = 2(x^4 + 12x^2 + 4).$$

§22.
$$\text{The value} = 2\left\{ {^5C_1}(x^2 - a^2)^2 x + {^5C_3}(x^2 - a^2)x^3 + {^5C_5}x^5 \right\}$$
$$= 2\left\{ 5x^5 - 10a^2 x^3 + 5a^4 x + 10x^5 - 10a^2 x^3 + x^5 \right\}$$
$$= 2(16x^5 - 20a^2 x^3 + 5a^4 x).$$

§23.
The value
$$= 2\left\{ 6(\sqrt{2})^5 + 20(\sqrt{2})^3 + 6\sqrt{2} \right\}$$
$$= 12 \times 4\sqrt{2} + 80\sqrt{2} + 12\sqrt{2}$$
$$= 140\sqrt{2}.$$

§24.
The value
$$= 2\left\{ 2^6 + 15.2^4(1-x) + 15.2^2(1-x)^2 + (1-x^3) \right\}$$
$$= 2\left\{ 64 + 240 - 240x + 60 - 120x + 60x^2 + 1 - 3x + 3x^2 - x^3 \right\}$$
$$= 2\{365 - 363x + 63x^2 - x^2\}.$$

§25. There are 11 terms in the series;
$$\therefore \text{ the middle term is the } 6^{th} = {^{10}C_5} = 252.$$

§26.
The 8th term
$$= {^{14}C_7} 7\left(-\dfrac{x^2}{2}\right)^7 = -\dfrac{14 \cdot 13 \cdot 12 \cdot 11 \cdot 10 \cdot 9 \cdot 8}{1 \cdot 2 \cdot 3 \cdot 4 \cdot 5 \cdot 6 \cdot 7} \cdot \dfrac{x^{14}}{128}$$
$$= -\dfrac{429}{16} x^{14}.$$

§27. The expression $= x^{30}\left(1 + \dfrac{3a}{x^3}\right)^{16};$

$\therefore$ in the expansion of $\left(1 + \dfrac{3a}{x^3}\right)^{16}$ we have to find the coefficient of x^{-12};
this is equal to $^{15}C_4(3a)^4 = 110565a^4$.
§28.
The expression

$$= a^9 x^{36}\left(1 - \dfrac{b}{ax^3}\right)^9, \text{ and the required coefficient}$$

$$= a^9 \times {}^9C_6\left(\dfrac{b}{a}\right)^6 = 84a^3b^6.$$

§29. Since the expression $= x^{60}\left(1 - \dfrac{1}{x^7}\right)^{15}$, we require the coefficients
of x^{-28} and x^{-77} in the expansion of $(1 - x^{-7})^{16}$, these are $^{15}C_4$ and $-^{15}C_{11}$
respectively.
Thus the coefficients required are 1365 and -1365.

§30. The 5th term $= {}^9C_4(3a)^5\left(-\dfrac{a^3}{6}\right)^4 = \dfrac{189}{8}a^{17}$.

The 6th term $= {}^9C_5(3a)^4\left(-\dfrac{a^3}{6}\right)^5 = -\dfrac{21}{16}a^{19}$.

§31. The expression $= \left(\dfrac{3}{2}x^2\right)^9\left(1 - \dfrac{2}{9x^3}\right)^9 = \dfrac{3^9 x^{18}}{2^9} \times \left(1 - \dfrac{2}{9x^3}\right)^9$;

$\therefore$ the term required $= {}^9C_6\left(-\dfrac{2}{9}\right)^6 \cdot \dfrac{3^9}{2^9} = \dfrac{9 \cdot 8 \cdot 7}{1 \cdot 2 \cdot 3} \cdot \dfrac{1}{3^2 \cdot 2^3} = \dfrac{7}{18}$.

§32. The 13th term

$$= {}^{16}C_{12} \cdot (9x)^6 \cdot \left(-\dfrac{1}{3\sqrt{x}}\right)^{12}$$

$$= \dfrac{18 \cdot 17 \cdot 16 \cdot 15 \cdot 14 \cdot 13}{1 \cdot 2 \cdot 3 \cdot 4 \cdot 5 \cdot 6}$$

$$= 18564.$$

§33. Let the $(p+1)^{th}$ term be the one required; then $^nC_p x^{n-p} \cdot \left(\dfrac{1}{x}\right)^p$, or
$^nC_p x^{n-2p}$ is the term containing x^r. Therefore $n - 2p = r$, or $p = \dfrac{n-r}{2}$;

$\therefore$ the coefficient $= {}^nC_p = \dfrac{\lfloor n}{\left\lfloor \frac{1}{2}(n-r) \right\rfloor \left\lfloor \frac{1}{2}(n+r) \right\rfloor}$.

§34. $\left(x - \dfrac{1}{x^2}\right)^{3n} = x^{8n}\left(1 - \dfrac{1}{x^3}\right)^{3n}$, and we require the coefficient of
$\left(\dfrac{1}{x^3}\right)^n$ in the expansion of $\left(1 - \dfrac{1}{x^3}\right)^{3n}$.

Hence the required term $= (-1)^n \dfrac{\lfloor 3n}{\lfloor n \lfloor 2n}$.

§35. Let the $(r+1)^{th}$ term contain x^p. Then $^{2n}C_r(x^2)^{2n-r} \cdot \left(\dfrac{1}{x}\right)^r$, or $^{2n}C_r x^{4n-3r}$
contains x^p; therefore $4n - 3r = p$, and $r = \frac{1}{3}(4n - p)$;

$$\therefore \quad {}^{2n}C_r = \dfrac{\lfloor 2n}{\left\lfloor \frac{1}{3}(4n-p) \right\rfloor \left\lfloor \frac{1}{3}(2n+p) \right\rfloor}.$$

EXAMPLES XIII b

§1. $(x - y)^{30} = x^{30}\left(1 - \dfrac{y}{x}\right)^{30}$. Let T_r and T_{r+1} denote consecutive terms of $\left(1 - \dfrac{y}{x}\right)^{30}$; then $T_{r+1} = \dfrac{30 - r + 1}{r} \cdot \dfrac{4}{11} \times T_r$, numerically; $\therefore T_{r+1} > T_r$, so long as $\dfrac{124 - 4r}{11r} > 1$; that is, $124 > 15r$; therefore $r = 8$ makes the 9th term greatest.

§2. $(2x - 3y)^{28} = (2x)^{28}\left(1 - \dfrac{3y}{2x}\right)^{28}$; $\therefore T_{r+1} = \dfrac{28 - r + 1}{r} \cdot \dfrac{3 \times 4}{2 \times 9} \times T_r$, numerically; $\therefore T_{r+1} > T_r$, so long as $58 - 3r > 3r$; that is, $r = 11$ makes the 12th term greatest.

§3. $(2a + b)^{14} = (2a)^{14}\left(1 + \dfrac{b}{2a}\right)^{14}$; $T_{r+1} = \dfrac{14 - r + 1}{r} \cdot \dfrac{5}{8} \times T_r$; $\therefore T_{r+1} > T_r$, so long as $75 > 13r$; that is, $r = 5$ makes the 6th term greatest.

§4. $(3 + 2x)^{15} = 3^{15}\left(1 + \dfrac{2x}{3}\right)^{15}$; $T_{r+1} = \dfrac{15 - r + 1}{r} \cdot \dfrac{5}{3} \times T_r$; $\therefore T_{r+1} > T_r$, so long as $80 > 8r$; that is, $r = 10$ makes the 10th and 11th terms equal, and greater than any other term.

§5. $T_{r+1} > T_r$, so long as $\dfrac{7 - r}{r} \cdot \dfrac{2}{3} > 1$; that is, $14 > 5r$. Therefore the 3rd term is the greatest, and its value $= \dfrac{6 \cdot 5}{1 \cdot 2} \cdot \dfrac{4}{9} = 6\frac{2}{3}$.

§6. $(a + x)^n = a^n\left(1 + \dfrac{x}{a}\right)^n$; and $T_{r+1} = \dfrac{10 - r}{r} \cdot \dfrac{2}{3} \times T_r$; $\therefore T_{r+1} > T_r$, so long as $20 - 2r > 3r$. Therefore the 4th and 5th terms are equal and greater than any other term. Their value

$$= \dfrac{9 \cdot 8 \cdot 7}{1 \cdot 2 \cdot 3} \cdot \left(\dfrac{1}{2}\right)^6 \left(\dfrac{1}{3}\right)^3 = \dfrac{7}{144}.$$

§7. We have to shew that $^{2n}C_n = {}^{2n-1}C_{n-1} + {}^{2n-1}C_n$.

Now $\qquad {}^{2n-1}C_{n-1} = {}^{2n-1}C_n = \dfrac{\underline{|2n - 1}}{\underline{|n - 1}\,\underline{|n}}$,

and $\qquad {}^{2n}C_n = \dfrac{\underline{|2n}}{\underline{|n}\,\underline{|n}} = \dfrac{2n\underline{|2n - 1}}{\underline{|n}\,\underline{|n}} = \dfrac{2 \cdot \underline{|2n - 1}}{\underline{|n - 1}\,\underline{|n}}$.

which proves the proposition.

§8. By Art. 165, we have $(x + a)^n = A + B$, and $(x - a)^n = A - B$; therefore by multiplication we get the required result.

§9. We have $nx^{n-1}y = 240$, $\dfrac{n(n - 1)}{1 \cdot 2}x^{n-2}y^2 = 720$, and

$$\dfrac{n(n - 1)(n - 2)}{1 \cdot 2 \cdot 3}x^{n-3}y^3 = 1080;$$

$\therefore$ by division, $\dfrac{n - 2}{3} \cdot \dfrac{y}{x} = \dfrac{3}{2}$ and $\dfrac{n - 1}{2} \cdot \dfrac{y}{x} = 3$. From these two equations we get $\dfrac{2(n - 2)}{3} = \dfrac{n - 1}{2}$; and $n = 5$; therefore it easily follows that $x = 2$, $y = 3$.

§10.

$$(1 + 2x - x^2)^4$$
$$= 1 + 4(2x - x^2) + 6(2x - x^2)^2 + 4(2x - x^2)^3 + (2x - x^2)^4$$
$$= 1 + 8x - 4x^2 + 24x^2 - 24x^3 + 6x^4 + 32x^3 - 48x^4$$

$$+ 24x^5 - 4x^6 + 16x^4 - 32x^5 + 24x^6 - 8x^7 + x^6$$
$$= 1 + 8x + 20x^2 + 8x^3 - 26x^4 - 8x^5 + 20x^6 - 8x^7 + x^6.$$

§11. $(3x^2 - 2ax + 3a^2)^3 = \{3x^2 - (2ax - 3a^2)\}^3$
$= 27x^6 - 3 \cdot 9x^4(2ax - 3a^2) + 3 \cdot 3x^2(2ax - 3a^2)^2 - (2ax - 3a^2)^3$
$= 27x^6 - 54ax^5 + 81a^2x^4 + 36a^2x^4 - 108a^3x^3 + 81a^4x^2 - 8a^3x^3 + 36a^4x^2 - 54a^5x + 27a^6$
$= 27x^6 - 54ax^5 + 117a^2x^4 - 116a^3x^3 + 117a^4x^2 - 54a^5x + 27a^6.$

§12. The rth term from the end is the $(n - r + 2)^{th}$ from the beginning

and is equal to $\dfrac{\lfloor n}{\lfloor r-1 \; \lfloor n-r+1}} \cdot x^{r-1}a^{n-r+1}.$

§13. There are $2n + 2$ terms in all, and the $(p + 2)^{th}$ term from the end has $2n + 2 - (p + 2)$ before it; therefore counting from the beginning it is the $(2n - p + 1)^{th}$ term which is

$$\frac{\lfloor 2n+1}{\lfloor p+1 \; \lfloor 2n-p}} \cdot x^{p+1}\left(-\frac{1}{x}\right)^{2n-p}, \text{ or } (-1)^p \frac{\lfloor 2n+1}{\lfloor p+1 \; \lfloor 2n-p}} \cdot x^{2p-2n-1}.$$

§14. We have $^{43}C_{2r} = {}^{43}C_{r+1}$; therefore $2r + r + 1 = 43$, or $r = 14$.

§15. We must have $^{2n}C_{3r-1} = {}^{2n}C_{r+1}$;
$$\therefore \; 3r - 1 + r + 1 = 2n, \text{ or } 2r = n.$$

§16. The middle term is the $(n + 1)^{th}$, which is $\dfrac{\lfloor 2n}{\lfloor n \; \lfloor n}} \cdot x^n.$

This may be written $\dfrac{1 \cdot 2 \cdot 3 \cdot 4 \ldots 2n}{\lfloor n \; \lfloor n}} \cdot x^n,$

and remembering that $2 \cdot 4 \cdot 6 \ldots 2n = 2^n \cdot \lfloor n$ this reduces to

$$\frac{1 \cdot 3 \cdot 5 \ldots (2n-1)}{\lfloor n}} \cdot 2^n x^n.$$

§17. This is solved in the first part of Art. 176.

§18. $(1+x)^{n+1} = 1 + (n+1)x + \dfrac{(n+1)n}{1 \cdot 2}x^2 + \dfrac{(n+1)n(n-1)}{1 \cdot 2 \cdot 3}x^3 + \ldots$
$$+ (n+1)x^n + x^{n+1};$$

$\therefore \dfrac{(1+x)^{n+1} - 1}{n+1} = x + \dfrac{n}{1 \cdot 2}x^2 + \dfrac{n(n-1)}{1 \cdot 2 \cdot 3}x^3 + \ldots + \dfrac{x^{n+1}}{n+1},$

$\therefore \dfrac{(1+x)^{n+1} - 1}{n+1} = c_0 x + \dfrac{c_1}{2}x^2 + \dfrac{c_2}{3}x^3 + \ldots + \dfrac{c_n x^{n+1}}{n+1}.$

Putting $x = 1$, we get the required result.

§19. Writing $c_0, c_1, c_2, \ldots$ in full we obtain

$$\frac{n}{1} + \frac{n(n-1)}{n} + \frac{n(n-1)(n-2)}{n(n-1)} + \ldots \text{ to } n \text{ terms,}$$

That is, $n + (n - 1) + (n - 2) + \ldots$ to n terms.

§20. We have $\dfrac{c_0 + c_1}{c_1} = 1 + \dfrac{1}{n} = \dfrac{n+1}{n}.$

$$\frac{c_1 + c_2}{c_2} = 1 + \frac{2}{n-1} = \frac{n+1}{n-1}, \quad \frac{c_2 + c_3}{c_3} = 1 + \frac{3}{n-2} = \frac{n+1}{n-2},$$

$\therefore$ by multiplication we get the required result.

§21. As in Ex. 18, we have

$$\frac{(1+x)^{n+1} - 1}{n+1} = c_0 x + \frac{c_1 x^2}{2} + \frac{c_2 x^3}{3} + \ldots + \frac{c_n x^{n+1}}{n+1}.$$

Putting $x = 2$, we easely get the required result.

§22. We have $(1+x)^n = c_0 + c_1 x + c_2 x^2 + \ldots + c_n x^n,$

$$\left(1+\frac{1}{x}\right)^n = c_0 + \frac{c_1}{x} + \frac{c_2}{x^2} + \ldots + \frac{c_n}{x^n};$$

$\therefore \dfrac{1}{x^n}(1+x)^{2n} = (c_0^2 + c_1^2 + c_2^2 + \ldots + c_n^2) +$ terms which contain x;

$\therefore c_0^2 + c_1^2 + c_2^2 + \ldots + c_n^2$ is equal to the term independent of x in $\dfrac{1}{x^n}(1+x)^{2n}$,

that is, the coefficient of x^n in the expansion of $(1+x)^{2n}$.

§23. $(1+x)^n = c_0 + c_1 x + c_2 x^2 + \ldots + c_r x^r + \ldots + c_n x^n$, also since terms equidistant from beginning and end have the same coefficient

$$(1+x)^n = c_n + c_{n-1}x + c_{n-2}x^2 + \ldots + c_{n-r}x^r + \ldots + c_0 x^n.$$

Now multiply these two series together and pick out the coefficient of x^{n+r}; then $c_0 c_r + c_1 c_{r+1} + c_2 c_{r+2} + \ldots + c_{n-r} c_n$ is equal to the coefficient of x^{n+r} in the expansion of $(1+x)^{2n}$.

EXAMPLES XIV : Binomial Theorem. Any Index

EXAMPLES XIV a

§1.

$$(1+x)^{\frac{1}{2}} = 1 + \frac{1}{2}x + \frac{\frac{1}{2}\left(\frac{1}{2}-1\right)}{1\cdot 2}x^2 + \frac{\frac{1}{2}\left(\frac{1}{2}-1\right)\left(\frac{1}{2}-2\right)}{\lfloor 3}x^3 + \ldots$$

$$= 1 + \frac{1}{2}x - \frac{1}{8}x^2 + \frac{1}{16}x^3 + \ldots$$

§2.

$$(1+x)^{\frac{3}{2}} = 1 + \frac{3}{2}x + \frac{\frac{3}{2}\left(\frac{3}{2}-1\right)}{1\cdot 2}x^2 + \frac{\frac{3}{2}\left(\frac{3}{2}-1\right)\left(\frac{3}{2}-2\right)}{\lfloor 3}x^3 + \ldots$$

$$= 1 + \frac{3}{2}x + \frac{3}{8}x^2 - \frac{1}{16}x^3 + \ldots$$

§3.

$$(1-x)^{\frac{2}{5}} = 1 - \frac{2}{5}x + \frac{\frac{2}{5}\left(\frac{2}{5}-1\right)}{1\cdot 2}x^2 - \frac{\frac{2}{5}\left(\frac{2}{5}-1\right)\left(\frac{2}{5}-2\right)}{1\cdot 2\cdot 3}x^3 + \ldots$$

$$= 1 - \frac{2}{5}x - \frac{3}{25}x^2 - \frac{8}{125}x^3 + \ldots$$

§4.

$$(1+x^2)^{-2} = 1 + (-2)x^2 + \frac{(-2)(-3)}{1\cdot 2}x^4 + \frac{(-2)(-3)(-4)}{\lfloor 3}x^8 + \ldots$$

$$= 1 - 2x^2 + 3x^4 - 4x^6 + \ldots$$

§5.

$$(1-3x)^{\frac{1}{3}}$$

$$= 1 - \frac{1}{3}3x + \frac{\frac{1}{3}\left(\frac{1}{3}-1\right)}{1\cdot 2}(3x)^2 - \frac{\frac{1}{3}\left(\frac{1}{3}-1\right)\left(\frac{1}{3}-2\right)}{\lfloor 3}(3x)^3 + \ldots$$

$$= 1 - x - x^2 - \frac{5}{3}x^3 + \ldots$$

§6.

$$(1-3x)^{-\frac{1}{3}} = 1 - \frac{1}{3}(-3x) + \frac{\left(-\frac{1}{3}\right)\left(-\frac{1}{3}-1\right)}{1\cdot 2}(-3x)^2$$

$$+ \frac{\left(-\frac{1}{3}\right)\left(-\frac{1}{3}-1\right)\left(-\frac{1}{3}-2\right)}{\lfloor 3}(-3x)^3 + \ldots$$

$$= 1 + x + 2x^2 + \frac{14}{3}x^3 + \ldots$$

§7.

$$(1+2x)^{-\frac{1}{2}} = 1 + \left(-\frac{1}{2}\right)2x + \frac{\left(-\frac{1}{2}\right)\left(-\frac{1}{2}-1\right)}{1\cdot 2}(2x)^2$$

$$+ \frac{\left(-\frac{1}{2}\right)\left(-\frac{1}{2}-1\right)\left(-\frac{1}{2}-2\right)}{\lfloor 3}(2x)^3 + \ldots$$

$$= 1 - x + \frac{3}{2}x^2 - \frac{5}{2}x^3 + \dots$$

§8.

$$\left(1 + \frac{x}{3}\right)^{-3}$$

$$= 1 + (-3)\frac{x}{3} + \frac{(-3)(-4)}{1 \cdot 2}\left(\frac{x}{3}\right)^2 + \frac{(-3)(-4)(-5)}{\lfloor 3}\left(\frac{x}{3}\right)^3 + \dots$$

$$= 1 - x + \frac{2}{3}x^2 - \frac{10}{27}x^3 + \dots$$

§9.

$$\left(1 + \frac{2x}{3}\right)^{\frac{3}{2}} = 1 + \frac{3}{2}\cdot\frac{2x}{3} + \frac{\frac{3}{2}\left(\frac{3}{2}-1\right)}{1 \cdot 2}\left(\frac{2x}{3}\right)^2$$

$$+ \frac{\frac{3}{2}\left(\frac{3}{2}-1\right)\left(\frac{3}{2}-2\right)}{\lfloor 3}\left(\frac{2x}{3}\right)^3 + \dots$$

$$= 1 + x + \frac{1}{6}x^2 - \frac{1}{54}x^3 + \dots$$

§10.

$$\left(1 + \frac{1}{2}a\right)^{-4} = 1 + (-4)\left(\frac{1}{2}a\right) + \frac{(-4)(-5)}{1 \cdot 2}\left(\frac{1}{2}a\right)^2$$

$$+ \frac{(-4)(-5)(-6)}{\lfloor 3}\left(\frac{1}{2}a\right)^3 + \dots$$

$$= 1 - 2a + \frac{5}{2}a^2 - \frac{5}{2}a^3 + \dots$$

§11.

$$(2 + x)^{-3} = 2^{-3}\left(1 + \frac{x}{2}\right)^{-3}$$

$$= \frac{1}{8}\left(1 - \frac{3}{2}x + \frac{3}{2}x^2 - \frac{5}{4}x^3 + \dots\right)$$

§12.

$$(9 + 2x)^{\frac{1}{2}} = 9^{\frac{1}{2}}\left(1 + \frac{2x}{9}\right)^{\frac{1}{2}}$$

$$= 3\left(1 + \frac{1}{9}x - \frac{1}{162}x^2 + \frac{1}{1458}x^3 + \dots\right)$$

§13.

$$(8 + 12a)^{\frac{2}{3}} = 8^{\frac{2}{3}}\left(1 + \frac{12a}{8}\right)^{\frac{2}{3}}$$

$$= 4\left(1 + \frac{3}{2}a\right)^{\frac{2}{3}}$$

$$= 4\left(1 + a - \frac{1}{4}a^2 + \frac{1}{6}a^3 - \dots\right)$$

§14.

$$(9 - 6x)^{-\frac{3}{2}} = 9^{-\frac{3}{2}}\left(1 - \frac{6x}{9}\right)^{-\frac{3}{2}} = \frac{1}{27}\left(1 - \frac{2}{3}x\right)^{-\frac{3}{2}}$$

$$= \frac{1}{27}\left(1 + x + \frac{5}{6}x^2 + \frac{35}{54}x^3 + \dots\right).$$

§15.

$$(4a - 8x)^{-\frac{1}{2}} = (4a)^{-\frac{1}{2}}\left(1 - \frac{2x}{a}\right)^{-\frac{1}{2}}$$

$$= \frac{1}{2a^{\frac{1}{2}}}\left(1 + \frac{x}{a} + \frac{3}{2}\frac{x^2}{a^2} + \frac{5}{2}\frac{x^3}{a^3} + \dots\right).$$

§16.

$$\frac{-\frac{1}{2}\left(-\frac{1}{2}-1\right)\dots\left(-\frac{1}{2}-7+1\right)}{\underline{|7}}(2x)^7 = -\frac{1\cdot3\cdot5\dots13}{2^7\underline{|7}}\cdot2^7x^7$$

$$= -\frac{429}{16}x^7.$$

§17. $\dfrac{\frac{11}{2}\left(\frac{11}{2}-1\right)\dots\left(\frac{11}{2}-10+1\right)}{\underline{|10}}(-2x^3)^{10}$

$$= \frac{11\cdot9\cdot7\cdot5\cdot3\cdot1\cdot(-1)(-3)(-5)(-7)}{2^{10}\underline{|10}}2^{10}\cdot x^{30} = \frac{77}{256}x^{30}.$$

§18. $\dfrac{\frac{16}{3}\left(\frac{16}{3}-1\right)\left(\frac{16}{3}-2\right)\dots\left(\frac{16}{3}-9+1\right)}{\underline{|9}}(3a^2)^9$

$$= \frac{16\cdot13\cdot10\cdot7\cdot4\cdot1\cdot(-2)(-5)(-8)}{3^9\underline{|9}}3^9a^{18} = -\frac{1040}{81}a^{18}.$$

§19. $(3a-2b)^{-1} = (3a)^{-1}\left(1-\dfrac{2b}{3a}\right)^{-1}$; therefore the 5th term

$$= \frac{(-1)(-2)(-3)(-4)}{\underline{|4}}(3a)^{-1}\left(-\frac{2b}{3a}\right)^4 = \frac{16}{243}\frac{b^4}{a^5}.$$

§20. $\dfrac{(-2)(-3)(-4)\dots(-1-r)}{\underline{|r}}(-x)^r = (r+1)x^r.$

§21. $\dfrac{(-4)(-5)\dots(-3-r)}{\underline{|r}} = \dfrac{4\cdot5\cdot6\dots(r+3)}{\underline{|r}}x^r$

$$= \frac{(r+1)(r+2)(r+3)}{\underline{|3}}x^r.$$

§22.

$$\frac{\frac{1}{2}\left(\frac{1}{2}-1\right)\left(\frac{1}{2}-2\right)\dots\left(\frac{1}{2}-r+1\right)}{\underline{|r}}x^r$$

$$= \frac{1(-1)(-3)\dots(-2r+3)}{2^r\underline{|r}}x^r$$

$$= (-1)^{r-1}\cdot\frac{1\cdot3\cdot5\dots(2r-3)}{2^r\underline{|r}}x^r.$$

§23.

$$\frac{\frac{11}{3}\left(\frac{11}{3}-1\right)\left(\frac{11}{3}-2\right)\dots\left(\frac{11}{3}-r+1\right)}{\underline{|r}}x^r$$

$$= \frac{11\cdot8\cdot5\dots(14-3r)}{3^r\underline{|r}}x^r$$

$$= (-1)^{r-4}\frac{11\cdot8\cdot5\cdot2\cdot1\cdot4\dots(3r-14)}{3^r\underline{|r}}x^r.$$

§24. The 14th term of $(2^{10})^{\frac{13}{2}}\left(1-\dfrac{x}{2^3}\right)^{\frac{13}{2}}$

$$= 2^{65} \frac{\frac{13}{2} \cdot \left(\frac{13}{2} - 1\right) \ldots \left(\frac{13}{2} - 13 + 1\right)}{\underline{|13}} \left(-\frac{x}{2^3}\right)^{13}$$

$$= -2^{65} \frac{13 \cdot 11 \cdot 9 \ldots (-11)}{2^{13} \underline{|13}} \frac{x^{13}}{2^{39}}$$

$$= -2^{13} \frac{(-1)(-3)(-5)(-7)(-9)(-11)}{12 \cdot 10 \cdot 8 \cdot 6 \cdot 4 \cdot 2} x^{13} = -1848 x^{13}.$$

§25. The 7th term of $(3^8)^{\frac{11}{4}} \left(1 + \frac{6^4 x}{3^8}\right)^{\frac{11}{4}}$

$$= 3^{22} \cdot \frac{\frac{11}{4} \left(\frac{11}{4} - 1\right) \left(\frac{11}{4} - 2\right) \ldots \left(\frac{11}{4} - 5\right)}{\underline{|6}} \left(\frac{2^4}{3^4} x\right)^6$$

$$= 3^{22} \cdot \frac{11 \cdot 7 \cdot 3(-1)(-5)(-9)}{2^{12} \underline{|6}} \cdot \frac{2^{24} x^6}{3^{24}}$$

$$= -\frac{11 \cdot 7 \cdot 3 \cdot 5 \cdot 9}{1 \cdot 2 \cdot 3 \cdot 4 \cdot 5 \cdot 6} \cdot \frac{2^{12}}{3^2} x^6, \text{ which reduces to } -\frac{19712}{3} x^6.$$

<h3 style="text-align:center">EXAMPLES XIV b</h3>

§1.
$$\frac{\frac{1}{2} \left(\frac{1}{2} + 1\right) \left(\frac{1}{2} + 2\right) \ldots \left(\frac{1}{2} + r - 1\right)}{\underline{|r}} (-x)^r$$
$$= (-1)^r \frac{1 \cdot 3 \cdot 5 \cdot 7 \ldots (2r - 1)}{2^r \underline{|r}} x^r.$$

§2. $\dfrac{5 \cdot 6 \cdot 7 \ldots (r + 4)}{\underline{|r}} x^r = \dfrac{(r + 1)(r + 2)(r + 3)(r + 4)}{\underline{|4}} x^r.$

§3.
$$\frac{\frac{1}{3} \left(\frac{1}{3} - 1\right) \left(\frac{1}{3} - 2\right) \ldots \left(\frac{1}{3} - r + 1\right)}{\underline{|r}} (3x)^r$$
$$= \frac{1 \cdot (-2)(-5) \ldots (4 - 3r)}{3^r \underline{|r}} 3^r x^r = (-1)^{r-1} \frac{1 \cdot 2 \cdot 5 \ldots (3r - 4)}{\underline{|r}} x^r.$$

§4.
$$\frac{\frac{2}{3} \left(\frac{2}{3} + 1\right) \left(\frac{2}{3} + 2\right) \ldots \left(\frac{2}{3} + r - 1\right)}{\underline{|r}} (-x)^r$$
$$= (-1)^r \cdot \frac{2 \cdot 5 \cdot 8 \ldots (3r - 1)}{3^r \underline{|r}} x^r.$$

§5. $\dfrac{3 \cdot 4 \cdot 5 \ldots (r + 2)}{\underline{|r}} (-x^2)^r = (-1)^r \dfrac{(r + 1)(r + 2)}{\underline{|2}} x^{2r}.$

§6.
$$\frac{\frac{3}{2} \left(\frac{3}{2} + 1\right) \left(\frac{3}{2} + 2\right) \ldots \left(\frac{3}{2} + r - 1\right)}{\underline{|r}} (2x)^r$$
$$= \frac{3 \cdot 5 \cdot 7 \ldots (2r + 1)}{\underline{|r}} x^r.$$

§7. $(a + bx)^{-1} = a^{-1} \left(1 + \dfrac{b}{a} x\right)^{-1}.$ Thus the $(r + 1)^{th}$ term is

$$(-1)^r \frac{b^r}{a^{r+1}} x^r, \text{ by Art. 186.}$$

§8. $(2 - x)^{-2} = \dfrac{1}{2^2} \left(1 - \dfrac{x}{2}\right)^{-2}.$

Thus the $(r+1)^{th}$ term $= \dfrac{(r+1)}{2^{r+2}} \cdot x^r$.

§9. $(a^3 - x^3)^{\frac{2}{3}} = a^2 \left(1 - \dfrac{x^3}{a^3}\right)^{\frac{2}{3}}$.

$\therefore (r+1)^{th}$ term

$$= a^2 \cdot \dfrac{\dfrac{2}{3}\left(\dfrac{2}{3}-1\right)\left(\dfrac{2}{3}-2\right)\ldots\left(\dfrac{2}{3}-r+1\right)}{\lfloor r}\left(-\dfrac{x^3}{a^3}\right)^r$$

$$= \dfrac{2(-1)(-4)\ldots(5-3r)}{3^r\lfloor r}(-1)^r\dfrac{x^{3r}}{a^{3r-2}}$$

$$= -\dfrac{2\cdot1\cdot4\ldots(3r-5)}{3^r\lfloor r}\cdot\dfrac{x^{3r}}{a^{3r-2}}.$$

§10. $(r+1)^{th}$ term of $(1+2x)^{-\frac{1}{2}}$

$$= \dfrac{\dfrac{1}{2}\left(\dfrac{1}{2}+1\right)\left(\dfrac{1}{2}+2\right)\ldots\left(\dfrac{1}{2}+r-1\right)}{\lfloor r}(-2x)^r$$

$$= (-1)^r \cdot \dfrac{1\cdot3\cdot5\ldots(2r-1)}{\lfloor r}x^r.$$

§11. $(r+1)^{th}$ term of $(1-3x)^{-\frac{2}{3}}$

$$= \dfrac{\dfrac{2}{3}\left(\dfrac{2}{3}+1\right)\left(\dfrac{2}{3}+2\right)\ldots\left(\dfrac{2}{3}+r-1\right)}{\lfloor r}(3x)^r$$

$$= \dfrac{2\cdot5\cdot8\ldots(3r-1)}{\lfloor r}x^r.$$

§12. $(r+1)^{th}$ term of $(a^n - nx)^{-\frac{1}{n}}$

$$= \dfrac{1}{a}\dfrac{\dfrac{1}{n}\left(\dfrac{1}{n}+1\right)\left(\dfrac{1}{n}+2\right)\ldots\left(\dfrac{1}{n}+r-1\right)}{\lfloor r}\left(\dfrac{nx}{a^n}\right)^r$$

$$= \dfrac{(n+1)(2n+1)(3n+1)\ldots(\overline{r-1}\cdot n+1)}{\lfloor r}\dfrac{x^r}{a^{nr+1}}.$$

§13. $T_{r+1} = \dfrac{7+r-1}{r}\cdot\dfrac{4}{15}\times T_r$ numerically;

$\therefore T_{r+1} > T_r$, so long as $24+4r > 15r$; $r=2$ makes the 3rd term greatest.

§14. $T_{r+1} = \dfrac{\dfrac{21}{2}-r+1}{r}\cdot\dfrac{2}{3}\times T_r$;

$\therefore T_{r+1} > T_r$, so long as $23 > 5r$; thus the 5th term is the greatest.

§15. $T_{r+1} = \dfrac{\dfrac{11}{4}+r-1}{r}\cdot\dfrac{7}{8}\times T_r$;

$\therefore T_{r+1} > T_r$, so long as $49+28r > 32r$; thus the 13th term is the greatest.

§16. $(2x+5y)^{12} = (2x)^{12}\left(1+\dfrac{5y}{2x}\right)^{12}$.

$$\therefore T_{r+1} = \dfrac{12-r+1}{r}\cdot\dfrac{5\times3}{2\times8}\times T_r;$$

that is, $T_{r+1} > T_r$, so long as $195 > 31r$; thus the 7th term is the greatest.

§17. $T_{r+1} = \dfrac{7+r-1}{r}\cdot\dfrac{2}{5}\times T_r$.

$\therefore T_{r+1} > T_r$, so long as $12+2r > 5r$; thus $r=4$ makes the 4th and 5th equal and greater than any other term.

§18. $(3x^2 + 4y^3)^{-n} = (3x^2)^{-n}\left(1 + \dfrac{4y^3}{3x^2}\right)^{-n}$.

$\therefore\ T_{r+1} = \dfrac{15 + r - 1}{r} \cdot \dfrac{4 \times 8}{3 \times 81} \times T_r$ numerically; that is, $T_{r+1} > T_r$, so long as $448 + 32r > 243r$; thus the 3rd term is the greatest.

§19.

$$\sqrt{98} = (100 - 2)^{\frac{1}{2}} = 10\left(1 - \frac{2}{100}\right)^{\frac{1}{2}}$$

$$= 10\left\{1 - \frac{1}{100} - \frac{1}{2}\left(\frac{1}{100}\right)^2 - \frac{1}{2}\left(\frac{1}{100}\right)^3 - \cdots\right\}$$

$$= 10(1 - {\cdot}01 - {\cdot}00005 - {\cdot}0000005)$$

$$= {\cdot}9899495 \times 10 = 9{\cdot}89949\ldots$$

§20.

$$\sqrt[3]{998} = (1000 - 2)^{\frac{1}{3}} = 10\left(1 - \frac{2}{1000}\right)^{\frac{1}{3}}$$

$$= 10\left\{1 - \frac{1}{3}\cdot\frac{2}{1000} - \frac{1}{9}\left(\frac{2}{1000}\right)^2 - \cdots\right\}$$

$$= 10(1 - {\cdot}0006666 - {\cdot}0000004)$$

$$= {\cdot}999333 \times 10 = 9{\cdot}99333.$$

§21.

$$\sqrt[3]{1003} = (10^3 + 3)^{\frac{1}{3}} = 10\left(1 + \frac{3}{10^3}\right)^{\frac{1}{3}}$$

$$= 10\left\{1 + \frac{1}{10^3} - \frac{1}{10^6} + \cdots\right\} = 10 + \frac{1}{10^2} - \frac{1}{10^5} + \cdots$$

$$= 10 + {\cdot}01 - {\cdot}00001 = 10{\cdot}00999.$$

§22.

$\sqrt[4]{2400}$

$$= (7^4 - 1)^{\frac{1}{4}} = 7\left(1 - \frac{1}{7^4}\right)^{\frac{1}{4}}$$

$$= 7\left\{1 - \frac{1}{4}\cdot\frac{1}{7^4} - \frac{3}{32}\cdot\left(\frac{1}{7^4}\right)^2 + \cdots\right\}$$

$$= 7\left\{1 - \frac{{\cdot}00041649}{4} - \frac{3}{32}({\cdot}00000017) + \cdots\right\}\ \text{[Ex. 3, p. 160.]}$$

$$= 7(1 - {\cdot}00010418) = 6{\cdot}99927.$$

§23.

$$(128)^{-\frac{1}{3}} = (5^3 + 3)^{-\frac{1}{3}} = \frac{1}{5}\left(1 + \frac{3}{5^3}\right)^{-\frac{1}{3}}$$

$$= \frac{1}{5}\left(1 - \frac{1}{5^3} - \frac{2}{5^6} - \frac{14}{3\cdot5^9} + \cdots\right)$$

$$= \frac{1}{5}\left(1 - \frac{2^3}{10^3} + \frac{2^7}{10^6} - \frac{14\cdot2^9}{3\cdot10^9} + \cdots\right)$$

$$= \frac{1}{5}(1 - {\cdot}008 + {\cdot}000128 - \ldots) = {\cdot}19842.$$

§24.

$$\left(1 + \frac{1}{250}\right)^{-\frac{1}{3}} = 1 + \frac{1}{3}\cdot\frac{1}{250} - \frac{1}{9}\left(\frac{1}{250}\right)^2 + \cdots$$

$$= 1 + \frac{4}{3 \times 10^3} - \frac{1}{9}\cdot\frac{16}{10^6} + \cdots$$

$$= 1 + \frac{\cdot004}{3} - \frac{\cdot000016}{9} + \dots$$
$$= 1 + \cdot00133, \text{ to five places,}$$
$$= 1\cdot00133.$$

§25.

$$(630)^{-\frac{3}{4}} = (5^4 + 5)^{-\frac{3}{4}} = 5^{-3}\left(1 + \frac{1}{5^3}\right)^{-\frac{3}{4}}$$
$$= \frac{1}{5^3}\left\{1 - \frac{3}{4}\cdot\frac{1}{5^3} + \frac{3\cdot7}{2^5}\left(\frac{1}{5^3}\right)^2 - \dots\right\}$$
$$= \frac{1}{5^3}\left(1 - \frac{6}{10^3} + \frac{42}{10^6} - \dots\right)$$
$$= \frac{1}{5^3}(1 - \cdot006 + \cdot000042 - \dots)$$
$$= \frac{8}{1000} \times \cdot994042 = \cdot00795.$$

§26.

$$\sqrt[5]{3128} = 5\left(1 + \frac{3}{5^5}\right)^{\frac{1}{5}} = 5\left(1 + \frac{3}{5^6} - \frac{18}{5^{12}} + \dots\right)$$
$$= 5\left(1 + \frac{3\cdot2^6}{10^6} - \frac{18\cdot2^{12}}{10^{12}} + \dots\right)$$
$$= 5(1 + \cdot000192), \text{ to five places,}$$
$$= 5\cdot00096.$$

§27.

$$(1 - 7x)^{\frac{1}{3}}(1 + 2x)^{-\frac{3}{4}}$$
$$= \left(1 - \frac{7}{3}x + \dots\right)\left(1 - \frac{3}{4}2x + \dots\right)$$
$$= 1 - \left(\frac{7}{3} + \frac{3}{2}\right)x + \dots$$
$$= 1 - \frac{23}{6}x, \text{ neglecting } x^2 \text{ and higher powers.}$$

§28.

$$\sqrt{4 - x}\left(3 - \frac{3}{2}\right)^{-1} = \left(1 - \frac{x}{4}\right)^{\frac{1}{2}} \times \frac{1}{3}\left(1 - \frac{x}{6}\right)^{-1}$$
$$= \frac{2}{3}\left(1 - \frac{x}{8}\right)\left(1 + \frac{x}{6}\right), \text{ neglecting } x^2;$$
$$= \frac{2}{3}\left(1 + \frac{x}{24}\right).$$

§29.

$$\frac{(8 + 3x)^{\frac{2}{3}}}{(2 + 3x)\sqrt{4 - 5x}} = \frac{4\left(1 + \frac{3}{8}x\right)^{\frac{2}{3}}}{2\left(1 + \frac{3}{2}x\right)2\left(1 - \frac{5x}{4}\right)^{\frac{1}{2}}}$$
$$= \left(1 + \frac{3}{8}x\right)^{\frac{2}{3}} \times \left(1 + \frac{3}{2}x\right)^{-1} \times \left(1 - \frac{5x}{4}\right)^{-\frac{1}{2}}$$
$$\approx \left(1 + \frac{1}{4}x\right)\left(1 - \frac{3}{2}x\right)\left(1 + \frac{5}{8}x\right)$$
$$= 1 - \frac{5x}{8}.$$

§30.

$$\frac{\left(1+\dfrac{2}{3}x\right)^{-5} \times (4+3x)^{\frac{1}{2}}}{(4+x)^{\frac{3}{2}}} = \frac{\left(1-\dfrac{10}{3}x\right) \times 2\left(1+\dfrac{3}{8}x\right)}{8\left(1+\dfrac{x}{4}\right)^{\frac{3}{2}}}$$

$$= \frac{1}{4}\left(1-\frac{10}{3}x\right)\left(1+\frac{3}{8}x\right)\left(1-\frac{3}{8}x\right)$$

$$= \frac{1}{4}\left(1-\frac{10}{3}x\right) = \frac{1}{4}-\frac{5}{6}x.$$

§31.

$$\text{Expression} = \frac{\left(1-\dfrac{3}{5}x\right)^{\frac{1}{4}}+\left(1+\dfrac{5}{6}x\right)^{-6}}{(1+2x)^{\frac{1}{3}}+\left(1-\dfrac{x}{2}\right)^{\frac{1}{5}}}$$

$$= \frac{1-\dfrac{3}{20}x+1-5x}{1+\dfrac{2}{3}x+1-\dfrac{1}{10}x} = \frac{2-\dfrac{103}{20}x}{2+\dfrac{17}{30}x} = \frac{1-\dfrac{103}{40}x}{1+\dfrac{17}{60}x}$$

$$= \left(1-\frac{103}{40}x\right)\left(1+\frac{17}{60}x\right)^{-1}$$

$$= \left(1-\frac{103}{40}x\right)\left(1-\frac{17}{60}x\right)$$

$$= 1-\frac{343}{120}x.$$

§32.

$$\text{Expression} = \frac{(8+3x)^{\frac{1}{3}}-(1-x)^{\frac{1}{5}}}{(1+5x)^{\frac{3}{5}}+2\left(1+\dfrac{x}{8}\right)^{\frac{1}{2}}}$$

$$= \frac{2\left(1+\dfrac{1}{8}x\right)-\left(1-\dfrac{1}{5}x\right)}{1+3x+2\left(1+\dfrac{x}{16}\right)} = \frac{1+\dfrac{9}{20}x}{3+\dfrac{25}{8}x}$$

$$= \frac{1+\dfrac{9}{20}x}{3\left(1+\dfrac{25}{24}x\right)} = \frac{1}{3}\left(1+\frac{9}{20}x\right)\left(1-\frac{25}{24}x\right)$$

$$= \frac{1}{3}-\frac{71}{360}x.$$

§33. Coefficient required $= \dfrac{\dfrac{1}{2}\left(\dfrac{1}{2}+1\right)\ldots\left(\dfrac{1}{2}+r-1\right)}{\underline{|r}}2^{2r}$

$$= \frac{1\cdot3\cdot5\ldots(2r-1)}{2^r\underline{|r}}2^{2r} = \frac{1\cdot3\cdot5\ldots(2r-1)}{\underline{|r}}2^r.$$

Multiply numerator and denominator by $\underline{|r}$; then, since

$$\underline{|r}\cdot2^r = 2\cdot4\cdot6\ldots2r, \text{ we get the required result.}$$

§34.

$$(1+x)^n = \left(\frac{1}{1+x}\right)^{-n}$$

$$= \left\{\frac{1}{2}\left(1+\frac{1-x}{1+x}\right)\right\}^{-n}$$

$$= 2^n \left(1 + \frac{1-x}{1+x}\right)^{-n} ; \&c.$$

§35.

$$(1+x)^{-2}(1+4x)^{-\frac{1}{2}} = (1 - 2x + 3x^2 - \ldots)(1 - 2x + 6x^2 - \ldots)$$
$$= 1 - 4x + x^2(4 + 3 + 6), \text{ neglecting } x^3,$$
$$= 1 - 4x + 13x^2.$$

§36.

$$\text{The expression } = \frac{\left(1 + \frac{3}{4}x - \frac{3}{32}x^2 \ldots\right) + \left(1 + \frac{5}{2}x - \frac{25}{8}x^2 \ldots\right)}{(1-x)^2}$$

$$= \frac{2 + \frac{13}{4}x - x^2\left(\frac{3}{32} + \frac{25}{8}\right)}{(1-x)^2}$$

$$= \left(2 + \frac{13}{4}x - \frac{103}{32}x^2\right)(1 + 2x + 3x^2)$$

$$= 2 + x\left(4 + \frac{13}{4}\right) + x^2\left(6 + \frac{13}{2} - \frac{103}{32}\right)$$

$$= 2 + \frac{29}{4}x + \frac{297}{32}x^2.$$

§37. The nth coefficient $= \dfrac{n(n+1)(n+2)\ldots(2n-2)}{\underline{n-1}}.$

The $(n-1)^{th}$ coefficient $= \dfrac{n(n+1)(n+2)\ldots(2n-3)}{\underline{n-2}};$

$$\therefore \frac{\text{the } n^{th} \text{ coefficient}}{\text{the } (n-1)^{th} \text{ coefficient}} = \frac{2n-2}{n-1} = 2.$$

EXAMPLES XIV c

§1. $(3 - 5x)(1 + 2x + 3x^2 + \ldots + 100x^{99} + 101x^{100} + \ldots);$
$\qquad \therefore$ the required coefficient $= 303 - 500 = -197.$

§2. See Example 1, Art. 193. With the same notation the required
coefficient $= 4p_{12} + 2p_{11} - p_{10} = 4 \cdot \dfrac{13 \cdot 14}{2} - 2 \cdot \dfrac{12 \cdot 13}{2} - \dfrac{11 \cdot 12}{2} = 142.$

§3. $\dfrac{3x^2 - 2}{x} \cdot (1+x)^{-1} = \dfrac{1}{x}(3x^2 - 2)(1 - x + x^2 - x^3 + \ldots);$
$\qquad \therefore$ the coefficient of $x^n = 3(-1)^{n-1} - 2(-1)^{n+1} = (-1)^{n-1}.$

§4. With the notation of Ex. 1, Art. 193, we have the coefficient of x^n
$$= 2p_n + p_{n-1} + p_{n-2}$$
$$= 2(-1)^n \frac{(n+1)(n+2)}{2} + (-1)^{n-1} \cdot \frac{n(n+1)}{2} + (-1)^{n-2}\frac{(n-1)n}{2}$$
$$= (-1)^n(n^2 + 2n + 2).$$

§5. Expansion $= \left(1 + \dfrac{1}{2}\right)^{-\frac{1}{2}} = \left(\dfrac{3}{2}\right)^{-\frac{1}{2}} = \left(\dfrac{2}{3}\right)^{\frac{1}{2}}.$

§6. $\sqrt{8} = 2^{\frac{3}{2}} = \left(\dfrac{1}{2}\right)^{-\frac{3}{2}} = \left(1 - \dfrac{1}{2}\right)^{-\frac{3}{2}}.$

§7. The first series $= \left(1 - \dfrac{2}{3}\right)^{-n} = \left(\dfrac{1}{3}\right)^{-n} = 2^n \cdot \dfrac{2^{-n}}{3^{-n}} = 2^n\left(1 - \dfrac{1}{3}\right)^{-n}.$

§8. The first series $= 7^n \left(1 + \dfrac{1}{7}\right)^n = 8^n = 4^n \cdot 2^n = 4^n \left(\dfrac{1}{2}\right)^{-n}$

$= 4^n \left(1 - \dfrac{1}{2}\right)^{-n}$.

§9. The expression

$$= \frac{3 \times \dfrac{2}{3} \left(1 + \dfrac{9x}{4}\right)^{\frac{1}{2}} \left(1 - \dfrac{3}{4}x^2\right)^{\frac{1}{3}}}{2\left(1 + \dfrac{9}{16}x\right)^2}$$

$$= \left\{1 + \dfrac{1}{2}\cdot\dfrac{9x}{4} - \dfrac{1}{8}\left(\dfrac{9x}{4}\right)^2\right\}\left(1 - \dfrac{1}{4}x^2\right)\left(1 + \dfrac{9}{16}x\right)^{-2}$$

$$= \left(1 + \dfrac{9}{8}x - \dfrac{81}{128}x^2\right)\left(1 - \dfrac{1}{4}x^2\right)\left(1 - \dfrac{9}{8}x + \dfrac{243}{256}x^2\right)$$

$$= 1 - \left(\dfrac{81}{128} + \dfrac{81}{64} + \dfrac{1}{4} - \dfrac{243}{256}\right)x^2, \text{ neglecting } x^2,$$

$$= 1 - \dfrac{307}{256}x^2.$$

10 and 11. See Ex.3, Art. 193.

§12. The expansion $= \left\{(1 + x)^{-2}\right\}^{-n} = (1 + x)^{2n}$;

$$\therefore \text{ the required coefficient } = \frac{\lfloor 2n}{\lfloor n \, \lfloor n}.$$

§13. The middle term of $\left(x + \dfrac{1}{x}\right)^{4n}$ is

$$\frac{\lfloor 4n}{\lfloor 2n \, \lfloor 2n} = \frac{2^{2n}\lfloor 2n \cdot 1 \cdot 3 \cdot 5 \dots (4n-1)}{\lfloor 2n \, \lfloor 2n}$$

$$= 2^{2n}\frac{1 \cdot 3 \cdot 5 \dots (4n-1)}{2^n\lfloor n \cdot 1 \cdot 3 \cdot 5 \dots (2n-1)}$$

$$= 2^n \cdot \frac{(2n+1)(2n+3)\dots(4n-1)}{\lfloor n}$$

$$= \frac{4^n\left(n + \dfrac{1}{2}\right)\left(n + \dfrac{3}{2}\right)\left(n + \dfrac{5}{2}\right)\dots \text{ to } n \text{ factors}}{\lfloor n}$$

$$= \text{ the coefficient of } x^n \text{ in } (1 - 4x)^{-\left(n + \frac{1}{2}\right)}.$$

§14. We have $(1 - x^3) = (1 - x)^3 + 3x - 3x^2$;

$$\therefore (1 - x^3)^n = \left\{(1 - x)^3 + 3x(1 - x)\right\}^n; \text{ \&c.}$$

§15. $\dfrac{1}{1 + x + x^2} = \dfrac{1 - x}{1 - x^3} = (1 - x)\{1 + x^3 + x^6 + x^9 + \dots\}$,

and in the series every index is a multiple of 3; therefore in the expansion of the given expression every index is of the form $3m$ or $3m + 1$. In the former case the coefficient is 1, and in the latter it is -1.

§16. (1) See Art. 191.

(2) The sum of the coefficients will be independent of a, b, and c; if these be each equal to 1, the whole expansion is the sum required, which is therefore equal to 3^8 or 6561.

§17. Multiply throughout by $\lfloor n$; then we have to shew that

$$^nC_1 + {}^nC_3 + {}^nC_5 + \dots + {}^nC_{n-1} = 2^{n-1},$$

which has been proved in Art. 174.

§18. (1) We have
$$(1+x)^n = c_0 + c_1 x + c_2 x^2 + \ldots + c_r x^r + \ldots + c_n x^n,$$
$$(1+x)^{-1} = 1 - x + x^2 - x^3 + \ldots$$

Multiply the two series together; the coefficient of x^r in the product on the right is
$$(-1)^r \{c_0 - c_1 + c_2 - c_3 + \ldots + (-1)^r c_r\}$$
which must be equal to the coefficient of x^r in $(1+x)^{n-1}$, that is to
$$\frac{\lfloor n-1}{\lfloor r \lfloor n-r-1}.$$

(2) $$(1+x)^n = c_0 + c_1 x + c_2 x^2 + \ldots + c_n x^n.$$
$$\left(1 + \frac{1}{x}\right)^{-2} = 1 - \frac{2}{x} + \frac{3}{x^2} - \ldots + (-1)^n (n+1) x^n + \ldots$$

$\therefore$ by multiplication, $(1+x)^n \left(1 + \dfrac{1}{x}\right)^{-2} = a$ series of terms in which the coefficient of x^0 is
$$c_0 - 2c_1 + 3c_2 - 4c_3 + \ldots + (-1)^n (n+1) c_n.$$

This expression is therefore equal to the coefficient of x^0 in $x^2(1+x)^{n-2}$, that is it is equal to zero.

(3) $$(1+x)^n = c_0 + c_1 x + c_2 x^2 + \ldots + c_n x^n,$$
$$\left(1 - \frac{1}{x}\right)^n = c_0 - \frac{c_1}{x} + \frac{c_2}{x^2} - \ldots + (-1)^n \frac{c_n}{x^n};$$

$\therefore$ by multiplication, $(1+x)^n \left(1 - \dfrac{1}{x}\right)^n = \{c_0^2 - c_1^2 + c_2^2 - \ldots + (-1)^n c_n^2\}$ together with terms involving x.

Hence $c_0^2 - c_1^2 + c_2^2 - \ldots + (-1)^n c_n^2$ is equal to the term independent of x in $\dfrac{(-1)^n}{x^n}(1-x^2)^n$.

This term is 0 when n is odd, and $(-1)^{\frac{n}{2}} c_{\frac{n}{2}}$ when n is even, since in the latter case we have only to consider the coefficient of x^n in $(1-x^2)^n$.

§19.

(1) $$(1-x)^{-3} = 1 + 3x + \frac{3 \cdot 4}{1 \cdot 2} x^2 + \frac{3 \cdot 4 \cdot 5}{1 \cdot 2 \cdot 3} x^3 + \ldots$$
$$= \frac{1 \cdot 2}{2} + \frac{2 \cdot 3}{2} x + \frac{3 \cdot 4}{2} x^2 + \frac{4 \cdot 5}{2} x^3 + \ldots$$
$$= s_1 + s_2 x + s_3 x^2 + s_4 x^3 + \ldots + s_r x^r + \ldots,$$

since $s_n = \dfrac{n(n+1)}{2}$.

(2) $$(1-x)^{-3} = s_1 + s_2 x + s_3 x^2 + \ldots + s_{2n} x^{2n-1} + \ldots$$
$$(1-x)^{-3} = s_1 + s_2 x + s_3 x^2 + \ldots + s_{2n} x^{2n-1} + \ldots$$

Multiply the two series together, and take the coefficient of x^{2n-1}; thus $s_1 s_{2n} + s_2 s_{2n-1} + \ldots$ to $2n$ terms = the coefficient of x^{2n-1} in the expansion of $(1-x)^{-6}$, & c.

§20. (1) It will be found that
$$(1-x)^{-\frac{1}{2}} = 1 + q_1 x + q_2 x^2 + q_3 x^3 + \ldots + q_{2n+1} x^{2n+1} + \ldots$$
$$(1-x)^{-\frac{1}{2}} = 1 + q_1 x + q_2 x^2 + q_3 x^3 + \ldots + q_{2n+1} x^{2n+1} + \ldots$$

$\therefore$ by multiplying these results together we see that
$$q_{2n+1} + q_1 q_{2n} + q_2 q_{2n-1} + \ldots \text{ to } 2n+2 \text{ terms} = \text{the coefficient of } x^{2n} \text{ in}$$

$(1-x)^{-1}$, which is unity;

$$\therefore q_{2n+1} + q_1 q_{2n} + q_2 q_{2n-1} + \ldots + q_{n-1} q_{n+2} + q_n q_{n+1} = \frac{1}{2}.$$

(2) We have $(1-x)^{-\frac{1}{2}} = 1 + q_1 x + q_2 x^2 + \ldots + q_{2n} x^{2n} + \ldots$;

and $\qquad (1+x)^{-\frac{1}{2}} = 1 - q_1 x + q_2 x^2 - \ldots + q_{2n} x^{2n} - \ldots$;

hence $\qquad (1-x^2)^{-\frac{1}{2}} = \{q_{2n} - q_1 q_{2n-1} + q_2 q_{2n-2} - \ldots + q_{2n}\} x^{2n}$,

together with terms containing other powers of x. Now the series in brackets consists of $2n + 1$ terms, those equidistant from the beginning and end being equal;

$\therefore 2\{q_{2n} - q_1 q_{2n-1} + q_2 q_{2n-2} - \ldots$ to n terms $\} + (-1)^n q_n^2$

$=$ the coefficient of x^{2n} in $(1-x^2)^{-\frac{1}{2}} = q_n$.

Transpose and we get the required result.

§21. We have

$$(c_0 + c_1 + c_2 + \ldots + c_n)^2 - 2(c_0 c_1 + c_0 c_2 + \ldots + c_1 c_2 + \ldots) = c_0^2 + c_1^2 + c_2^2 + \ldots + c_n^2;$$

$$\therefore 2(c_0 c_1 + c_0 c_2 + \ldots + c_1 c_2 + \ldots) = 2^{2n} - \frac{\underline{|2n}}{\underline{|n}\,\underline{|n}}.$$

See Ex. 22, XIII. b.

§22. $(7 + 4\sqrt{3})^n = p + \beta$, where $\beta < 1$.

Also $(7 + 4\sqrt{3})^n + (7 - 4\sqrt{3})^n = $ integer, and $(7 - 4\sqrt{3})^n$ is positive and less than 1; $\therefore (7 - 4\sqrt{3})^n$ must be equal to $1 - \beta$.

Now $\qquad (7 + 4\sqrt{3})^n \times (7 - 4\sqrt{3})^n = 1; \therefore (p + \beta)(1 - \beta) = 1.$

§23.

$$\text{Let } S_n = c_1 - \frac{c_2}{2} + \frac{c_3}{3} - \ldots + \frac{(-1)^{n-1} c_n}{n}$$

$$= n - \frac{n(n-1)}{2\underline{|2}} + \frac{n(n-1)(n-2)}{3\underline{|3}} - \ldots \text{ to } n \text{ terms.}$$

$$S_{n+1} = (n+1) - \frac{(n+1)n}{2\underline{|2}} + \frac{(n+1)n(n-1)}{3\underline{|3}}$$

$$- \ldots \text{ to } n+1 \text{ terms .}$$

$$\therefore S_{n+1} - S_n = 1 - \frac{n}{\underline{|2}} + \frac{n(n-1)}{\underline{|3}} - \ldots \text{ to } n+1 \text{ terms .}$$

$$= \frac{1}{n+1}\{1 - (1-1)^{n+1}\} = \frac{1}{n+1};$$

$$\therefore S_2 - S_1 = \frac{1}{2}; \text{ but } S_1 = 1, \text{ thus } S_2 = 1 + \frac{1}{2},$$

$$S_3 = S_2 + \frac{1}{3} = 1 + \frac{1}{2} + \frac{1}{3}; \text{ and so on.}$$

EXAMPLES XV : Multinomial Theorem

§1. As in Art. 194 the term involving $a^2b^3c^4d$ is

$$\frac{\lfloor 10}{\lfloor 2\ \lfloor 3\ \lfloor 4\ \lfloor 1}\,a^2(-b)^3(-c)^4 d;$$

that is, $-12600a^2b^3c^4d$.

§2. The term involving a^2b^5d is $\dfrac{\lfloor 8}{\lfloor 2\ \lfloor 5\ \lfloor 1}\,a^2b^5(-d)$; hence the coefficient is

-168.

§3. The term involving a^3b^3c is $\dfrac{\lfloor 7}{\lfloor 3\ \lfloor 3\ \lfloor 1}\,(2a)^3b^3(3c)$; hence the coefficient

is 3360.

§4. The term involving $x^2y^3z^4$ is $\dfrac{\lfloor 9}{\lfloor 2\ \lfloor 3\ \lfloor 4}\,(ax)^2(-by)^3(cz)^4$; hence the coef-

ficient is $-1260a^2b^3c^4$.

§5. The general term is $\dfrac{\lfloor 3}{\lfloor \alpha\ \lfloor \beta\ \lfloor \gamma}\,3^\beta(-2)^\gamma x^{\beta+2\gamma}$ where $\alpha+\beta+\gamma=3$, and

$\beta+2\gamma=3$.

$$\therefore \gamma=1,\ \beta=1, \alpha=1;\ \text{or } \gamma=0,\ \beta=3,\ \alpha=0;$$

$$\therefore \text{ the coefficient } = \sqrt{3}(3)(-2)+\frac{\lfloor 3}{\lfloor 3}3^3 = -36+27 = -9.$$

§6. The general term is $\dfrac{\lfloor 10}{\lfloor \alpha\ \lfloor \beta\ \lfloor \gamma}\,2^\beta 3^\gamma x^{\beta+2\gamma}$,

where $\qquad\qquad \alpha+\beta+\gamma=10,\ \beta+2\gamma=4.$

$$\therefore \gamma=2,\ \beta=0,\ \alpha=8;\ \gamma=1,\ \beta=2,\ \alpha=7; \gamma=0,\ \beta=4,\ \alpha=6;$$

$$\therefore \text{ the coefficient } = \frac{\lfloor 10}{\lfloor 8\ \lfloor 2}(3)^2+\frac{\lfloor 10}{\lfloor 7\ \lfloor 2}(2)^2(3)+\frac{\lfloor 10}{\lfloor 6\ \lfloor 4}(2)^4$$

$$= 405+4320+3360 = 8085.$$

§7. General term is $\dfrac{\lfloor 5}{\lfloor \alpha\ \lfloor \beta\ \lfloor \gamma}\,2^\beta(-1)^\gamma x^{\beta+2\gamma}$,

where $\alpha+\beta+\gamma=5,\ \beta+2\gamma=6.$

$$\therefore \gamma=3,\ \beta=0,\ \alpha=2;\ \gamma=2,\ \beta=2,\ \alpha=1;\ \gamma=1,\ \beta=4,\ \alpha=0.$$

$\gamma=0,\ \beta=6$ is not admissible, as it makes α negative;

$$\therefore \text{ the coefficient } = \frac{\lfloor 5}{\lfloor 2\ \lfloor 3}(-1)^3+\frac{\lfloor 5}{\lfloor 2\ \lfloor 2}(2)^2(-1)^2+\frac{\lfloor 5}{\lfloor 4}(2)^4(-1)$$

$$= -10+120-80 = 30.$$

§8. General term $= \dfrac{\lfloor 4}{\lfloor \alpha\ \lfloor \beta\ \lfloor \gamma\ \lfloor \delta}\,(-2)^\beta(3)^\gamma(-4)x^{\beta+2\gamma+3}$,

where $\qquad\qquad \alpha+\beta+\gamma+\delta=4,\ \beta+2\gamma+3\delta=8.$

Thus $\qquad \delta=2,\ \gamma=1,\ \beta=0,\ \alpha=1;\ \delta=2,\ \gamma=0,\ \beta=2,\ \alpha=0;$

$\qquad\qquad \delta=1,\ \gamma=2,\ \beta=1,\ \alpha=0;\ \delta=0,\ \gamma=4,\ \beta=0,\ \alpha=0;$

the other possible values make α negative and are not admissible;

∴ the coefficient

$$= \frac{\underline{4}}{\underline{2}}(3)^1(-4)^2 + \frac{\underline{4}}{\underline{2}\,\underline{2}}(-2)^2(-4)^2 + \frac{\underline{4}}{\underline{2}}(-2)^1(3)^2(-4)^1 + \frac{\underline{4}}{\underline{4}}(3)^4$$

$$= 576 + 384 + 864 + 81 = 1905.$$

§9. The expression

$$= (-x^5 - x^4 + 3x^2 - 2x + 1)^5$$
$$= (-x^5)^5 + 5(-x^5)^4(-x^4 + 3x^2 - 2x + 1)$$
$$+ 10(-x^5)^3(-x^4 + 3x^2 - 2x + 1)^2$$
$$+ 10(-x^5)^2(-x^4 + 3x^2 - 2x + 1)^3 + \ldots$$

The term containing x^{23} arises from $10(-x^5)^3(-x^4)^2$; hence the coefficient is -10.

§10. General term

$$= \frac{\left(-\frac{1}{2}\right)\left(-\frac{1}{2} - 1\right)\ldots\left(-\frac{1}{2} - p + 1\right)}{\underline{\beta}\,\underline{\gamma}}(-2)^\beta(3)^\gamma x^{\beta + 2\gamma},$$

where $\beta + 2\gamma = 5$, $p = \beta + \gamma$.

∴ $\gamma = 2$, $\beta = 1$, $p = 3$; $\gamma = 1$, $\beta = 3$, $p = 4$; $\gamma = 0$, $\beta = 5$, $p = 5$;

∴ the coefficient

$$= \frac{\left(-\frac{1}{2}\right)\left(-\frac{3}{2}\right)\left(-\frac{5}{2}\right)}{\underline{2}}(-2)^1(3)^2$$

$$+ \frac{\left(-\frac{1}{2}\right)\left(-\frac{3}{2}\right)\left(-\frac{5}{2}\right)\left(-\frac{7}{2}\right)}{\underline{3}}(-2)^3(3)$$

$$+ \frac{\left(-\frac{1}{2}\right)\left(-\frac{3}{2}\right)\left(-\frac{5}{2}\right)\left(-\frac{7}{2}\right)\left(-\frac{9}{2}\right)}{\underline{5}}(-2)^5$$

$$= \frac{135}{8} - \frac{105}{4} + \frac{63}{8} = -\frac{3}{2}.$$

§11. General term

$$= \frac{\frac{1}{2}\left(\frac{1}{2} - 1\right)\ldots\left(\frac{1}{2} - p + 1\right)}{\underline{\beta}\,\underline{\gamma}\,\underline{\delta}}(-2)^\beta(3)^\gamma(-4)x^{\beta + 2\gamma + 3\delta}$$

where $\beta + 2\gamma + 3\delta = 3$, $p = \beta + \gamma + \delta$.

∴ $\delta = 1$, $\gamma = 0$, $\beta = 0$, $p = 1$;

$\delta = 0$, $\gamma = 1$, $\beta = 1$, $p = 2$;

$\delta = 0$, $\gamma = 0$, $\beta = 3$, $p = 3$;

∴ the coefficient

$$= \frac{1}{2}(-4) + \frac{1}{2}\left(-\frac{1}{2}\right)(-2)(3) + \frac{\left(\frac{1}{2}\right)\left(-\frac{1}{2}\right)\left(-\frac{3}{2}\right)}{\underline{3}}(-2)^3$$

$$= -2 + \frac{3}{2} - \frac{1}{2} = -1.$$

§12. This is equivalent to finding the coefficient of x^4 in the expansion of

$$\left(1 - \frac{x}{3} + \frac{x^2}{9}\right)^{-2}.$$

General term $= \dfrac{(-2)(-3)\ldots(-2-p+1)}{\underline{|\beta}\ \underline{|\gamma}}\left(-\dfrac{1}{3}\right)^{\beta}\left(\dfrac{1}{9}\right)^{\gamma}x^{\beta+2\gamma},$

where $\beta + 2\gamma = 4,\ p = \beta + \gamma.$

$\therefore \gamma = 2,\ \beta = 0,\ p = 2;\ \gamma = 1, \beta = 2,\ p = 3;\ \gamma = 0,\ \beta = 4,\ \gamma = 4;$

$\therefore$ the coefficient $= \dfrac{(-2)(-3)}{\underline{|2}}\left(\dfrac{1}{9}\right)^{2} + \dfrac{(-2)(-3)(-4)}{\underline{|2}}\left(-\dfrac{1}{3}\right)^{2}\left(\dfrac{1}{9}\right)$

$+\dfrac{(-2)(-3)(-4)(-5)}{\underline{|4}}\left(-\dfrac{1}{3}\right)^{4} = \dfrac{1}{27} - \dfrac{4}{27} + \dfrac{5}{81} = -\dfrac{4}{81}.$

§13. $(2 - 4x + 3x^2)^{-2} = \dfrac{1}{4}\left(1 - 2x + \dfrac{3}{2}x^2\right)^{-2}.$

The general term of the expansion of $\left(1 - 2x + \dfrac{3}{2}x^2\right)^{-2}$ is

$\dfrac{(-2)(-3)\ldots(-2-p+1)}{\underline{|\beta}\ \underline{|\gamma}}(-2)^{\beta}\left(\dfrac{3}{2}\right)^{\gamma}x^{\beta+2\gamma},$

where $\beta + 2\gamma = 4,\ p = \beta + \gamma.$

$\therefore \gamma = 2,\ \beta = 0,\ p = 2;\ \gamma = 1, \beta = 2,\ p = 3;\ \gamma = 0, \beta = 4,\ p = 4;$

$\therefore$ the coefficient

$= \dfrac{(-2)(-3)}{\underline{|2}}\left(\dfrac{3}{2}\right)^{2} + \dfrac{(-2)(-3)(-4)}{\underline{|2}}(-2)^{2}\left(\dfrac{3}{2}\right)$

$+ \dfrac{(-2)(-3)(-4)(-5)}{\underline{|4}}(-2)^{4} = \dfrac{27}{4} - 72 + 80 = \dfrac{59}{4}.$

Thus the coefficient required $= \dfrac{59}{16}.$

§14. This is equivalent to finding the coefficient of x^3 in the expansion of

$$(1 + 4x + 10x^2 + 20x^3)^{-\frac{3}{4}}.$$

General term

$= \dfrac{\left(-\dfrac{3}{4}\right)\left(-\dfrac{7}{4}\right)\ldots\left(-\dfrac{3}{4}-p+1\right)}{\underline{|\beta}\ \underline{|\gamma}\ \underline{|\delta}}(4)^{\beta}(10)^{\gamma}(20)^{\delta}x^{\beta+2\gamma+3\delta}.$

Here $\beta + 2\gamma + 3\delta = 3,\ p = \beta + \gamma + \delta.$

Thus $\delta = 1,\ \gamma = 0,\ \beta = 0,\ p = 1;$

$\delta = 0, \gamma = 1,\ \beta = 1,\ p = 2;$

$\delta = 0,\ \gamma = 0, \beta = 3,\ p = 3;$

$\therefore$ the coefficient $= \left(-\dfrac{3}{4}\right)(20) + \left(-\dfrac{3}{4}\right)\left(-\dfrac{7}{4}\right)(4)(10)$

$+\dfrac{\left(-\dfrac{3}{4}\right)\left(-\dfrac{7}{4}\right)\left(-\dfrac{11}{4}\right)}{\underline{|3}}(4)^{3} = -15 + \dfrac{105}{2} - \dfrac{77}{2} = -1.$

§15. This is equivalent to finding the coefficient of x^4 in the expansion of

$$(3 - 15x + 18x^2)^{-1},\ \text{or}\ \dfrac{1}{3}(1 - 5x + 6x^2)^{-1}.$$

General term $= \dfrac{(-1)(-2)\ldots(-1-p+1)}{\underline{|\beta}\ \underline{|\gamma}}(-5)^{\beta}(6)^{\gamma}x^{\beta+2\gamma}.$

Here $\beta + 2\gamma = 4,\ p = \beta + \gamma.$

$\therefore \gamma = 2,\ \beta = 0,\ p = 2;\ \gamma = 1,\ \beta = 2,\ p = 3;\ \gamma = 0,\ \beta = 4,\ p = 4;$

$$\therefore \text{ the coefficient } = \frac{(-1)(-2)}{\underline{|2}}(6)^2 + \frac{(-1)(-2)(-3)}{\underline{|2}}(-5)^2(6)$$

$$+ \frac{(-1)(-2)(-3)(-4)}{\underline{|4}}(-5)^4 = 36 - 450 + 625 = 211.$$

Thus the coefficient required $= \dfrac{211}{3}$.

§16.

$$(1 - 2x - 2x^2)^{\frac{1}{4}} = 1 - \frac{1}{4}(2x + 2x^2) + \frac{\left(\frac{1}{4}\right)\left(-\frac{3}{4}\right)}{1 \cdot 2}(2x + 2x^2)^2 \dots$$

$$= 1 - \frac{1}{2}x - \frac{1}{2}x^2 - \frac{3}{8}x^2 = 1 - \frac{1}{2}x - \frac{7}{8}x^2.$$

§17. $(1 + 3x^2 - 6x^3)^{-\frac{2}{3}} = 1 - \dfrac{2}{3}(3x^2 - 6x^3) + \dfrac{\left(-\frac{2}{3}\right)\left(-\frac{5}{3}\right)}{1 \cdot 2}(3x^2 - 6x^3)^2 + \dots$

$= 1 - 2x^2 + 4x^3 + \dfrac{5}{9}(9x^4 - 36x^5 + \dots) + \dots = 1 - 2x^2 + 4x^3 + 5x^4 - 20x^5.$

§18. $(8 - 9x^3 + 18x^4)^{\frac{4}{3}} = 16\left(1 - \dfrac{9}{8}x^3 + \dfrac{9}{4}x^4\right)^{\frac{4}{3}}$

$$\left(1 - \frac{9}{8}x^3 + \frac{9}{4}x^4\right)^{\frac{4}{3}}$$

$$= 1 - \frac{4}{3}\left(\frac{9}{8}x^3 - \frac{9}{4}x^4\right) + \frac{\left(\frac{4}{3}\right)\left(\frac{1}{3}\right)}{1 \cdot 2}\left(\frac{9}{8}x^3 - \frac{9}{4}x^4\right)^2 + \dots$$

$$= 1 - \frac{3}{2}x^3 + 3x^4 + \frac{9}{32}(x^3 - 2x^4)^2 + \dots$$

$$= 1 - \frac{3}{2}x^3 + 3x^4 + \frac{9}{32}x^6 - \frac{9}{8}x^7 + \frac{9}{8}x^8;$$

and the required expansion is obtained by multiplying this result by 16.

§19. The first part is obtained by putting $x = 1$, for then

$$1 + x + x^2 + \dots + x^p = p + 1.$$

For the second part, change x into $1 + x$; thus

$$a_0 + a_1(1 + x) + a_2(1 + x)^2 + a_3(1 + x)^3 + \dots + a_{np}(1 + x)^{np}$$

$$= \{1 + (1 + x) + (1 + x)^2 + (1 + x)^3 + \dots + (1 + x)^p\}^n$$

$$= \{1 + 1 + 1 + \dots \text{ to } (p + 1) \text{ terms}$$

$$+ (1 + 2 + 3 + \dots \text{ to } p \text{ terms}) x + \text{ higher powers of } x\}^n$$

$$= \left\{p + 1 + \frac{p(p + 1)}{2}x + \text{higher powers of } x\right\}^n$$

$$= (p + 1)^n \left(1 + \frac{px}{2} + \dots\right)^n$$

$$= (p + 1)^n \left(1 + \frac{np}{2}x + \text{ higher powers of } x\right).$$

Hence by equating the coefficients of x

$$a_1 + 2a_2 + 3a_3 + \dots + np a_{np} = \frac{1}{2}np(p + 1)^n.$$

§20. In the expansion of $(1 + x + x^2)^n$, since the coefficient of x^2 is unity, it is evident that the coefficients of the terms equidistant from the beginning and end are equal; hence

$(1 + x + x^2)^n = a_0 + a_1 x + a_2 x^2 + a_3 x^3 + \dots + a_n x^n + \dots + a_3 x^{2n-3} + a_2 x^{2n-2} + a_1 x^{2n-1} + a_0 x^{2n}.$

Writing $-x$ for x,

$(1 - x + x^2)^n = a_0 - a_1 x + a_2 x^2 - a_3 x^3 + \ldots + (-1)^n a_n x^n + \ldots - a_3 x^{2n-3} + a_2 x^{2n-2} - a_1 x^{2n-1} + a_0 x^{2n}.$

Multiply together the two series on the right; the coefficient of x^{2n} is $a_0^2 - a_1^2 + a_2^2 - a_3^2 + \ldots + (-1)^{n-1} a_{n-1}^2 + (-1)^n a_n^2 + (-1)^{n-1} a_{n-1}^2 + \ldots - a_3^2 + a_2^2 - a_1^2 + a_0^2,$ or $2\{a_0^2 - a_1^2 + a_2^2 - a_3^2 + \ldots + (-1)^{n-1} a_{n-1}^2\} + (-1)^n a_n^2;$ and it is equal to the coefficient of x^{2n} in $(1 + x + x^2)^n (1 - x + x^2)^n$ or $(1 + x^2 + x^4)^n;$ that is, it is equal to a_n. Hence the result.

§21.

We have $\qquad (1 + x + x^2)^n = a_0 + a_1 x + a_2 x^2 + a_3 x^3 + \ldots \qquad (36)$

Denote the cube roots of unity by $1, \omega, \omega^2$.

By changing x into $\omega x, \omega^2 x$ successively, we have

$$(1 + \omega x + \omega^2 x^2)^n = a_0 + a_1 \omega x + a_2 \omega^2 x^2 + a_3 x^3 + a_4 \omega x^4 + \ldots \qquad (37)$$

$$(1 + \omega^2 x + \omega x^2)^n = a_0 + a_1 \omega^2 x + a_2 \omega x^2 + a_3 x^3 + a_4 \omega^2 x^4 + \ldots \qquad (38)$$

Put $x = 1$ in (36), (37), (38) and add the results; then since $1 + \omega + \omega^2 = 0$, we have $\qquad\qquad 3^n = 3(a_0 + a_3 + a_6 + \ldots);$

whence we have the first part of the question.

Multiply (36), (37), (38) by $1, \omega^2, \omega$ respectively, put $x = 1$, and add the results;

$$\therefore \; 3^n = 3(a_1 + a_4 + a_7 + \ldots),$$

which is the second part of the question.

Finally, by multiplying (36), (37), (38) by $1, \omega, \omega^2$ respectively, putting $x = 1$ and adding, we obtain the last part of the question.

EXAMPLES XVI : Logarithms

EXAMPLES XVI a

Examples 1 to 14 are too easy to require full solution; the following six solutions will suffice.

§1. (2) Let x be the required logarithm, then
$$(2\sqrt{3})^x = 1728 = 12^3 = (4 \cdot 3)^3 = (2\sqrt{3})^6; \quad \therefore \ x = 6.$$

§2. (2) $(4)^x = .25 = \dfrac{1}{4} = 4^{-1}; \ \therefore \ x = -1.$

§5. (2) $(9\sqrt{3})^x = .\dot{1} = \dfrac{1}{9} = 3^{-2};$

$$\therefore \ (3^2 \cdot 3^{\frac{1}{2}})^x = 3^{-2}; \ 3^{\frac{5}{2}x} = 3^{-2}; \quad \therefore \ x = -\dfrac{4}{5}.$$

§8. $\log\left(\sqrt{a^2 b^3}\right)^6 = \log(a^6 \cdot b^9) = 6\log a + 9\log b.$

§9. $\log(\sqrt[3]{a^2} \times \sqrt[2]{b^3}) = \log(a^{\frac{2}{3}} \times b^{\frac{3}{2}}) = \dfrac{2}{3}\log a + \dfrac{3}{2}\log b.$

§14. $\log\left\{\left(\dfrac{bc^{-2}}{b^{-4}c^3}\right)^{-3} \div \left(\dfrac{b^{-1}c}{b^2c^{-3}}\right)^5\right\} = \log\left(\dfrac{b^{-3}c^6}{b^{12}c^{-9}} \times \dfrac{b^{10}c^{-15}}{b^{-5}c^5}\right) = \log c^{-5} =$

$-5\log c.$

§15.

$$\log \dfrac{\sqrt[4]{5} \cdot \sqrt[10]{2}}{\sqrt[3]{18} \cdot \sqrt{2}} = \dfrac{1}{4}\log 5 + \dfrac{1}{10}\log 2 - \dfrac{1}{3}\log 18 - \dfrac{1}{6}\log 2$$

$$= \dfrac{1}{4}\log 5 + \dfrac{1}{10}\log 2 - \dfrac{1}{3}(\log 3^2 + \log 2) - \dfrac{1}{6}\log 2$$

$$= \dfrac{1}{4}\log 5 - \left(\dfrac{1}{3} + \dfrac{1}{6} - \dfrac{1}{10}\right)\log 2 - \dfrac{2}{3}\log 3$$

$$= \dfrac{1}{4}\log 5 - \dfrac{2}{5}\log 2 - \dfrac{2}{3}\log 3.$$

§16. $\log \sqrt[4]{729 \sqrt[3]{9^{-1} \cdot 27^{-\frac{4}{3}}}} = \log(3^6 \cdot 3^{-\frac{2}{3}} \cdot 3^{-\frac{4}{3}})^{\frac{1}{4}} = \log(3^6 \cdot 3^{-2})^{\frac{1}{4}} = \log 3.$

§17. $\log \dfrac{75}{16} - 2\log \dfrac{5}{9} + \log \dfrac{32}{243} = \log\left\{\dfrac{75}{16} \times \dfrac{32}{243} \times \left(\dfrac{9}{5}\right)^2\right\} = \log 2.$

§19. Taking logarithms we have
$2x\log a + 3x\log b = 5\log c; \ \therefore \ x(2\log a + 3\log b) = 5\log c;$

$$\therefore \ x = \dfrac{5\log c}{2\log a + 3\log b}.$$

§21.

Here
$$2x\log a + 3y\log b = 5\log m \tag{39}$$
$$3x\log a + 2y\log b = 10\log m \tag{40}$$

Multiply (40) by 3, and (39) by 2 and subtract, then $5x\log a = 20\log m$;

$$\therefore \ x = \dfrac{4\log m}{\log a}, \text{ and } y = -\dfrac{\log m}{\log b}.$$

§22. We have $2\log x + 3\log y = a$, and $\log x - \log y = b$.

Whence $\qquad \log x = \dfrac{1}{5}(a + 3b), \ \log y = \dfrac{1}{5}(a - 2b).$

§23. We have $b^{2x} = a^{x+5-(3-x)}; \ \therefore \ b^x = a^{x+1}; \ \therefore \ x\log b = (x+1)\log a;$

that is, $\qquad x(\log b - \log a) = \log a, \text{ or } x\log\left(\dfrac{b}{a}\right) = \log a.$

§24. We have
$$(a^2 - b^2)^{2x-2} = (a-b)^{2x}(a+b)^{-2};$$
$$\therefore \frac{(a+b)^{2x-2}}{(a+b)^{-2}} = \frac{(a-b)^{2x}}{(a-b)^{2x-2}};$$
that is,
$$(a+b)^{2x} = (a-b)^2;$$
$$\therefore x\log(a+b) = \log(a-b), \ x = \frac{\log(a-b)}{\log(a+b)}.$$

EXAMPLES XVI b

§7. $\log \cdot 128 = \log \dfrac{2^7}{10^3} = 7\log 2 - 3 = 2\cdot 10721 - 3 = \bar{1}\cdot 10721.$

§13.
$$\log(\cdot 0105)^{\frac{1}{4}} = \frac{1}{4}\log\left(\frac{5 \times 3 \times 7}{10^4}\right) = \frac{1}{4}(\log 5 + \log 3 + \log 7 - 4)$$
$$= \frac{1}{4}(\bar{2}\cdot 0211893) = \frac{1}{4}(\bar{4} + 2\cdot 0211893) = \bar{1}\cdot 5052973.$$

§14. $\log 324 = 2\log 2 + 4\log 3 = 2\cdot 5105452;$
$$\therefore \text{ if } x \text{ be the required } 7^{th} \text{ root, we have}$$
$$\log x = \frac{1}{7}\log(\cdot 00324) = \frac{1}{7}(\bar{3}\cdot 5105452)$$
$$= \frac{1}{7}(\bar{7} + 4\cdot 5105452) = \bar{1}\cdot 6443636;$$
$$\therefore x = \cdot 44092388.$$

§15.
$$\log x = \frac{2}{11}\log\frac{392}{10} = \frac{2}{11}\log\frac{7^2 \times 2^3}{10}$$
$$= \frac{2}{11}(2\log 7 + 3\log 2 - 1) = \cdot 28968836;$$
$$\therefore x = 1\cdot 948445.$$

§16. Let P be the product; then
$$\log P = \text{ sum of logs of its factors}$$
$$= 1\cdot 5705780 + \cdot 5705780 + \bar{3}\cdot 5705780 + 5\cdot 705780 = 5\cdot 2823120;$$
$$\therefore P = 191563\cdot 1.$$

§17.
$$\log x = \log\frac{3^{\frac{2}{3}} \cdot 5^{\frac{4}{3}}}{2^{\frac{1}{6}}} = \frac{2}{3}\log 3 + \frac{4}{3}\log 5 - \frac{1}{6}\log 2$$
$$= \frac{\cdot 9542426}{3} + \frac{4}{3}(\cdot 69897) - \frac{1}{6}(\cdot 30103) = 1\cdot 1998692.$$

§18.
$$\log x = \log\sqrt[3]{2^4 \cdot 3} + \log\sqrt[4]{3^3 \cdot 2^2} - \log\sqrt[12]{2 \cdot 3}$$
$$= \frac{4}{3}\log 2 + \frac{1}{2}\log 2 - \frac{1}{12}\log 2 + \frac{1}{3}\log 3 + \frac{3}{4}\log 3 - \frac{1}{12}\log 3$$
$$= \frac{7}{4}\log 2 + \log 3 = 1\cdot 0039238.$$

§19.
$$\log x = \log\left(\frac{7 \times 7 \times 6 \times 5^3}{7 \times 6 \times 2^5}\right)^{\frac{2}{3}}$$
$$= \frac{2}{3}(\log 7 + 3\log 5 - 5\log 2) = \cdot 9579053;$$
$$\therefore x = 9\cdot 076226.$$

§20.
$$\log x = \left(\frac{330}{49}\right)^4 - \log(22 \times 70)^{\frac{1}{3}}$$

$$= (4\log 3 + \log 11 + 1 - 2\log 7) - \frac{1}{3}(\log 2 + \log 11 + \log 7 + 1)$$

$$= 4\log 3 + \frac{11}{3}(1 + \log 11) - \frac{25}{3}\log 7 - \frac{1}{3}\log 2$$

$$= 9\cdot3935917 - 7\cdot1428266 = 2\cdot2507651;$$

$$\therefore \; x = 178\cdot141516.$$

§21. $\log x = 12\log 3 + 8\log 2 = 5\cdot7254556 + 2\cdot4082400 = 8\cdot1336956;$

$$\therefore \text{ contains 9 digits.}$$

§22.

Since
$$\left(\frac{21}{20}\right)^{100} = \left(\frac{3 \times 7}{2^2 \times 5}\right)^{100};$$

$$\therefore \; \log\left(\frac{21}{20}\right)^{100} = 100(\log 3 + \log 7 - 2\log 2 - 1 + \log 2)$$

$$= 132\cdot22193 - 130\cdot10300 = 2\cdot11893;$$

$\therefore \left(\dfrac{21}{20}\right)^{100}$ is a number which has 3 integral digits; that is it is greater than 100.

§23. $\log\left(\dfrac{1}{2}\right)^{1000} = 1000(-\log 2) = -301.03 = \overline{302}\cdot97;$

$\therefore$ there are 301 ciphers between the decimal point and the first significant digit.

§24. $3^{x-2} = 5;$ $\therefore (x-2)\log 3 = \log 5;$

$$\therefore \; x = \frac{\log 5 + 2\log 3}{\log 3} = \frac{1\cdot6532126}{\cdot4771213} = 3\cdot46\ldots$$

§25. $x\log 5 = 3\log 10,$ $\quad x = \dfrac{3}{\cdot69897} = 4\cdot29\ldots$

§26.

$$(5 - 3x)\log 5 = (x+2)\log 2; \quad \text{or } (5-3x)(1-\log 2) = (x+2)\log 2,$$
$$x(\log 2 + 3 - 3\log 2) = 5 - 7\log 2; \quad \therefore \; x(3 - 2\log 2) = 5 - 7\log 2;$$

$$\therefore \; x = \frac{2\cdot89279}{2\cdot39794} = 1\cdot206\ldots$$

§27.

Here
$$x\log 21 = (2x+1)\log 2 + x\log 5,$$
$$x(\log 3 + \log 7) = 2x\log 2 + \log 2 + x - x\log 2,$$
$$x(\log 3 + \log 7 - \log 2 - 1) = \log 2;$$

$$\therefore \; x = \frac{\log 2}{\log 3 + \log 7 - \log 2 - 1} = \frac{\cdot30103}{\cdot0211893} = 14\cdot206\ldots$$

§28. We have $2^x \cdot 6^{x-2} = 5^{2x} \cdot 7^{1-x},$

$$x\log 2 + (x-2)(\log 2 + \log 3) = 2x(1 - \log 2) + (1 - x)\log 7;$$
$$\therefore \; x(4\log 2 + \log 3 + \log 7 - 2) = 2\log 3 + 2\log 2 + \log 7;$$

$$\therefore \; x = \frac{24014006}{5263393} = 4\cdot562\ldots$$

§29. We have

$$\left.\begin{array}{l} 2^{x-y} = 6^y \\ 3^x = 3 \cdot 2^{y+1} \end{array}\right\}$$

$\therefore \; (x+y)\log 2 = y(\log 2 + \log 3),$ $x\log 3 = \log 3 + (y+1)\log 2;$

that is, $\qquad x\log 2 = y\log 3,$ $x\log 3 - y\log 2 = \log 2 + \log 3;$

$\therefore$ by substitution, $x\left\{(\log 3)^2 - (\log 2)^2\right\} = (\log 2 + \log 3)\log 3;$

$$\therefore \; x = \frac{\log 3}{\log 3 - \log 2}, \text{ and } y = \frac{\log 2}{\log 3 - \log 2}.$$

§30. Put $\log 3 = a$, $\log 2 = b$; then we have
$$ax + (a - 2b)y - a = 0, \quad (2b + a)x - 3ay - b = 0;$$
∴ by cross multiplication,

$$\frac{x}{b(2b - a) - 3a^2} = \frac{y}{-a(2b + a) + ab} = \frac{1}{-3a^2 - (a - 2b)(2b + a)};$$

or
$$\frac{x}{2b^2 - ab - 3a^2} = \frac{y}{-(ab + a^2)} = \frac{1}{4(b^2 - a^2)};$$

∴
$$\frac{x}{(b + a)(2b - 3a)} = \frac{y}{-(ab + a^2)} = \frac{1}{4(b + a)(b - a)};$$

∴
$$x = \frac{2b - 3a}{4(b - a)} = \frac{3\log 3 - 2\log 2}{4(\log 3 - \log 2)}, \text{ and}$$

$$y = \frac{a}{4(a - b)} = \frac{\log 3}{4(\log 3 - \log 2)}.$$

§31. Let $x = \log_{25} 200$, then $25^x = 200$,
$$2x \log 5 = 2 + \log 2, \quad x = \frac{2 \cdot 30103}{1 \cdot 39794} = 1 \cdot 6465.$$

§32. Let $x = \log_7 \sqrt{2}$, then $7^x = \sqrt{2}$,
$$x \log 7 = \frac{1}{2} \log 2, \quad \therefore \quad x = \frac{\cdot 150515}{\cdot 84509} = \cdot 1781.$$

Again, $2^{\frac{1}{2}x} = 7$; hence $\frac{1}{2}x \log 2 = \log 7$, and $x = \frac{2\log 7}{\log 2} = 5 \cdot 614.$

Or thus, $\log_7 \sqrt{2} \times \log_{\sqrt{2}} 7 = 1$;

$$\therefore \quad \log_{\sqrt{2}} 7 = \frac{1}{\log_7 \sqrt{2}} = \frac{1}{\cdot 1781} = 5 \cdot 614.$$

EXAMPLES XVII : Exponential and Logarithmic Series

§1. In the equation $\log_e(1+x) = x - \dfrac{x^2}{2} + \dfrac{x^3}{3} - \dfrac{x^4}{4} + \dots$

put $x = 1$; then $\qquad 1 - \dfrac{1}{2} + \dfrac{1}{3} - \dfrac{1}{4} + \dots = \log_e 2.$

§2. We have $\qquad \dfrac{1}{2} - \dfrac{1}{2 \cdot 2^2} + \dfrac{1}{3 \cdot 2^3} - \dfrac{1}{4 \cdot 2^4} + \dots = \log_e\left(1 + \dfrac{1}{2}\right) = \log_e \dfrac{3}{2} =$

$\log_e 3 - \log_e 2.$

§3.

$$\log_e(n+a) - \log_e(n-a)$$
$$= \log_e\left\{n\left(1 + \dfrac{a}{n}\right)\right\} - \log_e\left\{n\left(1 - \dfrac{a}{n}\right)\right\}$$
$$= \log_e n + \log_e\left(1 + \dfrac{a}{n}\right) - \log_e n - \log_e\left(1 - \dfrac{a}{n}\right)$$
$$= \log_e\left(1 + \dfrac{a}{n}\right) - \log_e\left(1 - \dfrac{a}{n}\right);$$

and the result follows by using the formulae for $\log_e(1+x)$, and $\log_e(1-x)$.

§4.

We have $\qquad\qquad\qquad y = \log_e(1+x),$

hence $\qquad\qquad 1 + x = e^y = 1 + y + \dfrac{y^2}{\underline{|2}} + \dfrac{y^3}{\underline{|3}} + \dots$

§5.

$$\text{The series on the left} = -\log_e\left(1 - \dfrac{a-b}{a}\right)$$
$$= -\log_e \dfrac{b}{a} = \log_e a - \log_e b.$$

§6. In the result of Example 3, put $n = 1000$, $a = 1$;

$$\therefore \log_e 1001 - \log_e 999 = 2\left(\dfrac{1}{1000} + \dfrac{1}{3} \cdot \dfrac{1}{1000^3} + \dfrac{1}{5} \cdot \dfrac{1}{1000^5}\right);$$

the term $\dfrac{1}{7} \cdot \dfrac{1}{1000^7}$ and subsequent terms may be omitted.

§7. In the series $e^x = 1 + x + \dfrac{x^2}{\underline{|2}} + \dfrac{x^2}{\underline{|3}} + \dots,$

put $x = -1$, then

$$e^{-1} = (1-1) + \left(\dfrac{1}{\underline{|2}} - \dfrac{1}{\underline{|3}}\right) + \left(\dfrac{1}{\underline{|4}} - \dfrac{1}{\underline{|5}}\right) + \dots$$
$$= \dfrac{3-1}{\underline{|3}} + \dfrac{5-1}{\underline{|5}} + \dfrac{7-1}{\underline{|7}} + \dots;$$

which gives the result.

§8.

$$\log_e(1+x)^{1+x}(1-x)^{1-x}$$
$$= (1+x)\log_e(1+x) + (1-x)\log_e(1-x)$$
$$= x\{\log_e(1+x) - \log_e(1-x)\} + \{\log_e(1+x) + \log_e(1-x)\}$$
$$= 2x\left(x + \dfrac{x^3}{3} + \dfrac{x^5}{5} + \dots\right) - 2\left(\dfrac{x^2}{2} + \dfrac{x^4}{4} + \dfrac{x^6}{6} + \dots\right);$$

whence the result.

§9. $x^2 - y^2 + \dfrac{1}{\underline{|2}}(x^4 - y^4) + \dfrac{1}{\underline{|3}}(x^6 - y^6) + \dots$

$$= x^2 + \frac{x^4}{\underline{2}} + \frac{x^6}{\underline{3}} + \dots - \left(y^2 + \frac{y^4}{\underline{2}} + \frac{y^6}{\underline{3}} + \dots \right) = (e^{x^2} - 1) - (e^{y^2} - 1) = e^{x^2} - e^{y^2}.$$

§10. Put $n = 50$ in the formula

$$\log_{10} n - \log_{10}(n-1) = \frac{\mu}{n} + \frac{\mu}{2n^2} + \frac{\mu}{3n^3} + \dots$$

The right side $= .00868589 + .00008686 + .00000116 + .00000001$
$$= .00877391,$$

thus $2 - \log 2 - 2\log 7 = .00877390$; whence $\log 7$ is found.

Put $n = 10$ in the formula

$$\log_{10}(n+1) - \log_{10} n = \frac{\mu}{n} + \frac{\mu}{2n^2} + \frac{\mu}{3n^3} + \dots$$

Positive terms.
.04342945
14476
87
1
———————
.04357509
218240
———————
.04139269

Negative terms.
.00217147
1086
7
———————
.00218240

Thus $\log 11 = 1.0413927$.

In the last formula put $n = 1000$;

$$\therefore \ \log_{10} 1001 - \log_{10} 1000 = .0004343 - .0000002;$$
$$\therefore \ \log_{10}(7 \times 11 \times 13) = 3.0004341, \text{ and}$$
$$\log_{10} 13 = 3.0004341 - \log 7 - \log 11.$$

§11.

The expression $= x^2 - \dfrac{x^4}{2} + \dfrac{x^6}{3} - \dots + \dfrac{1}{x^2} - \dfrac{1}{2x^4} + \dfrac{1}{3x^6} - \dots$

$$= \log(1 + x^2) + \log\left(1 + \frac{1}{x^2} \right)$$

$$= \log(1 + x^2)\left(1 + \frac{1}{x^2} \right) = \log\left(2 + x^2 + \frac{1}{x^2} \right).$$

§12. $\log_e(1 + 3x + 2x^2) = \log_e(1 + x)(1 + 2x) = \log_e(1 + x) + \log_e(1 + 2x)$

$$= \left(x - \frac{x^2}{2} + \frac{x^3}{3} - \frac{x^4}{4} + \dots \right) + \left(2x - \frac{4x^2}{2} + \frac{8x^3}{3} - \frac{16x^4}{4} + \dots \right);$$

whence the result.

The general term of the series is

$$\frac{(-1)^{r-1} x^n}{r} + \frac{(-1)^{r-1}(2x)^r}{r} \text{ or } (-1)^{r-1} \cdot \frac{2^r + 1}{r} x^r.$$

§13. The expression $= \log_e(1 + 3x) - \log_e(1 - 2x) =$

$$\left(3x - \frac{9x^2}{2} + \frac{27x^3}{3} - \frac{81x^4}{4} + \dots \right)$$

$$+ \left(2x + \frac{4x^2}{2} + \frac{8x^3}{3} + \frac{16x^4}{4} + \dots \right);$$

whence the result.

The general term is $\dfrac{(-1)^{r-1}(3x)^r}{r} + \dfrac{(2x)^r}{r}$ or $(-1)^{r-1}\dfrac{3^r + 2^r}{r} x^r.$

§14. $\dfrac{e^{5x} + e^x}{e^{3x}} = e^{2x} + e^{-2x} = \left\{ 1 + 2x + \dfrac{(2x)^2}{\underline{2}} + \dfrac{(2x)^3}{\underline{3}} + \dots \right\}$

$$+ \left\{ 1 - 2x + \frac{(2x)^2}{\underline{2}} - \frac{(2x)^3}{\underline{3}} + \dots \right\}.$$

§15.

$$e^{ix} + e^{ix} = \left(1 + ix + \frac{i^2 x^2}{\underline{2}} + \frac{i^3 x^3}{\underline{3}} + \dots\right)$$

$$+ \left(1 - ix + \frac{i^2 x^2}{\underline{2}} - \frac{i^3 x^3}{\underline{3}} + \dots\right)$$

$$= 2\left(1 + \frac{i^2 x^2}{\underline{2}} + \frac{i^4 x^4}{\underline{4}} + \frac{i^6 x^6}{\underline{6}} + \dots\right);$$

$$\therefore \text{ the expression} = 1 - \frac{x^2}{\underline{2}} + \frac{x^4}{\underline{4}} - \frac{x^6}{\underline{6}} + \dots$$

§16. $\log_e(x + 2h) + \log_e x - 2\log(x + h) = \log_e \dfrac{x(x + 2h)}{(x + h)^2}$

$$= \log_e\left\{1 - \frac{h^2}{(x + h)^2}\right\} = -\left\{\frac{h^2}{(x + h)^2} + \frac{h^4}{2(x + h)^4} + \dots\right\}.$$

§17. $\alpha + \beta = p$, $\alpha\beta = q$;

$$\therefore \log_e(1 + px + qx^2) = \log_e\{1 + (\alpha + \beta)x + \alpha\beta x^2\}$$
$$= \log_e(1 + \alpha x)(1 + \beta x)$$
$$= \log_e(1 + \alpha x) + \log_e(1 + \beta x); \&c.$$

§18.

$$S = (1 - 1)x + \left(1 - \frac{1}{2}\right)x^2 + \left(1 - \frac{1}{3}\right)x^3 + \left(1 - \frac{1}{4}\right)x^4 + \dots$$

$$= x + x^2 + x^3 + x^4 + \dots + \left(-x - \frac{1}{2}x^2 - \frac{1}{3}x^3 - \frac{1}{4}x^4 - \dots\right)$$

$$= \frac{x}{1 - x} + \log_e(1 - x).$$

§19.

$$\log_e\left(1 + \frac{1}{n}\right)^n = n\log_e \frac{n + 1}{n}$$

$$= -n\log_e \frac{n}{n + 1} = -n\log_e\left(1 - \frac{1}{n + 1}\right)$$

$$= -(n + 1)\log_e\left(1 - \frac{1}{n + 1}\right) + \log_e\left(1 - \frac{1}{n + 1}\right).$$

Hence by putting $k = n + 1$, we have

$$\log_e\left(1 + \frac{1}{n}\right)^n$$

$$= k\left(\frac{1}{k} + \frac{1}{2k^2} + \frac{1}{3k^3} + \dots\right) - \left(\frac{1}{k} + \frac{1}{2k^2} + \frac{1}{3k^3} + \dots\right)$$

$$= 1 - \left(1 - \frac{1}{2}\right)\frac{1}{k} - \left(\frac{1}{2} - \frac{1}{3}\right)\frac{1}{k^2} - \left(\frac{1}{3} - \frac{1}{4}\right)\frac{1}{k^3} - \dots$$

§20. $\log_e \dfrac{1}{1 + x + x^2 + x^3} = \log_e \dfrac{1 - x}{1 - x^4} = \log_e(1 - x) - \log_e(1 - x^4)$.

Unless n is a multiple of 4, the term involving x^n comes only from $\log_e(1 - x)$, and its coefficient is $-\dfrac{1}{n}$. If n is a multiple of 4, put $n = 4m$; then the coefficient of x^{4m} is

$$-\frac{1}{4m} + \frac{1}{m} = \frac{3}{4m} = \frac{3}{n}.$$

§21.

$$\text{The general term} = \frac{n^3}{\underline{n}} = \frac{n^2}{\underline{n - 1}}$$

$$= \frac{1 + 3(n-1) + (n-1)(n-2)}{\lfloor \underline{n-1}}$$

$$= \frac{1}{\lfloor \underline{n-1}} + \frac{3}{\lfloor \underline{n-2}} + \frac{1}{\lfloor \underline{n-3}}.$$

Thus the given series $= 1 + \left(\frac{1}{\lfloor \underline{1}} + 3\right) + \left(\frac{1}{\lfloor \underline{2}} + \frac{3}{\lfloor \underline{1}} + 1\right)$

$$+ \left(\frac{1}{\lfloor \underline{3}} + \frac{3}{\lfloor \underline{2}} + \frac{1}{\lfloor \underline{1}}\right) + \dots$$

$$= \left(1 + \frac{1}{\lfloor \underline{1}} + \frac{1}{\lfloor \underline{2}} + \frac{1}{\lfloor \underline{3}} + \dots\right)$$

$$+ 3\left(1 + \frac{1}{\lfloor \underline{1}} + \frac{1}{\lfloor \underline{2}} + \dots\right)$$

$$+ \left(1 + \frac{1}{\lfloor \underline{1}} + \frac{1}{\lfloor \underline{2}} + \dots\right)$$

$$= e + 3e + e = 5e.$$

§22. The result follows at once from Art. 224 by subtracting (1) from (2).

§23. $\dfrac{1}{n+1} + \dfrac{1}{2(n+1)^2} + \dfrac{1}{3(n+1)^3} + \dots = -\log\left(1 - \dfrac{1}{n+1}\right)$

$$= -\log \frac{n}{n+1} = \log \frac{n+1}{n} = \log\left(1 + \frac{1}{n}\right) = \frac{1}{n} - \frac{1}{2n^2} + \frac{1}{3n^3} - \dots$$

§24. We have (omitting the base)

$$\log 2 - 2\log 3 + \log 5 = a; \quad 3\log 2 + \log 3 - 2\log 5 = b;$$
$$-4\log 2 + 4\log 3 - \log 5 = c.$$

Solving these simultaneous equations, we obtain

$$\log 2 = 7a - 2b + 3c; \quad \log 3 = 11a - 3b + 5c; \quad \log 5 = 16a - 4b + 7c.$$

Now $a = -\log_e \dfrac{9}{10} = -\log_e\left(1 - \dfrac{1}{10}\right) = \dfrac{1}{10} + \dfrac{1}{2.10^2} + \dfrac{1}{3.10^3} + \dots$

$$= .105360516.$$

$$b = -\log_e\left(1 - \frac{4}{100}\right) = \frac{4}{100} + \frac{4^2}{2.100^2} + \frac{4^3}{3.100^3} + \dots$$

$$= .040821995.$$

$$c = \log_e\left(1 + \frac{1}{80}\right) = \frac{1}{80} - \frac{1}{2.80^2} + \frac{1}{3.80^3} - \dots = .012422520.$$

EXAMPLES XVIII a

§1. We have
$$M = 100 \left(\frac{21}{20}\right)^{50}.$$
$$\therefore \log M = 2 + 50(\log 21 - \log 20) = 2 + 50(\log 3 + \log 7 - 1 - \log 2)$$
$$= 2 + 1.059465; \therefore M = 1146.74.$$

§2. We have $Pnr = 90$, and $\dfrac{Pnr}{1+nr} = 80$. Substituting for nr, we obtain

$$\frac{90P}{P+90} = 80; \text{ whence } P = 720.$$

§3. Let n years be the time, then $P\left(\dfrac{21}{20}\right)^n = 2P$; whence

$$n(\log 21 - \log 20) = \log 2, \text{ and } n = \frac{.3010300}{.0211893} = 14.2.$$

§4. $V = 10000 \left(\dfrac{21}{20}\right)^{-8}$; whence $\log V = 4 - (8 \times .0211893) = 3.8304856$;

that is,
$$V = 6768.394.$$

§5. Here $2500 = 1000 \left(\dfrac{11}{10}\right)^n$; that is, $10 = 4\left(\dfrac{11}{10}\right)^n$;

$$\therefore 1 = 2\log 2 + n(\log 11 - 1); \text{ thus } n = \frac{.3979400}{.0413927} = 9.6.$$

§6. We have
$$D = \frac{Pnr}{1+nr}, \text{ and } I = Pnr.$$
$$\frac{1}{D} = \frac{1+nr}{Pnr} = \frac{1}{Pnr} + \frac{1}{P} = \frac{1}{I} + \frac{1}{P} = \frac{2}{H}, \text{ where } H \text{ is the harmonic mean}$$
between I and P; thus $D = \dfrac{H}{2}$.

§7. We have
$$M = P\left(\frac{21}{20}\right)^{100}.$$
$$\therefore \log M = \log P + 100(\log 21 - \log 20) = \log P + 2.11893;$$
$$\therefore M = P \times 10^{2.11893}; \text{ that is, } M \text{ is greater } P \times 100.$$

§8. The sum is the present worth of £1000;
hence $V = 1000(1.06)^{-12}$;
$$\therefore \log V = 3 - (12 \times .0253059) = 2.6963292.$$
Thus
$$V = 496.97.$$

§9. If n is the number of half-years, than $6000 = 600(1.18)^n$;

or
$$10 = (1.18)^n; \therefore 1 = n\log 1.18; \quad n = \frac{1}{.071882} = 13.9.$$

§10. $M = (1.06)^{200}$ farthings, $\log M = 200 \times .0253059 = 5.06118$;
$$M = 115027 \text{ farthings.}$$

EXAMPLES XVIII b

§1. In Art. 237, put $A = 120$, $n = 5$; then $672 = 600 + 10 \times 120r$, whence
$$r = \frac{72}{1200}, \text{ and } 100r = 6.$$

§2. $M = 100 \times \dfrac{(1.045)^{20} - 1}{1.045 - 1} = \dfrac{20000}{9}\{(1.045)^{20} - 1\}.$
If $x = (1.045)^{20}, \log x = 20 \log 1.045 = .382326; \therefore x = 2.4117;$
$$\therefore M = \frac{20000 \times 1.4117}{9} = 3137\frac{1}{9}.$$

§3. Here £2750 is the present value of a perpetual annuity of amount A say, interest being reckoned at 4 p.c. Hence by putting $r = \dfrac{1}{25}$ in Art. 240, we have $2750 = 25A$, and $A = 110$.

§4. Here $4000 = \dfrac{120}{r}$, whence $100r = \dfrac{120}{40} = 3$.

§5. By Art. 241, the number of years' purchase $= \dfrac{100}{3\frac{1}{2}} = 28\frac{4}{7}$.

§6. The rate per cent is 4; hence the amount at the end of the years
$$= £625 \times \frac{26}{25} \times \frac{26}{25} = £676.$$

§7. The rate of interest is 5 per cent. ; let A be the annuity; then
$$2522 = A\frac{1 - \left(\dfrac{21}{20}\right)^{-3}}{\dfrac{21}{20} - 1} = 20A\left\{1 - \left(\frac{20}{21}\right)^{3}\right\}; \qquad [\text{Art. 240.}]$$

whence $\qquad 2522 = 20A \times \dfrac{1261}{(21)^3}$ or $A = \dfrac{(21)^3}{10} = 926\dfrac{1}{10}.$

§8. This is equivalent to finding the present value of a perpetuity of £400 to commence after 10 years. Hence by Art. 242,
$$V = \frac{400 \times (1.04)^{-10}}{.04} = 10000 \times (1.04)^{10};$$
$$\therefore \ \log V = 4 - .170333 = 3.829667; \text{ and } V = 6755.65.$$

§9. Let P be be the sum, then using the formula $M = Pe^{nr}$, we have
$$500 = Pe, \text{ or } P = 500e^{-1} = 500 \times .3678.$$

§10. Equating the present value $\dfrac{A(1 - R^{-n})}{R - 1}$ [found in Art. 240] to mA,

we have $\qquad m = \dfrac{1 - R^{-n}}{R - 1}.$

Hence we have
$$25 = \frac{1 - R^{-n}}{R - 1}, \text{ and } 30 = \frac{1 - R^{-2n}}{R - 1};$$

whence by division
$$1 + R^{-n} = \frac{30}{25} = \frac{6}{5}, \text{ and } R^{-n} = \frac{1}{5}.$$

$$\therefore \ 25 = \frac{1 - \dfrac{1}{5}}{R - 1}, \text{ whence } R - 1 = \frac{4}{125}, \text{ that is, } r = \frac{4}{125},$$
$$\text{and } 100r = 3\frac{1}{5}.$$

§11. Let A be the number of pounds paid annually, then £5000 is the present value of an annuity to commence at once and to run 10 years;
$$\therefore \ 5000 = A\frac{(1 - 1.04^{-10})}{.04}; \ A = \frac{200}{1 - 1.04^{-10}}.$$

Now $\qquad \log(1.04)^{-10} = -.170333 = \bar{1}.829667;$
$$\therefore \ 1.04^{-10} = .676031.$$
$$\therefore \ A = \frac{200}{.323969} = 617.343.$$

§12. The present value of an annuity of £1800 is $\dfrac{1800(1 - R^{-n})}{R - 1}$; and the man will be ruined if this is greater than £20000. Now $R - 1 = \dfrac{1}{20}$; thus he will be ruined if $9(1 - R^{-n}) > 5$; that is if $R^{-n} < \dfrac{4}{9}$, or if $R^n > \dfrac{9}{4}$, when $n = 17$.

Now $\qquad\qquad \log R^{17} = 17 \log \dfrac{21}{20} = 17(\log 7 + \log 3 - 1 - \log 2)$;

and $\qquad\qquad\qquad\qquad \log \dfrac{9}{4} = 2 \log 3 - 2 \log 2$.

By comparing the values of these expressions we arrive at the required result.

§13. The fine $= \dfrac{500}{.06} \{(1.06)^{-13} - (1.06)^{-20}\} = \dfrac{50000}{6}(A - B)$, say.

$$\text{Now } \log A = -13 \times .0253059 = -.3289767 = \bar{1}.6710233;$$
$$\therefore A = .4688385.$$

$$\log B = -20 \times .0253059 = -.5061180 = \bar{1}.4938820;$$
$$\therefore B = .3118042.$$

$$\therefore \text{ the fine } = \dfrac{50000}{6} \times .1570343 = 1308.619.$$

§14. As in Example 10 we have

$$a = \dfrac{1 - R^{-n}}{R - 1}, \qquad b = \dfrac{1 - R^{-2n}}{R - 1}, \qquad c = \dfrac{1 - R^{-3n}}{R - 1}.$$

$$\therefore 1 + R^{-n} = \dfrac{b}{a}, \text{ and } 1 + R^{-n} + R^{-2n} = \dfrac{c}{a}$$

$$\therefore 1 + \left(\dfrac{b}{a} - 1\right) + \left(\dfrac{b}{a} - 1\right)^2 = \dfrac{c}{a}; \quad \therefore 1 - \dfrac{b}{a} + \dfrac{b^2}{a^2} = \dfrac{c}{a}.$$

§15. The present value of £10 due 1 year hence $= £\dfrac{10}{1.05}$,

$\dots\dots\dots\dots\dots\dots\dots\dots£20\dots2$ years $\dots\dots = £\dfrac{20}{(1.05)^2}$, and so on;

$\therefore$ the present value of the annuity in pounds

$$= \dfrac{10}{1.05} + \dfrac{20}{(1.05)^2} + \dfrac{30}{(1.05)^3} + \dfrac{40}{(1.05)^4} + \dots = a + 2ax + 3ax^2 + 4ax^3 + \dots;$$

where $a = \dfrac{10}{1.05}$, and $x = \dfrac{1}{1.05}$;

$\therefore$ present value in pounds

$$= \dfrac{a}{(1 - x)^2} = \dfrac{10 \times (1.05)^2}{1.05 \times .0025} = \dfrac{10 \times 1.05}{.0025};$$

$$\therefore \text{ present value } = £4200.$$

EXAMPLES XIX : Inequalities

EXAMPLES XIX a

§1. Multiply together the two inequalities,
$$ab + xy > 2\sqrt{abxy}, \text{ and } ax + by > \sqrt{axby}.$$

§2. Multiply together the three inequalities,
$$b + c > 2\sqrt{bc}, \quad c + a > 2\sqrt{ca}, \quad a + b > 2\sqrt{ab}.$$

§3. $\left(\sqrt{x} - \dfrac{1}{\sqrt{x}}\right)^2 > 0$; that is $x + \dfrac{1}{x} > 2.$

§4. We have $2ax < a^2 + x^2$, and $2by < b^2 + y^2$; hence by addition,
$$2ax + 2by < (a^2 + b^2) + (x^2 + y^2); \text{ that is } < 2.$$

§5. We have $2ax < a^2 + x^2$; $2by < b^2 + y^2$; $2cz < c^2 + z^2$; hence by addition,
$$2(ax + by + cz) < (a^2 + b^2 + c^2) + (x^2 + y^2 + z^2); \text{ that is } < 2.$$

§6. Here $a > b$; thus $a - b$ is positive, $\therefore a^{a-b} > b^{a-b}$;

or $\qquad\qquad \dfrac{a^a}{a^b} > \dfrac{b^a}{b^b}$; hence the result.

Again, $b < a$, and therefore $b + ab < a + ab$, that is, $b(1 + a) < a(1 + b)$;
$$\therefore \log b + \log(1 + a) < \log a + \log(1 + b);$$
$$\therefore \log b - \log a < \log(1 + b) - \log(1 + a),$$
and the result follows at once.

§7. By Art. 253, $\dfrac{x^2 y + y^2 z + z^2 x}{3} > (x^2 y \cdot y^2 z \cdot z^2 x)^{\frac{1}{3}}$, or $> xyz$;
$$\therefore x^2 y + y^2 z + z^2 x > 3xyz.$$

Similarly, $\qquad\qquad xy^2 + yz^2 + zx^2 > 3xyz.$

§8.
$$a^3 - 3ab^2 + 2b^3 = (a - b)(a^2 + ab - 2b^2)$$
$$= (a - b)(a - b)(a + 2b) = (a - b)^2(a + 2b);$$
which is always positive, hence $a^3 + 2b^3 > 3ab^2$.

§9. $a^4 - a^3 b - ab^3 + b^4 = (a - b)(a^3 - b^3) = (a - b)^2(a^2 + ab + b^2)$,
thus $\qquad\qquad a^4 + b^4 > a^3 b + ab^3.$

§10. $b^2 + c^2 > 2bc$; hence $(b^2 + c^2)a > 2abc$.

Similarly $\qquad\qquad (c^2 + a^2)b > 2abc, \text{ and } (a^2 + b^2)c > 2abc.$

By addition,
$$(b^2 + c^2)a + (c^2 + a^2)b + (a^2 + b^2)c > 6abc;$$
that is, $\qquad bc(b + c) + ca(c + a) + ab(a + b) > 6abc.$

§11. $b^2 + c^2 > 2bc$; hence $(b^2 + c^2)a^2 > 2a^2 bc.$

Similarly $\qquad (c^2 + a^2)b^2 > 2ab^2 c, \text{ and } (a^2 + b^2)c^2 > 2abc^2.$

By addition,
$$(b^2 + c^2)a^2 + (c^2 + a^2)b^2 + (a^2 + b^2)c^2 > 2a^2 bc + 2ab^2 c + 2abc^2;$$
that is $\qquad 2(b^2 c^2 + c^2 a^2 + a^2 b^2) > 2abc(a + b + c).$

§12. $x^3 - x^2 - x - 2 = (x - 2)(x^2 + x + 1)$, which is positive or negative according as x is greater or less than 2;
$$\therefore x^3 > \text{ or } < x^2 + x + 2, \text{ according as } x > \text{ or } < 2.$$

§13.
$$x^3 - 5ax^2 + 13a^2 x - 9a^3 = (x - a)(x^2 - 4ax + 9a^2)$$
$$= (x - a)\{(x - 2a)^2 + 5a^2\}.$$

By hypothesis the first factor is positive, and the second factor is always positive; hence the result.

§14. $11 - 17x + 7x^2 - x^3 = (1 - x)(11 - 6x + x^2) = (1 - x)\{2 + (3 - x)^2\}$.

The second of these factors is always positive, but the first is only positive so long as $x < 1$; hence the greatest value of x is 1.

§15. $x^2 - 12x + 40 = (x - 6)^2 + 4$, and is a minimum when $x = 6$; its value being 4.

$24x - 8 - 9x^2 = 8 - (4 - 3x)^2$, and is a maximum when $4 - 3x = 0$; its value being 8.

§16. It is easily seen that $r(n - r + 1) > n$; thus we have

$$1 \cdot n = n$$
$$2(n - 1) > n$$
$$3(n - 2) > n$$
$$\cdots\cdots\cdots$$
$$(n - 2)2 > n$$
$$n \cdot 1 = n.$$

By multiplication the required result is obtained.

Again, since the geometric mean is less than the arithmetic mean we have the n inequalities;

$$2 \cdot 2n < (n + 1)^2; \ 4(2n - 2) < (n + 1)^2;$$
$$6(2n - 4) < (n + 1)^2; \ldots\ldots\ldots\ldots$$
$$(2n - 2)4 < (n + 1)^2; \ 2n \cdot 2 < (n + 1)^2.$$

Hence by multiplication, $(2 \cdot 4 \cdot 6 \ldots 2n)^2 < (n + 1)^{2n}$.

§17. By Art. 253, $\dfrac{x + y + z}{3} > (xyz)^{\frac{1}{3}}$. Cube each side.

§18. The solution is similar to that of Ex. 16.

$$1 \cdot (2n - 1) < n^2; \ 3(2n - 3) < n^2; \ 5(2n - 5) < n^2; \ldots$$
$$\therefore \ \{1 \cdot 3 \cdot 5 \ldots (2n - 1)\}^2 < n^{2n}.$$

§19. By Art. 253, $\left(\dfrac{1 + 2 + 2^2 + \ldots + 2^{n-1}}{n}\right) > 1 \cdot 2 \cdot 2^2 \ldots 2^{n-1}$;

that is, $\left(\dfrac{2^n - 1}{n}\right)^n > 2^{1+2+3+\ldots+(n-1)}$, or $> 2^{\frac{n(n-1)}{2}}$;

hence $\dfrac{2^n - 1}{n} > 2^{\frac{n-1}{2}}$, that is $> \sqrt{2^{n-1}}$;

whence the result easily follows.

§20. $\dfrac{1^3 + 2^3 + 3^3 + \ldots + n^3}{n} > (1^3 \cdot 2^3 \cdot 3^3 \ldots n^3)^{\frac{1}{n}}; \quad \therefore \ \dfrac{n(n + 1)^2}{4} > \{(\underline{|n})^3\}^{\frac{1}{n}}$.

Raise each side to the nth power.

§21. In Ex. 17 we have proved that $(a + b + c)^3 > 27abc$.

Put $a = y + z - x$, $b = z + x - y$, $c = x + y - z$, so that $a + b + c = x + y + z$; we then obtain the result.

Again $(b + c)(c + a)(a + b) > 8abc$.   **[Ex. 2.]**

With the same substitutions, $b + c = 2x$, $c + a = 2y$, $a + b = 2z$; thus the result follows.

§22. The expression is a maximum when $\left(\dfrac{7 - x}{4}\right)^4 \left(\dfrac{2 + x}{5}\right)^5$ is a maximum.

But the sum of $4\left(\dfrac{7 - x}{4}\right)$ and $5\left(\dfrac{2 + x}{5}\right)$ is constant. Hence the maximum is when $\dfrac{7 - x}{4} = \dfrac{2 + x}{5}$, or $x = 3$.

§23. Put $u = \dfrac{(5+x)(2+x)}{1+x}$. Then if $1 + x = y$,

$$\therefore u = \dfrac{(4+y)(1+y)}{y} = \dfrac{4}{y} + y + 5; \quad \therefore u = \left(\dfrac{2}{\sqrt{y}} - \sqrt{y}\right)^2 + 9.$$

Hence u is a maximum when $\dfrac{2}{\sqrt{y}} - \sqrt{y} = 0$; that is when $y = 2$, $x = 1$; in this case the value of u is 9.

EXAMPLES XIX b

§1. We have $\dfrac{a^4 + b^4 + c^4}{3} > \left(\dfrac{a+b+c}{3}\right)^4.$ [Art. 258.]

By clearing of fractions we have the result.

§2. By Art. 258, $\dfrac{1^3 + 2^3 + 3^3 + \ldots + n^3}{n} > \left(\dfrac{1+2+3+\ldots+n}{n}\right)^3;$

and therefore $> \left(\dfrac{n+1}{2}\right)^3.$

By clearing of fractions we have the result.

§3. By Art. 258,

$$\dfrac{2^m + 4^m + 6^m + \ldots + (2n)^m}{n} > \left(\dfrac{2+4+6+\ldots+2n}{n}\right)^m;$$

and therefore $> (n+1)^m.$

By clearing of fractions we have the result.

§4. If $a > b$, $\left(1 + \dfrac{x}{a}\right)^a > \left(1 + \dfrac{x}{b}\right)^b.$ [Art. 259.]

Put $\dfrac{x}{a} = \dfrac{1}{\alpha}, \dfrac{x}{b} = \dfrac{1}{\beta}$, so that $a = \alpha x$, $b = \beta x$; $\therefore \left(1 + \dfrac{1}{\alpha}\right)^{\alpha x} > \left(1 + \dfrac{1}{\beta}\right)^{\beta x}.$

By taking the xth root we have the result; for since $a > b, \alpha$ must be greater than β. Also since x may be any positive quantity, α and β may be any positive quantities subject to the above restriction. Thus the expression $\left(1 + \dfrac{1}{n}\right)^n$ gradually increases as n increases. When $n = 1$ its value is 2, and when n is infinite its value is e.

§5. Put $c = ax$, and $c = by$, so that $a = \dfrac{c}{x}$ and $b = \dfrac{c}{y}$. Then

$$\left(\dfrac{a+c}{a-c}\right)^a < \left(\dfrac{b+c}{b-c}\right)^b, \text{ if } \left(\dfrac{1+x}{1-x}\right)^{\frac{c}{x}} < \left(\dfrac{1+y}{1-y}\right)^{\frac{c}{y}};$$

that is, if $\left(\dfrac{1+x}{1-x}\right)^{\frac{1}{x}} < \left(\dfrac{1+y}{1-y}\right)^{\frac{1}{y}}.$

Here $y > x$; hence the result follows from Art. 260.

§6. Consider the expression $a^a, b^b, c^c, \ldots k^k$.

If any two of the quantities, a and b suppose, are unequal, this expression is diminished when we replace a and b by the two equal quantities $\dfrac{a+b}{2} \cdot \dfrac{a+b}{2}.$ [Art. 261.]

Hence the least value of the expression is when all the quantities $a, b, c, \ldots k$ are equal; in this case each is equal to $\dfrac{a+b+c+\ldots+k}{n}.$

§7. $\dfrac{1}{m} \log(1 + a^m) < \dfrac{1}{n} \log(1 + a^n)$ if $n \log(1 + a^m) < m \log(1 + a^n)$; that is if $(1 + a^m)^n < (1 + a^n)^m.$

First suppose that $a < 1$; since $m > n$, therefore $a^n > a^m$, and $1 + a^n > 1 + a^m$; hence a fortiori $(1 + a^n)^m > (1 + a^m)^n$.

Dividing this inequality by a^{mn},

we have $\left(\dfrac{1}{a^n}+1\right)^m > \left(\dfrac{1}{a^m}+1\right)^n$.

If $a > 1$, then $\dfrac{1}{a} < 1$, and the inequality still holds.

§8.

$$\frac{1-x^{n+1}}{1-x^n} = 1 + \frac{x^n(1-x)}{1-x^n} = 1 + \frac{x^n}{1+x+x^2+\ldots+x^{n-1}}$$

$$= 1 + \frac{1}{\dfrac{1}{x}+\dfrac{1}{x^2}+\dfrac{1}{x^3}+\ldots+\dfrac{1}{x^n}}.$$

Since $x < 1$, each of the n terms $\dfrac{1}{x}$, $\dfrac{1}{x^2}$, ... $\dfrac{1}{x^n}$ is greater than 1, and their sum is greater than n;

$$\therefore \quad \frac{1-x^{n+1}}{1-x^n} < 1+\frac{1}{n}, \text{ that is } < \frac{n+1}{n};$$

$$\therefore \quad \frac{1-x^{n+1}}{n+1} < \frac{1-x^n}{n}.$$

§9. We have $\quad \dfrac{a^n+c^n}{2} > \left(\dfrac{a+c}{2}\right)^n.$ [Art. 257.]

But $\dfrac{a+c}{2}$ is the arithmetic mean of a and c, and is consequently greater than b the harmonic mean. [Art. 65.] Hence a fortiori $a^n + c^n > 2b^n$.

§10. $x^3(4a-x)^5$ is a maximum when $\left(\dfrac{x}{3}\right)^3\left(\dfrac{4a-x}{5}\right)^5$ is a maximum. This expression is the product of 8 factors whose sum is $3\left(\dfrac{x}{3}\right)+5\left(\dfrac{4a-x}{5}\right)$ or $4a$, which is constant.

Hence the maximum value is when $\dfrac{x}{3} = \dfrac{4a-x}{5}$, or $x = \dfrac{3}{2}$. Thus the maximum value is $\dfrac{3^3 \cdot 5^5}{2^8}a^8$.

The second expression is a maximum when its sixth power is a maximum; that is when $x^3(1-x)^2$ is a maximum. As in the preceding case, this is when $\dfrac{x}{3} = \dfrac{1-x}{2}$, or $x = \dfrac{3}{5}$. Thus the value required is $\sqrt{\dfrac{3}{5}} \cdot \sqrt[3]{\dfrac{2}{5}}$.

§11. $\log(1+x) < x$ if $(1+x) < e^x$; this is obviously the case since

$$e^x = 1 + x + \frac{x^2}{\lfloor 2} + \ldots$$

Again $\qquad \log(1+x) > \dfrac{x}{1+x}$, if $1+x > e^{\frac{x}{1+x}}$.

Now

$$1+x = \frac{1}{1-\dfrac{x}{1+x}} = 1 + \frac{x}{1+x} + \frac{x^2}{(1+x)^2} + \frac{x^3}{(1+x)^3} + \ldots$$

$$> 1 + \frac{x}{1+x} + \frac{x^2}{\lfloor 2\,(1+x)^2} + \frac{x^3}{\lfloor 3\,(1+x)^3} + \ldots > e^{\frac{x}{1+x}}.$$

§12. Consider the expression $\dfrac{1}{x}+\dfrac{1}{y}+\dfrac{1}{z}$, and suppose z constant, so that the sum of $x+y$ is also constant. Now $\dfrac{1}{x}+\dfrac{1}{y} = \dfrac{x+y}{xy} = \dfrac{\text{constant}}{xy}$, and the denominator is greatest when $x = y$; thus $\dfrac{1}{x}+\dfrac{1}{y}$ is least when $x = y$. Hence if any two of the quantities x, y, z are unequal, the expression $\dfrac{1}{x} + \dfrac{1}{y} + \dfrac{1}{z}$ can be diminished, and its value is a minimum when $x = y = z = \dfrac{1}{3}$.

Thus the minimum value of $\dfrac{1}{x} + \dfrac{1}{y} + \dfrac{1}{z}$ is 9.

Clearing of fractions, $yz + zx + xy > 9xyz$; and $1 - (x + y + z) = 0$;

$$\therefore\ 1 - (x + y + z) + (yz + zx + xy) - xyz > 8xyz;$$

that is,
$$(1 - x)(1 - y)(1 - z) > 8xyz.$$

§13. The expression $(a + b + c + d)(a^3 + b^3 + c^3 + d^3) - (a^2 + b^2 + c^2 + d^2)^2$
$= ab(a - b)^2 + ac(a - c)^2 + ad(a - d)^2 + bc(b - c)^2 + bd(b - d)^2 + cd(c - d)^2$,
and is therefore positive.

§14. Since both of the expressions involve the letters a, b, c symmetrically, we may suppose that a, b, c are in order of magnitude; let us suppose then that $a > b > c$. In this case $c(c - a)(c - b)$ is positive.

Also $a(a - b)(a - c) + b(b - c)(b - a) = (a - b)\{a^2 - ac - (b^2 - bc)\}$
$$= (a - b)^2(a + b - c),$$

and is therefore positive.

Again $c^2(c - a)(c - b)$ is positive.

Also $a^2(a - b)(a - c) + b^2(b - c)(b - a)$
$$= (a - b)\{a^3 - a^2c - (b^3 - b^2c)\}$$
$$= (a - b)^2(a^2 + ab + b^2 - ac - bc);$$

which is positive since $a^2 - ac$, and $b^2 - bc$ are positive.

§15. In Example 7 we have proved that if $m > n$, $(1 + a^n)^m > (1 + a^m)^n$.
Put $a = \dfrac{y}{x}$ and multiply both sides by x^{mn}, thus $(x^n + y^n)^m > (x^m + y^m)^n$.

§16. Let
$$P = (1 + x)^{1-x}(1 - x)^{1+x};$$
$$\log P = (1 - x)\log(1 + x) + (1 + x)\log(1 - x)$$
$$= \{\log(1 + x) + \log(1 - x)\} - x\{\log(1 + x) - \log(1 - x)\}$$
$$= -2\left(\frac{x^2}{2} + \frac{x^4}{4} + \frac{x^6}{6} + \dots\right) - 2x\left(x + \frac{x^3}{3} + \frac{x^5}{5} + \dots\right);$$

$$\therefore\ \log P \text{ is negative; } \therefore\ P < 1; \therefore\ (1 + x)^{1-x}(1 - x)^{1+x} < 1.$$

Now proceed exactly as in Art. 261.

§17. Let the three quantities p, q, r be in descending order of magnitude. Then the given expression
$$= a^2(p - q)(q - r) - b^2(p - q)(q - r) + c^2(p - r)(q - r),$$
and will consequently be least when the second term is greatest; this is when $b = a + c$, which is the extreme case when the triangle becomes a straight line. The expression then
$$= a^2(p - q)(q - r) - (a^2 + 2ac + c^2)(p - q)(q - r) + c^2(p - r)(q - r)$$
$$= a^2(p - q)^2 - 2ac(p - q)(q - r) + c^2(q - r)^2$$
$$= \{a(p - q) - c(q - r)\}^2, \text{ which is positive.}$$

Hence the expression is always positive.

(2) Substituting $z = -(x + y,)$ we have to shew that $-(a^2y + b^2x)(x + y) + c^2xy$ must be negative; that is (changing the signs) we must prove that
$$b^2x^2 + a^2y^2 + (a^2 + b^2 - c^2)xy$$
is positive. This expression is equal to $(bx - ay)^2 + \{(a + b)^2 - c^2\}xy$, which is positive; for $a + b > c$.

§18. Lemma. If $a + b = n$, a given quantity, then $\lfloor a\ \lfloor b$ becomes less and less the nearer a and b are to each other.

For $\lfloor n - r\ \lfloor r > \lfloor n - (r + 1)\ \lfloor r + 1$, if $n - r > r + 1$; that is if $n > 2r + 1$.

Hence
$$\lfloor n-1 \,\lfloor 1 \; > \; \lfloor n-2 \,\lfloor 2 \; > \; \lfloor n-3 \,\lfloor 3 \,,\ldots$$

Thus if $a+b = 2m$, the least value of $\lfloor a \,\lfloor b$ is $\lfloor m \,\lfloor m$; and if $a+b = 2m+1$, the least value of $\lfloor a \,\lfloor b$ is $\lfloor m+1 \,\lfloor m$.

By the preceding lemma,
$$\lfloor 2n-1 \,\lfloor 1 \; > \; \lfloor n \,\lfloor n \,, \quad \lfloor 2n-3 \,\lfloor 3 \; > \; \lfloor n \,\lfloor n \,, \quad \lfloor 2n-5 \,\lfloor 5 \; > \; \lfloor n \,\lfloor n \,,\ldots$$
$$\lfloor 3 \,\lfloor 2n-3 \; > \; \lfloor n \,\lfloor n \,, \quad \lfloor 1 \,\lfloor 2n-1 \; > \; \lfloor n \,\lfloor n \,.$$

Multiplying together these n inequalities, we have
$$(\lfloor 1 \,\lfloor 3 \,\lfloor 5 \,\ldots\, \lfloor 2n-1 \,)^2 \; > \; (\lfloor n \,)^{2n}.$$

§19. Consider the expression $\lfloor a \,\lfloor b \,\lfloor c \,\lfloor d \,\ldots$; then if any two of the quantities, a and b say, are unequal, we can without altering their sum diminish $\lfloor a \,\lfloor b$ by taking a and b equal. [This is proved in the lemma preceding Example 18.]

Hence the value $\lfloor a \,\lfloor b \,\lfloor c \,\lfloor d \,\ldots$ is least when all the quantities $a, b, c, d, \ldots$ are equal. If however n is not exactly divisible by p this will not be the case; suppose then that q is the quotient and r the remainder when n is divided by p; thus $n = pq + r = (p-r)q + r(q+1)$. Hence $p-r$ of the quantities $a, b, c, d \ldots$ will be equal to q, and the remaining r will be equal to $q+1$; thus the least value of the expression is $(\lfloor q \,)^{p-r}(\lfloor q+1 \,)^r$.

EXAMPLES XX : Limiting Values and Vanishing Fractions

§1. (1) Limit $= \dfrac{(2x)(-5x)}{7x^2} = -\dfrac{10}{7}.$ (2) Limit $= \dfrac{(-3)(3)}{4} = -\dfrac{9}{4}.$

§2. (2) Limit $= \dfrac{(3x^2)^2}{x^4} = 9.$ (2) Limit $= \dfrac{(-1)^2}{9} = \dfrac{1}{9}.$

§3. (1) Limit $= \dfrac{2x^3 \cdot x}{4x^3 \cdot x} = \dfrac{1}{2}.$ (2) Limit $= \dfrac{3 \cdot -5}{-9 \cdot 1} = \dfrac{5}{3}.$

§4. (1) Limit $= \dfrac{x \cdot -5x \cdot 3x}{(2x)^3} = -\dfrac{15}{8}.$ (2) Limit $= \dfrac{-3 \cdot 2 \cdot 1}{(-1)^3} = 6.$

§5. (1) Limit $= \dfrac{-x^2}{2x^3} \times \dfrac{2x^2}{-x} = 1.$ (2) Limit $= \dfrac{1}{-1} \times \dfrac{2x^2}{1} = 0.$

§6. (1) Limit $= \dfrac{-x \cdot x \cdot -7x}{7x(x)^3} = \dfrac{1}{x} = 0.$ (2) Limit $= \dfrac{3 \cdot 5 \cdot 2}{-1(1)^3} = -30.$

§7. $\dfrac{x^3+1}{x^2-1} = \dfrac{x^2-x+1}{x-1} = \dfrac{3}{-2} = -\dfrac{3}{2}.$

§8.

$$\frac{a^x - b^x}{x}$$

$$= \frac{1}{x}\left\{1 + x\log a + \frac{x^2(\log a)^2}{\lfloor 2} + \ldots - 1 - x\log b - \frac{x^2(\log b)^2}{\lfloor 2} - \ldots\right\}$$

$$= \log a - \log b, \text{ when } x = 0.$$

§9. $\dfrac{e^x - e^{-x}}{\log(1+x)} = \dfrac{1+x+\dfrac{x^2}{\lfloor 2}+\ldots - \left(1-x+\dfrac{x^2}{\lfloor 2}-\ldots\right)}{x - \dfrac{x^2}{2} + \dfrac{x^3}{3} - \ldots} = \dfrac{2x}{x} = 2.$

§10. By putting $x = a + h$, $\dfrac{e^{mx} - e^{ma}}{x-a} = \dfrac{e^{ma+mh} - e^{ma}}{h}$

$= \dfrac{e^{ma}}{h}(e^{mh} - 1) = \dfrac{e^{ma}}{h}\left(mh + \dfrac{m^2h^2}{\lfloor 2} + \ldots\right) = me^{ma}$, since $h = 0$ when $x = a$.

§11. Put $x = 2a + h$; then the expression

$$= \frac{(2a+h)^{\frac{1}{2}} - \sqrt{2a} + \sqrt{h}}{(4ah + h^2)^{\frac{1}{2}}} = \frac{\sqrt{2a}\left(1 + \dfrac{1}{2}\cdot\dfrac{h}{2a} - \ldots\right) - \sqrt{2a} + \sqrt{h}}{\sqrt{4ah}}$$

$$= \frac{\sqrt{h}}{\sqrt{4ah}} = \frac{1}{\sqrt{4a}}.$$

§12. $\dfrac{\log(1+x^2+x^4)}{3x^2(1-2x)} = \dfrac{(x^2+x^4) - \dfrac{1}{2}(x^2+x^4)^2 + \ldots}{3x^2(1-2x)} = \dfrac{x^2}{3x^2} = \dfrac{1}{3}.$

§13. Put $x = 1 + h$, then the expression

$$= \frac{-h + \log(1+h)}{1 - \sqrt{1 - h^2}} = \frac{-h + \left(h - \dfrac{1}{2}h^2 + \ldots\right)}{1 - \left(1 - \dfrac{1}{2}h^2 - \ldots\right)} = \frac{-\dfrac{1}{2}h^2}{\dfrac{1}{2}h^2} = -1.$$

§14. Put $x = a - h$; then the expression

$$= \frac{(2ah - h^2)^{\frac{1}{2}} + h^{\frac{3}{2}}}{(3a^2h - \ldots)^{\frac{1}{2}} + h^{\frac{1}{2}}}$$

$$= \frac{\sqrt{2a} \cdot h^{\frac{1}{2}}}{\sqrt{3a^2} \cdot h^{\frac{1}{2}} + h^{\frac{1}{2}}}, \text{ only keeping the lowest powers of } h,$$

$$= \frac{\sqrt{2a}}{\sqrt{3a^2 + 1}}.$$

§15. The expression $=$

$$\frac{(a^2 + ax + x^2) - (a^2 - ax + x^2)}{(a + x) - (a - x)} \cdot \frac{\sqrt{a + x} + \sqrt{a - x}}{\sqrt{a^2 + ax + x^2} + \sqrt{a^2 - ax + x^2}}$$

$$\frac{2ax}{2x} \cdot \frac{2\sqrt{a}}{2a} = \sqrt{a}.$$

§16. $\left(\dfrac{n + 1}{n}\right)^n = \left(1 + \dfrac{1}{n}\right)^n = e$, and $\dfrac{n + 1}{n} = 1 + \dfrac{1}{n} = 1.$

Hence the expression $= (e - 1)^{-n} = \dfrac{1}{(e - 1)^n} = \dfrac{1}{\infty} = 0$; since $e - 1$ is greater than 1.

§17. The expression

$$= n \left\{\log e - (n - 1) \log \left(1 + \frac{1}{n}\right)\right\}$$

$$= n \left\{1 - (n - 1) \left(\frac{1}{n} - \frac{1}{2n^2} + \frac{1}{3n^3} - \cdots\right)\right\}$$

$$= n \left\{1 - \left(1 - \frac{1}{2n} + \frac{1}{3n^2} - \frac{1}{n} + \frac{1}{2n^2} - \cdots\right)\right\}$$

$$= n \left\{1 - \left(1 - \frac{3}{2n} - \frac{5}{6n^2} + \cdots\right)\right\} = n \left(\frac{3}{2n} - \frac{5}{6n^2} + \cdots\right).$$

Hence the limit is $\dfrac{3}{2}$.

§18. Put $x = ay$; then the expression

$$= \sqrt[ay]{\frac{1 + y}{1 - y}} = \left\{\sqrt[y]{\frac{1 + y}{1 - y}}\right\}^{\frac{1}{a}} = \left(e^2\right)^{\frac{1}{a}}. \qquad\qquad \text{[Ex. 3, Art. 270.]}$$

Hence the limit is $e^{\frac{2}{a}}$.

EXAMPLES XXI : Convergency and Divergency of Series

EXAMPLES XXI a

§1. Convergent by Art. 280.

§2. The series $= \left(1 - \dfrac{1}{2}\right) + \left(\dfrac{1}{3} - \dfrac{1}{4}\right) + \left(\dfrac{1}{5} - \dfrac{1}{6}\right) + \ldots$; and is therefore convergent by Art. 280.

§3. Convergent by Art. 280.

§4. $\dfrac{u_n}{u_{n-1}} = \dfrac{x^n}{n(n+1)} \div \dfrac{x^{n-1}}{(n-1)n} = \dfrac{n-1}{n+1}x$. Hence if $x < 1$, the series is convergent; and if $x > 1$, divergent.

If $x = 1$, the series becomes

$$\left(1 - \frac{1}{2}\right) + \left(\frac{1}{2} - \frac{1}{3}\right) + \left(\frac{1}{3} - \frac{1}{4}\right) + \left(\frac{1}{4} - \frac{1}{5}\right)$$

$$+ \ldots + \left(\frac{1}{n} - \frac{1}{n+1}\right) + \ldots;$$

and the sum of the first n terms is $1 - \dfrac{1}{n+1}$. Hence the series is convergent.

§5. $\dfrac{u_n}{u_{n-1}} = \dfrac{x^n}{(2n-1)2n} \div \dfrac{x^{n-1}}{(2n-3)(2n-2)} = \dfrac{(2n-3)(2n-2)}{(2n-1)2n}x.$

If $x < 1$, the series is convergent; if $x > 1$, divergent.

If $x = 1$, the series $= \dfrac{1}{1 \cdot 2} + \dfrac{1}{3 \cdot 4} + \dfrac{1}{5 \cdot 6} + \ldots$, and is convergent; see Ex. 2.

§6. $\dfrac{u_n}{u_{n-1}} = \dfrac{n^2}{\lfloor n} \cdot \dfrac{(n-1)^2}{\lfloor n-1} = \dfrac{1}{n} \dfrac{n^2}{(n-1)^2}$; thus $Lim.\dfrac{u_n}{u_{n-1}} = 0$, and the series is convergent.

§7. Her $u_n = \sqrt{\dfrac{n}{n+1}}$, and is ultimately equal to unity; hence the series is divergent. [Art. 282.]

§8. $\dfrac{u_n}{u_{n-1}} = \dfrac{(2n-1)x^{n-1}}{(2n-3)x^{n-2}} = \dfrac{2n-1}{2n-3} \cdot x$. Hence if $x < 1$, the series is convergent; if $x > 1$, divergent. If $x = 1$, the series $= 1 + 3 + 5 + 7 + 9 + \ldots$, and is divergent.

§9. $\dfrac{u_n}{u_{n-1}} = \dfrac{(n+1)}{n^p} \div \dfrac{n}{(n-1)^p} = \dfrac{n+1}{n}\left(\dfrac{n-1}{n+1}\right)^p$, and thus is ultimately equal to unity.

But $u_n = \dfrac{n+1}{n^p} = \dfrac{1}{n^{p-1}}$ ultimately, hence we take for the auxiliary series the series whose n^{th} term is $\dfrac{1}{n^{p-1}}$, and this series is divergent except when $p - 1 > 1$. [Art. 290.]

Hence the given series is divergent except when $p > 2$.

§10. $\dfrac{u_n}{u_{n-1}} = \dfrac{x^{n-1}}{(n-1)^2 + 1} \div \dfrac{x^{n-2}}{(n-2)^2 + 1} = \dfrac{(n-2)^2 + 1}{(n-1)^2 + 1}x = x$, ultimately.

Hence if $x < 1$, the series is convergent; if $x > 1$, divergent.

If $x = 1$, the series $= 1 + \dfrac{1}{2} + \dfrac{1}{5} + \dfrac{1}{10} + \ldots + \dfrac{1}{n^2 + 1} + \ldots$

Now $\dfrac{1}{n^2 + 1} = \dfrac{1}{n^2}$ ultimately; but the series of which this is the general term is convergent. [Art. 290]

Hence the given series is convergent when $x = 1$.

§11. $\dfrac{u_n}{u_{n-1}} = \dfrac{n^2 - 1}{n^2 + 1} \cdot \dfrac{(n-1)^2 + 1}{(n-1)^2 - 1} x = x$, ultimately.

Hence if $x < 1$, the series is convergent; if $x > 1$, divergent.

If $x = 1$, then $u_n = \dfrac{n^{-1}}{n^2 + 1} = 1$, ultimately; hence the series is divergent.

[Art. 282.]

§12. $\dfrac{u_n}{u_{n-1}} = x$, ultimately. And when $x = 1$, $u_n = 1$, ultimately; hence the results are the same as in Ex. 11.

§13. $u_n = \dfrac{1}{(2n-1)^p} = \dfrac{1}{(2n)^p} = \dfrac{1}{2^p} \cdot \dfrac{1}{n^p}$, ultimately. But the series whose general term is $\dfrac{1}{n^p}$ is divergent except when $p > 1$; hence the given series is divergent except when $p > 1$.

§14. $\dfrac{u_n}{u_{n-1}} = x$, ultimately. Hence if $x < 1$, the series is convergent; if $x > 1$, divergent.

If $x = 1$, $u_n = \dfrac{n+1}{n^3} = \dfrac{1}{n^2}$, ultimately; and the series whose general term is $\dfrac{1}{n^2}$ is convergent. [Art. 290]

Hence the given series is convergent.

§15. $u_n = \left\{ \left(\dfrac{n+1}{n} \right)^n - \dfrac{n+1}{n} \right\}^{-n} = (e-1)^{-n}$,   [See Chap. XX. Ex. 16.]

Hence $\dfrac{u_n}{u_{n-1}} = \dfrac{(e-1)^{-n}}{(e-1)^{-(n-1)}} = \dfrac{1}{e-1}$, and since this is less than 1 the series is convergent.

§16. Here $u_n = \dfrac{(n-1)^{n-1}}{n^n} = \dfrac{1}{n} \left(\dfrac{n-1}{n} \right)^{n-1} = \dfrac{1}{n} \left(1 - \dfrac{1}{n} \right)^{n-1}$.

$\therefore u_n = \dfrac{1}{n} \cdot e^{-1} = \dfrac{1}{en}$, ultimately.   [Art. 220, Cor.]

Hence the series is divergent.   [Art. 290. Case II.]

§17. (1)

$$u_n = (n^2 + 1)^{\frac{1}{2}} - n = n \left(1 + \dfrac{1}{n^2} \right)^{\frac{1}{2}} - n$$

$$= n \left(1 + \dfrac{1}{2} \cdot \dfrac{1}{n^2} - \dots \right) - n = \dfrac{1}{2n}, \text{ ultimately.}$$

Hence the series is convergent.

(2)

$$\sqrt{n^4 + 1} - \sqrt{n^4 - 1} = \dfrac{(n^4 + 1) - (n^4 - 1)}{\sqrt{n^4 + 1} + \sqrt{n^4 - 1}}$$

$$= \dfrac{2}{n^2 + n^2} = \dfrac{1}{n^2}, \text{ ultimately.}$$

Hence the series is convergent.

§18. (1) Here $u_n = \dfrac{1}{x + n - 1} = \dfrac{1}{n}$, ultimately; hence the series is divergent.

(2) The series $= \dfrac{1}{x} + \left(\dfrac{1}{x-1} + \dfrac{1}{x+1} \right) + \left(\dfrac{1}{x-2} + \dfrac{1}{x+2} \right)$; the general term being $= \dfrac{1}{x-n} + \dfrac{1}{x+n} = \dfrac{2x}{x^2 - n^2}$. Thus the general term $= -\dfrac{2x}{n^2}$ ultimately, and the series is convergent.

§19. $\dfrac{u_n}{u_{n-1}} = \dfrac{n^p}{\lfloor n} \div \dfrac{(n-1)^p}{\lfloor n-1} = \dfrac{1}{n}\left(\dfrac{n}{n-1}\right)^p = \dfrac{1}{n}$, ultimately, whatever be
the value of p : and this being less than 1, the series is convergent.

§20. Let us compare the given series with the infinite geometrical progression $1 + r + r^2 + r^3 + \ldots$; then if $\dfrac{u_n}{r^n}$ be finite, the two series will be
both convergent, or both divergent. [Art. 288.]

Let $\dfrac{u_n}{r^n} = k$, so that $u_n = kr^n$; then $\sqrt[n]{u_n} = r\sqrt[n]{k} = rk^{\frac{1}{n}}$, so that $\sqrt[n]{u_n} = r$,
ultimately. Also if $r < 1$, the auxiliary series is convergent; if $r > 1$, the
auxiliary series is divergent; hence the proposition follows.

§21. The product, P suppose, consists of $2n - 1$ factors, and may be
written
$$P = u_1 u_2 u_3 \ldots u_{n-1} \times \dfrac{n}{n-1},$$
where
$$u_{n-1} = \dfrac{2n-2}{2n-3} \cdot \dfrac{2n-2}{2n-1}.$$

Proceeding as in Art. 296 we have,
$$\log u_{n-1} = \log\left\{1 + \dfrac{1}{(2n-3)(2n-1)}\right\}$$
$$= \log\left(1 + \dfrac{1}{4n^2}\right) = \dfrac{1}{4n^2}, \text{ ultimately};$$

hence $\log P$ is equal to the sum of a convergent series, and therefore
P is finite.

§22. When $x = 1$, the general term T_{r+1} in the expansion of
$$(1+x)^n \text{ is } \dfrac{n(n-1)(n-2)\ldots(n-r+1)}{1 \cdot 2 \cdot 3 \cdot r}.$$
Thus the numerical value of T_{r+1} is
$$\dfrac{r-n-1}{r} \cdot \dfrac{r-n-2}{r-1} \cdot \dfrac{r-n-3}{r-2} \ldots \text{ to } r \text{ factors}$$
$$= \left(1 - \dfrac{n+1}{r}\right)\left(1 - \dfrac{n+1}{r-1}\right)\left(1 - \dfrac{n+1}{r-2}\right)\ldots \text{ to } r \text{ factors};$$
$$\therefore \log T_{r+1} = \Sigma \log\left(1 - \dfrac{n+1}{q}\right),$$
where Σ denotes the sum for all values of q from 1 to r.

When r is finite, T_{r+1} is also finite; when r is infinite, $\log T_{r+1}$ is the
sum of an infinite series, which is convergent or divergent according as
the infinite series $-\Sigma\dfrac{n+1}{q}$ is convergent or divergent. [Art. 296.]

But this latter series is divergent; hence we may write
$$\log T_{r+1} = -(n+1) \times \infty.$$

If $n + 1$ is positive, $\log T_{r+1} = -\infty$, and $T_{r+1} = 0$; that is, the terms in
the expansion ultimately vanish.

If $n + 1$ is negative, $\log T_{r+1} = +\infty$, and $T_{r+1} = \infty$; that is, the terms in
the expansion become indefinitely great when n is negative and numerically greater than unity.

If $n + 1 = 0$, $n = -1$, and $(1+x)^n = (1+1)^{-1} = 1 - 1 + 1 - 1 + 1 - 1 + \ldots$,
which is an oscillating series.

EXAMPLES XXI b

§1.

Here
$$u_n = \dfrac{1 \cdot 3 \cdot 5 \ldots (4n - 7)}{2 \cdot 4 \cdot 6 \ldots (4n - 6)} \cdot \dfrac{x^{2(n-1)}}{4n - 4};$$

$$\therefore \frac{u_n}{u_{n+1}} = \frac{(4n-2)4n}{(4n-5)(4n-3)} \cdot \frac{1}{x^2} = \frac{1}{x^2}, \text{ ultimately.}$$

Hence if $x < 1$, the series is convergent; if $x > 1$, divergent.

If $x = 1$, then $n\left(\dfrac{u_n}{u_{n+1}} - 1\right) = \dfrac{n(24n - 15)}{(4n-5)(4n-3)}$, the limit of which is $\dfrac{3}{2}$;

hence the series is convergent.

§2.

Here
$$u_n = \frac{3 \cdot 6 \cdot 9 \ldots (3 \cdot \overline{n-1})}{7 \cdot 10 \cdot 13 \ldots (3n+1)} x^{n-1};$$

$$\therefore \frac{u_n}{u_{n+1}} = \frac{3n+4}{3n} \cdot \frac{1}{x} = \frac{1}{x}, \text{ ultimately.}$$

Hence if $x < 1$, the series is convergent; if $x > 1$, divergent.

If $x = 1$, $n\left(\dfrac{u_n}{u_{n+1}} - 1\right) = \dfrac{4n}{3n} = \dfrac{4}{3}$, and the series is convergent.

§3.

Here
$$u_n = \frac{2^2 \cdot 4^2 \cdot 6^2 \ldots (2n-2)^2}{3 \cdot 4 \cdot 5 \ldots (2n-1)2n} x^{2n};$$

$$\therefore \frac{u_n}{u_{n+1}} = \frac{(2n+1)(2n+2)}{(2n)^2} \cdot \frac{1}{x^2} = \frac{1}{x^2}, \text{ ultimately.}$$

Hence if $x < 1$, the series is convergent; if $x > 1$, divergent.

If $x = 1$, $n\left(\dfrac{u_n}{u_{n+1}} - 1\right) = \dfrac{n(6n+2)}{(2n)^2} = \dfrac{3}{2}$, and the series is convergent.

§4.

Here
$$u_n = \frac{n^{n-1} x^{n-1}}{\underline{|n}};$$

$$\therefore \frac{u_n}{u_{n+1}} = \frac{\underline{|n+1}}{\underline{|n}} \cdot \frac{n^{n-1}}{(n+1)^n} \cdot \frac{1}{x}$$

$$= \frac{n^{n-1}}{(n+1)^{n-1}} \cdot \frac{1}{x}$$

$$= \frac{1}{\left(1 + \dfrac{1}{n}\right)^{n-1}} \cdot \frac{1}{x} = \frac{1}{ex}, \text{ ultimately.}$$

Hence if $x < \dfrac{1}{e}$, the series is convergent; if $x > \dfrac{1}{e}$, divergent.

If $\quad x = \dfrac{1}{e}, \quad \dfrac{u_n}{u_{n+1}} = \dfrac{e}{\left(1 + \dfrac{1}{n}\right)^{n-1}},$

$$\therefore n \log \frac{u_n}{u_{n+1}} = \frac{3}{2}, \text{ ultimately.} \qquad \text{[Chap. XX. Ex. 17.]}$$

Hence the series is convergent. [Art. 302.]

§5.

Here
$$u_n = \frac{\underline{|n-1}}{n^{n-1}} x^{n-1};$$

$$\therefore \frac{u_n}{u_{n+1}} = \frac{\underline{|n-1}}{\underline{|n}} \cdot \frac{(n+1)^n}{n^{n-1}} \cdot \frac{1}{x} = \frac{(n+1)^n}{n^n} \cdot \frac{1}{x}$$

$$= \left(1 + \frac{1}{n}\right)^n \cdot \frac{1}{x} = \frac{e}{x}, \text{ ultimately.}$$

Hence if $x < e$, the series is convergent; if $x > e$, divergent.

If
$$x = e, \quad \frac{u_n}{u_{n+1}} = \left(1 + \frac{1}{n}\right)^n \cdot \frac{1}{e};$$
$$\therefore n \log \frac{u_n}{u_{n+1}} = n\left\{ n \log\left(1 + \frac{1}{n}\right) - 1 \right\}$$
$$= n\left\{ n\left(\frac{1}{n} - \frac{1}{2n^2} + \ldots\right) - 1 \right\},$$

the limit of which is $-\frac{1}{2}$; hence the series is divergent.

§6. $u_n = \dfrac{1^2 \cdot 3^2 \cdot 5^2 \ldots (2n-1)^2}{2^2 \cdot 4^2 \cdot 6^2 \ldots (2n)^2} x^{n-1}; \quad \therefore \dfrac{u_n}{u_{n+1}}$

$$= \frac{(2n+2)^2}{(2n+1)^2} \cdot \frac{1}{x} = \frac{1}{x}, \text{ ultimately.}$$

Hence if $x < 1$, the series is convergent; if $x > 1$, divergent.

If $x = 1$, $n\left(\dfrac{u_n}{u_{n+1}} - 1\right) = \dfrac{n(4n+3)}{(2n+1)^2}$, the limit of which is 1; we therefore pass to the next test.

$$\left\{ n\left(\frac{u_n}{u_{n+1}} - 1\right) - 1 \right\} \log n$$
$$= \frac{(-n-1)\log n}{(2n+1)^2} = -\frac{n\log n}{4n^2} = -\frac{\log n}{4n} = 0, \text{ ultimately.}$$

Hence the series is divergent.   [Art. 306.]

§7. $u_n =$
$$\frac{(n-2+a)(n-3+a)\ldots(1+a)a(1-a)\ldots(n-2-a)(n-1-a)}{1^2 \cdot 2^2 \cdot 3^2 \ldots (n-1)^2};$$
$$\therefore \frac{u_n}{u_{n+1}} = \frac{n^2}{(n-1+a)(n-a)} = 1, \text{ ultimately.}$$
$$n\left(\frac{u_n}{u_{n+1}} - 1\right) = \frac{n(n-a+a^2)}{(n-1+a)(n-a)} = 1, \text{ ultimately.}$$
$$\left\{ n\left(\frac{u_n}{u_{n+1}} - 1\right) - 1 \right\}\log n = \frac{(n-an+a^2n-a+a^2)\log n}{(n-1+a)(n-a)}$$
$$= \frac{(1-a+a^2)\log n}{n} = 0, \text{ ultimately.}$$

Hence the series is divergent.

§8.

Here
$$u_n = \frac{(a+nx)^n}{\lfloor n};$$
$$\therefore \frac{u_n}{u_{n+1}} = \frac{\lfloor n+1}{\lfloor n} \frac{(a+nx)^n}{(a+\overline{n+1}\cdot x)^{n+1}}$$
$$= \frac{(n+1)n^n x^n}{(n+1)^{n+1} x^{n+1}} = \frac{1}{\left(1 + \dfrac{1}{n}\right)^n x} = \frac{1}{ex}, \text{ ultimately.}$$

Hence if $x < \dfrac{1}{e}$, the series is convergent; if $x > \dfrac{1}{e}$, divergent.

This result it will be observed is quite independent of a, and if we put $a = 0$, we obtain $x + \dfrac{2^2 x^2}{\lfloor 2} + \dfrac{3^3 x^3}{\lfloor 3} + \ldots$, which is the series discussed in Art. 302; hence the conclusions obtained in that article hold for all values of a; thus when $x = \dfrac{1}{e}$, the series is divergent.

§9.

$$u_n = \frac{\alpha(\alpha+1)\dots(\alpha+n-2)\beta(\beta+1)\dots(\beta+n-2)}{1\cdot 2\cdot 3\dots(n-1)\gamma(\gamma+1)\dots(\gamma+n-2)};$$

$$\therefore \frac{u_n}{u_{n+1}} = \frac{n(\gamma+n-1)}{(\alpha+n-1)(\beta+n-1)} = 1, \text{ ultimately.}$$

$$n\left(\frac{u_n}{u_{n+1}}-1\right) = \frac{n^2(\gamma-\alpha-\beta+1)-n(\alpha-1)(\beta-1)}{(n+\alpha-1)(n+\beta-1)}$$

$$= \gamma-\alpha-\beta+1, \text{ ultimately.}$$

Hence if $\gamma-\alpha-\beta$ is positive, the series is convergent; if negative, divergent.

If $\gamma-\alpha-\beta=0$, then $n\left(\dfrac{u_n}{u_{n+1}}-1\right) = \dfrac{n^2-n(\alpha-1)(\beta-1)}{(n+\alpha-1)(n+\beta-1)} = 1$,

ultimately.

$$\left\{n\left(\frac{u_n}{u_{n+1}}-1\right)-1\right\}\log n$$

$$= -\frac{\{n(\alpha-1)(\beta-1)+n(\alpha+\beta-2)-(\alpha-1)(\beta-1)\}\log n}{(n+\alpha-1)(n+\beta-1)}$$

$$= -\frac{n\{(\alpha-1)(\beta-1)+(\alpha+\beta-2)\}\log n}{n^2} = 0, \text{ ultimately.}$$

Hence the series is divergent.

§10. $u_n = x^{n+1}\{\log(n+1)\}^q$; $\therefore \dfrac{u_n}{u_{n+1}} = \dfrac{\{\log(n+1)\}^q}{\{\log(n+2)\}^q}\cdot\dfrac{1}{x} = \dfrac{1}{x}$, ultimately.

Hence if $x<1$, the series is convergent; if $x>1$, divergent.

If $\quad x=1, \dfrac{u_n}{u_{n+1}} = \left\{\dfrac{\log(n+1)}{\log(n+2)}\right\}^q$

$$= \left\{\frac{\log n+\dfrac{1}{n}-\dots}{\log n+\dfrac{2}{n}-\dots}\right\}^q = \left\{\frac{1+\dfrac{1}{n\log n}}{1+\dfrac{2}{n\log n}}\right\}^q$$

$$= \left(1-\frac{1}{n\log n}\right)^q = 1-\frac{q}{n\log n};$$

$$\therefore n\left(\frac{u_n}{u_{n+1}}-1\right) = -\frac{q}{\log n} = 0, \text{ ultimately;}$$

hence the series is divergent whatever be the value of q.

§11.

Here $\quad u_n = \dfrac{a(a+1)(a+2)\dots(a+n-2)}{\lfloor n-1}$;

$$\therefore \frac{u_n}{u_{n+1}} = \frac{\lfloor n}{\lfloor n-1}\cdot\frac{1}{a+n-1} = \frac{n}{n+a-1} = 1, \text{ ultimately.}$$

$$n\left(\frac{u_n}{u_{n+1}}-1\right) = \frac{-n(a-1)}{n+a-1} = -a+1, \text{ ultimately.}$$

Hence if a be positive, the series is divergent; if negative, convergent. If a is zero, the series reduces to its first term 1 and is convergent.

§12. $\dfrac{u_n}{u_{n+1}} = 1$, ultimately;

$$n\left(\frac{u_n}{u_{n+1}}-1\right) = \frac{(A-a)n^k+(B-b)n^{k-1}+\dots}{n^k+an^{k-1}+bn^{k-2}+\dots}$$

$$= A-a, \text{ ultimately.}$$

Hence if $A-a>1$, the series is convergent; if $A-a<1$, divergent.

If $A - a = 1$, then $n\left(\dfrac{u_n}{u_{n+1}} - 1\right) = \dfrac{n^k + (B-b)n^{k-1} + \dots}{n^k + an^{k-1} + \dots}$;

$$\therefore \left\{ n\left(\dfrac{u_n}{u_{n+1}} - 1\right) - 1 \right\} \log n = \dfrac{(B-b-a)n^{k-1}\log n + \dots}{n^k + \dots}$$

$$= (B-b-a)\dfrac{\log n}{n} = 0, \text{ ultimately.}$$

Hence the series is divergent.

It should be noticed that the result is independent of B, b, C, c, $\dots$.

EXAMPLES XXII : Undetermined Coefficients

EXAMPLES XXII a

§1. Let $1^2 + 3^2 + 5^2 + \ldots + (2n-1)^2 = A + Bn + Cn^2 + Dn^3 + \ldots$
then $1^2 + 3^2 + 5^2 + \ldots + (2n-1)^2 + (2n+1)^2 = A + B(n+1) + C(n+1)^2 + D(n+1)^3 + \ldots$
$\therefore$ by subtraction, $(2n+1)^2 = B + C(2n+1) + D(3n^2 + 3n + 1) + \ldots$
$\therefore$ the coefficients after D all vanish, and on equating coefficients of like powers of n, we have $B + C + D = 1$, $2C + 3D = 4$, $3D = 4$;

$$\therefore D = \frac{4}{3}, \; C = 0, \; B = -\frac{1}{3}; \quad \therefore S = A - \frac{1}{3}n + \frac{4}{3}n^3.$$

Put $n = 1$; thus we find $A = 0$; hence $S = \frac{n}{3}(4n^2 - 1)$.

§2. Let $1 \cdot 2 \cdot 3 + 2 \cdot 3 \cdot 4 + 3 \cdot 4 \cdot 5 + \ldots + n(n+1)(n+2) = A + Bn + Cn^2 + Dn^3 + En^4$.
Then as in the last Example, we find
$(n+1)(n+2)(n+3) = B + C(2n+1) + D(3n^2 + 3n + 1) + E(4n^3 + 6n^2 + 4n + 1)$.
Equating coefficients, we find $E = \frac{1}{4}$, $D = \frac{3}{2}$, $C = \frac{11}{4}$, $B = \frac{3}{2}$;

$$\therefore S = A + \frac{3}{2}n + \frac{11}{4}n^2 + \frac{3}{2}n^3 + \frac{1}{4}n^4.$$

When $n = 1$, $A = 0$, and S reduces to $\dfrac{n(n+1)(n+2)(n+3)}{4}$.

§3. Let $1 \cdot 2^2 + 2 \cdot 3^2 + 3 \cdot 4^2 + \ldots + n(n+1)^2 = A + Bn + Cn^2 + Dn^3 + En^4$;
then as before
$(n+1)(n+2)^2 = B + C(2n+1) + D(3n^2 + 3n + 1) + E(4n^3 + 6n^2 + 4n + 1)$.
Equating coefficients, we find $E = \frac{1}{4}$, $D = \frac{7}{6}$, $C = \frac{7}{4}$, $B = \frac{5}{6}$;

$$\therefore S = A + \frac{5}{6}n + \frac{7}{4}n^2 + \frac{7}{6}n^3 + \frac{1}{4}n^4.$$

When $n = 1$, $A = 0$, and S reduces to $\dfrac{n(n+1)(n+2)(3n+5)}{12}$.

§4. Let $1^3 + 3^3 + 5^3 + \ldots + (2n-1)^3 = A + Bn + Cn^2 + Dn^3 + En^4$; then we find
$(2n+1)^3 = B + (2n+1)C + D(3n^2 + 3n + 1) + E(4n^3 + 6n^2 + 4n + 1)$.
Equating coefficients, we find $E = 2$, $D = 0$, $C = -1$, $B = 0$;
$$\therefore S = A - n^2 + 2n^4.$$
When $n = 1$, $A = 0$; hence $S = n^2(2n^2 - 1)$.

§5. Let $1^4 + 2^4 + 3^4 + \ldots + n^4 = A + Bn + Cn^2 + Dn^3 + En^4$.
Then in the usual way we get
$$(n+1)^4 = B + (2n+1)C + (3n^2 + 3n + 1)D$$
$$+ (4n^3 + 6n^2 + 4n + 1)E$$
$$+ F(5n^4 + 10n^3 + 10n^2 + 5n + 1);$$
whence by equating coefficients we obtain
$$F = \frac{1}{5}, \; E = \frac{1}{2}, \; D = \frac{1}{3}, \; C = 0, \; B = -\frac{1}{30};$$
$$\therefore S = A - \frac{1}{30}n + \frac{1}{3}n^3 + \frac{1}{2}n^4 + \frac{1}{5}n^5.$$
when $n = 1$, $A = 0$, and S reduces to $\dfrac{n}{30}(n+1)(2n+1)(3n^2 + 3n - 1)$.

§6. Assume $x^3 - 3px + 2q = (x+k)(x^2 + 2ax + a^2)$. Multiply out, and equate coefficients of like power of x; thus $k + 2a = 0$, $a^2k = 2q$, $-3p = 2ak + a^2$.

Eliminating k, we have $q = -a^3$, $p = a^2$; whence it follows that $p^3 = q^2$, which is the required condition.

§7. Assume $ax^3 + bx^2 + cx + d = (px+q)^3$. Equate coefficients of like powers of x, and we obtain $p^3 = a$, $3p^2q = b$, $3pq^2 = c$, $q^3 = d$, whence $\qquad\qquad b^3 = 27a^2d$, and $c^3 = 27ad^2$.

§8. Assume $a^2x^4 + bx^3 + cx^2 + dx + f^2 = (ax^2 + px + f)^2$; whence, by equating coefficients of like powers of x, we obtain $b = 2ap$, $c = p^2 + 2af$, $d = 2pf$;

$$\therefore ad = bf, \text{ and } c = \left(\frac{b}{2a}\right)^2 + 2af.$$

§9. Assume $ax^2 + 2bxy + cy^2 + 2dx + 2ey + f = (Ax + By + C)^2$. Then $\qquad A^2 = a$, $B^2 = c$, $AB = b$, $AC = d$, $BC = e$; whence the required conditions follow at once.

§10. Assume $ax^3 + bx^2 + cx + d = (x^2 + h^2)\left(ax + \dfrac{d}{h^2}\right)$; equate coefficients of like powers of x, and we obtain

$$b = \frac{d}{h^2}, \ c = ah^2; \quad \therefore \ \frac{c}{a} = \frac{d}{b}, \text{ or } bc = ad.$$

§11. Assume $x^5 - 5qx + 4r = (x^2 - 2xc + c^2)\left(x^3 + ax^2 + bx + \dfrac{4r}{c^2}\right)$.

Multiply out, and equate coefficients of like powers of x; then we have

$$a - 2c = 0; \ c^2 + b - 2ac = 0; \ ac^2 - 2bc + \frac{4r}{c^2} = 0, \ \frac{8r}{c} - bc^2 = 5q.$$

From the two first of these, $3c^2 = b$; substitute for b in the remaining equations, and we easily find $r = c^5$, $q = c^4$; $\quad \therefore r^4 = q^5$.

§12. (1) is an equation of the second degree satisfied by the three values a, b, c, d, as we easily find on trial. Therefore the equation is an identity.

(2) is solved in the same way.

§13. If $ax^2 + 2hxy + by^2 + 2gx + 2fy + c = (px + qy + r)(p'x + q'y + r')$. We have, by equating coefficients,

$$pp' = a, \ qq' = b, \ rr' = c, \ qr' + q'r = 2f,$$
$$rp' + r'p = 2g, \ pq' + p'q = 2h.$$

Multiply the last three results together; thus

$$2pp'qq'rr' + pp'(q^2r'^2 + q'^2r^2) + qq'(p^2r'^2 + p'^2r^2)$$
$$+ rr'(p^2q'^2 + p'^2q^2) = 8fgh.$$
$$2abc + a(4f^2 - 2bc) + b(4g^2 - 2ca) + c(4h^2 - 2ab) = 8fgh,$$

which reduces to

$$abc + 2fgh - af^2 - bg^2 - ch^2 = 0.$$

§14. We have $\qquad\qquad \xi = lx + my + nz$, also $\qquad x = l\xi + m\eta + n\zeta$, $y = n\xi + l\eta + m\zeta$, $z = m\xi + n\eta + l\zeta$; $\therefore$, by substitution, we have the identity

$$\xi = l(l\xi + m\eta + n\zeta) + m(n\xi + l\eta + m\zeta) + n(m\xi + n\eta + l\zeta).$$

Whence, by equating the coefficients of ξ, η, ζ on the two sides, we obtain the required relations.

§15. The sum of the products is the coefficient of x^r in the expansion of

$$(x + a)(x + a^2)(x + a^3) \ldots (x + a^n).$$

Let $(x+a)(x+a^2)\ldots(x+a^n) = x^n + A_1x^{n-1} + \ldots + A_{n-r-1}x^{r+1} + A_{n-r}x^r + \ldots$

Write $\dfrac{x}{a}$ for x, then since $\dfrac{x}{a} + a^r = \dfrac{1}{a}(x + a^{r+1})$, we have

$$\frac{1}{a^n}(x + a^2)(x + a^3)\ldots(x + a^{n+1})$$

$$= \left(\frac{x}{a}\right)^n + A_1\left(\frac{x}{a}\right)^{n-1} + A_2\left(\frac{x}{a}\right)^{n-2} + \ldots;$$

$$\therefore (x + a^2)(x + a^3) \ldots (x + a^{n+1}) = x^n + A_1 a x^{n-1} + \ldots$$
$$+ A_{n-r-1} a^{n-r-1} x^{r+1} + \ldots$$

$$\therefore (x + a)$$
$$\{x^n + A_1 a x^{n-1} + \ldots + A_{n-r-1} a^{n-r-1} x^{r+1} + A_{n-r} a^{n-r} x^r + \ldots\}$$
$$= (x + a^{n+1})$$
$$\{x^n + A_1 x^{n-1} + \ldots + A_{n-r-1} x^{r+1} + A_{n-r} x^r + \ldots\}.$$

Equate coefficients of x^{r+1}; then
$$A_{n-r} a^{n-r} + A_{n-r-1} a^{n-r} = A_{n-r-1} a^{n+1} + A_{n-r};$$
$$\therefore A_{n-r}(a^{n-r} - 1) = A_{n-r-1} a^{n-r}(a^{r+1} - 1);$$

that is, $A_{n-r} = A_{n-r-1} a^{n-r} \dfrac{a^{r+1} - 1}{a^{n-r} - 1};$

put $r + 1$ for r, then $A_{n-r-1} = A_{n-r-2} a^{n-r-1} \dfrac{a^{r+2} - 1}{a^{n-r-1} - 1};$

$$\ldots\ldots = \ldots\ldots\ldots\ldots\ldots\ldots$$

$$A_2 = A_1 a^2 \frac{a^{n-1} - 1}{a^2 - 1};$$

$$A_1 = a \frac{a^n - 1}{a - 1}, \text{ since } A_0 = 1.$$

Now multiply these results together and cancel like factors, and we easily obtain A_{n-r} in the required form.

EXAMPLES XXII b

§1. Let $\dfrac{1 + 2x}{1 - x - x^2} = a_0 + a_1 x + a_2 x^2 + a_3 x^3 + \ldots;$ then
$$1 + 2x = (1 - x - x^2)(a_0 + a_1 x + a_2 x^2 + a_3 x^3 + \ldots).$$
Then $1 = a_0$, $2 = a_1 - a_0$; whence $a_1 = 3$. The coefficients of higher powers of x are found in succession from the relation $a_n - a_{n-1} - a_{n-2} = 0$; hence $a_2 = 4$, and $a_3 = 7$; thus $\dfrac{1 + 2x}{1 - x - x^2} = 1 + 3x + 4x^2 + 7x^3 + \ldots$

§2. With the same notation we have
$$1 - 8x = (1 - x - 6x^2)(a_0 + a_1 x + a_2 x^2 + a_3 x^3 + \ldots).$$
Then $a_0 = 1$, $a_1 - a_0 = -8$; whence $a_1 = -7$. The other coefficients are determined in succession from the relation $a_n - a_{n-1} - 6a_{n-2} = 0$.

§3. We have $1 + x = (2 + x + x^2)(a_0 + a_1 x + a_2 x^2 + \ldots).$
Then $a_0 = \dfrac{1}{2}$, $2a_1 + a_0 = 1$; whence $a_1 = \dfrac{1}{4}$.
Also for values of $n > 1$, $2a_n + a_{n-1} + a_{n-2} = 0$;
$$\therefore 2a_2 = -\frac{1}{2} - \frac{1}{4}, \text{ or } a_2 = -\frac{3}{8}; \text{ and } 2a_3 = \frac{3}{8} - \frac{1}{4}; \text{ or } a_3 = \frac{1}{16};$$
$$\therefore \frac{1 + x}{2 + x + x^2} = \frac{1}{2} + \frac{1}{4}x - \frac{3}{8}x^2 + \frac{1}{16}x^3 + \ldots$$
Example 4 may be solved in a similar way.

§5. Let the required expansion be $b_0 + b_1 x + b_2 x^2 + b_3 x^3 + \ldots$

Then $1 = (1 + ax - ax^2 - x^3)(b_0 + b_1 x + b_2 x^2 + b_3 x^3 + \ldots);$
$$\therefore b_0 = 1, \; b_1 + b_0 a = 0; \text{ whence } b_1 = -a.$$
Also $b_2 + b_1 a - b_0 a = 0; \text{ whence } b_2 = a(a + 1).$
And $b_3 + b_2 a - b_1 a - b_0 = 0; \text{ whence } b_3 = 1 - 2a^2 - a^3;$
$$\therefore \frac{1}{1 + ax - ax^2 - x^3} = 1 - ax + a(a + 1)x^2 - (a^3 + 2a^2 - 1)x^3 + \ldots$$

§6. By putting $n = 1, 2, 3, \ldots$ successively we see that the required expansion will have coefficients $1, 4, 7, 10, \ldots$; that is,
$$a + bx = (1 - x)^2(1 + 4x + 7x^2 + 10x^3 + \ldots);$$
$\therefore$ by equating coefficients we have $a = 1, b = 4 - 2 = 2$.

§7. As in Example 6 we find that
$$a + bx + cx^2 = (1 - x)^3(1 + 2x + 5x^2 + 10x^3 + \ldots);$$
$\therefore a = 1, \ b = -3 + 2, \ c = 3 - 6 + 5;$ or $a = 1, \ b = -1, \ c = 2$.

§8. Since $y = 0$ when $x = 0$, we may assume $y = A_1 x + A_2 x^2 + A_3 x^3 + \ldots$; substitute this value for y in the given relation; thus
$$(A_1 x + A_2 x^2 + A_3 x^3 + \ldots)^2 + 2(A_1 x + A_2 x^2 + A_3 x^3 + \ldots) = x(1 + A_1 x + A_2 x^2 + \ldots).$$
Since this is an identity, we may equate the coefficients of powers of x; thus we obtain $2A_1 = 1$, or $A_1 = \dfrac{1}{2}$; $A_1^2 + 2A_2 = A_1$, whence $A_2 = \dfrac{1}{8}$; $2A_1 A_2 + 2A_3 = A_2$, whence $A_3 = 0$; $A_2^2 + 2A_1 A_3 + 2A_4 = A_3$, whence $A_4 = -\dfrac{1}{128}$;
$$\therefore y = \frac{1}{2}x + \frac{1}{8}x^2 - \frac{1}{128}x^4 + \ldots$$

§9. Here $y = 0$, when $x = 0$. Also y changes sign with x; therefore we may assume
$$x = A_1 y + A_3 y^3 + A_5 y^5 + A_7 y^7 + \ldots$$
Now proceed as in the last Example, and we get
$$c(A_1 y + A_3 y^3 + A_5 y^5 + \ldots)^3 + a(A_1 y + A_3 y^3 + \ldots) - y = 0;$$
$\therefore$ equating coefficients, $aA_1 - 1 = 0$; or $A_1 = \dfrac{1}{a}$;
$$cA_1^3 + aA_3 = 0; \text{ whence } A_3 = -\frac{c}{a^4}.$$
$$aA_5 + 3cA_1^2 A_3 = 0; \text{ whence } A_5 = \frac{3c^2}{a^5} \cdot \frac{1}{a^2} = \frac{3c^2}{a^7}.$$
$$aA_7 + 3cA_1^2 A_5 + 3cA_1 A_3^2 = 0;$$
$$\therefore A_7 = -\frac{3c}{a} \cdot \frac{1}{a^2} \cdot \frac{3c^2}{a^7} - \frac{3c}{a} \cdot \frac{1}{a} \cdot \frac{c^2}{a^8} = -\frac{12c^3}{a^{10}};$$
and so on; thus
$$x = \frac{y}{a} - \frac{cy^3}{a^4} + \frac{3c^2 y^5}{a^7} - \frac{12c^3 y^7}{a^{10}} \ldots$$
Now put $c = 1, \ y = 1, \ a = 100$; then $x = \dfrac{1}{100} - \dfrac{1}{(100)^4} + \dfrac{3}{(100)^7} - \ldots$
becomes the solution of
$$x^3 + 100x - 1 = 0;$$
$\therefore x = .01 - .00000001 + \ldots = .00999999$ approximately; also since the first term rejected is $\dfrac{3}{(100)^7}$, and this when expressed as a decimal begins with 13 ciphers, the value found for x is accurate to 13 places of decimals.

§10. Assume $(1 + x)(1 + ax)(1 + a^2 x) \ldots = 1 + A_1 x + A_2 x^2 + \ldots$
Change x into ax; then we have, as in XXII. a. 15,
$$(1 + x)(1 + A_1 ax + A_2 a^2 x^2 + \ldots) = 1 + A_1 x + A_2 x^2 + \ldots$$
Equate coefficients of x^r; thus $A_r a^r + A_{r-1} a^{r-1} = A_r$;
$$\therefore A_r(1 - a^r) = A_{r-1} a^{r-1};$$
$$\therefore A_r = \frac{a^{r-1}}{1 - a^r} A_{r-1}$$
$$= \frac{a^{r-1} \cdot a^{r-2}}{(1 - a^r)(1 - a^{r-1})} A_{r-2}$$
$$= \frac{a^{r-1} \cdot a^{r-2} \cdot a^{r-3}}{(1 - a^r)(1 - a^{r-1})(1 - a^{r-2})} A_{r-3}; \text{ and so on.}$$

Thus
$$A_r = \frac{a^{1+2+3+\dots+(r-1)}}{(1-a^r)(1-a^{r-1})\dots(1-a)} A_0$$
$$= \frac{a^{\frac{1}{2}r(r-1)}}{(1-a)(1-a^2)\dots(1-a^r)}, \text{ since } A_0 = 1.$$

§11. Let the expansion be $A_0 + A_1x + A_2x^2 + \dots + A_nx^n + \dots$
Multiply each side by $1 - ax$; thus
$$\frac{1}{(1-a^2x)(1-a^3x)\dots} = A_0 + (A_1 - A_0a)x + \dots + (A_n - A_{n-1}a)x^n + \dots$$
But by writing ax for x we see that the expression on the left
$$= A_0 + A_1ax + A_2a^2x^2 + \dots + A_na^nx^n + \dots;$$
$$\therefore A_na^n = (A_n - A_{n-1}a); \text{ thus } A_n = \frac{aA_{n-1}}{1-a^n}, \text{ &c.}$$

And finally
$$A_n = \frac{a^n}{(1-a)(1-a^2)\dots(1-a^n)}.$$

§12. (1) Proceed as in Ex.2, Art. 314, and we find
$$\frac{n^{n+1}}{\underline{|n+1}} - \frac{n(n-1)^{n+1}}{\underline{|n+1}} + \frac{n(n-1)}{\underline{|2}} \cdot \frac{(n-2)^{n+1}}{\underline{|n+1}} - \dots \text{ to } n \text{ terms} = \text{ the coefficient}$$
of x^{n+1} in $\left(x + \dfrac{x^2}{2} + \dfrac{x^3}{\underline{|3}} + \dots\right)^n = \dfrac{1}{2}n.$

(2) $(e^x - 1)^{n+1} = \left(x + \dfrac{x^2}{\underline{|2}} + \dfrac{x^3}{\underline{|3}} + \dots\right)^{n+1} = x^{n+1} + \text{ terms containing}$

higher powers of x. Expand the left-hand side and multiply all through
by e^{-x}; then
$$e^{nx} - (n+1)e^{(n-1)x} + \frac{(n+1)n}{1 \cdot 2}e^{(n-2)x}$$
$$- \dots \text{ to } n + 2 \text{ terms}$$
$$= e^{-x}(x^{n+1} + \dots).$$
The last two terms of the series on the left are
$$(-1)^n(n+1)e^0 + (-1)^{n+1}e^{-x};$$
$\therefore$ the coefficient of x^n in the series is
$$\frac{n^2}{\underline{|n}} - (n+1)\frac{(n-1)^n}{\underline{|n}} + \frac{(n+1)n}{1 \cdot 2} \cdot \frac{(n-2)^n}{\underline{|n}}$$
$$- \dots \text{ to } n \text{ terms } + (-1)^{n+1}\frac{(-1)^n}{\underline{|n}};$$
and this is equal to zero, since on the right-hand side there is no term
containing x^n. Transpose and multiply up by $\underline{|n}$.

(3) We have $e^x(1 - e^x)^n = e^x - ne^{2x} + \dfrac{n(n-1)}{1 \cdot 2}e^{3x} - \dots$ to $n+1$ terms;
$\therefore$ the coefficient of x^n in the expression on the right is
$$\frac{1}{\underline{|n}} - n \cdot \frac{2^n}{\underline{|n}} + \frac{n(n-1)}{1 \cdot 2} \cdot \frac{3^n}{\underline{|n}} - \dots \text{ to } n+1 \text{ terms.}$$
Again the expression on the left is $(-1)^ne^x(e^x - 1)^n$, which may be
written $(-1)^n(1 + x + \dots)(x^n + \dots)$; thus the coefficient of x^n is $(-1)^n$.
Equate the two coefficients, multiply up by $\underline{|n}$, and the required result
follows.

(4)
$$e^{px}(e^x - 1)^n = e^{px}\left\{e^{nx} - ne^{(n-1)x} + \frac{n(n-1)}{1 \cdot 2}e^{(n-2)x} - \dots\right\}$$
$$= e^{(n+p)x} - ne^{(n+p-1)x} + \frac{n(n-1)}{1 \cdot 2}e^{(n+p-2)x} - \dots$$

Equate the coefficients of x^n, and the required result follows.

EXAMPLES XXIII : Partial Fractions

§1. Assume $\dfrac{7x-1}{1-5x+6x^2} = \dfrac{A}{1-3x} + \dfrac{B}{1-2x}$.

Then $\qquad 7x-1 = A(1-2x) + B(1-3x);$

∴ equating coefficients, $A+B = -1$, $2A+3B = -7$; whence $A = 4$, $B = -5$;

$$\therefore \dfrac{7x-1}{1-5x+6x^2} = \dfrac{4}{1-3x} - \dfrac{5}{1-2x}.$$

§2. Assume $\dfrac{46+13x}{12x^2-11x-15} = \dfrac{A}{4x+3} + \dfrac{B}{3x-5}$.

Then $\qquad 46+13x = A(3x-5) + B(4x+3);$

∴ $3A+4B = 13$, $-5A+3B = 46$; whence $A = -5$, $B = 7$;

$$\therefore \dfrac{46+13x}{12x^2-11x-15} = \dfrac{7}{3x-5} - \dfrac{5}{4x+3}.$$

§3. Here $\dfrac{1+3x+2x^2}{(1-2x)(1-x^2)} = \dfrac{(1+2x)(1+x)}{(1-2x)(1-x^2)} = \dfrac{1+2x}{(1-2x)(1-x)}.$

Assume $\qquad \dfrac{1+2x}{(1-2x)(1-x)} = \dfrac{A}{1-2x} + \dfrac{B}{1-x}.$

$$\therefore 1+2x = A(1-x) + B(1-2x).$$

Put $x = 1$; then $3 = -B$; also $A + B = 1$; thus $A = 4$.

§4. Assume $\dfrac{x^2-10x+13}{(x-1)(x-2)(x-3)} = \dfrac{A}{x-1} + \dfrac{B}{x-2} + \dfrac{C}{x-3}$; whence by

putting $x = 1$, $x = 2$, $x = 3$ successively, we get $A = 2$, $B = 3$, $C = -4$;

$$\therefore \text{ the expression } = \dfrac{2}{x-1} + \dfrac{3}{x-2} - \dfrac{4}{x-3}.$$

§5. Dividing out we have $\dfrac{2x^3+x^2-x-3}{x(x-1)(2x+3)} = 1 + \dfrac{2x-3}{x(x-1)(2x+3)}.$

Now assume $\dfrac{2x-3}{x(x-1)(2x+3)} = \dfrac{A}{x} + \dfrac{B}{x-1} + \dfrac{C}{2x+3}$; we find

$$A = 1, \ B = -\dfrac{1}{5}, \ C = -\dfrac{8}{5};$$

$$\therefore \text{ the expression } = 1 + \dfrac{1}{x} - \dfrac{1}{5(x-1)} - \dfrac{8}{5(2x+3)}.$$

§7. By division we find the given expression

$$= x - 2 - \dfrac{7x-4}{(x+1)^2(x-3)}.$$

Assume $\dfrac{7x-4}{(x+1)^2(x-3)} = \dfrac{A}{x+1} + \dfrac{B}{(x+1)^2} + \dfrac{C}{x-3}$; we find

$$A = -\dfrac{17}{16}, \ B = \dfrac{11}{4}, \ C = \dfrac{17}{16}.$$

$$\therefore \text{ the expression } = x - 2 - \dfrac{17}{16(x-3)} - \dfrac{11}{4(x+1)^2} + \dfrac{17}{16(x+1)}.$$

§8. The expression $= \dfrac{26x(x+8)}{(x^2+1)(x+5)}$. Now assume

$$\dfrac{x+8}{(x^2+1)(x+5)} = \dfrac{A}{x+5} + \dfrac{Bx+C}{x^2+1}; \text{ then } x+8 = A(x^2+1) + (Bx+C)(x+5);$$

whence $\qquad A = \dfrac{3}{26}, \ B = -\dfrac{3}{26}, \ C = \dfrac{41}{26}.$

$$\therefore \dfrac{26x^2+208x}{(x^2+1)(x+5)} = x\left(\dfrac{3}{x+5} - \dfrac{3x-41}{x^2+1}\right).$$

§9. Assume $\dfrac{2x^2 - 11x + 5}{(x-3)(x^2+2x-5)} = \dfrac{A}{x-3} + \dfrac{Bx+C}{x^2+2x-5}$.

Then $\qquad 2x^2 - 11x + 5 = (A+B)x^2 + (2A - 3B + C)x - (5A+3C)$;

whence by equating coefficients $A = -1$, $B = 3$, $C = 0$;

$$\therefore \text{ the expression } = \dfrac{3x}{x^2+2x-5} - \dfrac{1}{x-3};$$

§10. Put $x - 1 = z$; then the expression

$$= \dfrac{3(z+1)^3 - 8(z+1)^2 + 10}{z^4}$$

$$= \dfrac{3z^3 + z^2 - 7z + 5}{z^4} = \dfrac{3}{z} + \dfrac{1}{z^2} - \dfrac{7}{z^3} + \dfrac{5}{z^4}$$

$$= \dfrac{3}{x-1} + \dfrac{1}{(x-1)^2} - \dfrac{7}{(x-1)^3} + \dfrac{5}{(x-1)^4}.$$

§11. Put the expression $= \dfrac{A}{x-1} + \dfrac{f(x)}{(x+1)^4}$, and proceed as in Art. 317.

We thus find $A = 1$, and $f(x) = 1 - x^3$.

Now $\qquad \dfrac{1-x^3}{(1+x)^4} = \dfrac{2 - 3z + 3z^2 - z^3}{z^4}$, if $1 + x = z$,

$$= \dfrac{2}{z^4} - \dfrac{3}{z^3} + \dfrac{3}{z^2} - \dfrac{1}{z};$$

$$\therefore \text{ the expression } = \dfrac{1}{x-1} - \dfrac{1}{x+1} + \dfrac{3}{(x+1)^2}$$

$$- \dfrac{3}{(x+1)^3} + \dfrac{2}{(x+1)^4}.$$

§12. The expression $= \dfrac{4}{3(1+7x)} - \dfrac{1}{3(1+4x)}$, and the general term is

$$\dfrac{(-1)^r}{3}\{4.7^r - 4^r\}x^r.$$

§13. The expression $= \dfrac{11}{3(1-x)} - \dfrac{4}{3(2+x)}$, and the general term is

$$\dfrac{1}{3}\left\{11 - \dfrac{(-1)^r}{2^{r-1}}\right\}x^r.$$

§14. $\dfrac{x^2 + 7x + 3}{x^2 + 7x + 10} = 1 - \dfrac{7}{x^2 + 7x + 10} = 1 + \dfrac{7}{3(x+5)} - \dfrac{7}{3(x+2)}$. The general

term is $(-1)^r \cdot \dfrac{7}{3}\left\{\dfrac{1}{5^{r+1}} - \dfrac{1}{2^{r+1}}\right\}x^r.$

§15. The expression $= -\dfrac{1}{1+x} + \dfrac{1}{1-x} - \dfrac{4}{1-2x}$, and the general term is

$\{1 - (-1)^r - 2^{r+2}\}x^r$ or $\{1 + (-1)^{r-1} - 2^{r+2}\}x^r.$

§16. The expression $= \dfrac{4}{3(1+2x)} - \dfrac{1}{3(1-x)} + \dfrac{3}{(1-x)^2}$, and the general

term is $\dfrac{1}{3}\{9r + 8 + (-1)^r 2^{r+2}\}x^r.$

§17. The expression $= \dfrac{1}{4(1-4x)} + \dfrac{11}{4(1-4x)^2}$, and the general term is

$4^{r-1}(12 + 11r)x^r.$

§18. The expression $= \dfrac{2}{1+x} + \dfrac{3}{(1+x)^2} - \dfrac{6}{2+3x}$. The general term is

$$(-1)^r\left\{3r + 5 - \dfrac{3^{r+1}}{2^r}\right\}x^r.$$

§19. The expression $= \dfrac{3}{2(x-1)} + \dfrac{1-3x}{2(1+x^2)}$. The general term is

$\frac{1}{2}\left\{(-1)^{\frac{r}{2}} - 3\right\} x^r$ if r is even and $-\frac{3}{2}\left\{1 + (-1)^{\frac{r-1}{r}}\right\} x^r$ if r is odd.

§20. If $z = 1 - x$,

$$\text{the expression} = \frac{z + 2(1 - z)^2}{z^3} = \frac{2 - 3z + 2z^2}{z^3}$$

$$= \frac{2}{z^3} - \frac{3}{z^2} + \frac{2}{z}$$

$$= 2(1 - x)^{-3} - 3(1 - x)^{-2} + 2(1 - x)^{-1};$$

$\therefore$ the coefficient of x^r is $(r + 1)(r + 2) - 3(r + 1) + 2$, or $r^2 + 1$.

§21. Assume $\dfrac{1}{(1 - ax)(1 - bx)(1 - cx)} = \dfrac{A}{1 - ax} + \dfrac{B}{1 - bx} + \dfrac{C}{1 - cx};$

$\therefore 1 = A(1 - bx)(1 - cx) + B(1 - ax)(1 - cx) + C(1 - ax)(1 - bx).$

By putting in succession $1 - ax = 0$, $1 - bx = 0$, $1 - cx = 0$, we find that

$A = \dfrac{a^2}{(a - b)(a - c)}$, and similar expressions for B and C;

$\therefore$ the required term is the $(r + 1)^{th}$ term in the expansion of

$$\frac{a^2}{(a - b)(a - c)}(1 - ax)^{-1} + \ldots + \ldots$$

$$= \left\{\frac{a^{r+2}}{(a - b)(a - c)} + \frac{b^{r+2}}{(b - c)(b - a)} + \frac{c^{r+2}}{(c - a)(c - b)}\right\} x^r.$$

§22. Put $\dfrac{3 - 2x^2}{(2 - 3x + x^2)^2} = \dfrac{A}{(2 - x)^2} + \dfrac{B}{2 - x} + \dfrac{C}{(1 - x)^2} + \dfrac{D}{1 - x},$

$\therefore 3 - 2x^2 = A(1 - x)^2 + B(2 - x)(1 - x)^2$

$$+ C(2 - x)^2 + D(2 - x)^2(1 - x).$$

Then $A + 2B + 4C + 4D = 3; -B - D = 0$. Also $x = 2$ gives $A = -5$; and $x = 1$ gives $C = 1$.

$\therefore 2B + 4D = 4$, and $B = -D$, whence $D = 2, B = -2$;

$$\therefore \frac{3 - 2x^2}{(2 - 3x + x^2)^2} = -\frac{5}{4\left(1 - \frac{x}{2}\right)^2} - \frac{1}{1 - \frac{x}{2}} + \frac{1}{(1 - x)^2} + \frac{2}{1 - x};$$

$\therefore$ the coefficient of $x^r = -\dfrac{5}{4} \cdot \dfrac{r + 1}{2^r} - \dfrac{1}{2^r} + (r + 1) + 2$

$$= 3 + r - \frac{5r + 9}{2^{r+2}}.$$

§23. (1) the nth term is $\dfrac{x^{n-1}}{(1 + x^n)(1 + x^{n+1})}$, and this may be put in the form

$$\frac{1}{x(1 - x)}\left\{\frac{1}{1 + x^{n+1}} - \frac{1}{1 + x^n}\right\}.$$

Similarly each term of the series may be decomposed, and on addition we find that all the terms disappear except one at the beginning and one at the end; thus

$$S = \frac{1}{x(1 - x)}\left\{\frac{1}{1 + x^{n+1}} - \frac{1}{1 + x}\right\}.$$

(2)The nth term is $\dfrac{a^{n-1}x(1 - a^n x)}{(1 + a^{n-1}x)(1 + a^n x)(1 + a^{n+1}x)}.$

Assume this $= \dfrac{A}{1 + a^{n-1}x} + \dfrac{B}{1 + a^n x} + \dfrac{C}{1 + a^{n+1}x}$; then in the usual way

we find $A = -\dfrac{1}{(a - 1)^2} = C, B = \dfrac{2}{(a - 1)^2}$. Thus the nth term

$$= \frac{1}{(a - 1)^2}\left\{-\frac{1}{1 + a^{n-1}x} + \frac{2}{1 + a^n x} - \frac{1}{1 + a^{n+1}x}\right\}.$$

If we decompose each term of the series in this way, we find on addition that all the terms disappear except two at the beginning and two at the end.

Thus the sum

$$= \frac{1}{(a-1)^2} \left\{ \frac{1}{1+a^n x} - \frac{1}{1+a^{n+1} x} - \frac{1}{1+x} + \frac{1}{1+ax} \right\}.$$

§24. The nth term

$$= \frac{x^{2n-2}}{(1-x^{2n-1})(1-x^{2n+1})} = \frac{1}{x(1-x^2)} \left\{ \frac{1}{1-x^{2n-1}} - \frac{1}{1-x^{2n+1}} \right\}.$$

Thus the series may be written

$$\frac{1}{x(1-x^2)} \left\{ \frac{1}{1-x} - \frac{1}{1-x^3} + \frac{1}{1-x^3} - \frac{1}{1-x^5} + \dots \text{ to inf. } \right\};$$

and this reduces to $\dfrac{1}{x(1-x)(1-x^2)}$.

§25. The nth term can be put in the form

$$\frac{1}{(1-x)^2} \left[\frac{1}{1-x^n} - \frac{2x^{n+1}}{1-x^{n+1}} + \frac{x^{n+2}}{1-x^{n+2}} \right];$$

and, as in Ex. 23, the sum is found to be

$$\frac{1}{(1-x)^2} \left\{ \frac{x}{1-x} - \frac{x^2}{1-x^2} - \frac{x^{n+1}}{1-x^{n+1}} + \frac{x^{n+2}}{1-x^{n+2}} \right\}.$$

§26. We have $\dfrac{1}{(1-ax)(1-bx)(1-cx)} = 1 + S_1 x + S_2 x^2 + \dots + S_n x^n + \dots,$

where $S_n =$ the sum of the homogeneous products of n dimensions which can be formed of a, b, c and their powers.   [Art. 190]

Assume $\dfrac{1}{(1-ax)(1-bx)(1-cx)} = \dfrac{A}{1-ax} + \dfrac{B}{1-bx} + \dfrac{C}{1-cx}$; then by putting $1-ax$, $1-bx$, $1-cx$ equal to zero successively, we find $A = \dfrac{a^2}{(a-b)(a-c)}$, and similar values for B and C.

$$\therefore \frac{1}{(1-ax)(1-bx)(1-cx)} = \frac{a^2}{(a-b)(a-c)} (1-ax)^{-1} + \dots + \dots;$$

∴ the coefficient of x^n is

$$\frac{a^{n+2}}{(a-b)(a-c)} + \frac{b^{n+2}}{(b-c)(b-a)} + \frac{c^{n+2}}{(c-a)(c-b)},$$ which may easily be thrown into the required form.

EXAMPLES XXIV : Recurring Series

§1. Let $1-px-qx^2$ be the scale of relation; then $13-9p+5q = 0, 9-5p+q = 0$; whence $p = 2, q = 1$.

Now let
$$S = 1 + 5x + 9x^2 + 13x^3 + \ldots,$$

then
$$-2xS = \quad -2x - 10x^2 - 18x^3 - \ldots,$$
$$x^2 S = \quad\quad x^2 + 5x^3 + \ldots;$$
$$\therefore S(1 - 2x + x^2) = 1 + 3x;$$

$$\therefore S = \frac{1 + 3x}{1 - 2x + x^2} = (1 + 3x)(1 - x)^{-2}$$
$$= (1 + 3x)\{1 + 2x + 3x^2 + \ldots + (r + 1)x^r + \ldots\};$$
$$\therefore \text{ the general term } = (3r + r + 1)x^r = (4r + 1)x^r.$$

§2. Here $p = -1, q = -2$, $S = \dfrac{2 + x}{1 + x - 2x^2} = \dfrac{1}{1 + 2x} + \dfrac{1}{1 - x}$;
$$\therefore \text{ the general term } = \{1 + (-2)^r\} x^r$$

§3. Here $p = 3, q = -2$, $S = \dfrac{2 - 3x}{1 - 3x + 2x^2} = \dfrac{1}{1 - x} + \dfrac{1}{1 - 2x}$;
$$\therefore \text{ the general term } = (1 + 2^r)x^r.$$

§4. Here the term involving x^3 is absent;
$$\therefore 27 - 0p - 9q = 0, \ 0 - 9p + 6q = 0; \ \therefore p = 2, q = 3.$$

Let
$$S = 7 - 6x + 9x^2 + 0x^3 + \ldots,$$

then
$$-2xS = \quad -14x + 12x^2 - 18x^3 - \ldots$$
$$-3x^2 S = \quad\quad -21x^2 + 18x^3 - \ldots;$$
$$\therefore S(1 - 2x - 3x^2) = 7 - 20x,$$

$$\therefore S = \frac{7 - 20x}{1 - 2x - 3x^2} = \frac{1}{4(1 - 3x)} + \frac{27}{4(1 + x)};$$
$$\therefore \text{ the general term } = \frac{1}{4} \{3^r + 27(-1)^r\} x^r.$$

§5. Let $(1 - px - qx^2 - rx^3)$ be the scale of relation; then
$276 - 98p - 36q - 14r = 0, \ 98 - 36 - 14q - 6r = 0, \ 36 - 14p - 6q - 3r = 0,$
whence $p = 6, q = -11, r = 3$;
$\therefore$ the scale of relation is $1 - 6x + 11x^2 - 6x^3$, and the generating function

$$= \frac{3 - 12x + 11x^2}{1 - 6x + 11x^2 - 6x^3} = \frac{1}{1 - 3x} + \frac{1}{1 - 2x} + \frac{1}{1 - x};$$

and the general term $= (3^r + 2^r + 1)x^r$.

§6. Proceed as in Art. 329; we find the scale of relation $= 1 - 5x + 6x^2$.

The generating function $= \dfrac{2 - 5x}{1 - 5x + 6x^2} = \dfrac{1}{1 - 3x} + \dfrac{1}{2x}$.

The $(r + 1)^{th}$ or general term of the given series is $3^r + 2^r$.

The nth term $= 3^{n-1} + 2^{n-1}$.

The sum of n terms $= \Sigma 3^{n-1} + \Sigma 2^{n-1} = \dfrac{1}{2}(3^n - 1) + (2^n - 1)$.

§7. The scale of relation is $1 - 5x + 6x^2$; and the generating function

$$= \frac{5x - 1}{1 - 5x + 6x^2} = \frac{2}{1 - 3x} - \frac{3}{1 - 2x}.$$

The nth term $= 2(3x)^{n-1} - 3(2x)^{n-1}$.

The sum of n terms $= \dfrac{2(1 - 3^n x^n)}{1 - 3x} - \dfrac{3(1 - 2^n x^n)}{1 - 2x}$.

§8. The scale of relation is $1 - 7x + 12x^2$. The generating function

$$= \frac{2 - 7x}{1 - 7x + 12x^2} = \frac{1}{1 - 4x} + \frac{1}{1 - 3x}.$$

The nth term $= (4^{n-1} + 3^{n-1})x^{n-1}$;

$\therefore$ the sum of n terms $= \Sigma 4^{n-1}x^{n-1} + \Sigma 3^{n-1}x^{n-1} = \dfrac{1 - 4^n x^n}{1 - 4x} + \dfrac{1 - 3^n x^n}{1 - 3x}$.

§9. The scale of relation is $1 - 6x + 11x^2 - 6x^3$. The generating function

$$= \frac{1 - 4x + 5x^2}{(1 - x)(1 - 2x)(1 - 3x)} = \frac{1}{1 - x} - \frac{1}{1 - 2x} + \frac{1}{1 - 3x}.$$

The nth term $= (1 - 2^{n-1} + 3^{n-1})x^{n-1}$;

$\therefore$ the sum of n terms $= \dfrac{1 - x^n}{1 - x} - \dfrac{1 - 2^n x^n}{1 - 2x} + \dfrac{1 - 3^n x^n}{1 - 3x}$.

§10. The scale of relation of the series $-\dfrac{3}{2} + 2x + 0x^2 + 8x^3 + \ldots$ is $1 - 3x - 4x^2$, and the generating function is

$$\frac{13x - 3}{2(1 - 3x - 4x^2)}, \text{ or } \frac{1}{10}\left\{\frac{1}{1 - 4x} - \frac{16}{1 + x}\right\};$$

$\therefore$ the nth term $= \dfrac{1}{10}\{4^{n-1} - (-1)^{n-1}16\}x^{n-1}$

$$= \left\{\frac{8}{5}(-1)^n + \frac{2^{2n-3}}{5}\right\}x^{n-1}.$$

Put $x = 1$, then the nth term of the given series $= \dfrac{1}{5}\{8(-1)^n + 2^{2n-3}\}$;

and the sum to n terms $= \dfrac{4}{5}\{(-1)^n - 1\} + \dfrac{1}{30}(2^{2n} - 1)$.

§11. If we denote the series by $u_1 + u_2 + \ldots + u_n$, we have in the first case

$$u_n - u_{n-1} = 2n - 1, \quad u_{n-1} - u_{n-2} = 2n - 3;$$

$$\therefore u_n - 2u_{n-1} + u_{n-2} = 2 = u_{n-1} - 2u_{n-2} + u_{n-3};$$

$$\therefore u_n - 3u_{n-1} + 3u_{n-2} - u_{n-3} = 0,$$

which is a relation connecting any four consecutive terms.

In the second case we have $u_n - u_{n-1} = 3n^2 - 3n + 1$;

$$\therefore u_n - 2u_{n-1} + u_{n-2} = 3(2n - 1) - 3 = 6n - 6;$$

$$\therefore u_n - 3u_{n-1} + 3u_{n-2} - u_{n-3} = 6 = u_{n-1} - 3u_{n-2} + 3u_{n-3} - u_{n-4};$$

$$\therefore u_n - 4u_{n-1} + 6u_{n-2} - 4u_{n-3} + u_{n-4} = 0,$$

which is a relation connecting any five consecutive terms

§12. The sum to infinity is $\dfrac{a_0 + (a_1 - pa_0)x}{1 - px - qx^2}$ by Art. 326.

Also the sum to infinity beginning with the $(n + 1)^{th}$ term is

$$\frac{a_n x^n + (a_{n+1} - pa_n)x^{n+1}}{1 - px - qx^2};$$

$\therefore$ by subtraction the sum to n terms

$$\frac{a_0 + (a_1 - pa_0)x}{1 - px - qx^2} - \frac{a_n x^n + (a_{n+1} - pa_n)x^{n+1}}{1 - px - qx^2};$$

and since $a_n = pa_{n-1} + qa_{n-5}$, and $a_{n+1} - pa_n = qa_{n-1}$, this result agrees with that in Art. 325.

§13. The scale of relation is $1 - 3x^2 + 2x^3$, and this $= (1 + 2x)(1 - x)^2$;

$\therefore$ we find the generating function

$$= \frac{3 - x + 4x^2}{1 - 3x^2 + 2x^3} = \frac{2}{(1 - x)^2} - \frac{1}{1 - x} + \frac{2}{1 + 2x}.$$

Hence the mth term $= \{2m - 1 + (-1)^{m-1}2^m\}x^m$.

Put $x = 1$; then the sum of m terms

$= (1 + 3 + 5 + \ldots + 2m - 1) + \{2 - 2^2 + 2^3 - \ldots + (-1)^{m-1}2^m\}$

$$= m^2 + 2 \cdot \frac{1 - (-2)^m}{1 + 2} = m^2 + \frac{2}{3}\{1 - (-2)^m\};$$

$$\therefore \text{ the sum of } 2n + 1 \text{ terms} = (2n+1)^2 + \frac{2}{3}(2^{2n+1} + 1).$$

§14. Let the generating functions of the two series be

$\dfrac{A}{1 + px + qx^2}$ and $\dfrac{B}{1 + rx + sx^2}$ respectively; then the sum of the two infinite series is

$$\frac{A}{1 + px + qx^2} + \frac{B}{1 + rx + sx^2}.$$

This is therefore the generating function of the series whose general term is $(a_n + b_n)x^n$, and on reduction we find the generating function has for its denominator $1 + (p + r)x + (q + s + pr)x^2 + (qr + ps)x^3 + qsx^4$, which is therefore the scale of relation of the new series.

§15. Let the given series be $u_0 + u_1 + u_2 + \ldots$, and let the scale of relation contain k constants; so that

$$u_n = p_1 u_{n-1} + p_2 u_{n-2} + \ldots + p_k u_{n-k}.$$

Let $\qquad\qquad\qquad S_n = u_1 + u_2 + u_3 + \ldots + u_n;$

then $S_n - S_{n-1} = u_n$

$$= p_1 u_{n-1} + p_2 u_{n-2} + \ldots + p_k u_{n-k}$$
$$= p_1(S_{n-1} - S_{n-2}) + p_2(S_{n-2} - S_{n-3}) + \ldots$$
$$+ p_k(S_{n-k} - S_{n-k-1});$$
$$\therefore S_n = (p_1 + 1)S_{n-1} - (p_1 - p_2)S_{n-2} - (p_2 - p_3)S_{n-3} - \ldots$$
$$- (p_{k-1} - p_k)S_{n-k} - p_k S_{n-k-1}.$$

Thus if u_n is formed from the preceding k terms of the series $u_1 + u_2 + u_3 + \ldots$, S_n is formed from the preceding $k + 1$ terms of the series $S_1, S_2, S_3 \ldots$.

EXAMPLES XXV a

Examples $4-11$ may be worked as in Art. 333. It will be sufficient here to give the two following solutions.

§6.

3	1189	3927	3
3	109	360	3
	10	33	3
3	1	3	3

∴ the successive quotients are $3, 3, 3, 3, 3, 3, 3$; and

$$\frac{1189}{3927} = \frac{1}{3+} \frac{1}{3+} \frac{1}{3+} \frac{1}{3+} \frac{1}{3+} \frac{1}{3+} \frac{1}{3};$$

and the first four convergents are $\dfrac{1}{3}, \dfrac{3}{10}, \dfrac{10}{33}, \dfrac{33}{109}$.

§10. $\cdot 3029 = \dfrac{3029}{10000}$.

3	3029	10000	3
6	290	913	3
2	32	43	1
10	10	11	1
		1	

∴ the successive quotients are $3, 3, 3, 6, 1, 2, 1, 10$; and

$$\cdot 3029 = \frac{1}{3+} \frac{1}{3+} \frac{1}{3+} \frac{1}{6+} \frac{1}{1+} \frac{1}{2+} \frac{1}{1+} \frac{1}{10};$$

and the first four convergents are $\dfrac{1}{3}, \dfrac{3}{10}, \dfrac{10}{33}, \dfrac{63}{208}$.

§12. A metre $= 39.37079 \times \dfrac{1}{36}$ yards $= 1.0936$ yards.

$$\text{Also } 1.0936 = 1 + \frac{1}{10+} \frac{1}{1+} \frac{1}{2+} \frac{1}{6+} \frac{1}{6}.$$

The convergents are $1, \dfrac{11}{10}, \dfrac{12}{11}, \dfrac{35}{32}, \dots$ Thus 32 metres are nearly equal to 35 yards.

§13. The continued fraction corresponding to .24226 is

$$\frac{1}{4+} \frac{1}{7+} \frac{1}{1+} \frac{1}{4+} \frac{1}{1+} \cdots$$

and the first five convergents are $\dfrac{1}{4}, \dfrac{7}{29}, \dfrac{8}{33}, \dfrac{39}{161}, \dfrac{47}{194}$.

§14. The continued fraction corresponding to .62138 is

$$\frac{1}{1+} \frac{1}{1+} \frac{1}{1+} \frac{1}{1+} \frac{1}{1+} \frac{1}{3+} \frac{1}{1+} \frac{1}{2+} \cdots$$

and the convergents are $1, \dfrac{1}{2}, \dfrac{2}{3}, \dfrac{3}{5}, \dfrac{5}{8}, \dfrac{18}{29}, \dfrac{23}{37}, \dfrac{64}{103}, \dots$.

§15. 162 parts of the first scale are equal to 209 parts of the second scale.

∴ 1 part of the first $= \dfrac{209}{162}$ parts of the second.

Convert $\dfrac{209}{162}$ into a continued fraction and we find

$$1 + \frac{1}{3+} \frac{1}{2+} \frac{1}{4+} \frac{1}{5};$$ and the fourth convergent is $\dfrac{40}{31}$.

In other words, 1 part of the first scale is nearly equal to $\frac{40}{31}$ parts of the second; that is the 31st division of the first nearly coincides with the 40th of the second.

§16.

$$
\begin{array}{c|ll|l}
n+1 & n^3+n^2+n+1 & n^4+n^2-1 & n-1 \\
 & n^3+n^2 & n^4+n^3+n^2+n & \\
\cline{2-3}
n+1 & n+1 & -n^3-n-1 & \\
 & n+1 & -n^3-n^2-n-1 & \\
\cline{2-3}
 & & n^2 & n-1 \\
 & & n^2-1 & \\
\cline{3-3}
 & & 1 &
\end{array}
$$

Thus the continued fraction is $n - 1 + \cfrac{1}{(n+1)+} \cfrac{1}{(n-1)+} \cfrac{1}{n+1}$, and the convergents are

$$\frac{n-1}{n}, \frac{n^2}{n+1}, \frac{n^3-n^2+n-1}{n^2}, \frac{n^4+n^2-1}{n^3+n^2+n+1}.$$

§17. (1) The expression on the left $= \dfrac{(a_n p_n + p_{n-1}) - p_{n-1}}{(a_n q_n + q_{n-1}) - q_{n-1}} = \dfrac{a_n p_n}{a_n q_n} = \dfrac{p_n}{q_n}$.

(2) $\dfrac{p_{n+2}}{p_n} - 1 = \dfrac{a_{n+2} p_{n+1} + p_n - p_n}{p_n} = \dfrac{a_{n+2} p_{n+1}}{p_n}$.

$1 - \dfrac{p_{n-1}}{p_{n+1}} = \dfrac{a_{n+1} p_n + p_{n-1} - p_{n-1}}{p_{n+1}} = \dfrac{a_{n+1} p_n}{p_{n+1}}$,

$\therefore \left(\dfrac{p_{n+2}}{p_n} - 1 \right) \left(1 - \dfrac{p_{n-1}}{p_{n+1}} \right)$

$$= a_{n+2} a_{n+1}$$

$$= \left(\frac{q_{n+2}}{q_n} - 1 \right) \left(1 - \frac{q_{n-1}}{q_{n+1}} \right), \text{ similarly}$$

§18. We have $p_{n-1} q_{n-2} - p_{n-2} q_{n-1} = (-1)^{n-1}$. [Art. 338.]

Again,

$$p_n q_{n-2} - p_{n-2} q_n = (a_n p_{n-1} + p_{n-2}) q_{n-2}$$
$$- (a_n q_{n-1} + q_{n-2}) p_{n-2}$$
$$= a_n (p_{n-1} q_{n-2} - p_{n-2} q_{n-1}) = (-1)^{n-1} a_n.$$

Similarly,

$$p_{n+1} q_{n-2} - p_{n-2} q_{n+1} = a_{n+1} (p_n q_{n-2} - p_{n-2} q_n)$$
$$+ (p_{n-1} q_{n-2} - p_{n-2} q_{n-1})$$
$$= (-1)^{n-1} (a_{n+1} a_n + 1).$$

Finally,

$$p_{n+2} q_{n-2} - p_{n-2} q_{n+2} = a_{n+2} (p_{n+1} q_{n-2} - p_{n-2} q_{n+1})$$
$$+ (p_n q_{n-2} - p_{n-2} q_n)$$
$$= (-1)^{n-1} (a_{n+2} a_{n+1} a_n + a_{n+2} + a_n).$$

EXAMPLES XXV b

§1. The convergents are $1, \dfrac{11}{10}, \dfrac{12}{11}, \dfrac{35}{32}, \dfrac{222}{203}, \dfrac{1367}{1250}$. [XXV. a. 12.]

$\therefore$ in taking $\dfrac{222}{203}$, the error is $< \dfrac{1}{(203)^2}$, and $> \dfrac{1}{2(1250)^2}$.

§2. The convergents are $1, \dfrac{4}{3}, \dfrac{21}{16}, \dfrac{151}{115}, \dots$; the fourth convergent $\dfrac{151}{115}$ differs from the true value by $< \dfrac{1}{(115)^2}$, which is $< \dfrac{1}{(100)^2}$, or .0001.

§3. $1.41421 = 1 + \dfrac{1}{2+}\dfrac{1}{2+}\dfrac{1}{2+}\dfrac{1}{2+}\dfrac{1}{2+}\dfrac{1}{2+}\dots$; the convergents are

$$1, \frac{3}{2}, \frac{7}{5}, \frac{17}{12}, \frac{41}{29}, \frac{99}{70}, \frac{239}{169} \dots;$$

$\therefore$ the error in taking $\dfrac{99}{70}$ as an approximation $< \dfrac{1}{70 \times 169}$.   [Art. 340.]

§4.

$$
\begin{array}{ll}
a+1 \,\big|\, a^3 + 6a^2 + 13a + 10 & a^4 + 6a^3 + 14a^2 + 15a + 7 \,\big|\, a \\
\quad\quad a^3 + 5a^2 + 7a & a^4 + 6a^3 + 13a^2 + 10a \\
\hline
\quad\quad a^2 + 6a + 10 & a^2 + 5a + 7 \,\big|\, a+2 \\
\quad\quad a^2 + 5a + 7 & a^2 + 5a + 6 \\
\hline
a+3 \,\big|\, a+3 & 1 \\
\quad\quad a+3
\end{array}
$$

$\therefore$ the fraction $= \dfrac{1}{a+}\dfrac{1}{(a+1)+}\dfrac{1}{(a+2)+}\dfrac{1}{a+3}$, and the first 3 convergents

$\dfrac{1}{a}, \dfrac{a+1}{a^2+a+1}, \dfrac{a^2+3a+3}{a^3+3a^2+4a+2}$.

§5. $\dfrac{p_2}{q_2} - \dfrac{p_1}{q_1} = \dfrac{1}{q_1 q_2}, \dfrac{p_3}{q_3} - \dfrac{p_2}{q_2} = -\dfrac{1}{q_2 q_3}, \dots \dfrac{p_n}{q_n} - \dfrac{p_{n-1}}{q_{n-1}} = \dfrac{(-1)^n}{q_n q_{n-1}}$; add these results together.

§6. We have $\dfrac{p_n}{p_{n-1}} = \dfrac{a_n p_{n-1} + p_{n-2}}{p_{n-1}} = a_n + \dfrac{p_{n-2}}{p_{n-1}}$;

and $\dfrac{p_{n-1}}{p_{n-2}} = a_{n-1} + \dfrac{p_{n-3}}{p_{n-2}}$; $\dfrac{p_{n-2}}{p_{n-3}} = a_{n-2} + \dfrac{p_{n-4}}{p_{n-3}}$; and finally $\dfrac{p_2}{p_1} = a_2 + \dfrac{1}{a_1}$.

Thus $\dfrac{p_n}{p_{n-1}} = a_n + \dfrac{1}{a_{n-1}+}\dfrac{1}{a_{n-2}+}\dots\dfrac{1}{a_2+}\dfrac{1}{a_1}$.

Similarly the second result follows, for $\dfrac{q_2}{q_1} = a_2$.

§7. (1) We have to prove that

$$p_n p_{n+2} - p_n^2 = p_{n+1}^2 - p_{n+1}p_{n-1}, \text{ or}$$
$$p_n(p_{n+2} - p_n) = p_{n+1}(p_{n+1} - p_{n-1}).$$

Now $p_{n+2} = ap_{n+1} + p_n$, so that $p_{n+2} - p_n = ap_{n+1}$;

$p_{n+1} = ap_n + ap_{n-1}$, so that $p_{n+1} - p_{n-1} = ap_n$,

whence the required result easily follows.

(2) is the particular case of Ex.8, when $b = a$.

§8. By trial we find the required results hold in the case of the first few convergents. Assume that

$$q_{2n-2} = p_{2n-1}, \quad q_{2n-3} = \frac{a}{b}p_{2n-2},$$

$\therefore q_{2n} = bq_{2n-1} + q_{2n-2} = b(aq_{2n-2} + q_{2n-3}) + q_{2n-2}$
$= (ab+1)q_{2n-2} + bq_{2n-3} = (ab+1)p_{2n-1} + ap_{2n-2}$
$= a(bp_{2n-1} + p_{2n-2}) + p_{2n-1} = ap_{2n} + p_{2n-1} = p_{2n+1}$;

therefore, by induction, the result follows. Similarly we may whew that

$$q_{2n-1} = \frac{a}{b}p_{2n}.$$

§9. We have $p_{2n+1} = ap_{2n} + p_{2n-1}$, and $p_{2n} = bp_{2n-1} + p_{2n-2}$;

$\therefore p_{2n+1} = (ab+1)p_{2n-1} + ap_{2n-2}$; and $p_{2n-1} = ap_{2n-2} + p_{2n-3}$;

whence, by substitution, $p_{2n+1} = (ab+2)p_{2n-1} - p_{2n-3}$.

Similarly we may shew that $p_{2n} = (ab+2)p_{2n-2} - p_{2n-4}$;

$\therefore$ generally $p_n = (ab+2)p_{n-2} - p_{n-4}$.

§10. The first expression $= ax_1 + \dfrac{a}{ax_2+}\dfrac{1}{x_3+}\dfrac{1}{ax_4+}\dots$

$$= ax_1 + \cfrac{1}{x_2+} \cfrac{1}{ax_3+} \cfrac{a}{ax_4+} \ldots = ax_1 + \cfrac{1}{x_2+} \cfrac{1}{ax_3+} \cfrac{1}{x_4+}; \text{ and so on. } \quad \text{[Compare}$$

Art. 448.]

§11. We have
$$\frac{M}{N} = \cfrac{1}{a_1+} \cfrac{P}{Q} = \frac{Q}{a_1 Q + P};$$

and
$$\frac{P}{Q} = \cfrac{1}{a_2+} \cfrac{R}{S} = \frac{S}{a_2 S + R}.$$

But the fractions $\dfrac{M}{N}, \dfrac{P}{Q}, \dfrac{R}{S}$ are in their lowest terms; hence $\dfrac{Q}{a_1 Q + P}$

and $\dfrac{S}{a_2 S + R}$ are in their lowest terms. [See Art. 338, Cor. 1.]

Thus $M = Q, N = a_1 Q + P, P = S, Q = a_2 S + R$; whence
$$M = a_2 S + R = a_2 P + R;$$

and
$$N = a_1(a_2 P + R) + P = (a_1 a_2 + 1)P + a_1 R.$$

§12. We have $p_n = a p_{n-1} + p_{n-2}$, $q_n = a q_{n-1} + q_{n-2}$; thus the numerators and denominators of the successive convergents are the coefficients of the terms of a recurring series whose scale of relation is $1 - ax - x^2$.

Let
$$S = p_1 x + p_2 x^2 + p_3 x^3 + p_4 x^4 + \ldots;$$

then as in Art. 326, we have $\quad S = \dfrac{p_1 x + (p_2 - a p_1)x^2}{1 - ax - x^2};$

put $p_1 = 1, p_2 = a$; hence $\quad S = \dfrac{x}{1 - ax - x^2}.$

Now α, β are the roots of the equation $t^2 - at - 1 = 0$;

hence
$$S = \frac{x}{(1 - \alpha x)(1 - \beta x)} = \frac{1}{\alpha - \beta}\left(\frac{1}{1 - \alpha x} - \frac{1}{1 - \beta x}\right);$$

$\therefore p_n$, which is the coefficient of x^n in S, is equal to $\dfrac{\alpha^n - \beta^n}{\alpha - \beta}.$

Similarly, if $S' = q_1 x^2 + q_2 x^3 + q_3 x^4 + \ldots$, we find
$$S' = \frac{ax^2 + x^3}{1 - ax - x^2} = \frac{x}{1 - ax - x^2}; \text{ hence } q_{n-1} = \frac{\alpha^n - \beta^n}{\alpha - \beta}.$$

§13. As in Example 9 we have
$$p_n = (ab + 2)p_{n-2} - p_{n-4}, \quad q_n = (ab + 2)q_{n-2} - q_{n-4}.$$

Hence the numerators and denominators of the successive convergents each form a recurring series whose scale of relation is $1 - (ab + 2)x^2 + x^4$.

Let
$$S = p_1 x + p_2 x^2 + p_3 x^3 + p_4 x^4 + p_5 x^5 + \ldots + p_n x_n + \ldots,$$
$$S' = q_1 x^2 + q_2 x^3 + q_3 x^4 + q_4 x^5 + \ldots + q_{n-1} x^n + \ldots;$$

then $S\{1 - (ab+2)x^2 + x^4\} = p_1 x + p_2 x^2 + \{p_3 - (ab+2)p_1\}x^3 + \{p_4 - (ab+2)p_2\}x^4;$ all the other terms vanishing in virtue of the scale of relation. A similar result holds for S'.

But
$$p_1 = 1, p_2 = b, p_3 = ab + 1, p_4 = ab^2 + 2b;$$
$$q_1 = a, q_2 = ab + 1, q_3 = a^2 b + 2a, q_4 = a^2 b^2 + 3ab + 1;$$

hence
$$S = \frac{x + bx^2 - x^3}{1 - (ab + 2)x^2 + x^4} \tag{41}$$

and
$$S' = \frac{ax^2 + (ab + 1)x^3 - x^5}{1 - (ab + 2)x^2 + x^4} = \frac{x + ax^2 - x^3}{1 - (ab + 2)x^2 + x^4} - x \tag{42}$$

Now α and β are the roots of the reciprocal equation $1 - (ab+2)x^2 + x^4 = 0$;

hence
$$1 - (ab + 2)x^2 + x^4 = (1 - \alpha x^2)(1 - \beta x^2);$$
$$\therefore \frac{1}{1 - (ab + 2)x^2 + x^4} = \frac{1}{(1 - \alpha x^2)(1 - \beta x^2)}$$

$$= \frac{1}{\alpha - \beta} \left(\frac{\alpha}{1 - \alpha x^2} - \frac{\beta}{1 - \beta x^2} \right)$$

that is, $$\frac{1}{1 - (ab + 2)x^2 + x^4} = \Sigma \frac{\alpha^{n+1} - \beta^{n+1}}{\alpha - \beta} x^{2n} \qquad (43)$$

Hence from (41), $p_{2n} = b \times$ coefficient of x^{2n-2} in (43) $= b \dfrac{\alpha^n - \beta^n}{\alpha - \beta}$.

Similarly from (42), $q_{2n-1} = a \dfrac{\alpha^n - \beta^n}{\alpha - \beta}$.

Again, from (41) and (42), it is obvious that $p_{2n+1} = q_{2n}$; and also that $p_{2n+1} =$ coefficient of x^{2n} in (43) $-$ coefficient of x^{2n-2} in (43)

$$= \frac{1}{\alpha - \beta} \left\{ (\alpha^{n+1} - \beta^{n+1}) - (\alpha^n - \beta^n) \right\}.$$

EXAMPLES XXVI : Indeterminate Equations of the First Degree

§1. The convergents to $\dfrac{775}{711}$ are $\dfrac{1}{1}, \dfrac{12}{11}, \dfrac{109}{100}$;

thus $775 \times 100 - 711 \times 109 = 1$, and $775x - 711y = 1$;

$$\therefore 775(x - 100) = 711(y - 109); \text{ hence } \frac{x - 100}{711} = \frac{y - 109}{775} = t,$$
$$x = 711t + 100, \quad y = 775t + 109.$$

§2. The convergents to $\dfrac{519}{455}$ are $\dfrac{1}{1}, \dfrac{8}{7}, \dfrac{73}{64}$;

thus $455 \times 73 - 519 \times 64 = -1$, and $455x - 519y = 1$;

$$\therefore 455(x + 73) = 519(y + 64); \text{ hence } x + 73 = 519t, \quad y + 64 = 455t.$$

§3. The convergents to $\dfrac{436}{393}$ are $\dfrac{1}{1}, \dfrac{10}{9}, \dfrac{71}{64}$;

thus $436 \times 64 - 393 \times 71 = 1$, and therefore $436 \times 320 - 393 \times 355 = 5$;

whence
$$436(x - 320) = 393(y - 355);$$
$$\therefore x - 320 = 393t, \quad y - 355 = 436t.$$

§4. Let x, y be the number of florins and half-crowns respectively; then
$$4x + 5y = 79.$$
One solution is $x = 1$, $y = 15$; thus the general solution is
$$x = 1 + 5t, \ y = 15 - 4t.$$
Here t can have the values $0, 1, 2, 3$; hence there are 4 ways.

§5. By trial $x = 1$, $y = 78$ is a solution; hence the general solution is $x = 1 + 15t$, $y = 78 - 11t$; and t can have the values $0, 1, 2, \ldots 7$; thus there are 8 solutions.

§6. Let x, y be the numerators; then $\dfrac{x}{7} + \dfrac{y}{9} = \dfrac{73}{63}$, or $9x + 7y = 73$, the only solution of which is $x = 5$, $y = 4$.

§7. Let x, y be the numerators; then $\dfrac{x}{12} - \dfrac{y}{8} = \dfrac{1}{24}$; that is $2x - 3y = 1$, or $2x - 3y = -1$.

(i) The general solution of $2x - 3y = 1$ is $x = 3t + 2$, $y = 2t + 1$, and since $y < 8$, the values of t are restricted to $0, 1, 2, 3$. Thus
$$x = 2, 5, 8, 11; \ y = 1, 3, 5, 7.$$

(ii) The general solution of $2x - 3y = -1$ is $x = 3t + 1$, $y = 2t + 1$; thus $x = 1, 4, 7, 10; \ y = 1, 3, 5, 7$.

§8. x pounds y shillings is equivalent to $20x + y$ shillings; hence
$$20x + y = \frac{1}{2}(20y + x); \text{ that is, } 39x = 18y, \text{ or } 13x = 6y.$$
The general solution is $x = 6t$, $y = 13t$; and as x, y are both restricted to values less than 20, it follows that t can only have the value 1; thus $x = 6$, $y = 13$.

§9. Eliminating z, we have $40x + 37y = 656$. By trial one solution is $y = 8$, $x = 9$; hence the general solution is $x = 9 + 37t, y = 8 - 40t$; thus t can only have the value 0, and $x = 9$, $y = 8$ is the only solution. By substitution we find $z = 3$.

§10. Eliminating x, we have $4y + 7z = 73$; the general solution is
$$y = 13 - 7t, \ z = 3 + 4t.$$
Thus t can only have the values 0 and 1. When $t = 0$, $y = 13, z = 3$, but the value of x is fractional; when $t = 1$, $y = 6$, $z = 7$, $x = 5$.

§11. The general solution of $3y + 4z = 34$ is $y = 10 - 4t$, $z = 1 + 3t$. Thus $y = 10, 6, 2$; $z = 1, 4, 7$.

From the equation $20x - 21y = 38$, we see that when $y = 10$, or $y = 6$, the value of x is fractional; and when $y = 2, x = 4, z = 7$.

§12. The general solution of $13x + 11z = 103$ is $x = 2 + 11t$, $z = 7 - 13t$; thus $x = 2$, $z = 7$ is the only solution.

From $7z - 5y = 4$, we have $y = 9$.

§13. Put $z = 1$, then $7x + 4y = 65$; the solutions are $x = 3, y = 11$; $x = 7, y = 4$.

Put $z = 2$, then $7x + 4y = 46$; here the solutions are $x = 6, y = 1$; $x = 2, y = 8$.

Put $z = 3$, then $7x + 4y = 27$; here $x = 1, y = 5$ is the only solution.

Put $z = 4$, then $7x + 4y = 8$, which has no integral solution.

§14. Put $x = 1$, then $17y + 11z = 107$; solution $y = 5$, $z = 2$;

put $x = 2$, then $17y + 11z = 84$; solution $y = 3$, $z = 3$;

put $x = 3$, then $17y + 11z = 61$; solution $y = 1$, $z = 4$;

put $x = 4$, then $17y + 11z = 38$; no solution;

put $x = 5$, then $17y + 11z = 15$; no solution.

§15. Let N denote the number, x, y, z the quotients when N is divided by $5, 7, 8$ respectively;

then $N = 5x + 3 = 7y + 2 = 8z + 5$; hence $7y - 5x = 1$ and $7y - 8z = 3$.

The general solution of $7y - 5x = 1$ is $x = 4 + 7s$, $y = 3 + 5s$.

Substituting this value of y in $7y - 8z = 3$, we have $35s - 8z = -18$, the general solution of which is $s = 2 + 8t$, $z = 11 + 35t$.

Substituting for s, we obtain

$$x = 56t + 18, \quad y = 40t + 13, \quad z = 35t + 11, \quad N = 280t + 93.$$

§16. With the notation of the preceding example, we have

$$N = 3x + 1 = 7y + 6 = 11z + 5; \quad \text{hence } 3x - 7y = 5, 3x - 11z = 4.$$

Thus $x = 4 + 7s$, $y = 1 + 3s$, and substituting for x, we have $11z - 21s = 8$; when $z = 16 + 21t, s = 8 + 11t$; thus

$$x = 77t + 60, \quad y = 33t + 25, \quad z = 21t + 16, \quad N = 231t + 181.$$

By putting $t = 0, t = 1$, we find that the two smallest values of N are 181 and 412.

§17. In the septenary scale let the number be denoted by $x0y$; then in the nonary scale it is denoted by $y0x$.

In the septenary scale $x0y$ represents the denary number $y + 0.7 + x.7^2$, or $y + 49x$. Similarly in the nonary scale $y0x$ represents $x + 81y$; hence

$$y + 49x = x + 81y, \text{ or } 3x = 5y$$

The general solution of this equation is $x = 5t, y = 3t$; but x and y are both less than 7; hence $x = 5, y = 3$ is the only solution. Thus $y + 49x$, the value of the number in the denary scale, is equal to 248.

§18. By hypothesis $\dfrac{2}{a} = \dfrac{1}{6} + \dfrac{1}{b}$; hence $b = \dfrac{6a}{12 - a}$. By ascribing to a the values $1, 2, 3 \ldots 11$, we get the corresponding values of b.

§19. Since 250 and 243 have no common factor, no two divisions will be coincident. If a is the length of the two rods, then the distance from the zero end of the xth division of the first is $\dfrac{xa}{250}$; and of the yth division of the second is $\dfrac{ya}{243}$. Hence the distance between these divisions is

$$\left(\frac{x}{250} - \frac{y}{243} \right) a, \text{ or } \frac{243x - 250y}{250 \cdot 243} a.$$

As the numerator cannot be equal to zero, this fraction will be least when $243x - 250y = \pm 1$.

The penultimate convergent to $\dfrac{250}{243}$ is $\dfrac{107}{104}$, and $243 \times 107 - 250 \times 104 = 1$;

also $243(250 - 107) - 250(243 - 104) = -1$;

thus the values of x are $107, 143$; and the values of y are $104, 139$.

§20. Let x, y, z denote the required number of times; then the three bells tolled for $23x, 29y, 34z$ times, excluding the first of each. Hence
$$29y = 23x + 39, \quad 34z = 23x + 40; \quad \text{therefore } 34z - 29y = 1.$$
The general solution of this equation is $z = 6 + 29t; y = 7 + 34t$.

Now since the bells cease in less than 20 minutes, $29y$, or $203 + 29 \times 34t$, must be less than 1200; that is $t < \dfrac{997}{29 \times 34} < 2$.

When $t = 0, y = 7$, but the value of x is not integral; when $t = 1, y = 41, x = 50, z = 35$.

§21. Let a, b be a solution of the equation $7x + 9y = c$, and let a be the smallest value of x for any particular value of c, so that b is the greatest value of y; then the general solution is $x = 9t + a, y = b - 7t$. Since there are to be 6 solutions, t is restricted to the values $0, 1, 2, 3, 4, 5$.

Also $c = 7a + 9b$, and will therefore have its greatest value when a and b have their greatest values. Now $b - 7t$ is a positive integer; hence $b > 7t$; thus $b > 35$; and the greatest value of b is 41, for if $b = 42$, then $t = 6$ would be an admissible value. The greatest value of a is 8, for if $a = 9$, then $t = -1$ would be an admissible value; thus $c = (7 \times 8) + (9 \times 41) = 425$.

§22. As in the preceding example, $x = 11t + a$, $y = b - 14t$; where t may have the values $0, 1, 2, 3, 4$. Thus the greatest value of a is 10, and since b must be greater than 4×14 and less than 5×14, the greatest value of b is 69; hence $c = 14a + 11b = 14 \times 10 + 11 \times 69 = 899$.

§23. The general solution of $19x + 14y = c$ is $x = a + 14t$, $y = b - 19t$; where t may have the values $0, 1, 2, 3, 4, 5$.

Since zero solutions are inadmissible, a must lie between 1 and 13, and b must be greater than 5×19 and less than 6×19.

Now $c = 19a + 14b$, and is greatest when $a = 13$ and $b = 113$, in which case $c = 1829$; also c has its least value when $a = 1, b = 96$, in which case $c = 1363$.

§24. Let $x = h$, $y = k$ be a particular solution of $ax + by = c$, and let h be the smallest value that x can have for any particular value of c, so that k is the greatest value of y; then the general solution is $x = h + bt$, $y = k - at$, where t is restricted to the values $0, 1, 2, \ldots (n - 1)$.

Since zero solutions are inadmissible, h must lie between 1 and $b - 1$, while k must lie between $1 + a(n - 1)$ and $a - 1 + a(n - 1)$.

Now $c = ah + bk$, and the greatest values of h and k are $b - 1$ and $a - 1 + a(n - 1)$ respectively; hence the greatest value of $c = (n + 1)ab - a - b$.

The least values of h and k are 1 and $1 + a(n - 1)$ respectively; hence the least value of $\quad c = (n - 1)ab + a + b$.

This Example includes Examples $21 - 23$ as particular cases.

EXAMPLES XXVII : Recurring Continued Fractions

EXAMPLES XXVII a

§1. $\sqrt{3} = 1 + \sqrt{3} - 1 = 1 + \dfrac{2}{\sqrt{3}+1}$; $\dfrac{\sqrt{3}+1}{2} = 1 + \dfrac{\sqrt{3}-1}{2} = 1 + \dfrac{1}{\sqrt{3}+1}$;

$\dfrac{\sqrt{3}+1}{1} = 2 + \sqrt{3} - 1$;

$\therefore$ the continued fraction $= 1 + \dfrac{1}{1+}\dfrac{1}{2+}\dfrac{1}{1+}\dfrac{1}{2+}\cdots$;

and the convergents are $1, 2, \dfrac{5}{3}, \dfrac{7}{4}, \dfrac{19}{11}, \dfrac{26}{15}, \cdots$

§2. $\sqrt{5} = 2 + \sqrt{5} - 2 = 2 + \dfrac{1}{\sqrt{5}+2}$; $\sqrt{5} + 2 = 4 + \sqrt{5} - 2$;

$\therefore$ the continued fraction $= 2 + \dfrac{1}{4+}\dfrac{1}{4+}\cdots$;

and the convergents are $\dfrac{2}{1}, \dfrac{9}{4}, \dfrac{38}{17}, \dfrac{161}{72}, \dfrac{682}{305}, \dfrac{2889}{1292}, \cdots$;

§3. $\sqrt{6} = 2 + \sqrt{6} - 2 = 2 + \dfrac{2}{\sqrt{6}+2}$; $\sqrt{6} + 2 = 2 + \dfrac{\sqrt{6}-2}{2} = 2 + \dfrac{1}{\sqrt{6}+2}$;

$\sqrt{6} + 2 = 4 + \sqrt{6} - 2$;

$\therefore$ the continued fraction $= 2 + \dfrac{1}{2+}\dfrac{1}{4+}\dfrac{1}{2+}\dfrac{1}{4+}$;

and the convergents are $\dfrac{2}{1}, \dfrac{5}{2}, \dfrac{22}{9}, \dfrac{49}{20}, \dfrac{218}{89}, \dfrac{485}{198}, \cdots$

§4. $\sqrt{8} = 2 + \sqrt{8} - 2 = 2 + \dfrac{4}{\sqrt{8}+2}$; $\dfrac{\sqrt{8}+2}{4} = 1 + \dfrac{\sqrt{8}-2}{4} = 1 + \dfrac{1}{\sqrt{8}+2}$;

$\sqrt{8} + 2 = 4 + \sqrt{8} - 2$;

$\therefore$ the continued fraction $= 2 + \dfrac{1}{1+}\dfrac{1}{4+}\dfrac{1}{1+}\dfrac{1}{4+}\cdots$;

and the convergents are $\dfrac{2}{1}, \dfrac{3}{1}, \dfrac{14}{5}, \dfrac{17}{6}, \dfrac{82}{29}, \dfrac{99}{35}\cdots$

§5. $\sqrt{11} = 3 + \sqrt{11} - 3 = 3 + \dfrac{2}{\sqrt{11}+3}$; $\dfrac{\sqrt{11}+3}{2} = 3 + \dfrac{\sqrt{11}-3}{2} = 3 + \dfrac{1}{\sqrt{11}+3}$;

$\sqrt{11} + 3 = 6 + \sqrt{11} - 3$;

$\therefore$ the continued fraction $= 3 + \dfrac{1}{3+}\dfrac{1}{6+}\dfrac{1}{3+}\dfrac{1}{6+}\cdots$;

and the convergents are $\dfrac{3}{1}, \dfrac{10}{3}, \dfrac{63}{19}, \dfrac{199}{60}, \dfrac{1257}{379}, \dfrac{3970}{1197}, \cdots$

§6. $\sqrt{13} = 3 + \sqrt{13} - 3 = 3 + \dfrac{4}{\sqrt{13}+3}$; $\dfrac{\sqrt{13}+3}{4} = 1 + \dfrac{\sqrt{3}-1}{4} = 1 + \dfrac{3}{\sqrt{13}+1}$;

$\dfrac{\sqrt{13}+1}{3} = 1 + \dfrac{\sqrt{13}-2}{3} = 1 + \dfrac{3}{\sqrt{13}+2}$; $\dfrac{\sqrt{13}+2}{3} = 1 + \dfrac{\sqrt{13}-1}{3} = 1 + \dfrac{4}{\sqrt{13}+1}$;

$\dfrac{\sqrt{13}+1}{4} = 1 + \dfrac{\sqrt{13}-3}{4} = 1 + \dfrac{1}{\sqrt{13}+3}$; $\sqrt{13} + 3 = 6 + \sqrt{13} - 3$;

$\therefore$ the continued fraction $= 3 + \dfrac{1}{1+}\dfrac{1}{1+}\dfrac{1}{1+}\dfrac{1}{1+}\dfrac{1}{6+}\cdots$

and the convergents are $\dfrac{3}{1}, \dfrac{4}{1}, \dfrac{7}{2}, \dfrac{11}{3}, \dfrac{18}{5}, \dfrac{119}{33}, \cdots$

§7. $\sqrt{14} = 3 + \sqrt{14} - 3 = 3 + \dfrac{5}{\sqrt{14}+3}$; $\dfrac{\sqrt{14}+3}{5} = 1 + \dfrac{\sqrt{14}-2}{5} = 1 + \dfrac{2}{\sqrt{14}+2}$;

$$\frac{\sqrt{14}+2}{2} = 2 + \frac{\sqrt{14}-2}{2} = 2 + \frac{5}{\sqrt{14}+2}; \quad \frac{\sqrt{14}+2}{5} = 1 + \frac{\sqrt{14}-3}{5} = 1 + \frac{1}{\sqrt{14}+3};$$

$$\sqrt{14}+3 = 6 + \sqrt{14} - 3;$$

∴ the continued fraction $= 3 + \dfrac{1}{1+}\ \dfrac{1}{2+}\ \dfrac{1}{1+}\ \dfrac{1}{6+}\ \cdots;$

and the convergents are $\dfrac{3}{1},\ \dfrac{4}{1},\ \dfrac{11}{3},\ \dfrac{15}{4},\ \dfrac{101}{27},\ \dfrac{116}{31},\ \cdots$

§8. $\sqrt{22} = 4 + \sqrt{22} - 4 = 4 + \dfrac{6}{\sqrt{22}+4};\quad \dfrac{\sqrt{22}+4}{6} = 1 + \dfrac{\sqrt{22}-2}{6} = 1 + \dfrac{3}{\sqrt{22}+2};$

$$\frac{\sqrt{22}+2}{3} = 2 + \frac{\sqrt{22}-4}{3} = 2 + \frac{2}{\sqrt{22}+4};\quad \frac{\sqrt{22}+4}{2} = 4 + \frac{\sqrt{22}-4}{2} = 4 + \frac{3}{\sqrt{22}+4};$$

$$\frac{\sqrt{22}+4}{3} = 2 + \frac{\sqrt{22}-2}{3} = 2 + \frac{6}{\sqrt{22}+2};\quad \frac{\sqrt{22}+2}{6} = 1 + \frac{\sqrt{22}-4}{6} = 1 + \frac{1}{\sqrt{22}+4};$$

$$\sqrt{22}+4 = 8 + \sqrt{22} - 4;$$

∴ the continued fraction $= 4 + \dfrac{1}{1+}\ \dfrac{1}{2+}\ \dfrac{1}{4+}\ \dfrac{1}{2+}\ \dfrac{1}{1+}\ \dfrac{1}{8+}\ \cdots$

and the convergents are $\dfrac{4}{1},\ \dfrac{5}{1},\ \dfrac{14}{3},\ \dfrac{61}{13},\ \dfrac{136}{29},\ \dfrac{197}{42},\ \cdots$

§9. $2\sqrt{3} = \sqrt{12} = 3 + \sqrt{12} - 3 = 3 + \dfrac{3}{\sqrt{12}+3};\quad \dfrac{\sqrt{12}+3}{3} = 2 + \dfrac{\sqrt{12}-3}{3} =$

$2 + \dfrac{1}{\sqrt{12}+3};$

$$\sqrt{12}+3 = 6 + \sqrt{12} - 3;$$

∴ the continued fraction $= 3 + \dfrac{1}{2+}\ \dfrac{1}{6+}\ \dfrac{1}{2+}\ \dfrac{1}{6+}\ \cdots;$

and the convergents are $\dfrac{3}{1},\ \dfrac{7}{2},\ \dfrac{45}{13},\ \dfrac{97}{28},\ \dfrac{627}{181},\ \dfrac{1351}{390},\ \cdots$

§10. $\sqrt{32} = 5 + \sqrt{32} - 5 = 5 + \dfrac{7}{\sqrt{32}+5};\quad \dfrac{\sqrt{32}+5}{7} = 1 + \dfrac{\sqrt{32}-2}{7} = 1 + \dfrac{4}{\sqrt{32}+2};$

$$\frac{\sqrt{32}+2}{4} = 1 + \frac{\sqrt{32}-2}{4} = 1 + \frac{7}{\sqrt{32}+2};\quad \frac{\sqrt{32}+2}{7} = 1 + \frac{\sqrt{32}-5}{7} = 1 + \frac{1}{\sqrt{32}+5};$$

$$\sqrt{32}+5 = 10 + \sqrt{32} - 5;$$

∴ the continued fraction $= 5 + \dfrac{1}{1+}\ \dfrac{1}{1+}\ \dfrac{1}{1+}\ \dfrac{1}{10+}\ \cdots;$

and the convergents are $\dfrac{5}{1},\ \dfrac{6}{1},\ \dfrac{11}{2},\ \dfrac{17}{3},\ \dfrac{181}{32},\ \dfrac{198}{35},\ \cdots$

§11. $3\sqrt{5} = 6 + \sqrt{45} - 6 = 6 + \dfrac{9}{\sqrt{45}+6};\quad \dfrac{\sqrt{45}+6}{9} = 1 + \dfrac{\sqrt{45}-3}{9} = 1 + \dfrac{4}{\sqrt{45}+3};$

$$\frac{\sqrt{45}+3}{4} = 2 + \frac{\sqrt{45}-5}{4} = 2 + \frac{5}{\sqrt{45}+5};\quad \frac{\sqrt{45}+5}{5} = 2 + \frac{\sqrt{45}-5}{5} = 2 + \frac{4}{\sqrt{45}+5};$$

$$\frac{\sqrt{45}+5}{4} = 2 + \frac{\sqrt{45}-3}{4} = 2 + \frac{9}{\sqrt{45}+3};\quad \frac{\sqrt{45}+3}{9} = 1 + \frac{\sqrt{45}-6}{9} = 1 + \frac{1}{\sqrt{45}+6};$$

$$\sqrt{45}+6 = 12 + \sqrt{45} - 6;$$

∴ the continued fraction $= 6 + \dfrac{1}{1+}\ \dfrac{1}{2+}\ \dfrac{1}{2+}\ \dfrac{1}{2+}\ \dfrac{1}{1+}\ \dfrac{1}{12+}\ \cdots$

and the convergents are $\dfrac{6}{1},\ \dfrac{7}{1},\ \dfrac{20}{3},\ \dfrac{47}{7},\ \dfrac{114}{17},\ \dfrac{161}{24},\ \cdots$

§12. $\sqrt{160} = 12 + \sqrt{160} - 12 = 12 + \dfrac{4}{\sqrt{10}+3};\quad \dfrac{\sqrt{10}+3}{4} = 1 + \dfrac{\sqrt{10}-1}{4} =$

$1 + \dfrac{9}{4\sqrt{10}+4};$

$$\frac{4\sqrt{10}+4}{9} = 1 + \frac{4\sqrt{10}-5}{9} = 1 + \frac{15}{4\sqrt{10}+5};\quad \frac{4\sqrt{10}+5}{15} = 1 + \frac{4\sqrt{10}-10}{15} =$$

$$1 + \frac{2}{2\sqrt{10}+5};$$

$$\frac{2\sqrt{10}+5}{2} = 5 + \frac{2\sqrt{10}-5}{2} = 5 + \frac{15}{4\sqrt{10}+10};\quad \frac{4\sqrt{10}+10}{15} = 1 + \frac{4\sqrt{10}-5}{15} =$$

$$1 + \frac{9}{4\sqrt{10}+5};$$

$$\frac{4\sqrt{10}+5}{9} = 1 + \frac{4\sqrt{10}-4}{9} = 1 + \frac{4}{\sqrt{10}+1};\quad \frac{\sqrt{10}+1}{4} = 1 + \frac{\sqrt{10}-3}{4} = 1 +$$

$$\frac{1}{4(\sqrt{10}+3)};$$

$$4(\sqrt{10}+3) = 24 + 4\sqrt{10} - 12;$$

∴ the continued fraction

$$= 12 + \frac{1}{1+}\ \frac{1}{1+}\ \frac{1}{1+}\ \frac{1}{5+}\ \frac{1}{1+}\ \frac{1}{1+}\ \frac{1}{1+}\ \frac{1}{24+}\ \cdots;$$

and the convergents are $\dfrac{12}{1}, \dfrac{13}{1}, \dfrac{25}{2}, \dfrac{38}{3}, \dfrac{215}{17}, \dfrac{253}{20}, \cdots$

§13. $\sqrt{21} = 4 + \sqrt{21} - 4 = 4 + \dfrac{5}{\sqrt{21}+4};\quad \dfrac{\sqrt{21}+4}{5} = 1 + \dfrac{\sqrt{21}-1}{5} = 1 + \dfrac{4}{\sqrt{21}+1};$

$$\frac{\sqrt{21}+1}{4} = 1 + \frac{\sqrt{21}-3}{4} = 1 + \frac{3}{\sqrt{21}+3};\quad \frac{\sqrt{21}+3}{3} = 2 + \frac{\sqrt{21}-3}{3} = 2 + \frac{4}{\sqrt{21}+3};$$

$$\frac{\sqrt{21}+3}{4} = 1 + \frac{\sqrt{21}-1}{4} = 1 + \frac{5}{\sqrt{21}+1};\quad \frac{\sqrt{21}+1}{5} = 1 + \frac{\sqrt{21}-4}{5} = 1 + \frac{1}{\sqrt{21}+4};$$

$$\sqrt{21} + 4 = 8 + \sqrt{21} - 4;$$

∴ the continued fraction $= \dfrac{1}{4+}\ \dfrac{1}{1+}\ \dfrac{1}{1+}\ \dfrac{1}{2+}\ \dfrac{1}{1+}\ \dfrac{1}{1+}\ \dfrac{1}{8+}\ \cdots;$

and the convergents are $\dfrac{1}{4}, \dfrac{1}{5}, \dfrac{2}{9}, \dfrac{5}{23}, \dfrac{7}{32}, \dfrac{12}{55}, \cdots;$

§14. $\sqrt{33} = 5 + \sqrt{33} - 5 = 5 + \dfrac{8}{\sqrt{33}+5};\quad \dfrac{\sqrt{33}+5}{8} = 1 + \dfrac{\sqrt{33}-3}{8} = 1 + \dfrac{3}{\sqrt{33}+3};$

$$\frac{\sqrt{33}+3}{3} = 2 + \frac{\sqrt{33}-3}{3} = 2 + \frac{8}{\sqrt{33}+3};\quad \frac{\sqrt{33}+3}{8} = 1 + \frac{\sqrt{33}-3}{8} = 1 + \frac{1}{\sqrt{33}+5};$$

$$\sqrt{33} + 5 = 10 + \sqrt{33} - 5;$$

∴ the continued fraction $= \dfrac{1}{5+}\ \dfrac{1}{1+}\ \dfrac{1}{2+}\ \dfrac{1}{1+}\ \dfrac{1}{10+}\ \cdots;$

and the convergents are $\dfrac{1}{5}, \dfrac{1}{6}, \dfrac{3}{17}, \dfrac{4}{23}, \dfrac{43}{247}, \dfrac{219}{1258}, \cdots;$

§15. $\sqrt{\dfrac{6}{5}} = \dfrac{\sqrt{30}}{5} = 1 + \dfrac{\sqrt{30}-5}{5} = 1 + \dfrac{1}{\sqrt{30}+5};\quad \sqrt{30} + 5 = 10 + \sqrt{30} - 5 =$

$$10 + \frac{5}{\sqrt{30}+5};$$

$$\frac{\sqrt{30}+5}{5} = 2 + \frac{\sqrt{30}-5}{5};$$

∴ the continued fraction $= 1 + \dfrac{1}{10+}\ \dfrac{1}{2+}\ \cdots;$

and the convergents are $\dfrac{1}{1}, \dfrac{11}{10}, \dfrac{23}{21}, \dfrac{241}{220}, \dfrac{505}{461}, \dfrac{5291}{4830}, \cdots;$

§16. $\sqrt{\dfrac{7}{11}} = \dfrac{7}{\sqrt{77}};$ and $\dfrac{\sqrt{77}}{7} = 1 + \dfrac{\sqrt{77}-7}{7} = 1 + \dfrac{4}{\sqrt{77}+7};$

$$\frac{\sqrt{77}+7}{4} = 3+\frac{\sqrt{77}-5}{4} = 3+\frac{13}{\sqrt{77}+5}; \quad \frac{\sqrt{77}+5}{13} = 1+\frac{\sqrt{77}-8}{13} = 1+\frac{1}{\sqrt{77}+8};$$

$$\sqrt{77}+8 = 16+\sqrt{77}-8 = 16+\frac{13}{\sqrt{77}+8}; \quad \frac{\sqrt{77}+8}{13} = 1+\frac{\sqrt{77}-5}{13} = 1+\frac{4}{\sqrt{77}+5};$$

$$\frac{\sqrt{77}+5}{4} = 3+\frac{\sqrt{77}-7}{4} = 3+\frac{7}{\sqrt{77}+7}; \quad \frac{\sqrt{77}+7}{7} = 2+\frac{\sqrt{77}-7}{7};$$

∴ the continued fraction

$$= \frac{1}{1+}\ \frac{1}{3+}\ \frac{1}{1+}\ \frac{1}{16+}\ \frac{1}{1+}\ \frac{1}{3+}\ \frac{1}{2+}\ \frac{1}{3+}\ \frac{1}{1+} \cdots;$$

and the convergents are $\dfrac{1}{1}, \dfrac{3}{4}, \dfrac{4}{5}, \dfrac{67}{84}, \dfrac{71}{89}, \dfrac{280}{351}, \cdots$

§17. $\sqrt{17} = 4+\sqrt{17}-4 = 4+\dfrac{1}{\sqrt{17}+4}; \quad \sqrt{17}+4 = 8+\sqrt{17}-4;$

∴ the continued fraction $= 4+\dfrac{1}{8+}\ \dfrac{1}{8+} \cdots;$

and the convergents are $\dfrac{4}{1}, \dfrac{33}{8}, \dfrac{268}{65}, \dfrac{2177}{528}, \cdots$

∴ the error in taking $\dfrac{268}{65}$ is less than $\dfrac{1}{(65)^2}$ and greater than $\dfrac{1}{2(528)^2}$.

§18. $\sqrt{23} = 4+\dfrac{1}{1+}\ \dfrac{1}{3+}\ \dfrac{1}{1+}\ \dfrac{1}{8+} \cdots;$ and the convergents are

$$\frac{4}{1}, \frac{5}{1}, \frac{19}{4}, \frac{24}{5}, \frac{211}{44}, \frac{235}{49}, \frac{916}{191}, \frac{1151}{240}, \cdots;$$

∴ the error in taking $\dfrac{916}{191}$ is less than $\dfrac{1}{(191)^2}$ and greater than $\dfrac{1}{2(240)^2}$.

§19. $\sqrt{101} = 10+\dfrac{1}{20+}\ \dfrac{1}{20+} \cdots;$ and the convergents are

$$\frac{10}{1}, \frac{201}{20}, \frac{4030}{401}, \cdots$$

The third convergent differs from $\sqrt{101}$ by less than $\dfrac{1}{(401)^2}$, and is therefore correct to five places of decimals.

§20. $\sqrt{15} = 3+\dfrac{1}{1+}\ \dfrac{1}{6+} \cdots;$ and the convergents are

$$\frac{3}{1}, \frac{4}{1}, \frac{27}{7}, \frac{31}{8}, \frac{213}{55}, \frac{244}{63}, \frac{1677}{433}, \cdots$$

The seventh convergent differs from $\sqrt{15}$ by less than $\dfrac{1}{(433)^2}$, and is therefore correct to five places of decimals.

§21. The positive root of $x^2 + 2x - 1 = 0$ is $\sqrt{2} - 1$.

Now $\sqrt{2} - 1 = \dfrac{1}{\sqrt{2}+1}; \quad \sqrt{2}+1 = 2+\sqrt{2} - 1;$

∴ the continued fraction $= \dfrac{1}{2+}\ \dfrac{1}{2+}\ \dfrac{1}{2+} \cdots$

§22. The positive root of $x^2 - 4x - 3 = 0$ is $\sqrt{7} + 2$.

Now $\sqrt{7}+2 = 4+\sqrt{7}-2 = 4+\dfrac{3}{\sqrt{7}+2};$

$$\frac{\sqrt{7}+2}{3} = 1+\frac{\sqrt{7}-1}{3} = 1+\frac{2}{\sqrt{7}+1};$$

$$\frac{\sqrt{7}+1}{2} = 1+\frac{\sqrt{7}-1}{2} = 1+\frac{3}{\sqrt{7}+1};$$

$$\frac{\sqrt{7}+1}{3} = 1 + \frac{\sqrt{7}-2}{3} = 1 + \frac{1}{\sqrt{7}+2}; \quad \sqrt{7}+2 = 4 + \sqrt{7}-2;$$

∴ the continued fraction $= 4 + \cfrac{1}{1+}\cfrac{1}{1+}\cfrac{1}{1+}\cfrac{1}{4+}\cdots$

§23. The positive root of $7x^2 - 8x - 3 = 0$ is $\dfrac{\sqrt{37}+4}{7}$.

Now

$$\frac{\sqrt{37}+4}{7} = 1 + \frac{\sqrt{37}-3}{7} = 1 + \frac{4}{\sqrt{37}+3};$$

$$\frac{\sqrt{37}+3}{4} = 2 + \frac{\sqrt{37}-5}{4} = 2 + \frac{3}{\sqrt{37}+5};$$

$$\frac{\sqrt{37}+5}{3} = 3 + \frac{\sqrt{37}-4}{3} = 3 + \frac{7}{\sqrt{37}+4};$$

∴ the continued fraction $= 1 + \cfrac{1}{2+}\cfrac{1}{3+}\cfrac{1}{1+}\cdots$

§24. The roots of $x^2 - 5x + 3 = 0$ are $\dfrac{5 \pm \sqrt{13}}{2}$.

Now $\dfrac{5+\sqrt{13}}{2} = 4 + \dfrac{\sqrt{13}-3}{2} = 4 + \dfrac{2}{\sqrt{13}+3}; \quad \dfrac{\sqrt{13}+3}{2} = 3 + \dfrac{\sqrt{13}-3}{2} =$

$3 + \dfrac{2}{\sqrt{13}+3};$

∴ the continued fraction $= 4 + \cfrac{1}{3+}\cfrac{1}{3+}\cdots$

Again, $\dfrac{5-\sqrt{13}}{2} = \dfrac{6}{5+\sqrt{13}}; \quad \dfrac{\sqrt{13}+5}{6} = 1 + \dfrac{\sqrt{13}-1}{6} = 1 + \dfrac{2}{\sqrt{13}+1};$

$\dfrac{\sqrt{13}+1}{2} = 2 + \dfrac{\sqrt{13}-3}{2} = 1 + \dfrac{2}{\sqrt{13}+3}; \quad \dfrac{\sqrt{13}+3}{2} = 3 + \dfrac{\sqrt{13}-3}{2};$

∴ the continued fraction $= \cfrac{1}{1+}\cfrac{1}{2+}\cfrac{1}{3+}\cfrac{1}{3+}\cdots$

§25. Let $x = 3 + \cfrac{1}{6+}\cfrac{1}{6+}\cdots$; then $x - 3 = \dfrac{1}{6 + (x-3)}$, whence $x = \sqrt{10}$.

§26. Let $x = \cfrac{1}{1+}\cfrac{1}{3+}\cfrac{1}{1+}\cfrac{1}{3+}\cdots$; then $x = \cfrac{1}{1+}\cfrac{1}{3+x}$.

Therefore $x = \dfrac{3+x}{4+x}$, or $x^2 + 3x - 3 = 0$; and the given continued fraction
is the positive root of this quadratic.

§27. Here $x = 3 + \cfrac{1}{1+}\cfrac{1}{2+}\cfrac{1}{3+(x-3)}; \quad x - 3 = \cfrac{1}{1+}\cfrac{x}{2x+1} = \dfrac{2x+1}{3x+1};$

whence we obtain $3x^2 - 10x - 4 = 0$, and the continued fraction is the
positive root of this quadratic.

§28. Here $x - 5 = \cfrac{1}{1+}\cfrac{1}{1+}\cfrac{1}{1+}\cfrac{1}{x+5}$; which reduces to $3x^2 = 96$. Therefore

$x = 4\sqrt{2}$.

§29. By the method of the preceding examples

$$3 + \cfrac{1}{1+}\cfrac{1}{6+}\cdots = \sqrt{15}, \text{ and } 1 + \cfrac{1}{3+}\cfrac{1}{2+}\cdots = \sqrt{\frac{5}{3}};$$

whence the required result follows.

Or it may be proved thus;

$$3\left(1 + \cfrac{1}{3+}\cfrac{1}{2+}\cdots\right) = 3 + \cfrac{3}{3+}\cfrac{1}{2+}\cdots = 3 + \cfrac{1}{1+}\cfrac{1}{6+}\cdots$$

§30. The expressions are equal to $\dfrac{-9+\sqrt{145}}{4}$, and $\dfrac{-11+\sqrt{145}}{4}$.

The difference of these values $= \dfrac{1}{2}$.

EXAMPLES XXVII b

§1. $\sqrt{a^2+1} = a + \left(\sqrt{a^2+1} - a\right) = a + \dfrac{1}{\sqrt{a^2+1}+a}$;

$$\sqrt{a^2+1} + a = 2a + \left(\sqrt{a^2+1} - a\right) = \ldots$$

Thus $\quad \sqrt{a^2+1} = a + \dfrac{1}{2a+}\ \dfrac{1}{2a+}\ \ldots,$

and the convergents are $\dfrac{a}{1}, \dfrac{2a^2+1}{2a}, \dfrac{4a^3+3a}{4a^2+1}, \dfrac{8a^4+8a^2+1}{8a^3+4a}$.

§2. $\sqrt{a^2-a} = (a-1) + \left(\sqrt{a^2-a} - \overline{a-1}\right) = (a-1) + \dfrac{a-1}{\sqrt{a^2-a}+a-1}$;

$$\dfrac{\sqrt{a^2-a}+a-1}{a-1} = 2 + \dfrac{\sqrt{a^2-a}-(a-1)}{a-1} = 2 + \dfrac{1}{\sqrt{a^2-a}+a-1};$$

$$\dfrac{\sqrt{a^2-a}+a-1}{1} = 2(a-1) + \left(\sqrt{a^2-a} - \overline{a-1}\right) = \ldots$$

Thus $\quad \sqrt{a^2-1} = a-1 + \dfrac{1}{2+}\ \dfrac{1}{2(a-1)+}\ \ldots$

The convergents are $\dfrac{a-1}{1}, \dfrac{2a-1}{2}, \dfrac{4a^2-5a+1}{4a-3}, \dfrac{8a^2-8a+1}{8a-4}$.

§3. $\sqrt{a^2-1} = a-1 + \sqrt{a^2-1} - (a-1) = (a-1) + \dfrac{2a-2}{\sqrt{a^2-1}+a-1}$;

$$\dfrac{\sqrt{a^2-1}+a-1}{2a-2} = 1 + \dfrac{\sqrt{a^2-1}-(a-1)}{2a-2} = 1 + \dfrac{1}{\sqrt{a^2-1}+a-1};$$

$$\dfrac{\sqrt{a^2-1}+a-1}{1} = 2(a-1) + \sqrt{a^2-1} - (a-1) = \ldots$$

Thus $\quad \sqrt{a^2-1} = a-1 + \dfrac{1}{1+}\ \dfrac{1}{2(a-1)+}\ \dfrac{1}{1+}\ \dfrac{1}{2(a-1)}\ +\ldots;$

and the convergents are $\dfrac{a-1}{1}, \dfrac{a}{1}, \dfrac{2a^2-a-1}{2a-1}, \dfrac{2a^2-1}{2a}$.

§4. $\sqrt{1+\dfrac{1}{a}} = \dfrac{\sqrt{a^2+a}}{a} = 1 + \dfrac{\sqrt{a^2+a}-a}{a} = 1 + \dfrac{1}{\sqrt{a^2+a}+a}$;

$$\dfrac{\sqrt{a^2+a}+a}{1} = 2a + \dfrac{\sqrt{a^2+a}-a)}{1} = 2a + \dfrac{a}{\sqrt{a^2+a}+a};$$

$$\dfrac{\sqrt{a^2+a}+a}{a} = 2 + \dfrac{\sqrt{a^2+a}-a)}{a} = \ldots$$

Thus $\quad \sqrt{1+\dfrac{1}{a}} = 1 + \dfrac{1}{2a+}\ \dfrac{1}{2+}\ \dfrac{1}{2a+}\ \dfrac{1}{2+}\ \ldots;$

and the convergents are $\dfrac{1}{1}, \dfrac{2a+1}{2a}, \dfrac{4a+3}{4a+1}, \dfrac{8a^2+8a+1}{8a^2+4a}$.

§5. $\sqrt{a^2+\dfrac{2a}{b}} = \dfrac{\sqrt{a^2b^2+2ab}}{b} = a + \dfrac{\sqrt{a^2b^2+2ab}-ab}{b} = a + \dfrac{2a}{\sqrt{a^2b^2+2ab}+ab}$;

$$\dfrac{\sqrt{a^2b^2+2ab}+ab}{2a} = b + \dfrac{\sqrt{a^2b^2+2ab}-ab}{2a} = b + \dfrac{b}{\sqrt{a^2b^2+2ab}+ab};$$

$$\dfrac{\sqrt{a^2b^2+ab}+ab}{b} = 2a + \dfrac{\sqrt{a^2+b^2}-ab}{b} = \ldots$$

Thus $\quad \sqrt{a^2+\dfrac{2a}{b}} = a + \dfrac{1}{b+}\ \dfrac{1}{2a+}\ \dfrac{1}{b+}\ \dfrac{1}{2a+}\ \ldots;$

and the convergents are $\dfrac{a}{1},\ \dfrac{ab+1}{b},\ \dfrac{2a^2b+3a}{2ab+1},\ \dfrac{2a^2b^2+4ab+1}{2ab^2+2b}.$

§6.

$$\sqrt{a^2-\frac{a}{n}}=\frac{\sqrt{a^2n^2-an}}{n}=a-1+\frac{\sqrt{a^2n^2-an}-(a-1)n}{n}$$

$$=a-1+\frac{2an-a-n}{\sqrt{a^2n^2-an}+(a-1)n};$$

$$\frac{\sqrt{a^2n^2-an}+(a-1)n}{2an-a-n}=1+\frac{\sqrt{a^2n^2-an}-(an-a)}{2an-a-n}$$

$$=1+\frac{a}{\sqrt{a^2n^2-an}+an-a};$$

$$\frac{\sqrt{a^2n^2-an}+(an-a)}{a}=2(n-1)+\frac{\sqrt{a^2n^2-an}-(an-a)}{a}$$

$$=2(n-1)+\frac{2an-a-n}{\sqrt{a^2n^2-an}+an-a};$$

$$\frac{\sqrt{a^2n^2-an}+an-a}{2an-a-n}=1+\frac{\sqrt{a^2n^2-an}-(an-n)}{2an-a-n}$$

$$=1+\frac{n}{\sqrt{a^2n^2-an}+an-n};$$

$$\frac{\sqrt{a^2n^2-an}+an-n}{n}=2(a-1)+\frac{\sqrt{a^2n^2-an}-(a-1)n}{n}=\dots$$

Thus $\sqrt{a^2-\dfrac{a}{n}}=(a-1)+\dfrac{1}{1+}\ \dfrac{1}{2(n-1)+}\ \dfrac{1}{1+}\ \dfrac{1}{2(a-1)+}\cdots;$

and the convergents are $\dfrac{a-1}{1},\ \dfrac{a}{1},\ \dfrac{2an-a-1}{2n-1},\ \dfrac{2an-1}{2n}.$

§7.

$$\sqrt{9a^2+3}=3a+\left(\sqrt{9a^2+3}-3a\right)=3a+\frac{3}{\sqrt{9a^2+3}+3a};$$

$$\frac{\sqrt{9a^2+3}+3a}{3}=2a+\frac{\sqrt{9a^2+3}-3a}{3}=2a+\frac{1}{\sqrt{9a^2+3}+3a};$$

$$\frac{\sqrt{9a^2+3}+3a}{1}=6a+\left(\sqrt{9a^2+3}-3a\right)=\dots$$

Thus $\qquad\sqrt{9a^2+3}=3a+\dfrac{1}{2a+}\ \dfrac{1}{6a+}\ \dfrac{1}{2a+}\ \dfrac{1}{6a+}\cdots;$

and the convergents are

$$\frac{3a}{1},\ \frac{6a^2+1}{2a},\ \frac{36a^3+9a}{12a^2+1},\ \frac{72a^4+24a^2+1}{24a^3+4a},\ \frac{432a^5+180a^3+15a}{144a^4+36a^2+1}.$$

§8.

We have $\qquad\qquad\qquad x=p+\dfrac{2}{1+y}\qquad\qquad$ (44)

where $\qquad\qquad\qquad y=\dfrac{1}{p+}\ \dfrac{1}{1+y}\qquad\qquad$ (45)

From (45), $\qquad (1+y)(1-py)=y.$

From (44), $\qquad (1+y)(x-p)=2$ $\qquad\qquad$ (46)

$\qquad\qquad\qquad\therefore\ 2(1-py)=y(x-p)$ $\qquad\qquad$ (47)

From (46), $\qquad 1+y=\dfrac{2}{x-p};$ from (47), $y=\dfrac{2}{x+p}.$

By subtraction, $1=\dfrac{2}{x-p}-\dfrac{2}{x+p}=\dfrac{4p}{x^2-p^2};$

whence $x^2=p^2+4p.$

§9. $p\left(a_1 + \dfrac{1}{pqa_2 + \dfrac{1}{R_2}}\right) = pa_1 + \dfrac{p}{pqa_2 + \dfrac{1}{R_2}} = pa_1 + \dfrac{1}{qa_2 + \dfrac{1}{pR_2}};$

where $\qquad\qquad\qquad R_2 = a_3 + \dfrac{1}{pqa_4 +} \cdots$

Similarly $pR_2 = p\left(a_3 + \dfrac{1}{pqa_4 + \dfrac{1}{R_4}}\right) = pa_3 + \dfrac{1}{qa_4 + \dfrac{1}{pr_4}};$ and so on.

§10. From Ex. 1 we see that the complete quotient at any stage is always $\sqrt{a^2 + 1} + a$; hence, as in Art. 358,

$$\sqrt{a^2 + 1} = \frac{\left(\sqrt{a^2 + 1} + a\right)p_n + p_{n-1}}{\left(\sqrt{a^2 + 1} + a\right)q_n + q_{n-1}}.$$

Multiplying up, and equating rational and irrational parts,

$$(a^2 + 1)q_n = ap_n + p_{n-1} \qquad\qquad (48)$$
$$aq_n + q_{n-1} = p_n \qquad\qquad (49)$$

Now $\qquad\qquad \sqrt{a^2 + 1} = a + \dfrac{1}{2a+}\ \dfrac{1}{2a+}\ \dfrac{1}{2a+} \cdots;$

$\therefore\ p_{n+1} = 2ap_n + p_{n-1}, \quad q_{n+1} = 2aq_n + q_{n-1}.$

From (48), $\qquad 2(a^2 + 1)q_n = 2ap_n + 2p_{n-1} = p_{n+1} + p_{n-1}.$
From (49), $\qquad 2p_n = 2aq_n + 2q_{n-1} = q_{n+1} + q_{n-1}.$

§11. We have $x = \dfrac{1}{a_1+}\ \dfrac{1}{a_2 + x},$

whence $\qquad\qquad\qquad a_1x^2 + a_1a_2x - a_2 = 0 \qquad\qquad (50)$

Similarly, $\qquad\qquad 2a_1y^2 + 4a_1a_2y - 2a_2 = 0 \qquad\qquad (51)$

$$3a_1z^2 + 9a_1a_2z - 3a_2 = 0 \qquad\qquad (52)$$

From (50) and (51), $2a_1(x^2 - y^2) + 2a_1a_2(x - 2y) = 0;$
From (50) and (52), $3a_1(x^2 - z^2) + 3a_1a_2(x - 3z) = 0;$

$$\therefore\ \frac{2(x^2 - y^2)}{3(x^2 - z^2)} = \frac{2(x - 2y)}{3(x - 3z)};$$

that is $(x^2 - y^2)(x - 3z) = (x^2 - z^2)(x - 2y).$

§12. If x and y denote the two continued fractions,

$$x = a + \frac{1}{b+}\ \frac{1}{x}, \text{ or } bx^2 - abx - a = 0;$$

and $\qquad\qquad y = \dfrac{1}{b+}\ \dfrac{1}{a + y}, \text{ or } by^2 + aby - a = 0;$

Hence $x,\ -y$ are the roots of the equation $bt^2 - abt - a = 0,$

$$\therefore\ x(-y) = -\frac{a}{b}; \text{ or } xy = \frac{a}{b}.$$

§13. We have

$$x - a = \frac{1}{b+}\ \frac{1}{b+}\ \frac{1}{a+}\ \frac{1}{a+} \cdots = \frac{1}{b+}\ \frac{1}{b + y - b} = \frac{1}{b+}\ \frac{1}{y};$$

$$\therefore\ x - a = \frac{y}{by + 1}, \text{ or } bxy + x - (ab + 1)y - a = 0.$$

Similarly $\qquad y - b = \dfrac{1}{a+}\ \dfrac{1}{x}, \text{ or } axy + y - (ab + 1)x - b = 0.$

Eliminating xy, we obtain $ax - (a^2b + a)y - by + (ab^2 + b)x = a^2 - b^2$, which is the result required.

§14.

Since
$$\sqrt{a^2+1} = a + \cfrac{1}{2a+} \cfrac{1}{2a+} \cfrac{1}{2a+} \cdots,$$

we have
$$p_{n+2} = 2ap_{n+1} + p_n,$$

Similarly
$$p_{n+1} = 2ap_n + p_{n-1};$$
$$p_n = 2ap_{n-1} + p_{n-2};$$
$$\cdots\cdots\cdots\cdots\cdots$$
$$p_3 = 2ap_2 + p_1.$$

Multiply these equations by p_{n+1}, p_n, $\ldots p_2$ respectively, add, and erase the terms $p_{n+1}p_n$, $p_n p_{n-1}, \ldots p_3 p_2$ from each side of the sum; we obtain
$$p_{n+2}p_{n+1} = 2a(p_{n+1}^2 + p_n^2 + \ldots + p_2^2) + p_1 p_2,$$

or
$$p_{n+2}p_{n+1} - p_1 p_2 = 2a(p_2^2 + p_3^2 + \ldots + p_{n+1}^2);$$

and similarly for the $q's$; hence the result.

§15. Denote the continued fractions by x and y; then
$$x = \cfrac{1}{a+} \cfrac{1}{b+} \cfrac{1}{c+x} = \frac{(c+x)b+1}{(c+x)(ab+1)+a};$$

that is,
$$(1+ab)x^2 + (abc+c+a-b)x - (1+bc) = 0.$$

Again,
$$y = c + \cfrac{1}{b+} \cfrac{1}{a+} \cfrac{1}{y} = \frac{(abc+a+c)y+bc+1}{(ab+1)y+b};$$

that is,
$$(1+ab)y^2 - (abc+a+c-b)y - (1+bc) = 0.$$

Hence, x, $-y$ are the roots of the equation
$$(1+ab)t^2 + (abc+a+c-b)t - (1+bc) = 0;$$
$$\therefore x(-y) = -\frac{1+bc}{1+ab}; \text{ or } xy = \frac{1+bc}{1+ab}.$$

§16.

We have
$$\frac{\sqrt{5}+1}{2} = 1 + \cfrac{1}{1+} \cfrac{1}{1+} \cfrac{1}{1+} \cdots;$$

hence
$$p_{n+1} = p_n + p_{n-1}, \text{ or } p_{n+1} - p_{n-1} = p_n;$$
$$\therefore p_8 + p_5 + p_7 + \ldots + p_{2n-1} = (p_4 - p_2) + (p_6 - p_4) + (p_8 - p_6)$$
$$+ \ldots + (p_{2n} - p_{2n-2})$$
$$= p_{2n} - p_2.$$

Similarly for the other result.

§17.

Let
$$x = \cfrac{1}{a+} \cfrac{1}{b+} \cfrac{1}{c+} \cfrac{1}{a+} \cfrac{1}{b+} \cfrac{1}{c+} \cdots;$$

then
$$x = \cfrac{1}{a+} \cfrac{1}{b+} \cfrac{1}{c+x} = \frac{b(c+x)+1}{(ab+1)(c+x)+a};$$

on reduction we obtain $(1+ab)x^2 + (abc+a-b+c)x - (bc+1) = 0.$

Denoting the value of the second continued fraction by y, we have by interchanging a and b,
$$(1+ab)y^2 + (abc-a+b+c)y - (ac+1) = 0.$$

Subtracting and rearranging, we have
$$(1+ab)(x^2-y^2) + (ab+1)c(x-y) + (a-b)(x+y) + c(a-b) = 0;$$

that is,
$$(1+ab)(x-y)(x+y+c) + (a-b)(x+y+c) = 0;$$

now $x+y+c$ is positive, hence $(1+ab)(x-y) + (a-b) = 0$; which proves the result.

§18. In Art. 364, we have proved that the $2n^{th}$ convergent
$$= \frac{1}{2}\left(\frac{p_n}{q_n} + \frac{Nq_n}{p_n}\right) = \frac{p_n^2 + Nq_n^2}{2p_n q_n};$$

and this we denote by $\dfrac{p_{2n}}{q_{2n}}$; hence

$$q_{2n} = 2p_n q_n, \text{ and } p_{2n} = p_n^2 + N q_n^2.$$

Also from Art. 364,

$$a_1 p_n + p_{n-1} = N q_n, \quad a_1 q_n + q_{n-1} = p_n;$$

$$\therefore \ p_n^2 - N q_n^2 = p_n q_{n-1} - p_{n-1} q_n = (-1)^n; \qquad [\text{Art. 338}]$$

$$\therefore \ N q_n^2 = p_n^2 + (-1)^{n+1}; \text{ and therefore } p_{2n} = 2p_n^2 + (-1)^{n+1}.$$

§19. As in Art. 364, we have $\sqrt{N} = \dfrac{\left(a_1 + \sqrt{N}\right) p_{2n} + p_{2n-1}}{\left(a_1 + \sqrt{N}\right) q_{2n} + q_{2n-1}}$,

$$\therefore \ a_1 p_{2n} + p_{2n-1} = N q_{2n}, \qquad a_1 q_{2n} + q_{2n-1} = p_{2n}.$$

Again,
$$\frac{p_{3n}}{q_{3n}} = \frac{\left(a_1 + \dfrac{p_n}{q_n}\right) p_{2n} + p_{2n-1}}{\left(a_1 + \dfrac{p_n}{q_n}\right) q_{2n} + q_{2n-1}} = \frac{N q_{2n} + \dfrac{p_n}{q_n} p_{2n}}{p_{2n} + \dfrac{p_n}{q_n} q_{2n}} \qquad (53)$$

From formula (53) of Art. 364, we have

$$n_2 = \frac{1}{2}\left(n_1 + \frac{N}{n_1}\right) = \frac{n_1^2 + N}{2n_1}; \text{ hence } \frac{n_2}{\sqrt{N}} = \frac{n_1^2 + N}{2n_1\sqrt{N}}.$$

Componendo and dividendo, $\dfrac{n_2 + \sqrt{N}}{n_2 - \sqrt{N}} = \left(\dfrac{n_1 + \sqrt{N}}{n_1 - \sqrt{N}}\right)^2$.

The proof may be completed by induction: for suppose

$$\frac{n_k + \sqrt{N}}{n_k - \sqrt{N}} = \left(\frac{n_1 + \sqrt{N}}{n_1 - \sqrt{N}}\right)^k;$$

then proceeding in the way by which equation (53) was obtained, it is easy to shew that

$$\frac{p_{(k+1)n}}{q_{(k+1)n}} = \frac{\left(a_1 + \dfrac{p_n}{q_n}\right) p_{kn} + p_{kn-1}}{\left(a_1 + \dfrac{p_n}{q_n}\right) q_{kn} + q_{kn-1}} = \frac{N q_{kn} + \dfrac{p_n}{q_n} p_{kn}}{p_{kn} + \dfrac{p_n}{q_n} q_{kn}};$$

that is,
$$n_{k+1} = \frac{N + n_1 n_k}{n_k + n_1}; \text{ hence } \frac{n_{k+1}}{\sqrt{N}} = \frac{N + n_1 n_k}{\sqrt{N}(n_k + n_1)};$$

$$\therefore \ \frac{n_{k+1} + \sqrt{N}}{n_{k+1} - \sqrt{N}} = \frac{(n_k + \sqrt{N})(n_1 + \sqrt{N})}{(n_k - \sqrt{N})(n_1 - \sqrt{N})} = \left(\frac{n_1 + \sqrt{N}}{n_1 - \sqrt{N}}\right)^{k+1};$$

EXAMPLES XXVIII : Indeterminate Equations of the Second Degree

§1. Solving for x, we obtain
$$5x = 5y \pm \sqrt{385 - 10y^2} = 5y \pm \sqrt{5(77 - 2y^2)};$$
hence y cannot be greater than 6.

When $\quad y = 4, \qquad 5x = 20 \pm 15, \qquad$ that is, $x = 7$, or 1;

when $\quad y = 6, \qquad 5x = 30 \pm 5, \qquad$ that is, $x = 7$, or 5.

§2. Solving for x, we obtain $7x = y \pm \sqrt{189 - 20y^2}$.

When $\qquad\qquad y = 1,\ 7x = 1 \pm 13,$ that is $x = 2$;
$$y = 3,\ 7x = 3 \pm 3,$$ and there is no solution.

§3. Solving for y, we obtain
$$y = 2x \pm \sqrt{4 + 10x - x^2} = 2x \pm \sqrt{29 - (x-5)^2};$$
hence $x - 5$ cannot be greater than 5.

When $\qquad\qquad x = 3,\ y = 6 \pm 5 = 11$ or 1;
$$x = 7,\ y = 14 \pm 5 = 19 \text{ or } 9;$$
$$x = 10, y = 20 \pm 2 = 22 \text{ or } 18.$$

§4. Expressing y in terms of x, we have $y = \dfrac{8 + 2x}{x - 1} = 2 + \dfrac{10}{x - 1}$;

hence $\qquad\qquad x - 1 = \pm 1,\ \pm 2,\ \pm 5,\ \pm 10.$

$x = 2$ gives $y = 12$; $x = 3$ gives $y = 7$; $x = 6$ gives $y = 4$; $x = 11$ gives $y = 3$.

§5. Expressing y in terms of x, we have $y = \dfrac{14 - 3x}{3x - 4} = \dfrac{10}{3x - 4} - 1$;

hence $\qquad\qquad 3x - 4 = \pm 1,\ \pm 2,\ \pm 5,\ \pm 10.$

Hence $\qquad\qquad x = 2,\quad y = 4;\quad x = 3,\quad y = 1.$

§6. We have $(2x + y)(2x - y) = 315.$

The factors of 315 are $1, 315;\ 3, 105;\ 5, 63;\ 7, 45;\ 9, 35;\ 15, 21.$

Thus solutions are obtained from
$$2x + y = 315, 105, 63, 45, 35, 21;$$
$$2x - y = \quad 1, 3, 5, 7, 9, 15.$$

§7. $\sqrt{14} = 3 + \dfrac{1}{1+}\dfrac{1}{2+}\dfrac{1}{1+}\dfrac{1}{6+}$; hence the penultimate convergent is $\dfrac{15}{4}$; thus
$x = 15$, $y = 4$ is the smallest solution.

§8. $\sqrt{19} = 4 + \dfrac{1}{2+}\dfrac{1}{1+}\dfrac{1}{3+}\dfrac{1}{1+}\dfrac{1}{2+}\dfrac{1}{8+} \dots$; [Art. 355]

here the penultimate convergent is $\dfrac{170}{39}$; thus $x = 170$, $y = 39$ is the
smallest solution.

§9. $\sqrt{41} = 6 + \dfrac{1}{2+}\dfrac{1}{2+}\dfrac{1}{12+} \dots$; here the penultimate is $\dfrac{32}{5}$, and since the
number of quotients in the period is odd, $x = 32$, $y = 5$ is a solution. [Art. 370.]

§10.

As in Art. 355, $\qquad \sqrt{61} = 7 + (\sqrt{61} - 7) = 7 + \dfrac{12}{\sqrt{61} + 7};$

$$\dfrac{\sqrt{61} + 7}{12} = 1 + \dfrac{\sqrt{61} - 5}{12} = 1 + \dfrac{3}{\sqrt{61} + 5};$$

$$\dfrac{\sqrt{61} + 5}{3} = 4 + \dfrac{\sqrt{61} - 7}{3} = 4 + \dfrac{4}{\sqrt{61} + 7};$$

$$\dfrac{\sqrt{61} + 7}{4} = 3 + \dfrac{\sqrt{61} - 5}{4} = 3 + \dfrac{9}{\sqrt{61} + 5};$$

$$\frac{\sqrt{61}+5}{9} = 1 + \frac{\sqrt{61}-4}{9} = 1 + \frac{5}{\sqrt{61}+4};$$

$$\frac{\sqrt{61}+4}{5} = \dots$$

Thus 5 is the denominator of one of the complete quotients which occur in the process of converting $\sqrt{61}$ into a continued fraction; and the convergent preceding this quotient is $\frac{164}{21}$; thus $x = 164, y = 21$ is a solution.

§11. Put $x = 3x'$, $y = 3y'$, then $x'^2 - 7y'^2 = 1$; and

$$\sqrt{7} = 2 + \frac{1}{1+}\frac{1}{1+}\frac{1}{1+}\frac{1}{4+}\dots;$$

the penultimate convergent is $\frac{8}{3}$; thus $x' = 8, y' = 3$; and $x = 24, y = 9$.

§12. $\sqrt{3} = 1 + \frac{1}{1+}\frac{1}{2+}\dots$; thus $x = 2, y = 1$ is a solution; hence

$x^2 - 3y^2 = (2^2 - 3)^n$; that is, $(x + \sqrt{3}y)(x - \sqrt{3}y) = (2 + \sqrt{3})^n(2 - \sqrt{3})^n$; thus as in Art. 371,

$$2x = (2 + \sqrt{3})^n + (2 - \sqrt{3})^n; \quad 2y\sqrt{3} = (2 + \sqrt{3})^n - (2 - \sqrt{3})^n.$$

§13. $\sqrt{5} = 2 + \frac{1}{4+}\frac{1}{4+}\dots$; hence $x = 9$, $y = 4$ is a solution;

thus $\qquad x^2 - 5y^2 = (9^2 - 5 \cdot 4^2)^n = (9 + 4\sqrt{5})^n(9 - 4\sqrt{5})^n$;

$$\therefore 2x = (9 + 4\sqrt{5})^n + (9 - 4\sqrt{5})^n; \quad 2y\sqrt{5} = (9 + 4\sqrt{5})^n - (9 - 4\sqrt{5})^n.$$

§14. $\sqrt{17} = 4 + \frac{1}{8+}\frac{1}{8+}\dots$; and $x = 4$, $y = 1$ is a solution. Thus

$$(x + y\sqrt{17})(x - y\sqrt{17}) = (4 + \sqrt{17})^n(4 - \sqrt{17})^n,$$

where n is any positive integer.

$$\therefore 2x = (4 + \sqrt{17})^n + (4 - \sqrt{17})^n; \quad 2y\sqrt{17} = (4 + \sqrt{17})^n - (4 - \sqrt{17})^n.$$

§15. Put $x^2 - 3xy + 3y^2 = z^2$; then $x^2 - z^2 = 3y(x - y)$.
Put $m(x + z) = 3ny$, $n(x - z) = m(x - y)$; then by cross multiplication

$$\frac{x}{3n^2 - m^2} = \frac{y}{-m^2 + 2mn} = \frac{z}{m^2 - 3mn + 3n^2}.$$

§16. We have $(x + y)^2 = z^2 - y^2$.
Put $\qquad m(x + y) = n(z + y); \quad n(x + y) = m(z - y);$

then $\qquad \frac{x}{-m^2 + 2mn + n^2} = \frac{y}{m^2 - n^2} = \frac{z}{m^2 + n^2}.$

§17. We have $5x^2 = z^2 - y^2$.
Put $\qquad 5mx = n(z + y); \quad nx = m(z - y);$

then $\qquad \frac{x}{2mn} = \frac{y}{5m^2 - n^2} = \frac{z}{5m^2 + n^2}.$

§18.. If x and y represent the two numbers, $x^2 - y^2 = 105$. The factors of 105 are $1, 105; 3, 35; 5, 21; 7, 15$; the solution may easily be completed as in Art. 377.

§19. Denote the lengths of the two sides and hypotenuse by x, y, z respectively; then $x^2 + y^2 = z^2$, or $x^2 = z^2 - y^2$.
Put $\qquad mx = n(z + y)$, and $nx = m(z - y);$

then $\qquad \frac{x}{2mn} = \frac{y}{m^2 - n^2} = \frac{z}{m^2 + n^2}.$

§20. Let x, y be the integers; then

$$x^2 + xy + y^2 = \text{perfect square} = z^2 \text{ say; thus } x(x + y) = z^2 - y^2.$$

Put $\qquad mx = n(z + y), \quad n(x + y) = m(z - y);$

then $\qquad \frac{x}{2mn + n^2} = \frac{y}{m^2 - n^2} = \frac{z}{m^2 + mn + n^2}.$

§21. Let x denote the number of hogs bought by any one of the men, then since x shillings is the price of each hog, x^2 shillings represents the value of the hogs bought by this man. Similarly if y^2 shillings be taken to represent the value of the hogs bought by the wife of this man, we have $x^2 - y^2 = 63$.

Proceeding as in Art. 377, we find for the solution
$$x = 32, 12, 8; \quad y = 31, 9, 1.$$
Thus the men bought $32, 12, 8$ hogs, and the women $31, 9, 1$.

Further, Hendriek bought 23 more than Catriin; thus Hendriek bought 32 hogs, and Catriin 9 hogs; also Class bought 11 more than Geertruij; thus Class bought 12 and Geetruij 1 hog. Hence Cornelius bought 8, and Anna 31 hogs; therefore we have the following arrangement

$$\begin{array}{llllll} \text{Hendriek} & 32\} & \text{Class} & 12\} & \text{Cornelius} & 8\} \\ \text{Anna} & 31\} & \text{Catriin} & 9\} & \text{Geertuij} & 1\} \end{array} .$$

§22. The sum of the first n natural numbers is $\dfrac{n(n+1)}{2}$. This expression is a perfect square when $n = k^2$, provided that
$$\frac{k^2+1}{2} = \text{a perfect square} = x^2 \text{ say; that is, } k^2 - 2x^2 = -1.$$

Since $\sqrt{2} = 1 + \dfrac{1}{2+}\dfrac{1}{2+}\ldots$, the number of quotients in the period is odd, and the values of k are the numerators of the odd convergents.   [Art. 370.]

Again, the expression $\dfrac{n(n+1)}{2}$ is a perfect square when $n + 1 = k^2$, provided that $\dfrac{k'^2-1}{2} = \text{a perfect square} = x^2 \text{ say; that is, } k'^2 - 2x^2 = 1.$

In this case the values of k' are the numerators of the even convergents.   [Art. 369.]

EXAMPLES XXIX : Summation of Series

EXAMPLES XXIX a

§1. Here $u_n = n(n+1)(n+2)$, and $S_n = \dfrac{n(n+1)(n+2)(n+3)}{4} + C$;

when $n = 1$, we find $C = 0$; thus $S_n = \dfrac{n(n+1)(n+2)(n+3)}{4}$.

§3.

Here
$$u_n = (3n-2)(3n+1)(3n+4),$$
$$\therefore S_n = C + \frac{(3n-2)(3n+1)(3n+4)(3n+7)}{4 \times 3};$$

when $n = 1$, we have $28 = C + \dfrac{1 \cdot 4 \cdot 7 \cdot 10}{12}$; thus $C = \dfrac{14}{3}$.

$$\therefore S_n = \frac{1}{12}(3n-2)(3n+1)(3n+4)(3n+7) + \frac{56}{12}.$$

§4. Here $u_n = n(n+3)(n+6) = n(n+1)(n+2) + 6n(n+1) + 10n.$

$$\therefore S_n = C + \frac{n(n+1)(n+2)(n+3)}{4} + 2n(n+1)(n+2) + 5n(n+1);$$

when $n = 1$, we have $28 = C + 6 + 12 + 10$; thus $C = 0$;

$$\therefore S_n = \frac{n(n+1)}{4}\{n^2 + 5n + 6 + 8n + 16 + 20\}$$

$$= \frac{n(n+1)(n+6)(n+7)}{4}.$$

§5. Here $u_n = n(n+4)(n+8) = n(n+1)(n+2) + 9n(n+1) + 21n;$

$$\therefore S_n = C + \frac{n(n+1)(n+2)(n+3)}{4} + 3n(n+1)(n+2)$$

$$+ \frac{21n(n+1)}{2},$$

and by putting $n = 1$, we find $C = 0$;

$$\therefore S_n = n(n+1)\left\{\frac{(n+2)(n+3)}{4} + 3(n+2) + \frac{21}{2}\right\}$$

$$= \frac{n(n+1)}{4}(n^2 + 17n + 72) = \frac{1}{4}n(n+1)(n+8)(n+9).$$

§6. By Art. 386, $S_n = C - \dfrac{1}{n+1}$, and it will be found that $C = 1$;

$$\therefore S_n = \frac{n}{n+1}; \text{ and clearly } s_\infty = 1.$$

§7. Here $u_n = \dfrac{1}{(3n-2)(3n+1)}$; and $S_n = C - \dfrac{1}{3(3n+1)}$.

Put $n = 1$, then $\dfrac{1}{4} = -\dfrac{1}{3 \cdot 4} + C$. thus $C = \dfrac{1}{3}, S_n = \dfrac{n}{3n+1}$, and $S_\infty = \dfrac{1}{3}$.

§8. Here $u_n = \dfrac{1}{(2n-1)(2n+1)(2n+3)}$; and

$$S_n = C - \frac{1}{4(2n+1)(2n+3)}.$$

Put $n = 1$, then $\dfrac{1}{15} = C - \dfrac{1}{4 \cdot 3 \cdot 5}$.

$$\therefore C = \frac{1}{12}, \text{ and } S_n = \frac{1}{12} - \frac{1}{4(2n+1)(2n+3)}.$$

§9.

Here
$$u_n = \frac{1}{(3n-2)(3n+1)(3n+4)}.$$

$$\therefore S_n = C - \frac{1}{6(3n+1)(3n+4)}; \ \&c.$$

§10.. Here

$$u_n = \frac{n+3}{n(n+1)(n+2)}$$

$$= \frac{1}{(n+1)(n+2)} + \frac{3}{n(n+1)(n+2)}.$$

$$\therefore S_n = C - \frac{1}{n+2} - \frac{3}{2(n+1)(n+2)}.$$

when $n = 1$, we find $C = \frac{5}{4}$.

$$\therefore S_n = \frac{5}{4} - \frac{1}{n+2} - \frac{3}{2(n+1)(n+2)} = \frac{5}{4} - \frac{2n+5}{2(n+1)(n+2)}.$$

§11. Here $u_n = \dfrac{n}{(n+2)(n+3)(n+4)} = \dfrac{(n+2)-2}{(n+2)(n+3)(n+4)}$

$$= \frac{1}{(n+3)(n+4)} - \frac{2}{(n+2)(n+3)(n+4)}; \ \&c.$$

§12. Here

$$u_n = \frac{2n-1}{n(n+1)(n+2)} = \frac{2}{(n+1)(n+2)} - \frac{1}{n(n+1)(n+2)}.$$

§13. Here $u_n = n(n+1)(n+2)(n+1) = n(n+1)(n+2)(n+3) - 2n(n+1)(n+2);$

$$\therefore S_n = C - \frac{1}{5}n(n+1)(n+2)(n+3)(n+4) - \frac{1}{2}n(n+1)(n+2)(n+3); \ \&c.$$

§14. Here $S_n = n^2(1+2+3+\ldots+n) - (1^3+2^3+\ldots+n^3) = \frac{1}{2}n^3(n+1) -$

$\left\{\dfrac{n(n+1)}{2}\right\}^2 = \dfrac{1}{4}n^2(n^2-1).$

§15. Here $u_n = (n-1)n(n+1)n = (n-1)n(n+1)(n+2) - 2(n-1)n(n+1); \ \&c.$

§16. Here $u_n = (n+1)(n+4)\{(n+2)(n+3)+2\} = (n+1)(n+2)(n+3)(n+4) + 2(n+1)(n+2) + 4(n+1);$

$\therefore S_n = C + \dfrac{1}{5}(n+1)(n+2)(n+3)(n+4)(n+5) + \dfrac{2}{3}(n+1)(n+2)(n+3) + 2(n+1)(n+2).$

When $n = 1$, we find $C = -32$, and S_n reduces to

$$\frac{1}{15}(n+1)(n+2)(3n^3+36n^2+151n+240) - 32.$$

§17.

Here $u_n = \dfrac{n^2}{4} \cdot \dfrac{4n^2-4}{4n^2-1} = \dfrac{n^2}{4} - \dfrac{3}{4} \cdot \dfrac{n^2}{4n^2-1} = \dfrac{n^2}{4} - \dfrac{3}{16} \cdot \dfrac{4n^2}{4n^2-1}$

$$= \frac{n^2}{4} - \frac{3}{16} - \frac{3}{16(4n^2-1)}.$$

$$\therefore S_n = \frac{n(n+1)(2n+1)}{24} - \frac{3n}{16} - \frac{3}{16}\left\{\frac{1}{2} - \frac{1}{2(2n+1)}\right\}$$

$$= \frac{n(n+1)(2n+1)}{24} - \frac{3}{16} \cdot \frac{4n^2+4n}{2(2n+1)}$$

$$= \frac{n(n+1)}{8}\left\{\frac{2n+1}{3} - \frac{3}{2n+1}\right\}$$

$$= \frac{n(n+1)(4n^2+4n-8)}{24(2n+1)} = \frac{(n-1)n(n+1)(n+2)}{6(2n+1)}.$$

§18. Here $u_n = \dfrac{n^2(n+1)^2-1}{n(n+1)} = n(n+1) - \dfrac{1}{n(n+1)}.$

$$\therefore S_n = \frac{n(n+1)(n+2)}{3} - \frac{n}{n+1}.$$

§19. Here $u_n = \dfrac{(n+1)(n^2+2n)+2}{n^2+2n} = n+1 + \dfrac{2(n+1)}{n(n+1)(n+2)}$

$$= n+1 + \dfrac{2}{(n+1)(n+2)} + \dfrac{2}{n(n+1)(n+2)};$$

$$\therefore S_n = C + \dfrac{n(n+1)}{2} + n - \dfrac{2}{n+2} - \dfrac{1}{(n+1)(n+2)};$$

by putting $n = 1$, we find $C = \dfrac{3}{2}$.

§20. Here $u_n = \dfrac{(n^2+n+1)(n^2-n+1)}{n(n+1)(n^2-n+1)} = \dfrac{n^2+n+1}{n^2+n} = 1 + \dfrac{1}{n(n+1)}$;

$$\therefore S_n = C + n - \dfrac{1}{n+1}; \text{ by putting } n = 1, \text{ we find } C = 1.$$

§21. The nth term of the rth order $= \dfrac{\lfloor n+r-2}{\lfloor n-1 \; \lfloor r-1}$, and the rth term of

the nth order $= \dfrac{\lfloor r+n-2}{\lfloor r-1 \; \lfloor n-1}$.

§22. The nth term of the rth order $= \dfrac{\lfloor n+r-2}{\lfloor n-1 \; \lfloor r-1}$, and the $(n+2)^{th}$ term

of the $(r-2)^{th}$ order $= \dfrac{\lfloor n+r-2}{\lfloor n+1 \; \lfloor r-3}$;

$$\therefore (r-1)(r-2) = n(n+1); \text{ whence } r-2 = n.$$

§23. The sum of the first n terms of the rth order of polygonal numbers

$$= \dfrac{1}{6}n(n+1)\{(r-2)(n-1)+3\}; \text{ [Art. 390]}$$

Hence the required sum

$$= \dfrac{1}{6}(n-1)n(n+1) \sum_{r=2}^{r=r}(r-2) + \dfrac{1}{2}n(n+1)(r-1)$$

$$= \dfrac{1}{12}(n-1)n(n+1)(r-2)(r-1) + \dfrac{1}{2}n(n+1)(r-1)$$

$$= \dfrac{(r-1)n(n+1)}{12}\{6 + (r-2)(n-1)\}$$

$$= \dfrac{(r-1)n(n+1)}{12}\{rn - 2n - r + 8\}.$$

EXAMPLES XXIX b

§1. The successive orders of differences are

4,	14,	30,	52,	80,	114,	$\cdots$
	10,	16,	22,	28,	34,	$\cdots$
	6,	6,	6,	6,		$\cdots$

Assume $u^n = A + Bn + Cn^2$; whence by putting for n the values $1, 2, 3$ successively, we get

$$4 = A + B + C,$$
$$14 = A + 2B + 4C,$$
$$30 = A + 3B + 9C;$$

and from these equaltions we find $A = 0, B = 1, C = 3$.

$$\therefore \text{ the } n^{th} \text{ term} = 3n^2 + n.$$

$$\therefore \text{ the sum of } n \text{ terms} = 3\Sigma n^2 + \Sigma n = n(n+1)^2.$$

§2. We have

$$8, \quad 26, \quad 54, \quad 92, \quad 140, \quad \ldots$$
$$18, \quad 28, \quad 38, \quad 48, \quad \ldots$$
$$10, \quad 10, \quad 10, \quad \ldots$$

$$\therefore u_n = 8 + 18(n-1) + \frac{10(n-1)(n-2)}{\lfloor 2} = 5n^2 + 3n;$$

$$\therefore S_n = 5\Sigma n^2 + 3\Sigma n = \frac{1}{3}n(n+1)(5n+7).$$

§3. We have

$$2, \quad 12, \quad 36, \quad 80, \quad 150, \quad 252, \quad \ldots$$

$$10, \quad 24, \quad 44, \quad 70, \quad 102, \quad \ldots$$

$$14, \quad 20, \quad 26, \quad 32, \quad \ldots$$

$$6, \quad 6, \quad 6, \quad \ldots$$

Assume $u_n = A + Bn + Cn^2 + Dn^3$; then by the method of Art. 397, we find $u_n = n^3 + n^2$.

$$\therefore S_n = \Sigma n^3 + \Sigma n^2 = \frac{1}{12}n(n+1)(n+2)(3n+1).$$

§4. We have

$$8, \quad 16, \quad 0, \quad -64, \quad -200, \quad -432, \quad \ldots$$

$$8, \quad -16, \quad -64, \quad -136, \quad -232, \quad \ldots$$

$$-24, \quad -48, \quad -72, \quad -96, \quad \ldots$$

$$-24, \quad -24, \quad -24 \quad \ldots$$

$$\therefore u_n = 8 + 8(n-1) - \frac{24(n-1)(n-2)}{\lfloor 2} - \frac{24(n-1)(n-2)(n-3)}{\lfloor 3}$$

$$= 8n - \frac{24(n-1)(n-2)}{\lfloor 2}\left\{1 + \frac{n-3}{3}\right\} = 8n - 4n(n-1)(n-2)$$

$$= -4n^2(n-3);$$

$\therefore$ by the method of Art. 396, we find $S_n = -n(n+1)(n^2 - 3n - 2)$.

§5. We have

$$30, \quad 144, \quad 420, \quad 960, \quad 1890, \quad 3360, \quad \ldots$$

$$114, \quad 276, \quad 540, \quad 930, \quad 1470, \quad \ldots$$

$$162, \quad 264, \quad 390, \quad 540, \quad \ldots$$

$$102, \quad 126, \quad 150, \quad \ldots$$

$$24, \quad 24, \quad \ldots$$

$\therefore u_n = 30 + 114(n-1) + 81(n-1)(n-2) + 17(n-1)(n-2)(n-3) + (n-1)(n-2)(n-3)(n-4);$

$$\therefore u_n = n^4 + 7n^3 + 14n^2 + 8n = n(n^3 + 7n^2 + 14n + 8)$$

$$= n(n+1)(n^2+6n+8) = n(n+1)(n+2)(n+4).$$

And, by Art. 383, $S_n = \dfrac{1}{20}n(n+1)(n+2)(n+3)(4n+21)$.

§6. By the method of Art. 398, we have

1,	3,	7,	13,	21,	31,	...
	2,	4,	6,	8,	10,	...
	2,	2,	2,	2,		...

Thus the scale of relation is $(1-x)^3$.

$$S = 1 + 3x + 7x^2 + 13x^3 + 21x^4 + \dots$$
$$-3xS = \quad -3x - 9x^2 - 21x^3 - 39x^4 - \dots$$
$$3x^2 S = \quad\quad\quad 3x^2 + 9x^3 + 21x^4 + \dots$$
$$-x^3 S = \quad\quad\quad\quad -x^3 - 3x^4 - \dots$$

By addition, $\quad\quad S(1-x)^3 = 1 + x^2;$

$$\therefore S = \frac{1+x^2}{(1-x)^3}.$$

Examples 7, 8, 9 may be solved in the same way.

§10. We have

1,	16,	81,	256,	625,	1296,	...
	15,	65,	175,	369,	671,	...
		50,	110,	194,	302,	...
			60,	84,	108,	...
				24,	24,	...

$\therefore$ the scale of relation $= (1-x)^5$.

Now $\quad\quad S = 1 + 16x + 81x^2 + 256x^3 + 625x^4 + 1296x^5 + \dots$
$$-5xS = \quad -5x - 80x^2 - 405x^3 - 1280x^4 - 3125x^5 - \dots$$
$$10x^2 S = \quad\quad 10x^2 + 160x^3 + 810x^4 + 2560x^5 + \dots$$
$$-10x^3 S = \quad\quad\quad -10x^3 - 160x^4 - 810x^5 - \dots$$
$$5x^4 S = \quad\quad\quad\quad 5x^4 + 80x^5 + \dots$$
$$-x^5 S = \quad\quad\quad\quad\quad -x^5 - \dots$$

$$\therefore S(1-x)^5 = 1 + 11x + 11x^2 + x^3.$$
$$\therefore S = \frac{1+11x+11x^2+x^3}{(1-x)^5}.$$

§11. The general term of the series is $n(n+1)x^n$, where $x = \dfrac{1}{3}$; hence the series is a recurring series whose scale of relation is $(1-x)^3$. [Art. 398.]

Now $\quad\quad\quad S = 2x + 6x^2 + 12x^3 + 20x^4 + \dots$
$$-3xS = \quad -6x^2 - 18x^3 - 36x^4 - \dots$$
$$3x^2 S = \quad\quad\quad 6x^3 + 18x^4 + \dots$$
$$-x^3 S = \quad\quad\quad\quad -2x^4 - \dots$$

$\therefore S(1-x)^3 = 2x$; that is, $S = \dfrac{2x}{(1-x)^3} = \dfrac{9}{4}$, since $x = \dfrac{1}{3}$.

§12. Put $x = \dfrac{1}{5}$, then we have

$$S = 1 - 4x + 9x^2 - 16x^3 + 25x^4 - 36x^5 + \ldots$$
$$Sx = \quad x - 4x^2 + 9x^3 - 16x^4 + 25x^5 - \ldots$$

$$\therefore S(1+x) = 1 - 3x + 5x^2 - 7x^2 + 9x^4 - 11x^5 + \ldots = \frac{1-x}{(1+x)^2};$$

$$\therefore S = \frac{1-x}{(1+x)^3} = \frac{25}{54}, \text{ since } x = \frac{1}{5}.$$

§13. We have

9,	16,	29,	54,	103,	...
7,	13,	25,	49,		...
6,	12,	24,		...	

$\therefore$, as in Art. 401, we assume $u_n = a \cdot 2^{n-1} + bn + c$.
Put $n = 1, 2, 3$ successively; thus we obtain $a = 6, b = 1, c = 2$.

$$\therefore u_n = 3 \cdot 2^n + n + 2.$$

$$\therefore S_n = 6(2^n - 1) + \frac{n(n+1)}{2} + 2n = 6(2^n - 1) + \frac{n(n+5)}{2}.$$

§14. We have

2,	12,	28,	50,	78,	...
10,	16,	22,	28,		...
6,	6,	6,		...	

We may therefore assume $u_n = A + Bn + Cn^2 + Dn^3$. And as in Art. 397, we find $u_n = n^3 - (n+1)^2 = n^3 - n - (n^2 + n + 1) = (n-1)n(n+1) - n(n+1) - 1$.
Whence S_n is easily found.

§15. We have

2,	5,	12,	31,	86,	...
3,	7,	19,	55,		...
4,	12,	36,		...	

$$\therefore u_n = a \cdot 3^{n-1} + bn + c,$$

and as in Art. 401 we obtain $a = 1, b = 1, c = 0$;

$$\therefore u_n = 3^{n-1} + n,$$

and $\qquad S_n = \dfrac{1}{2}(3^n - 1) + \dfrac{n(n+1)}{2} = \dfrac{3^n + n^2 + n - 1}{2}.$

§16. We have

1,	0,	1,	8,	29,	80,	193,	...
−1,	1,	7,	21,	51,	113,		...
2,	6,	14,	30,	62,			...
4,	8,	16,	32,				...

$\therefore u_n = a \cdot 2^{n-1} + bn^2 + cn + d$; and as before we find
$$a = 4, b = -1, c = -2, d = 0.$$

$$\therefore u_n = 4 \cdot 2^{n-1} - n^2 - 2n = 2^{n+1} - n^2 - 2n.$$

$$\therefore S_n = (2^2 + 2^3 + \ldots + 2^{n+1}) - \Sigma n^2 - 2\Sigma n$$

$$= 4(2^n - 1) - \frac{1}{6}n(n+1)(2n+1) - n(n+1)$$

$$= 2^{n+2} - 4 - \frac{1}{6}n(n+1)(2n+7).$$

§17. We have

$$4, \quad 13, \quad 35, \quad 94, \quad 262, \quad 755, \quad \ldots$$

$$9, \quad 22, \quad 59, \quad 168, \quad 493, \quad \ldots$$

$$13, \quad 37, \quad 109, \quad 325, \quad \ldots$$

$$24, \quad 72, \quad 216, \quad \ldots$$

$\therefore u_n = a \cdot 3^{n-1} + bn^2 + cn + d$; and as before we find $a = 3, a = \dfrac{1}{2}, c = \dfrac{3}{2}, d = -1$.

$$\therefore u_n = 3 \cdot 3^{n-1} + \frac{1}{2}n^2 + \frac{3}{2}n - 1 = 3^n - 1 + \frac{1}{2}n(n+3);$$

$$\therefore S_n = \frac{3(3^n - 1)}{2} - n + \frac{n(n+1)(2n+1)}{12} + \frac{3n(n+1)}{4}$$

$$= \frac{1}{2}(3^{n+1} - 3) + \frac{n(n+1)(n+5)}{6} - n.$$

§18. The series is the expansion of $\dfrac{1}{(1-x)^2}$;

$$\therefore S_n = 1 + 2x + 3x^2 + 4x^3 + \ldots nx^{n-1},$$

$$-2xS_n = \quad -2x - 4x^2 - 6x^3 - \ldots - 2(n-1)x^{n-1} - 2nx^n,$$

$$x^2 S_n = \quad \quad x^2 + 2x^3 + \ldots + (n-2)x^{n-1} + (n-1)x^n + nx^{n+1};$$

$$\therefore S_n(1-x)^2 = 1 - (n+1)x^n + nx^{n+1};$$

$$\therefore S_n = \frac{1 - x^n}{(1-x)^2} - \frac{nx^n}{1-x}.$$

§19.

$$S = 1 + 3x + 6x^2 + 10x^3 + \ldots + \frac{n(n+1)}{2}x^{n-1},$$

$$xS = \quad x + 3x^2 + 6x^3 + \ldots + \frac{(n-1)n}{2}x^{n-1} + \frac{n(n+1)}{2}x^n;$$

$$\therefore (1-x)S = (1 + 2x + 3x^2 + 4x^3 + \ldots \text{ to } n \text{ terms }) - \frac{n(n+1)}{2}x^n$$

by the previous Example,

$$= \frac{1 - x^n}{(1-x)^2} - \frac{nx^n}{1-x} - \frac{n(n+1)}{2}x^n.$$

§20.

We have
$$u_n = \frac{n+2}{n(n+1)} \cdot \frac{1}{2^n}.$$

Assume
$$\frac{n+2}{n(n+1)} = \frac{A}{n} + \frac{B}{n+1}; \text{ then } A = 2, B = -1.$$

$$\therefore u_n = \left(\frac{2}{n} - \frac{1}{n+1}\right)\frac{1}{2^n} = \frac{1}{n \cdot 2^{n-1}} - \frac{1}{(n+1)2^n}.$$

Similarly
$$u_{n-1} = \frac{1}{(n-1)2^{n-2}} - \frac{1}{n \cdot 2^{n-1}},$$

$$\cdots\cdots\cdots\cdots\cdots\cdots\cdots\cdots$$

$$u_1 = 1 - \frac{1}{2 \cdot 2};$$

$$\therefore S_n = 1 - \frac{1}{(n+1)2^n}.$$

§21. We have

$$u_n = \frac{n^2}{(n+1)(n+2)} \cdot .4^n.$$

Now

$$\frac{n^2}{(n+1)(n+2)} = \frac{(n+1)(n+2) - (3n+2)}{(n+1)(n+2)} = 1 - \frac{3n+2}{(n+1)(n+2)}$$

$$= 1 - \frac{4}{n+2} + \frac{1}{n+1}.$$

$$\therefore u_n = 4^n - \frac{4^{n+1}}{n+2} + \frac{4^n}{n+1},$$

$$u_{n-1} = 4^{n-1} - \frac{4^n}{n+1} + \frac{4^{n-1}}{n},$$

$$\cdots\cdots\cdots\cdots\cdots\cdots\cdots\cdots$$

$$u_2 = 4^2 - \frac{1}{4} \cdot 4^3 + \frac{1}{3} \cdot 4^2,$$

$$u_1 = 4 - \frac{1}{3} \cdot 4^2 + \frac{1}{2} \cdot 4.$$

$$\therefore S_n = (4 + 4^2 + \ldots + 4^n) - \frac{4^{n+1}}{n+2} + 2$$

$$= \frac{4^{n+1} - 4}{3} - \frac{4^{n+1}}{n+2} + 2 = \frac{n-1}{n+2} \cdot \frac{4^{n+1}}{3} + \frac{2}{3}.$$

§22. We have

3,		8,		15,		24,		35,		$\ldots$
	5,		7,		9,		11,		$\ldots$	
		2,		2,		2,		$\ldots$		

$$\therefore \text{ the } n^{th} \text{ term } = 3 + 5(n-1) + \frac{2(n-1)(n-2)}{2} = n^2 + 2n.$$

Again, we have

4,		11,		20,		31,		44,		$\ldots$
	7,		9,		11,		13,		$\ldots$	
		2,		2,		2,		$\ldots$		

$$\therefore \text{ the } n^{th} \text{ term } = 4 + 7(n-1) + (n-1)(n-2) = n^2 + 4n - 1;$$

$\therefore$ in the given series we have

$$u_n = n(n+2)(n^2 + 4n - 1) = n(n+2)\{(n+1)(n+3) - 4\}$$
$$= n(n+1)(n+2)(n+3) - 4n(n+2)$$
$$= n(n+1)(n+2)(n+3) - 4n(n+1) - 4n.$$

$$\therefore S_n = \frac{n(n+1)(n+2)(n+3)(n+4)}{5}$$

$$- \frac{4n(n+1)(n+2)}{3} - 2n(n+1)$$

$$= \frac{n(n+1)(3n^2 + 27n^2 + 58n + 2)}{15}, \text{ on reduction .}$$

§23. The series is $1^2 \cdot 3 + 2^2 \cdot 7 + 3^2 \cdot 13 + 4^2 \cdot 21 + 5^2 \cdot 31 + \ldots$.

Consider the successive orders of differences of $3, 7, 13, 21, 31, \ldots$

We have

3,		7,		13,		21,		31,		$\ldots$
	4,		6,		8,		10,		$\ldots$	
		2,		2,		2,		$\ldots$		

$$\therefore n^{th} \text{ term } = 3 + 4(n-1) + (n-1)(n-2) = n^2 + n + 1;$$

$$\therefore u_n = n^2(n^2 + n + 1) = \{(n-1)(n+1) + 1\}\{n(n+2) - n + 1\}$$
$$= (n-1)n(n+1)(n+2) - (n-1)(n+1)n$$

$$+ (n-1)(n+1) + 1 + n(n+2) - n$$
$$= (n-1)n(n+1)(n+2) - (n-1)n(n+1) + n(2n+1);$$
$$\therefore S_n = \frac{1}{5}(n-1)n(n+1)(n+2)(n+3) - \frac{1}{4}(n-1)n(n+1)(n+2)$$
$$+ \frac{2}{3}n(n+1)(n+2) - \frac{1}{2}n(n+1)$$
$$= \frac{n(n+1)}{60}$$
$$\{12(n^3 + 4n^2 + n - 6) - 15(n^2 + n - 2) + 40(n+2) - 30\}$$
$$= \frac{n(n+1)(12n^2 + 33n^2 + 37n + 8)}{60}$$

§24. As in the last example we find $u_n = n(1 + n + 3n^2) = n + n^2 + 3n^3$,
$$\therefore S_n = \Sigma n + \Sigma n^2 + 3\Sigma n^3 = \frac{n(n+1)}{2}\left\{1 + \frac{2n+1}{3} + \frac{3n(n+1)}{2}\right\}$$
$$= \frac{n(n+1)}{2} \cdot \frac{6 + 4n + 2 + 9n^2 + 9n}{6} = \frac{n(n+1)(9n^2 + 13n + 8)}{12}.$$

§25.
$$u_n = \frac{n}{1 \cdot 3 \cdot 5 \ldots (2n-1)(2n+1)}$$
$$= \frac{A(n+1) + B}{1 \cdot 3 \ldots (2n+1)} - \frac{An + B}{1 \cdot 3 \ldots (2n-1)}, \text{ say.}$$
$$\therefore n = An + A + B - (An + B)(2n+1);$$

whence, by equating coefficients, $A = 0$, $B = -\dfrac{1}{2}$

$$\therefore u_n = \frac{1}{2} \cdot \frac{1}{1 \cdot 3 \ldots (2n-1)} - \frac{1}{2} \cdot \frac{1}{1 \cdot 3 \ldots (2n+1)},$$
$$u_{n-1} = \frac{1}{2} \cdot \frac{1}{1 \cdot 3 \ldots (2n-3)} - \frac{1}{2} \cdot \frac{1}{1 \cdot 3 \ldots (2n-1)},$$
$$\ldots\ldots\ldots\ldots\ldots\ldots\ldots\ldots\ldots\ldots\ldots\ldots$$
$$u_2 = \frac{1}{2} \cdot \frac{1}{1 \cdot 3} - \frac{1}{2} \cdot \frac{1}{1 \cdot 3 \cdot 5},$$
$$u_1 = \frac{1}{2} \cdot \frac{1}{1} - \frac{1}{2} \cdot \frac{1}{1 \cdot 3};$$
$$\therefore S_n = \frac{1}{2} - \frac{1}{2} \cdot \frac{1}{1 \cdot 3 \cdot \ldots (2n+1)}.$$

§26.

Since
$$\frac{n}{\lfloor n+2} = \frac{(n+2) - 2}{\lfloor n+2} = \frac{1}{\lfloor n+1} - \frac{2}{\lfloor n+2};$$
$$\therefore u_n = \frac{n \cdot 2^n}{\lfloor n+2} = \frac{2^n}{\lfloor n+1} - \frac{2^{n+1}}{\lfloor n+2};$$
$$\therefore S_n = 1 - \frac{2^{n+1}}{\lfloor n+2}.$$

§27. The nth term of the series $2, 4, 7, 11, 16, \ldots$ is $\dfrac{n^2 + n + 2}{2}$;
$$\therefore u_n = (n^2 + n + 2)2^{n-1} = (An^2 + Bn + C)2^{n-1} - \{A(n-1)^2 + B(n-1) + C\}2^{n-2},$$
say.
As in Ex. 4, Art. 403, we find $A = 2, B = -2, C = 8$.
$$\therefore u_n = (2n^2 - 2n + 8)2^{n-1} - \{2(n-1)^2 - 2(n-1) + 8\}2^{n-2}.$$
$$\therefore S_n = (n^2 - n + 4)2^n - 4.$$

§28. Here $u_n = (2n-1)3^n = (An + B)3^n - \{A(n-1) + B\}3^{n-1}$, suppose.

Divide by 3^{n-1}, and equate coefficients; thus we obtain $A = 3, B = -3$.

$$\therefore u_n = (3n-3)3^n - \{3(n-1)-3\}3^{n-1} = (n-1)3^{n+1} - (n-2)3^n;$$

$$\therefore S_n = (n-1)3^{n+1} + 3.$$

§29. We have $u_n = \dfrac{1 \cdot 3 \cdot 5 \ldots (2n-1)}{2 \cdot 4 \cdot 6 \ldots (2n+2)}$

$$= \frac{1 \cdot 3 \cdot 5 \ldots (2n-1)}{2 \cdot 4 \cdot 6 \ldots 2n} - \frac{1 \cdot 3 \cdot 5 \ldots (2n+1)}{2 \cdot 4 \cdot 6 \ldots (2n+1)};$$

$$\therefore S_n = \frac{1}{2} - \frac{1 \cdot 3 \cdot 5 \ldots (2n+1)}{2 \cdot 4 \cdot 6 \ldots (2n+1)}.$$

§30. The coefficient of 2^{n-1} is $\dfrac{n^2+1}{n(n+1)}$.

$$\text{Now } \frac{n^2+1}{n(n+1)} = 1 - \frac{2}{n+1} + \frac{1}{n}$$

$$= \left(2 - \frac{2}{n+1}\right) - \left(1 - \frac{1}{n}\right)$$

$$= \frac{2n}{n+1} - \frac{n-1}{n};$$

$$\therefore u_n = \frac{(n^2+1)2^{n-1}}{n(n+1)}$$

$$= \frac{n \cdot 2^n}{n+1} - \frac{(n-1)2^{n-1}}{n};$$

$$\therefore S_n = \frac{n \cdot 2^n}{n+1}.$$

§31.

We have

$$\frac{n+3}{n(n+2)} = \frac{3}{2n} - \frac{1}{2(n+2)};$$

$$\therefore u_n = \frac{n+3}{n(n+1)(n+2)} \cdot \frac{1}{3^n}$$

$$= \frac{1}{2n(n+1)} \cdot \frac{1}{3^{n-1}} - \frac{1}{2(n+1)(n+2)} \cdot \frac{1}{3^n};$$

$$\therefore S_n = \frac{1}{4} - \frac{1}{2(n+1)(n+2)} \cdot \frac{1}{3^n}.$$

§32. The nth term of the series $1, 5, 11, 19, \ldots$ is $n^2 + n - 1$.

Now

$$u_n = \frac{n^2+n-1}{\lfloor n+2} = \frac{n(n+2)-(n+1)}{\lfloor n+2} = \frac{n}{\lfloor n+1} - \frac{n+1}{\lfloor n+2};$$

$$\therefore S_n = \frac{1}{2} - \frac{n+1}{\lfloor n+2}.$$

§33. The nth term of the series $19, 28, 39, 52, \ldots$ is $n^2 + 6n + 12$;

Now

$$\frac{n^2+6n+12}{n(n+2)} = 1 + \frac{6}{n} - \frac{2}{n+2} = \left(2 + \frac{6}{n}\right) - \left(1 + \frac{2}{n+2}\right)$$

$$= \frac{2(n+3)}{n} - \frac{n+4}{n+2};$$

$$\therefore u_n = \frac{n^2+6n+12}{n(n+1)(n+2)} \cdot \frac{1}{2^{n+1}}$$

$$= \frac{n+3}{n(n+1)} \cdot \frac{1}{2^n} - \frac{n+4}{(n+1)(n+2)} \cdot \frac{1}{2^{n+1}};$$

$$\therefore S_n = 1 - \frac{n+4}{(n+1)(n+2)} \cdot \frac{1}{2^{n+1}}.$$

EXAMPLES XXIX c

§1.

We have
$$e^x = 1 + x + \frac{x^2}{\underline{|2}} + \frac{x^3}{\underline{|3}} + \frac{x^4}{\underline{|4}} + \frac{x^5}{\underline{|5}} + \dots,$$

$$e^{-x} = 1 - x + \frac{x^2}{\underline{|2}} - \frac{x^3}{\underline{|3}} + \frac{x^4}{\underline{|4}} - \frac{x^5}{\underline{|5}} + \dots,$$

By subtraction, $e^x - e^{-x} = 2x + 2S.$

§2.

$$S = \left(\frac{x}{1} - \frac{x}{2}\right) + \left(\frac{x^2}{2} - \frac{x^2}{3}\right) + \left(\frac{x^3}{3} - \frac{x^4}{4}\right) + \dots$$

$$= \left(\frac{x}{1} + \frac{x^2}{2} + \frac{x^3}{3} + \dots\right) - \frac{1}{x}\left(\frac{x^2}{2} + \frac{x^3}{3} + \frac{x^4}{4} + \dots\right)$$

$$= -\log(1-x) - \frac{1}{x}\{-\log(1-x) - x\} = \frac{1-x}{x}\log(1-x) + 1.$$

§3. By writing down the series for e^x, e^{-x}, e^{ix} and e^{-ix} the result is easily obtained.

§4.

Here
$$u_n = \frac{\underline{|n}}{\underline{|r+n-1}} = \frac{\underline{|n}}{\underline{|r+n-2}} \cdot \frac{1}{r+n-1}.$$

$$\therefore (r-2)u_n = \frac{\underline{|n}}{\underline{|r+n-2}}\left(1 - \frac{n+1}{r+n-1}\right)$$

$$= \frac{\underline{|n}}{\underline{|r+n-2}} - \frac{\underline{|n+1}}{\underline{|r+n-1}}.$$

Thus $(r-2)u_1 = \dfrac{1}{\underline{|r-1}} - \dfrac{\underline{|2}}{\underline{|r}}$; and $(r-2)S = \dfrac{1}{\underline{|r-1}}.$

§5.

$$S = 1 + 2x + \frac{3}{\underline{|2}}x^2 + \frac{4}{\underline{|3}}x^3 + \frac{5}{\underline{|4}}x^4 + \dots$$

$$= 1 + 2x + \frac{2+1}{\underline{|2}}x^2 + \frac{3+1}{\underline{|3}}x^3 + \frac{4+1}{\underline{|4}}x^4 + \dots$$

$$= 1 + x + \frac{x^2}{\underline{|2}} + \frac{x^3}{\underline{|3}} + \frac{x^4}{\underline{|4}} + \dots + x\left(1 + x + \frac{x^2}{\underline{|2}} + \frac{x^3}{\underline{|3}} + \dots\right)$$

$$= e^x + xe^x.$$

§6.

We have
$$e^{px} = 1 + px + \frac{p^2 x^2}{\underline{|2}} + \dots + \frac{p^{r-1}x^{r-1}}{\underline{|r-1}} + \frac{p^r x^r}{\underline{|r}} + \dots$$

$$e^{qx} = 1 + qx + \frac{q^2 x^2}{\underline{|2}} + \dots + \frac{q^{r-1}x^{r-1}}{\underline{|r-1}} + \frac{q^r x^r}{\underline{|r}} + \dots;$$

$$\therefore S = \text{coefficient of } x^r \text{ in } e^{px} \times e^{qx} \text{ or } e^{(p+q)x} = \frac{(p+q)^r}{\underline{|r}}.$$

§7. The given series may be expressed as the sum of the series

$$\frac{n}{1+nx} - \frac{n(n-1)}{\lfloor 2} \cdot \frac{1}{(1+nx)^2} + \frac{n(n-1)(n-2)}{\lfloor 3} \cdot \frac{1}{(1+nx)^3} - \dots$$

and

$$\frac{nx}{1+nx}\left\{1 - (n-1)\cdot\frac{1}{1+nx} + \frac{(n-1)(n-2)}{\lfloor 2}\cdot\frac{1}{(1+nx)^2} - \dots\right\};$$

$$\therefore S = 1 - \left(1 - \frac{1}{1+nx}\right)^n + \frac{nx}{1+nx}\left(1 - \frac{1}{1+nx}\right)^{n-1}$$

$$= 1 - \left(\frac{nx}{1+nx}\right)^n + \left(\frac{nx}{1+nx}\right)^n = 1.$$

§8. The scale of relation of the recurring series $1+3x+5x^2+\dots$ is $(1-x)^2$. [Art. 398.]

$$S_n = 1 + 3x + 5x^2 + \dots + (2n-1)x^{n-1},$$
$$-2xS_n = \quad -2x - 6x^2 - \dots - (4n-6)x^{n-1} - (4n-2)x^n,$$
$$x^2 S_n = \qquad\qquad x^2 + \dots + (2n-5)x^{n-1} + (2n-3)x^n$$
$$\qquad\qquad + (2n-1)x^{n+1};$$
$$\therefore (1-x)^2 S_n = 1 + x - (2n+1)x^n + (2n-1)x^{n+1}$$
$$= 1 + x - x^n\{(2n+1) - (2n-1)x\}.$$

When $x = \dfrac{2n+1}{2n-1}$, we have $\dfrac{4}{(2n-1)^2}S_n = \dfrac{4n}{2n-1}$; that is, $S_n = n(2n-1)$.

§9. $(1+x)^n = 1 + nx + \dfrac{n(n-1)}{\lfloor 2}x^2 + \dfrac{n(n-1)(n-2)}{\lfloor 3}x^3 + \dots,$

$$\left(1 - \frac{1}{x}\right)^n = 1 - \frac{n}{x} + \frac{n(n-1)}{\lfloor 2}\cdot\frac{1}{x^2} - \frac{n(n-1)(n-2)}{\lfloor 3}\cdot\frac{1}{x^3} + \dots,$$

$\therefore S = $ the term independent of x in $(1+x)^n\left(1 - \dfrac{1}{x}\right)^n$

$= $ the term containing x^n in $(x+1)^n(x-1)^n$, the is in $(x^2-1)^n$;

$\therefore S = 0$, if n is odd; and $S = \dfrac{\lfloor n}{\left\lfloor\dfrac{n}{2}\right.\left\lfloor\dfrac{n}{2}\right.}(-1)^{\frac{n}{2}}$, if n is even.

§10. The given series is the sum of the two series $e^{\log_e 2} - 1$ and $e^{2\log_e 2} - 1$. Now $N = e^{\log_e N}$, therefore $e^{\log_e 2} = 2$, and $e^{2\log_e 2} = e^{\log_e 4} = 4$; thus
$$S = (2-1) + (4-1) = 4.$$

§11.

$$u_n = \frac{1}{(2n-1)2n(2n+1)} = \frac{1}{2}\left(\frac{1}{2n-1} - \frac{2}{2n} + \frac{1}{2n+1}\right).$$

$$\therefore 2S = \left(\frac{1}{1} - \frac{2}{2} + \frac{1}{3}\right) + \left(\frac{1}{3} - \frac{2}{4} + \frac{1}{5}\right)\left(\frac{1}{5} - \frac{2}{6} + \frac{1}{7}\right)$$

$$+ \left(\frac{1}{7} - \frac{2}{8} + \frac{1}{9}\right) + \dots$$

$$\therefore 1 + 2S = 2\left(1 - \frac{1}{2} + \frac{1}{3} - \frac{1}{4} + \frac{1}{5} - \frac{1}{6} + \dots\right) = 2\log_e 2.$$

§12. The general term of $2, 3, 6, 11, 18, \dots$ is $n^2 - 2n + 3$;

$$\therefore u_n = \frac{n^2 - 2n + 3}{\lfloor n} = \frac{n(n-1) - n + 3}{\lfloor n} = \frac{1}{\lfloor n-2} - \frac{1}{\lfloor n-1} + \frac{3}{\lfloor n}.$$

Thus $\qquad u_1 = \dfrac{3}{\lfloor 1} - 1; \quad u_2 = \dfrac{3}{\lfloor 2} - \dfrac{1}{\lfloor 1} + 1; \quad \dfrac{3}{\lfloor 3} - \dfrac{1}{\lfloor 2} + \dfrac{1}{\lfloor 1};$

hence as in Art. 404, Ex.1, $S = 3(e-1) - e + e = 3(e-1)$.

§13.

$$S = 1 + (1-1)x + \frac{\lfloor 1 + 1}{\lfloor 2}x^2 - \frac{\lfloor 2 - 1}{\lfloor 3}x^3 + \frac{\lfloor 3 + 1}{\lfloor 4}x^4 - \frac{\lfloor 4 - 1}{\lfloor 5}x^5$$

$$+ \frac{\lfloor 5 + 1}{\lfloor 6}x^6 - \ldots$$

$$= 1 + x + \frac{x^2}{\lfloor 2} + \frac{x^3}{\lfloor 3} + \frac{x^4}{\lfloor 4} + \frac{x^5}{\lfloor 5} + \frac{x^6}{\lfloor 6} + \ldots$$

$$- \left(x - \frac{x^2}{2} + \frac{x^3}{3} - \frac{x^4}{4} + \frac{x^5}{5} - \frac{x^6}{6} + \ldots\right)$$

$$= e^x - \log(1+x).$$

§14. (1) Assume $1^6 + 2^6 + 3^6 + \ldots = A_0 n^7 + A_1 n^6 + A_2 n^5 + \ldots + A_7$, then as in Art. 405, $(n+1)^6 = A_0\{(n+1)^7 - n^7\} + A_1\{(n+1)^6 - n^6\} + \ldots$

Equate coefficients of $n^6, n^5, \ldots$ [the various coefficients are given on p. 320].

Then
$$1 = 7A_0; \quad A_0 = \frac{1}{7}.$$

$$6 = 21A_0 + 6A_1; \quad A_1 = \frac{1}{2}.$$

$$15 = 35A_0 + 15A_1 + 5A_2; \quad A_2 = \frac{1}{2}.$$

$$20 = 35A_0 + 20A_1 + 10A_2 + 4A_3; \quad A_3 = 0.$$

$$15 = 21A_0 + 15A_1 + 10A_2 + 3A_4; \quad A_4 = -\frac{1}{6}.$$

$$6 = 7A_0 + 6A_1 + 5A_2 + 3A_4 + 2A_5; \quad A_5 = 0.$$

$$1 = A_0 + A_1 + A_2 + A_3 + A_4 + A_8; \quad A_6 = \frac{1}{42}.$$

And by putting $n = 1$, we have
$$1 = A_0 + A_1 + A_2 + A_3 + A_4 + A_6 + A_7; \quad A_7 = 0.$$

(2) Assume $1^7 + 2^7 + 3^7 + \ldots + n^7 = A_0 n^8 + A_1 n^7 + A_2 n^6 + \ldots + A_7 n + A_8;$

then
$$(n+1)^7 = A_0(n+1)^8 - n^8 + A_1\{(n+1)^7 - n^7\} + \ldots + A_7.$$

$$\therefore 1 = 8A_0; \quad A_0 = \frac{1}{8}.$$

$$7 = 28A_0 + 7A_1; \quad A_1 = \frac{1}{2}.$$

$$21 = 56A_0 + 21A_1 + 6A_2; \quad A_2 = \frac{7}{12}.$$

$$35 = 70A_0 + 35A_1 + 15A_2 + 5A_3; \quad A_3 = 0.$$

$$35 = 56A_0 + 35A_1 + 20A_2 + 4A_4; \quad A_4 = -\frac{7}{24}.$$

$$21 = 28A_0 + 21A_1 + 15A_2 + 6A_4 + 3A_5; \quad A_5 = 0.$$

$$7 = 8A_0 + 7A_1 + 6A_2 + 4A_4 + 2A_8; \quad A_8 = \frac{1}{12}.$$

$$1 = A_0 + A_1 + A_2 + A_4 + A_8 + A_7; \quad A_7 = 0.$$

Putting $n = 1$, $\quad 1 = A_0 + A_1 + A_2 + A_4 + A_8 + A_8; \quad A_8 = 0.$

§15.

$$u_{n+1} = \frac{(n+1)^3}{\lfloor n} = \frac{n(n-1)(n-2) + 6n(n-1) + 7n + 1}{\lfloor n}$$

$$= \frac{1}{\underline{n-3}} + \frac{6}{\underline{n-2}} + \frac{7}{\underline{n-1}} + \frac{1}{\underline{n}}.$$

Thus $u_1 = 1$; $u_2 = \dfrac{1}{\underline{1}} + 7$; $u_3 = \dfrac{1}{\underline{2}} + \dfrac{7}{\underline{1}} + 6$; $\dfrac{1}{\underline{3}} + \dfrac{7}{\underline{2}} + \dfrac{6}{\underline{1}} + 1$;

hence as in Ex.1, Art. 404, $S = (1 + 7 + 6 + 1)e = 15e$.

§16. The required coefficient is equal to the coefficient of x^{n-1} in

$$\frac{1}{(1-x)^2 - cx} \quad \text{or} \quad \frac{1}{(1-x)^2}\left\{1 - \frac{cx}{(1-x)^2}\right\}^{-1}.$$

Expanding by the Binomial Theorem, the last expression becomes

$$\frac{1}{(1-x)^2} + \frac{cx}{(1-x)^4} + \frac{c^2 x^2}{(1-x)^6} + \frac{c^3 x^3}{(1-x)^8} + \dots$$

and we have to pick out from the expansions of

$$(1-x)^{-2}, (1-x)^{-4}, (1-x)^{-6}, (1-x)^{-8}, \dots$$

the terms involving $x^{n-1}, x^{n-2}, x^{n-3}, x^{n-4}, \dots$ respectively, and multiply them by $1, c, c^2, c^3, \dots$

§17. (1) This is the particular case of Ex.3, Art. 404, in which $a = 1$, $b = 2$.

(2) By putting $a = 1$, $b = 1$ in Ex.3, Art. 404, and proceeding as in that example, we have

$$S = \text{ the coefficient of } x^n \text{ in } \frac{1}{1 - x + x^2}$$

$$= \text{ the coefficient of } x^n \text{ in } \frac{1+x}{1+x^3}$$

$$= \text{ the coefficient of } x^n \text{ in } (1+x)(1+x^3)^{-1};$$

and since n is a multiple of 3 the coefficient of x^n is unity and is negative when n is odd, and positive when n is even. Hence $S = (-1)^n$.

§18.

If

$$(x+1)^n = x^n + c_1 x^{n-1} + c_2 x^{n-2} + c_3 x^{n-3} + \dots;$$

then

$$(e^x + 1)^n = e^{nx} + c_1 e^{(n-1)x} + c_2 e^{(n-2)x} + \dots,$$

$$(e^x - 1)^n = e^{nx} - c_1 e^{(n-1)x} + c_2 e^{(n-2)x} + \dots;$$

$$\therefore 2\{e^{nx} + c_2 e^{(n-2)x} + c_4 e^{(n-4)x} + \dots\} = (e^x + 1)^n + (e^x - 1)^n.$$

Equating coefficients of x^3, we have

$$\frac{2S}{\underline{3}} = \text{ the coefficient of } x^3 \text{ in } (e^x + 1)^n + (e^x - 1)^n$$

$$= \text{ the coefficient of } x^3 \text{ in } \left(2 + x + \frac{x^2}{2} + \frac{x^3}{6} + \dots\right)^n;$$

that is, in $n \cdot 2^{n-1}\left(x + \frac{x^2}{2} + \frac{x^3}{6}\right) + \dfrac{n(n-1)}{\underline{2}} \cdot 2^{n-2}\left(x + \frac{x^2}{2}\right)^2$

$$+ \frac{n(n-1)(n-2)}{\underline{3}} \cdot 2^{n-3} \cdot x^3;$$

$$\therefore \frac{2S}{6} = \frac{n \cdot 2^{n-1}}{6} + \frac{n(n-1)}{2} 2^{n-2} + \frac{n(n-1)(n-2)}{\underline{3}} 2^{n-3}$$

$$= \frac{2^{n-3}}{6}\{4n + 6n(n-1) + n(n-1)(n-2)\};$$

$$\therefore S = n^2(n+3)2^{n-4}.$$

§19. (1) $u_n = \dfrac{n}{1 + n^2 + n^4} = \dfrac{1}{2}\left(\dfrac{1}{1 - n + n^2} - \dfrac{1}{1 + n + n^2}\right)$;

$$\therefore S_n = \frac{1}{2}\left(1 - \frac{1}{1 + n + n^2}\right).$$

(2) For the odd terms, $u_n = \dfrac{2n+3}{n(n+1)} = \dfrac{3}{n} - \dfrac{1}{n+1}$; and for the even

terms, $u_n = \dfrac{2n-1}{n(n+1)} = \dfrac{3}{n+1} - \dfrac{1}{n}$. Hence

$$S_{2m} = \left(\frac{3}{1} - \frac{1}{2}\right) + \left(\frac{1}{2} - \frac{3}{3}\right)$$
$$+ \left(\frac{3}{3} - \frac{1}{4}\right) + \left(\frac{1}{4} - \frac{3}{5}\right) + \ldots + \left(\frac{1}{2m} - \frac{3}{2m+1}\right);$$

and

$$S_{2m+1} = \left(\frac{3}{1} - \frac{1}{2}\right) + \left(\frac{1}{2} - \frac{3}{3}\right)$$
$$+ \left(\frac{3}{3} - \frac{1}{4}\right) + \left(\frac{1}{4} - \frac{3}{5}\right) + \left(\frac{3}{2m+1} - \frac{1}{2m+2}\right);$$

that is, $\qquad S_{2m} = 3 - \dfrac{3}{2m+1}$, and $S_{2m+1} = 3 - \dfrac{1}{2m+2}$;

$$\therefore S_n = 3 - \frac{2 + (-1)^n}{n+1}.$$

§20. $\dfrac{1}{n(n+1)(n+2)} = \dfrac{1}{2}\left(\dfrac{1}{n} - \dfrac{2}{n+1} + \dfrac{1}{n+2}\right);$

$$\therefore 2S = \left(\frac{1}{1} - \frac{2}{2} + \frac{1}{3}\right)x - \left(\frac{1}{2} - \frac{2}{3} + \frac{1}{4}\right)x^2 + \left(\frac{1}{3} - \frac{2}{4} + \frac{1}{5}\right)x^3 + \ldots$$
$$= \left(x - \frac{x^2}{2} + \frac{x^3}{3} - \ldots\right) - \frac{2}{x}\left(\frac{x^2}{2} - \frac{x^3}{3} + \frac{x^4}{4} - \ldots\right)$$
$$+ \frac{1}{x^2}\left(\frac{x^3}{3} - \frac{x^4}{4} + \frac{x^5}{5} - \ldots\right)$$
$$= \log(1+x) - \frac{2}{x}\{x - \log(1+x)\} + \frac{1}{x^2}\left\{-x + \frac{x^2}{2} + \log(1+x)\right\}$$
$$= \left(1 + \frac{2}{x} + \frac{1}{x^2}\right)\log(1+x) - \frac{3}{2} - \frac{1}{x}.$$

§21.
$$(e^x + 1)^n = e^{nx} + c_1 e^{(n-1)x} + c_2 e^{(n-2)x} + c_3 e^{(n-3)x} + \ldots$$
$$(e^x - 1)^n = e^{nx} - c_1 e^{(n-1)x} + c_3 e^{(n-2)x} - c_3 e^{(n-3)x} + \ldots$$
$$\therefore 2\left\{c_1 e^{(n-1)x} + c_3 e^{(n-3)x} + c_5 e^{(n-5)x} + \ldots\right\} = (e^x + 1)^n - (e^x - 1)^n$$

Equate coefficients of x^2; then, if S denote the required series, we have

$$\frac{2S}{\lfloor 2} = \text{ the coefficient of } x^2 \text{ in } (e^x + 1)^n - (e^x - 1)^n;$$

that is, in

$$\left(2 + x + \frac{x^2}{2}\right) \text{ or in } 2^{n-1}n\left(x + \frac{x^2}{2}\right) + \frac{n(n-1)}{\lfloor 2}2^{n-2}x^2;$$

$$\therefore S = n \cdot 2^{n-2} + n(n-1)2^{n-3} = n(n+1)2^{n-3}.$$

§22. (1) When n is odd,

$$u_n = \frac{2^n}{(2^n - 1)(2^{n+1} + 1)} = \frac{1}{3}\left(\frac{1}{2^n - 1} + \frac{1}{2^{n+1} + 1}\right);$$

and when n is even,

$$u_n = \frac{2^n}{(2^n + 1)(2^{n+1} - 1)} = \frac{1}{3}\left(\frac{1}{2^n + 1} + \frac{1}{2^{n+1} - 1}\right);$$

$$\therefore 3S_{2m} = \left(\frac{1}{1} + \frac{1}{5}\right) - \left(\frac{1}{5} + \frac{1}{7}\right) + \left(\frac{1}{7} + \frac{1}{17}\right) - \left(\frac{1}{17} + \frac{1}{31}\right) + \ldots -$$

$$\left(\frac{1}{2^{2m}+1}+\frac{1}{2^{2m+1}-1}\right);$$

$$\text{and } 3S_{2m+1} = \left(\frac{1}{1}+\frac{1}{5}\right)-\left(\frac{1}{5}+\frac{1}{7}\right)+\left(\frac{1}{7}+\frac{1}{17}\right)-\left(\frac{1}{17}+\frac{1}{31}\right)+\dots$$

$$+\left(\frac{1}{2^{2m+1}-1}+\frac{1}{2^{2m+2}+1}\right);$$

that is, $\qquad 3S_{2m} = 1 - \dfrac{1}{2^{2m+1}-1}, \quad 3S_{2m+1} = 1 + \dfrac{1}{2^{2m+2}+1};$

$$\therefore 3S_n = 1 + \frac{(-1)^{n+1}}{2^{n+1}+(-1)^{n+1}}.$$

(2) The general term of the series $7, 17, 31, 49, 71, \dots$ is

$$7 + 10(n-1) + 2(n-1)(n-2) \text{ or } 2n^2 + 4n - 1;$$

$$\therefore u_n = \frac{2n^2 + 4n + 1}{n(n+1)(n+2)} = \frac{1}{2}\left(\frac{1}{n}+\frac{2}{n+1}+\frac{1}{n+2}\right);$$

$$\therefore 2S_n = \left(\frac{1}{1}+\frac{2}{2}+\frac{1}{3}\right)-\left(\frac{1}{2}+\frac{2}{3}+\frac{1}{4}\right)+\left(\frac{1}{3}+\frac{2}{4}+\frac{1}{5}\right)-\dots$$

$$+(-1)^{n-2}\left(\frac{1}{n-1}+\frac{2}{n}+\frac{1}{n+1}\right)$$

$$+(-1)^{n-1}\left(\frac{1}{n}+\frac{2}{n+1}+\frac{1}{n+2}\right)$$

$$= 1 + \frac{1}{2} + (-1)^{n-1}\frac{1}{n+1} + (-1)^{n-1}\frac{1}{n+2}.$$

§23. Assume $(1+ax)(1+a^3x)(1+a^5x)\dots = 1 + A_1x + A_2x^2 + A_3x^3 + \dots;$ change x into a^2x; then

$$(1+a^3x)(1+a^5x)(1+a^7x)\dots$$

$$= 1 + A_1a^2x + A_2a^4x^2 + A_3a^6x^3 + \dots;$$

$$\therefore (1+ax)(1 + A_1a^2x + A_2a^4x^2 + A_3a^6x^3 + \dots)$$

$$= 1 + A_1x + A_2x^2 + A_3x^3 + \dots;$$

$$\therefore a + A_1a^2 = A_1; \text{ hence } A_1 = \frac{a}{1-a^2}.$$

$$A_1a^3 + A_2a^4 = A_2; \text{ hence } A_2 = \frac{A_1a^3}{1-a^4} = \frac{a^4}{(1-a^2)(1-a^4)}.$$

$$A_2a^5 + A_3a^6 = A_3; \text{ hence } A_3 = \frac{A_2a^5}{1-a^6} = \frac{a^9}{(1-a^2)(1-a^4)(1-a^6)}.$$

§24. Here $(1+x)^2\left(1+\dfrac{x}{2}\right)^2\left(1+\dfrac{x}{2^2}\right)^2\dots$

$$= 1 + A_1x + A_2x^2 + \dots + A_{r-2}x^{r-2} + A_{r-1}x^{r-1} + A_rx^r + \dots$$

Change x into $\dfrac{x}{2}$; then

$$\left(1+\frac{x}{2}\right)^2\left(1+\frac{x}{2^2}\right)^2\left(1+\frac{x}{2^3}\right)^2\dots$$

$$= 1 + A_1\frac{x}{2} + \dots + A_{r-2}\frac{x^{r-2}}{2^{r-2}} + A_{r-1}\frac{x^{r-1}}{2^{r-1}} + A_r\frac{x^r}{2^r} + \dots$$

$$\therefore (1+x)^2\left\{1 + A_1\frac{x}{2} + \dots + A_{r-2}\frac{x^{r-2}}{2^{r-2}} + A_{r-1}\frac{x^{r-1}}{2^{r-1}} + A_r\frac{x^r}{2^r} + \dots\right\}$$

$$= 1 + A_1x + \dots + A_{r-1}x^{r-1} + A_rx^r + \dots;$$

$$\therefore A_r\frac{1}{2^r} + 2A_{r-1}\frac{1}{2^{r-1}} + A_{r-2}\frac{1}{2^{r-2}} = A_r; \text{ hence } A_r(2^r - 1) = 4(A_{r-1} + A_{r-2}).$$

Now $A_0 = 1$, and $A_1 = $ sum of coefficients of x in the component factors [Art. 133]

$$= 2\left(1 + \frac{1}{2} + \frac{1}{2^2} + \frac{1}{2^3} + \ldots\right) = 4.$$

Put $r = 2$, then $A_2 = \frac{4}{3}(4+1) = \frac{20}{3}$;

similarly, $A_3 = \frac{4}{7}\left(\frac{20}{3} + 4\right) = \frac{128}{21}$ and $A_4 = \frac{4}{15}\left(\frac{128}{21} + \frac{20}{3}\right) = \frac{1072}{315}$.

§25.

Let $\qquad (1+x)^n = 1 + c_1 x + c_2 x^2 + c_3 x^3 + \ldots;$

then $\qquad (1+ix)^n = 1 + ic_1 x - c_2 x^2 - ic_3 x^3 + c_4 x^4 + ic_5 x^5 - \ldots;$

$\qquad (1-ix)^n = 1 - ic_1 x - c_2 x^2 + ic_3 x^3 + c_4 x^4 - ic_5 x^5 - \ldots;$

$\qquad \therefore 2ix(c_1 - c_3 x^2 + c_5 x^4 - \ldots) = (1+ix)^n - (1-ix)^n.$

Put $x^2 = 3$, so that $x = \lfloor 3$, and let S_1 denote the value of the first series; also as usual let ω, ω^2 be the imaginary cube roots of unity; so that

$$\omega = \frac{-1 + \sqrt{-3}}{2}, \quad \omega^2 = \frac{-1 - \sqrt{-3}}{2}.$$

We have $2i\lfloor 3 \cdot S_1 = (1+\sqrt{-3})^n - (1-\sqrt{-3})^n = (-2\omega^2)^n - (-2\omega)^n = (2)^n - (2)^n = 0$, when n is a multiple of 6, for then $(-\omega)^n = 1, (-\omega^2)^n = 1$.

Put $x^2 = \frac{1}{3}$, and let S_2 denote the value of the second series; then

$$\frac{2i}{\sqrt{3}} S_2 = \left(1 + \frac{\sqrt{-1}}{\sqrt{3}}\right)^n - \left(1 - \frac{\sqrt{-1}}{\sqrt{3}}\right)^n$$

$$= \left(\frac{\sqrt{-3}-1}{\sqrt{-3}}\right)^n - \left(\frac{\sqrt{-3}+1}{\sqrt{-3}}\right)^n$$

$$= \left(\frac{2\omega}{\sqrt{-3}}\right)^n - \left(\frac{-2\omega^2}{\sqrt{-3}}\right)^n = 0, \text{ if } n \text{ is a multiple of 6.}$$

§26. As in Example 3, Art. 404, we may shew that the given series is equal to the coefficient of x^n in $\dfrac{1}{1 - bx + ax^2}$, where $b = p + q$, $a = pq$. In this case

$$\frac{1}{1 - bx + ax^2} = \frac{1}{1 - (p+q)x + pqx^2} = \frac{1}{p-q}\left\{\frac{p}{1 - px} - \frac{q}{1 - qx}\right\};$$

whence the result at once follows.

§27. $P_r = \lfloor p \times$ the coefficient of x^{n-r-1} in $(1-x)^{-(p+1)}$,

$\qquad Q_r = \lfloor q \times$ the coefficient of x^{r-1} in $(1-x)^{-(q+1)}$.

$\qquad \therefore (1-x)^{-(q+1)} = \dfrac{1}{\lfloor q}\{Q_1 + Q_2 x + Q_3 x^2 + \ldots + Q_{n-1}x^{n-2} + \ldots\},$

$(1-x)^{-(p+1)} = \dfrac{1}{\lfloor p}\{P_{n-1} + P_{n-2}x + P_{n-3}x^2 + \ldots + P_1 x^{n-2} + \ldots\},$

as far as terms not higher than x^{n-2}.

$\qquad \therefore \dfrac{S}{\lfloor p \lfloor q} =$ the coefficient of x^{n-2} in $(1-x)^{-(p+1)} \times (1-x)^{-(q+1)};$

$\qquad \therefore S = \lfloor p \lfloor q \times$ the coefficient of x^{n-2} in $(1-x)^{-(p+q+2)}$

$$= \frac{\lfloor p \lfloor q \lfloor n + p + q - 1}{\lfloor p + q + 1 \lfloor n - 2}.$$

§28. Here $\dfrac{n-3}{2}$ is the coefficient of x^{n-4} in $\dfrac{1}{2}(1-x)^{-2}$;

$\dfrac{(n-5)(n-4)}{\lfloor 3}$ is the coefficient of x^{n-6} in $\dfrac{1}{3}(1-x)^{-3}$;

$\dfrac{(n-7)(n-6)(n-5)}{\lfloor 4}$ is the coefficient of x^{n-8} in $\dfrac{1}{4}(1-x)^{-4}$;

and so on. Hence $S=$ the coefficient of x^n in

$$x^2(1-x)^{-1} - \frac{1}{2}x^4(1-x)^{-2} + \frac{1}{3}x^6(1-x)^{-3} - \frac{1}{4}x^8(1-x)^{-4} + \dots$$

$$= \text{ the coefficient of } x^n \text{ in } \log\{1+x^2(1-x)^{-1}\}.$$

But $\qquad 1+x^2(1-x)^{-1} = 1 + \dfrac{x^2}{1-x} = \dfrac{1-x+x^2}{1-x} = \dfrac{1+x^3}{1-x^2};$

$$\therefore S = \text{ coefficient of } x^n \text{ in } \log(1+x^3) - \log(1-x^2).$$

If $n=6r$, the coefficient of x^n is $-\dfrac{1}{2r}$ from the first series, and $\dfrac{1}{3r}$ from the second, $\therefore S = -\dfrac{1}{6r} = -\dfrac{1}{n}.$

If $n = 6r+3$, the coefficient of x^n is $-\dfrac{1}{2r+1}$ from the first series, and zero from the second, thus $S = \dfrac{3}{n}.$

§29. $\dfrac{x}{1-x^2} - \dfrac{x^3}{1-x^6} + \dfrac{x^5}{1-x^{10}} - \dfrac{x^7}{1-x^{14}} + \dots$

$$= x + x^3 + x^5 + x^7 + x^9 + x^{11} + \dots$$
$$- x^3 - x^9 - x^{15} - x^{21} - x^{27} - x^{33} - \dots$$
$$+ x^5 + x^{15} + x^{25} + x^{35} + x^{45} + x^{55} + \dots$$
$$- x^7 - x^{21} - x^{35} - x^{49} - x^{63} - x^{77} - \dots$$

By adding the vertical columns, we obtain

$$\dfrac{x}{1+x^2} + \dfrac{x^3}{1+x^6} + \dfrac{x^5}{1+x^{10}} + \dfrac{x^7}{1+x^{14}} + \dots$$

EXAMPLES XXX : Theory of Numbers

EXAMPLES XXX a

§1.

$3675 = 3 \cdot 5^2 \cdot 7^2$; thus the multiplier is 3.

$4374 = 2 \cdot 3^7$; thus the multiplier is $2 \cdot 3$ or 6.

$18375 = 3 \cdot 5^3 \cdot 7^2$; thus the multiplier is $3 \cdot 5$ or 15.

$74088 = 2^3 \cdot 3^3 \cdot 7^3$; thus the multiplier is $2 \cdot 3 \cdot 7$ or 42.

§2.

$7623 = 3^2 \cdot 7 \cdot 11^2$; thus the multiplier is $3 \cdot 7^2 \cdot 11$ or 1617.

$109350 = 2 \cdot 5^2 \cdot 3^7$; thus the multiplier is $2^2 \cdot 3^2 \cdot 5$ or 180.

$539539 = 7^3 \cdot 11^2 \cdot 13$; thus the multiplier is $11 \cdot 13^2$ or 1859.

§3. If $x - y$ is even, then $x - y + 2y$ or $x + y$ is also even; hence $x - y$ and $x + y$ are both divisible by 2, and therefore their product is divisible by 4.

§4. Let n be the number; then the difference $= n^2 - n = n(n-1)$; and one of the numbers n, $n - 1$ must be even; hence the result.

§5. $4x^2 + 7xy - 2y^2 = (4x - y)(x + 2y)$; since $4x - y$ is a multiple of 3, it follows that $4x - y - 3(x - y)$ or $x + 2y$ is also a multiple of 3; thus the expression is divisible by 3×3 or 9.

§6. $8064 = 2^7 \cdot 3 \cdot 7^2$; hence by Art. 412,

the number of divisors $= (7 + 1)(1 + 1)(2 + 1) = 48$.

§7. $7056 = 2^4 \cdot 3^2 \cdot 7^2$; hence by Art. 413,

the number of ways $= \dfrac{1}{2}\{5 \cdot 3 \cdot 3 + 1\} = 23$.

§8. $2^{4n} - 1 = (2^4)^n - 1^n = 16^n - 1^n$, and is divisible by $16 - 1$, or 15.

§9. $n(n + 1)(n + 5) = n(n + 1)(\overline{n - 1} + 6) = (n - 1)n(n + 1) + 6n(n + 1)$; and each of the terms of this last expression is divisible by 6.

§10. The difference between a number n and its cube
$$= n^3 - n = n(n - 1)(n + 1) = (n - 1)n(n + 1),$$
and this being the product of three consecutive integers is divisible by 6; hence n^3 and n when divided by 6 must leave the same remainder.

§11. $n(n^2 + 20) = n(n^2 - 4 + 24) = n(n - 2)(n + 2) + 24n$.

Now $(n - 2)n(n + 2)$ is the product of three consecutive even integers and therefore must be divisible by $2 \cdot 4 \cdot 6$ or 48; also $24n$ is divisible by 48; hence the result.

§12.

$$n(n^2 - 1)(3n + 2) = n(n + 1)(n - 1)(\overline{n + 2} + 2n)$$
$$= (n - 1)n(n + 1)(n + 2) + 2n(n - 1)n(n + 1).$$

This last expression consists of two parts, the first of which is divisible by $\lfloor 4$ or 24. [Art. 418.]

The second part is divisible by 3; it is also divisible by 8, for if n is even, $2n^2$ is divisible by 8, and if n is odd $2(n - 1)(n + 1)$ is divisible by 8; thus the second part is also divisible by 24; hence the whole expression is divisible by 24.

§13. $n^5 - 5n^3 + 4n = n(n^2 - 4)(n^2 - 1) = (n - 2)(n - 1)n(n + 1)(n + 2)$, which being the product of five consecutive integers is divisible by $\lfloor 5$, or 120.

§14. $3^{2n} + 7 = (3^2)^n - 1 + 8 = 9^n - 1^n + 8$; now $9^n - 1^n$ is divisible by $9 - 1$ or 8; hence the result.

§15. Since n is prime to 3, $n^2 - 1$ is also divisible by 3 [Art. 421]. Also since $n^2 - 1 = (n - 1)(n + 1)$, it is the product of two consecutive even integers, since n is prime. Thus the expression is divisible by $2 \cdot 4 \cdot 3$ or 24.

§16. $n^5 - n$ is divisible by 5 [Art. 422].

Again $n^5 - n = n(n^4 - 1) = n(n - 1)(n + 1)(n^2 + 1)$; and this expression is divisible by $\lfloor 3$ or 6. Thus $n^5 - n$ is divisible by $5 \cdot 6$ or 30.

Again if n is odd, the expression $n(n - 1)(n + 1)(n^2 + 1)$ is divisible by 240; for the product of $n - 1$ and $n + 1$ is divisible by $2 \cdot 4$ or 8; one of the first three factors is divisible by 3; and $n^2 + 1$ is even, since n is odd. As in the first part of the question the expression is divisible by 5; thus it divisible by $2 \cdot 4 \cdot 3 \cdot 2 \cdot 5$ or 240.

§17. Let m and n be any two prime numbers greater than 6. Then $m^2 - n^2 = (m^2 - 1) - (n^2 - 1)$; and each part of this expression is divisible by 3. [Art. 421.] Also each part is the product of two consecutive even numbers, and therefore divisible by 8. Thus $m^2 - n^2$ is divisible by 24.

§18. If possible suppose $N^2 = 3n - 1$, then $N^2 + 1 = 3n$, a multiple of 3. But this is impossible for $N^2 + 1 = (N^2 - 1) + 2$, and by Fermat's Theorem $N^2 - 1$ is divisible by 3 when N is prime to 3; thus $N^2 + 1$ exceeds a multiple of 3 by 2; and therefore N^2 is of the form $3n + 1$. If N is not prime to 3, it is clear that N^2 must be of the form $3n$.

§19. Every number x is one of the forms $3q$, $3q \pm 1$.

If $x = 3q$, then $x^3 = 27q^3$, and is of the form $9n$.

If $x = 3q \pm 1$, then $x^3 = 27q^3 \pm 27q^2 + 9q \pm 1$, and is of the form $9n \pm 1$.

§20. N is either equal to $7n$, or else is prime to 7; in the latter case $N^6 - 1$ is a multiple of 7, and therefore either $N^3 - 1$ or $N^3 + 1$ is a multiple of 7.

Thus every cube number is of the form $7n$ or $7n \pm 1$.

Also $7n - 1 = 7(n - 1) + 6$; therefore if N^3 is divided by 7, the remainder is 0, 1, or 6.

§21. Let the number be N^6; then if N is a multiple of 7, $N = 7n$; if N is a prime to 7, $N^6 - 1 = 7n$, or $N^6 = 7n + 1$.

§22. Let $\dfrac{x(x + 1)}{2}$ be the triangular number. Then this is a multiple of 3 if either x or $x + 1$ is divisible by 3. If neither x nor $x + 1$ is divisible by 3, x must be of the form $3n + 1$; in this case $\dfrac{1}{2}x(x + 1) = \dfrac{9}{2}n(n + 1) + 1$ and is therefore of the form $3r + 1$. Thus the form $3n - 1$ is inadmissible.

§23. Let r, s represent any two of the numbers 1, 2, 3, ..., n; also suppose that $r^2 - s^2$ is divisible by $2n + 1$. Now $2n + 1$ is prime; hence either $r + s$ or $r - s$ must be divisible by $2n + 1$; but r and s are each less than n, so that $r + s$ and $r - s$ are each less than $2n + 1$; hence $r^2 - s^2$ cannot be divisible by $2n + 1$, that is r^2 and s^2 cannot leave the same remainder when divided by $2n + 1$.

§24. If a is odd, then a^x is odd; hence $a^x + a$ and $a^x - a$ are both even. If a is even, then a^x is even; hence $a^x + a$ and $a^x - a$ are both even.

§25. $(2x + 1)^{2n} = (4x^2 + 4x + 1)^n = \{4x(x + 1) + 1\}^n = (8m + 1)^n$, because $x(x + 1)$ is even; but $(8m + 1)^n = 8r + 1$; hence the result.

§26. From Fermat's Theorem, by putting $p = 13$, $N^{12} - 1 = M(13) = 13n$, when N is prime to 13; thus $N^{12} = 13n + 1$. If N is a multiple of 13 then evidently $N^{12} = 13n$.

§27. If N is not prime to 17, then $N^8 = 17n$.

If N is prime to 17, then by Fermat's Theorem $N^{16} - 1 = M(17)$; that is, $(N^8 + 1)(N^8 - 1) = M(17)$; hence $N^8 \pm 1 = 17n$; or $N^8 = 17n \pm 1$.

§28. We have $n^4 - 1 = (n + 1)(n - 1)(n^2 + 1)$; and $(n + 1)(n - 1)$ is divisible by 8, being the product of two consecutive even numbers.

By Fermat's Theorem $n^2 - 1$ is divisible by 3, and $n^4 - 1$ by 5; also $n^2 + 1$ is even. Hence $n^4 - 1$ is divisible by $8 \cdot 3 \cdot 5 \cdot 2$ or 240.

§29. $n^8 - 1$ is divisible by $n^2 - 1$ and therefore by 8, when n is prime.

Also $n^2 - 1$ is divisible by 3, from Fermat's Theorem; and $n^6 - 1$ is divisible by 7, except when $n = 7$; thus $n^8 - 1$ is divisible by $8 \cdot 3 \cdot 7$ or 168.

§30. $n^{36} - 1 = (n^{18} + 1)(n^9 + 1)(n^9 - 1)$, and each of these 3 factors is even; and of the last two factors one is divisible by 2 and the other by 4; thus $n^{36} - 1$ is divisible by $2 \cdot 2 \cdot 4$ or 16.

Again $n^2 - 1$ is divisible by 3, $n^{18} - 1$ is divisible by 19, and $n^{36} - 1$ is divisible by 37. [Art. 421.]

Thus $n^{36} - 1$ is divisible by $16 \cdot 3 \cdot 19$, 37 or 33744.

§31. Since x is odd, $x^{2p} - 1$ or $(x^p + 1)(x^p - 1)$ is divisible by 8. Again by Fermat's Theorem $x^p - 1$ is divisible by $p + 1$, and $x^{2p} - 1$ is divisible by $2p + 1$; whence the result follows at once.

§32. By Fermat's Theorem $x^{p-1} - 1 = M(p)$; hence $x^{p-1} = 1 + kp$;

$$\therefore (x^{p-1})^{p^{r-1}} = (1 + kp)^{p^{r-1}} = 1 + kpM(p^{r-1}).$$

$$\therefore x^{p^r - p^{r-1}} = 1 + M(p^r);$$

which proves the proposition.

§33. Here both a and b are prime to m; hence by Fermat's Theorem, $a^{m-1} - 1$ and $b^{m-1} - 1$ are both multiples of m; hence their difference $a^{m-1} - b^{m-1}$ must also be a multiple of m. Since a and b are both less than m, their difference $a - b$ is less than m and therefore prime to m; hence $(a^{m-1} - b^{m-1}) \div (a - b)$ must be a multiple of m; that is,

$$a^{m-2} + a^{m-1}b + a^{m-2}b^2 + \ldots + b^{m-2}$$

is a multiple of m.

EXAMPLES XXX b

§1. Let $f(n) = 10^n + 3 \cdot 4^{n+2} + 5$; then $f(n+1) = 10^{n+1} + 3 \cdot 4^{n+3} + 5$;

$$\therefore f(n+1) - f(n) = 10^n(10 - 1) + 3 \cdot 4^{n+2}(4 - 1)$$
$$= 9 \cdot 10^n + 9 \cdot 4^{n+2} = M(9).$$

And $\qquad f(1) = 10 + 3 \cdot 4^3 + 5 = 207 = M(9)$.

§2. Let $f(n) = 2 \cdot 7^n + 3 \cdot 5^n - 5$; then $f(n+1) = 2 \cdot 7^{n+1} + 3 \cdot 5^{n+1} - 5$;

$$\therefore f(n+1) - f(n) = 2 \cdot 7^n \cdot 6 + 3 \cdot 5^n \cdot 4$$
$$= M(24), \text{ for } 7^n + 5^n \text{ is even;}$$

and $\qquad f(1) = 2 \cdot 7 + 3 \cdot 5 - 5 = 24$.

§3. This will follow if we shew that $4 \cdot 6^n + 5^{n+1} - 9$ is divisible by 20.

Let

$$f(n) = 4 \cdot 6^n + 5^{n+1} - 9; \text{ then } f(n+1) = 4 \cdot 6^{n+1} + 5^{n+2} - 9;$$
$$\therefore f(n+1) - f(n) = 4 \cdot 6^n \cdot 5 + 5^n(5^2 - 5) = M(20).$$

and $\qquad f(1) = 4 \cdot 6 + 5^2 - 9 = 40 = M(20)$.

§4. $8 \cdot 7^n + 4^{n+2} = 8 \cdot 7^n + 2^{2n+4} = 8(7^n + 2^{2n+1})$.

Let $\qquad f(n) = 7^n + 2^{2n+1}$; then $f(n+1) = 7^{n+1} + 2^{2n+3}$;

$$f(n+1) - f(n) = 7^n \cdot 6 + 2^{n+1} \cdot 3 = M(3);$$

and $\qquad f(1) = 7 + 2^3 = 15 = M(3)$;

thus $7^n + 2^{2n+1}$ is divisible by 3; but 7^n is odd and 2^{2n+1} is even, hence the quotient must be odd, and therefore of the form $2r - 1$.

Thus $\qquad 8 \cdot 7^n + 4^{n+2} = 24(2r - 1)$.

§5.

By Wilson's Theorem, $\qquad 1 + \underline{|p - 1} = M(p)$;

$$\therefore 1 + (p - 1)(p - 2)\underline{|p - 3} = M(p);$$

$$\therefore 1 + (p^2 - 3p + 2)\big|p - 3 = M(p);$$

$$\therefore 1 + M(p) + 2\big|p - 3 = M(p);$$

whence the result follows at once.

§6. $a^{4b+1} - a = a(a^{4b} - 1)$, and is therefore divisible by $a(a^4 - 1)$ or $a^5 - a$; hence by Art. 422 the given expression is divisible by 5. Similarly it is divisible by $a(a^2 - 1)$ or $a^3 - a$ and therefore by 3.

If a is even, the expression is clearly divisible by 2, and if a is odd, $a^{4b} - 1$ is even and the expression is again divisible by 2. Thus the given expression is divisible by $2 \cdot 3 \cdot 5$ or 30.

§7. The highest power is the sum of the integral parts of the expressions

$$\frac{2^r - 1}{2}, \ \frac{2^r - 1}{2^2}, \ \frac{2^r - 1}{2^3}, \ \frac{2^r - 1}{2^{r-1}},$$

and is therefore equal to

$$(2^{r-1} - 1) + (2^{r-2} - 1) + (2^{r-3} - 1) + \ldots + (2 - 1)$$

$$= \frac{2^r - 2}{2 - 1} - (r - 1) = 2^r - r - 1.$$

§8.

Let $\qquad f(n) = 3^{4n+2} + 5^{2n+1}$; then $f(n + 1) = 3^{4n+6} + 5^{2n+3}$;

$$\therefore f(n + 1) - 25f(n) = 3^{4n+2}(81 - 25) = M(56) = M(14);$$

also $\qquad f(1) = 3^6 + 5^3 = 729 + 125 = 854 = M(14)$;

§9.

Let $\qquad f(n) = 3^{2n+5} + 160n^2 - 56n - 243$;

then $\qquad f(n + 1) = 3^{2n+7} + 160(n + 1)^2 - 56(n + 1) - 243$;

$$\therefore 9f(n) - f(n + 1) = 160(8n^2 - 2n - 1) - 56(8n - 1) - 1944$$

$$= 1280n^2 - 768n - 2048 = 256(5n^2 - 3n - 8)$$

$$= 256(5n - 8)(n + 1);$$

and it is easy to shew that $(5n - 8)(n + 1)$ is divisible by 2;

$$\therefore 9f(n) - f(n + 1) = M(512);$$

Also $\qquad f(1) = 3^7 + 160 - 56 - 243 = 2048 = M(512)$.

§10.

Let $\qquad (1 + x + x^2 + x^3 + x^4)^{n-1} = 1 + c_1 x + c_2 x^2 + c_3 x^3 + \ldots$

then $\qquad (1 - x + x^2 - x^3 + x^4)^{n-1} = 1 - c_1 x + c_2 x^2 - c_3 x^3 + \ldots$

Let $\qquad S = c_1 + c_3 + c_5 + \ldots$;

by subtracting and putting $x = 1$, we have $2S = 5^{n-1} - 1 = M(n)$, by Fermat's Theorem; hence S is divisible by n, since n is prime and greater than 2.

§11. $n^6 - 1 = M(7)$, by Fermat's Theorem.

Again $n^6 - 1$ is divisible by $n^2 - 1$ and therefore by 8. Since n must be one of the forms $3q + 1$ or $3q - 1$, it is easy to see that $n^6 - 1$ is divisible by 9; hence the given expression is divisible by $7 \cdot 8 \cdot 9$ or 504.

§12. $n^6 + 3n^4 + 7n^2 - 11 = (n^2 - 1)(n^4 + 4n^2 + 11) = M(8) \cdot (n^4 + 4n^2 + 11)$.

And $n^4 + 4n^2 + 11 = (n^2 - 1)(n^2 - 11) + 16n^2 = M(16)$, for $n^2 - 11$ is even. Thus the given expression is divisible by 8×16 or 128.

§13. Let the coefficients be denoted by $c_0, c_1, c_2, \ldots, c_r, \ldots$

Then $\qquad c_r = \dfrac{(p - 1)(p - 2)(p - 3) \ldots (p - r)}{\lfloor r}$

$$= \frac{M(p) + (-1)^r \lfloor r}{\lfloor r} = \frac{M(p)}{\lfloor r} + (-1)^r.$$

Now c_r is a positive integer; hence $\dfrac{M(p)}{\lfloor r}$ must be a positive integer,

and since p is a prime number, it must be a multiple of p; therefore
$$c_r = M(p) + (-1)^r.$$
§14. In Art. 426, Cor., it is proved that if the p terms of a series in A. P. are divided by p the remainders will be 0, 1, 2, 3, ..., $p-1$.

Hence disregarding the order of the terms, the series may be represented by ap, $bp+1$, $cp+2$, $dp+3$, ..., $kp+(p-1)$; a, b, c, d, ...k being the various quotients. With the exception of the first term all the terms of the series are prime to p; hence by Fermat's Theorem, their $(p-1)^{th}$ powers are all of the form $M(p)+1$, whilst that of the first term is of the form $M(p)$.

Thus the sum of the $(p-1)^{th}$ powers $= M(p) + p - 1 = M(p) - 1$.

§15. $(a^{12} - 1) - (b^{12} - 1) = M(13)$, by Fermat's Theorem.

Similarly, $$(a^6 - 1) - (b^6 - 1) = M(7);$$

hence $a^{12} - b^{12}$ is divisible both by 13 and 7, and therefore by 91.

§16.

By Wilson's Theorem,
$$1 + \underline{|p-1} = M(p);$$
$$\therefore 1 + (p-1)(p-2)\ldots(p-\overline{2r-1})\underline{|p-2r} = M(p);$$
$$\therefore 1 + \{M(p) - \underline{|2r-1}\}\underline{|p-2r} = M(p);$$
$$\therefore 1 + M(p) - \underline{|p-2r}\,\underline{|2r-1} = M(p);$$

whence the result follows.

§17. Since $(n-2)(n-1)n(n+1)(n+2)$ is divisible by $\underline{|5}$ or 120; and $n-1$ and $n+1$ are both prime and greater than 5,
$$\therefore\ n(n-2)(n+2) \text{ is divisible by 120, that is, } n(n^2-4) \text{ is divisible by}$$
120. Again $(n-1)n(n+1)$ is divisible by 6, and therefore n is divisible by 6 since $n-1$ and $n+1$ are prime and greater than 5.
$$\therefore\ n^2(n^2-4) \text{ is divisible by 720; also } 20n^2 \text{ is divisible by 720;}$$
$$\therefore\ n^2(n^2-4) + 20n^2, \text{ or } n^2(n^2+16) \text{ is divisible by 720.}$$

Lastly $n = 6s$, and one of the three numbers $n+2$, n, $n-2$ is divisible by 5.

(1) If $n = 6s = 5r - 2$; $s + \dfrac{s+2}{5} = r$; $s = 5t - 2$; $\therefore n = 30t - 12$.

(2) If $n = 6s = 5r$; $n = 30t$.

(3) If $n = 6s = 5r + 2$; $s + \dfrac{s-2}{5} = r$; $s = 5t + 2$; $\therefore n = 30t + 12$.

§18. The highest power required is equal to the sum of the integral parts of
$$\frac{n^r - 1}{n}, \frac{n^r - 1}{n^2}, \frac{n^r - 1}{n^3}, \ldots \frac{n^r - 1}{n^{r-1}}; \qquad \text{[Art. 416]}$$
that is,
$$= (n^{r-1} - 1) + (n^{r-2} - 1) + (n^{r-3} - 1) + \ldots + (n - 1)$$
$$= \frac{n^r - n}{n-1} - (r-1) = \frac{n^r - nr + r - 1}{n-1}.$$

§19. We have $c^2 - a = kp$, so that $a = c^2 - kp$;
$$\therefore\ a^{\frac{1}{2}(p-1)} - 1 = (c^2 - kp)^{\frac{p-1}{2}} - 1 = c^{p-1} - M(p) - 1,$$
since $p - 1$ is an even integer.

Also a is prime to p, so that c must be prime to p; hence $c^{p-1} - 1 = M(p)$, by Fermat's Theorem, and the result at once follows.

§20. The congruence $98x - 1 \equiv 0 \pmod{139}$ means that $98x - 1$ is divisible by 139; if y is the quotient, then $98x - 1 = 139y$, or $98x - 139y = 1$.

If $\dfrac{139}{98}$ is converted into a continued fraction the convergent just preceding the fraction is $\dfrac{61}{43}$. Hence the general solution of the equation is

$$x = 61 + 139t, \quad y = 43 + 98t.$$

§21. The numbers less than N and not prime to it are given by

$$\Sigma\frac{N}{a} - \Sigma\frac{N}{ab} + \Sigma\frac{N}{abc} - \dots \qquad \text{[See Art. 432.]}$$

Let us first find the sum of the squares of all numbers less than N and not prime to it.

These are given by the sum of

$$a^2 + (2a)^2 + (3a)^2 + \dots + \left(\frac{N}{a} \cdot a\right)^2$$

$$+ b^2 + (2b)^2 + (3b)^2 + \dots + \left(\frac{N}{b} \cdot b\right)^2$$

$$\dots\dots\dots\dots\dots\dots\dots\dots$$

$$-(ab)^2 - (2ab)^2 - (3ab)^2 - \dots - \left(\frac{N}{ab} \cdot ab\right)^2$$

$$-(bc)^2 - (2bc)^2 - (3bc)^2 - \dots - \left(\frac{N}{bc} \cdot bc\right)^2$$

$$\dots\dots\dots\dots\dots\dots\dots\dots\dots\dots$$

$$+(abc)^2 + (2abc)^2 + (3abc)^2 + \dots + \left(\frac{N}{abc} \cdot abc\right)^2$$

$$\dots\dots\dots\dots\dots\dots\dots\dots$$

Now

$$a^2 + (2a)^2 + (3a)^2 + \dots + \left(\frac{N}{a} \cdot a\right)^2$$

$$= \frac{1}{6}a^2\frac{N}{a}\left(\frac{N}{a} + 1\right)\left(\frac{2N}{a} + 1\right)$$

$$= \frac{N^3}{3a} + \frac{N^2}{2} + \frac{Na}{6};$$

$\therefore$ the sum of the squares of all numbers less than N and not prime to it is

$$\frac{N^3}{3}\left\{\frac{1}{a} + \frac{1}{b} + \frac{1}{c} + \dots - \frac{1}{ab} - \frac{1}{ac} - \dots + \frac{1}{abc} + \dots\right\}$$

$$+ \frac{N^2}{2}\left\{m - \frac{m(m-1)}{1\cdot 2} + \frac{m(m-1)(m-2)}{1\cdot 2\cdot 3} - \dots\right\}$$

$$+ \frac{N}{6}\{a + b + c + \dots - ab - ac - \dots + abc + \dots\},$$

where m is the number of prime factors in N.

Thus the coefficient of $\dfrac{N^2}{2} = 1 - (1-1)^m = 1$.

$\therefore$ the sum of the squares of all numbers less than N and prime to it is obtained by subtracting the above expression from

$$\frac{N}{6}(N+1)(2N+1), \quad \text{or} \quad \frac{N^3}{3} + \frac{N^2}{2} + \frac{N}{6}.$$

$\therefore$ the sum required is

$$\frac{N^3}{3}\left\{1 - \frac{1}{a} - \frac{1}{b} - \frac{1}{c} - \dots + \frac{1}{ab} + \frac{1}{ac} + \dots - \frac{1}{abc} - \dots\right\}$$

$$+ \frac{N}{6}\{1 - a - b - c - \dots + ab + ac + \dots - abc - \dots\}$$

$$= \frac{N^3}{3}\left(1 - \frac{1}{a}\right)\left(1 - \frac{1}{b}\right)\left(1 - \frac{1}{c}\right)\ldots + \frac{N}{6}(1 - a)(1 - b)(1 - c)\ldots$$

To find the sum of the cubes we may conveniently use the following method.

As in Art. 431 let the integers less than N and prime to it be denoted by

$$1, \ p, \ q, \ r, \ldots N - r, \ N - q, \ N - p, \ N - 1.$$

If x stands for any one of these integers, then $\Sigma x^3 = \Sigma(N - x)^3$; for each of these expressions denotes the sum of the same series of terms, only the order in one is the reverse of that in the order.

Hence $\qquad\qquad \Sigma x^3 = \Sigma N^3 - 3\Sigma N^2 x + 3\Sigma N x^2 - \Sigma x^3;$

that is, $\qquad\qquad 2\Sigma x^3 = N^3\phi(n) - 3N^2\Sigma x + 3N\Sigma x^2,$

for the number of terms is $\phi(N)$.

Also, by Art. 431, $\Sigma x = \dfrac{1}{2}N\phi(N)$; and we have shewn that

$$\Sigma x^2 = \frac{N^2}{3}\phi(N) + \frac{N}{6}(1 - a)(1 - b)(1 - c)\ldots$$

$$\therefore 2\Sigma x^3 = \frac{N^3}{2}\phi(N) + \frac{N^2}{2}(1 - a)(1 - b)(1 - c)\ldots$$

§22. These results are easily established by Induction. Suppose that $\lfloor p(q - 1)$ is divisible by $\left(\lfloor p\right)^{q-1}\lfloor q - 1$. Now

$$\frac{\lfloor pq}{\left(\lfloor p\right)^{q}\lfloor q} \div \frac{\lfloor p(q - 1)}{\left(\lfloor p\right)^{q-1}\lfloor q - 1} = \frac{\lfloor pq}{\lfloor pq - p} \times \frac{\lfloor q - 1}{\lfloor p \lfloor q}$$

$$= \frac{pq(pq - 1)(pq - 2)\ldots \text{ to } p \text{ factors}}{pq\lfloor q - 1}$$

$$= \frac{(pq - 1)(pq - 2)\ldots \text{ to } p - 1 \text{ factors}}{\lfloor p - 1}$$

$$= \text{ integer.}$$

But $\dfrac{\lfloor p}{\lfloor p \lfloor 1}$ is an integer; hence $\dfrac{\lfloor 2p}{\left(\lfloor p\right)^{2}\lfloor 2}$ is an integer; and so on.

§23. From Art. 389, triangular numbers are of the form $\dfrac{1}{2}n(n + 1)$; if these numbers are also square numbers, we have $\dfrac{1}{2}n(n + 1) = k^2$; it remains to shew that k is the coefficient of any power of x in the expansion of $\dfrac{1}{1 - 6x + x^2}$.

From the equation $n^2 + n = 2k^2$, it follows that $2n + 1 = \sqrt{8k^2 + 1} = t$ say; so that $t^2 - 8k^2 = 1$.

Also $\sqrt{8} = 2 + \dfrac{1}{1+}\dfrac{1}{4+}\dfrac{1}{1+}\dfrac{1}{4+}\dfrac{1}{1+}\dfrac{1}{4+}\ldots$ Hence by Art. 369, the values of k are the denominators of the even convergents of the above continued fraction.

Now $q_{2n+2} = q_{2n+1} + q_{2n}$, $\quad q_{2n+1} = 4q_{2n} + q_{2n-1}$; $\quad q_{2n} = q_{2n-1} + q_{2n-2}$. Eliminating q_{2n+1} and q_{2n-1}, we have $q_{2n+2} - 6q_{2n} + q_{2n-2} = 0$.

And since $q_2 = 1$, $q_4 = 6$, the sum of the recurring series $q_2 + q_4 x + q_6 x^2 + \ldots$, in which the scale of relation is $1 - 6x + x^2$, is $\dfrac{1}{1 - 6x + x^2}$.

All pentagonal numbers are of the form $\frac{1}{2}n(3n - 1)$. Proceeding as in the former case, we have $3n^2 - n = 2k^2$; hence

$$(6n - 1)^2 = 24k^2 + 1 = t^2; \text{ that is } t^2 - 24k^2 = 1.$$

Also

$$\sqrt{24} = 4 + \frac{1}{1+} \frac{1}{8+} \frac{1}{1+} \frac{1}{8+} \cdots$$

Here

$$q_{2n+2} = q_{2n+1} + q_{2n}; \quad q_{2n+1} = 8q_{2n} + q_{2n-1}; \quad q_{2n} = q_{2n-1} + q_{2n-2};$$

whence

$$q_{2n+2} - 10q_{2n} + q_{2n-2} = 0.$$

Also $q_2 = 1$, $q_4 = 10$, and thus the sum of the series $q_2 + q_4 x + q_6 x^2 + \ldots$

is $\dfrac{1}{1 - 10x + x^2}$.

§24. Proceeding as in Example 21, we may shew that the sum of the rth powers of all integers less than N and prime to it is

$$S_N - a^r S_{\frac{N}{a}} - b^r S_{\frac{N}{b}} - \ldots + (ab)^r S_{\frac{N}{ab}} + \ldots \tag{54}$$

where

$$S_p = 1^r + 2^r + 3^r + \ldots + p_r$$

Now by Art. 406,

$$S_N = \frac{N^{r+1}}{r+1} + \frac{1}{2}N^r + B_1 \frac{r}{\lfloor 2} N^{r-1} - B_3 \frac{r(r-1)(r-2)}{\lfloor 4} N^{r-3} + \ldots$$

$$\therefore x^r S_{\frac{N}{x}} = \frac{N^{r+1}}{r+1} \cdot \frac{1}{x} + \frac{1}{2}N^r + B_1 \frac{r}{\lfloor 2} N^{r-1} x$$

$$- B_3 \frac{r(r-1)(r-2)}{\lfloor 4} N^{r-3} x^3 + \ldots$$

Therefore, by substituting in (54), we see that the sum of the rth powers of all integers less than N and prime to it

$$= \frac{N^{r+1}}{r+1} \left\{ 1 - \frac{1}{a} - \frac{1}{b} - \frac{1}{c} - \ldots + \frac{1}{ab} + \ldots \right\}$$

$$+ \frac{1}{2}N^r \left\{ 1 - m + \frac{m(m-1)}{\lfloor 2} - \ldots \right\}$$

$$+ B_1 \frac{r}{\lfloor 2} N^{r-1} \{ 1 - a - b - c - \ldots + ab + \ldots \}$$

$$- B_3 \frac{r(r-1)(r-2)}{\lfloor 4} N^{r-3} \{ 1 - a^3 - b^3 - \ldots + a^3 b^3 + \ldots \}$$

$$+ \ldots\ldots\ldots\ldots\ldots\ldots\ldots\ldots\ldots\ldots\ldots ;$$

where m is the number of prime factors in N.

Thus the coefficient of N^r is zero, and the sum required

$$= \frac{N^{r+1}}{r+1} \left(1 - \frac{1}{a} \right) \left(1 - \frac{1}{b} \right) \left(1 - \frac{1}{c} \right) \ldots$$

$$+ B_1 \frac{r}{\lfloor 2} N^{r-1} (1 - a)(1 - b)(1 - c) \ldots$$

$$- B_3 \frac{r(r-1)(r-2)}{\lfloor 4} N^{r-3} (1 - a^3)(1 - b^3)(1 - c^3) \ldots$$

$$+ \ldots$$

If $r = 2$, $B_1 \dfrac{r}{\lfloor 2} = \dfrac{1}{6}$.   [Compare Ex. 21.]

If $r = 3$, $B_1 \dfrac{r}{\lfloor 2} = \dfrac{1}{4}$; and since $\Sigma n^3 = \dfrac{n^4}{4} + \dfrac{n^3}{2} + \dfrac{n^2}{4}$. there is no term in S_3 which involves B_3.

Again, if $r = 4$, $B_1 \dfrac{r}{\underline{|2}} = \dfrac{1}{3}$, $B_3 \dfrac{r(r-1)(r-2)}{\underline{|4}} = \dfrac{1}{30}$; and by substituting these values we obtain the result given in Ex. 24.

§25. Let 1, a, b, c,... $(N-1)$ denote the $\phi(N)$ numbers less than N and prime to it; also let x represent any one of these numbers. Then $1x$, ax, bx,... $(N-1)x$ are all different and all prime to N. There are $\phi(N)$ of such products, and, as in Art. 426, it is easily shewn that when these products are divided by N the remainders are all different and all prime to N : thus the $\phi(N)$ remainders must be 1, a, b, c,... $(N-1)$, though not necessarily in this order. Hence $x \cdot ax \cdot bx \dots (N-1)x$ must differ from $1 \cdot a \cdot b \cdot c \dots (N-1)$ by a multiple of N.

$$\therefore \ \left\{ x^{\phi(N)} - 1 \right\} abc \dots (N-1) = \text{a multiple of } N;$$

but the product $abc\dots (N-1)$ is prime to N;

$$\therefore \ x^{\phi(N)} - 1 \equiv 0 \pmod{N}.$$

§26. Let $N = a^p b^q c^r \dots$, then d_1, d_2, d_3,... are the terms of the product
$$(1 + a + a^2 + \dots + a^p)(1 + b + b^2 + \dots + b^q)(1 + c + c^2 + \dots + c^r)\dots$$
Consider any divisor d, and suppose $d = a^f b^g c^h \dots$; then
$$\phi(d) = \phi(a^f) \cdot \phi(b^g) \cdot \phi(c^h) \dots; \quad [\text{Art. 430.}]$$

$$\therefore \ \{1 + \phi(a) + \phi(a^2) + \dots + \phi(a^p)\}\{1 + \phi(b) + \phi(b^2) + \dots + \phi(b^p)\}\dots$$
$$= \phi(d_1) + \phi(d_2) + \phi(d_3) + \dots$$

But
$$1 + \phi(a) + \phi(a^2) + \dots + \phi(a^p)$$
$$= 1 + a\left(1 - \frac{1}{a}\right) + a^2\left(1 - \frac{1}{a}\right) + \dots + a^p\left(1 - \frac{1}{a}\right) \quad [Art.431.]$$
$$= 1 + a\left(1 - \frac{1}{a}\right) \cdot \frac{a^p - 1}{a - 1} = a^p;$$

$$\therefore \ \Sigma\phi(d) = a^p b^q c^r \dots = N.$$

If the terms of the series
$$\phi(1)\frac{y}{1 - y^2} + \phi(3)\frac{y^3}{1 - y^6} + \phi(5)\frac{y^5}{1 - y^{10}} + \phi(7)\frac{y^7}{1 - y^{14}} + \dots$$
be expanded by the Binomial Theorem, the coefficient of y^N will be
$$\phi(d_1) + \phi(d_2) + \phi(d_3) + \dots,$$
where d_1, d_2, d_3,... are the divisors of N, including unity and N itself; for a term involving y^N can arise from the expansion of $\phi(d)\dfrac{y^d}{1 - y^{2d}}$ only when d is a divisor of N.

But $\phi(d_1) + \phi(d_2) + \phi(d_3) + \dots = N$; hence the above series
$$= y + 3y^3 + 5y^5 + 7y^7 + 9y^9 + \dots$$
$$= y(1 + y^2)(1 + 2y^2 + 3y^4 + 4y^6 + 5y^8 + \dots)$$
$$= y(1 + y^2)(1 - y^2)^{-2}.$$
Writing ix for y, the required theorem at once follows.

EXAMPLES XXXI : The General Theory of Continued Fractions

EXAMPLES XXXI a

§1. Assume $p_n = a_n p_{n-1} + b_n p_{n-2}, q_n = a_n q_{n-1} + b_n q_{n-2}$; then, as in Art. 438, the $(n+1)^{th}$ convergent

$$= \frac{\left(a_n - \dfrac{b_{n+1}}{a_{n+1}}\right)p_{n-1} - b_n p_{n-2}}{\left(a_n - \dfrac{b_{n+1}}{a_{n+1}}\right)q_{n-1} - b_n q_{n-2}}$$

$$= \frac{p_n - \dfrac{b_{n+1}}{a_{n+1}}p_{n-1}}{q_n - \dfrac{b_{n+1}}{a_{n+1}}q_{n-1}} = \frac{a_{n+1}p_n - b_{n+1}p_{n-1}}{a_{n+1}q_n - b_{n+1}q_{n-1}}.$$

Then by induction the required result follows.

§2. We have $\left(\dfrac{2x+1}{2x}\right)^2 = \dfrac{4x^2 + 4x + 1}{4x^2} = 1 + \dfrac{4x+1}{4x^2}$;

$$\frac{4x^2}{4x+1} = x - \frac{x}{4x+1}; \qquad \frac{4x+1}{x} = 4 + \frac{1}{x};$$

$$\therefore \left(\frac{2x+1}{2x}\right)^2 = 1 + \frac{1}{x-} \frac{1}{4+} \frac{1}{x};$$

§3. (1) Since $\sqrt{a^2+b} = a + (\sqrt{a^2+b} - a) = a + \dfrac{b}{\sqrt{a^2+b}+a}$;

and $\qquad \sqrt{a^2+b} + a = 2a + (\sqrt{a^2+b} - a) = 2a + \dfrac{b}{\sqrt{a^2+b}+a}$;

$$\therefore \sqrt{a^2+b} = a + \frac{b}{2a+} \frac{b}{2a+} \cdots$$

(2) Again, $\sqrt{a^2-b} = a - (a - \sqrt{a^2-b}) = a - \dfrac{b}{a+\sqrt{a^2-b}}$;

$$a + \sqrt{a^2-b} = 2a - (a - \sqrt{a^2-b}) = 2a - \frac{b}{a+\sqrt{a^2-b}};$$

$$\therefore \sqrt{a^2-b} = a - \frac{b}{2a-} \frac{b}{2a-} \cdots$$

§4. As in Example 1, $\quad p_n = a_n p_{n-1} - b_n p_{n-1}$;

$$\therefore p_n - p_{n-1} = (a_n - 1)p_{n-1} - b_n p_{n-2}.$$

Now $a_n - 1$ is at least as great as b_n; therefore $p_n - p_{n-1}$ is at least as great as $b_n(p_{n-1} - p_{n-2})$; therefore $p_n > p_{n-1}$ if $p_{n-1} > p_{n-2}$; and so on. But p_2 is clearly greater than p_1; hence $p_n > p_{n-1}$. Similarly $q_n > q_{n-1}$.

§5. By definition, $\quad \dfrac{1}{a_n} + \dfrac{1}{a_{n-2}} = \dfrac{2}{a_{n-1}}$;

$$\therefore \frac{a_{n-1}}{a_n} = 2 - \frac{a_{n-1}}{a_{n-2}}; \text{ that is, } \frac{a_n}{a_{n-1}} = \frac{1}{2 - \dfrac{a_{n-1}}{a_{n-2}}};$$

similarly,

$$\frac{a_{n-1}}{a_{n-2}} = \frac{1}{2 - \dfrac{a_{n-2}}{a_{n-3}}}; \ldots; \text{ and finally } \frac{a_3}{a_2} = \frac{1}{2 - \dfrac{a_2}{a_1}}.$$

§6. Denote the continued fractions by x and y;

then
$$x - a = \frac{1}{2a + x - a} = \frac{1}{x + a}; \text{ whence } x^2 - a^2 = 1, \text{ or } x = \sqrt{a^2 + 1}.$$

Again,
$$y - a = -\frac{1}{2a + y - a} = \frac{-1}{y + a};$$
$$\therefore y^2 - a^2 = -1, \text{ or } y = \sqrt{a^2 - 1};$$
$$\text{whence } x^2 + y^2 = a^2 + 1 + a^2 - 1 = 2a^2.$$

Finally,
$$xy = \sqrt{a^4 - 1} = a^2 - (a^2 - \sqrt{a^4 - 1}) = a^2 - \frac{1}{a^2 + \sqrt{a^4 - 1}};$$
$$a^2 + \sqrt{a^4 - 1} = 2a^2 - (a^2 - \sqrt{a^4 - 1}) = 2a^2 - \frac{1}{a^2 + \sqrt{a^4 - 1}};$$

thus
$$xy = a^2 - \frac{1}{2a^2 -} \frac{1}{2a^2 -} \cdots$$

§7.
The series
$$P = p_1 + p_2 x + p_3 x^2 + p_4 x^3 + \ldots,$$
and
$$Q = q_1 + q_2 x + q_3 x^2 + q_4 x^3 + \ldots,$$
are both recurring series in which the scale of relation is $1 - ax - bx^2$.
Also
$$p_1 = b, p_2 = ab; \quad q_1 = a, q_2 = a^2 + b;$$
$$\therefore P = \frac{p_1 + (p_2 - ap_1)x}{1 - ax - bx^2} = \frac{b}{1 - ax - bx^2};$$
$$Q = \frac{q_1 + (q_2 - aq_1)x}{1 - ax - bx^2} = \frac{a + bx}{1 - ax - bx^2};$$
$$\therefore xQ = \frac{ax + bx^2}{1 - ax - bx^2} = \frac{1}{1 - ax - bx^2} - 1.$$

Thus $\frac{p_{n+1}}{b} = q_n = $ the coefficient of x^n in $\dfrac{1}{1 - ax - bx^2}$.

Again $\quad bq_{n+1} - ap_{n+1} = p_{n+2} - ap_{n+1} = bp_n = b^2 q_{n-1}$.

§8. Proceeding as in Example 7, we see that $p_x = $ coefficient of y^{x-1} in $\dfrac{b}{1 - ay - by^2}$, and $q_x = $ coefficient of y^x in $\dfrac{1}{1 - ay - by^2}$.

If α, β are the roots of $k^2 - ak - b = 0$, then $\alpha + \beta = \alpha, \alpha\beta = -b$, and
$$\frac{1}{1 - ay - by^2} = \frac{1}{1 - (\alpha + \beta)y + \alpha\beta y^2} = \frac{1}{(1 - \alpha y)(1 - \beta y)}$$
$$= \frac{1}{\alpha - \beta}\left(\frac{\alpha}{1 - \alpha y} - \frac{\beta}{1 - \beta y}\right).$$
$$\therefore p_x = \frac{b}{\alpha - \beta}(\alpha^x - \beta^x); \quad q_x = \frac{1}{\alpha - \beta}(\alpha^{x+1} - \beta^{x+1}).$$

§9.
Let
$$x = a + \frac{1}{b+} \frac{1}{c+} \frac{1}{d+} \frac{1}{a+} \ldots;$$
$$\therefore x = a + \frac{1}{b+} \frac{1}{c+} \frac{1}{d+} \frac{1}{x}.$$

The convergents are
$$\frac{a}{1}, \frac{ab+1}{b} \cdot \frac{abc+c+a}{bc+1}, \frac{abcd+cd+ad+ab+1}{bcd+d+b};$$
$$\therefore x = \frac{(abcd+cd+ad+ab+1)x + abc+c+a}{(bcd+d+b)x + bc+1};$$
$$\therefore (bcd + b + d)x^2 - (abcd + ab + ad - bc + cd)x - (abc + c + a) = 0.$$

If $y = -d + \dfrac{1}{-c+} \dfrac{1}{-b+} \dfrac{1}{-a+} \dfrac{1}{-d+} \ldots$, by writing $-d, -c, -b, -a$ for a, b, c, d

respectively, we have

$$(-abc - c - a)y^2 - (abcd + cd + ad - bc + ab)y - (-bcd - b - d) = 0;$$

or

$$(abc + c + a)y^2 + (abcd + ab + ad - bc + cd)y - (bcd + b + d) = 0.$$

Now y is the negative root of this equation; by putting $y = -\dfrac{1}{z}$ we have

$$(bcd + b + d)z^2 - (abcd + ab + ad - bc + cd)z - (abc + c + a) = 0,$$

and therefore $z = x$; that is $y = -\dfrac{1}{x}$ or $xy = -1$.

§10. Here $\dfrac{(n^2 - 1)^2}{n^2 + (n+1)^2}$ is the $(n+1)^{th}$ component.

$$\therefore u_{n+1} = \{n^2 + (n+1)^2\}u_n - (n^2 - 1)^2 u_{n-1};$$

or

$$u_{n+1} - n^2 u_n = (n+1)^2\{u_n - (n-1)^2 u_{n-1}\},$$

similarly

$$u_n - (n-1)^2 u_{n-1} = n^2\{u_{n-1} - (n-2)^2 u_{n-2}\};$$

$$\dotfill$$

$$u_3 - 2^2 u_2 = 3^2(u_2 - u_1).$$

Hence, by multiplication we obtain

$$u_{n+1} - n^2 u_n = 3^2 \cdot 4^2 \ldots (n+1)^2(u_2 - u_1).$$

Now

$$p_1 = 1, \; q_1 = 1; \; p_2 = 5, \; q_2 = 1.$$

Thus

$$q_{n+1} - n^2 q_n = 0, \text{ or } q_{n+1} = n^2 q_n.$$

Hence

$$q_{n+1} = n^2(n-1)^2(n-2)^2 \ldots 1^2 = (\lfloor n)^2.$$

Again

$$p_{n+1} - n^2 p_n = 3^2 \cdot 4^2 \ldots (n+1)^2 \cdot 4 = (\lfloor n+1)^2.$$

$$\therefore \frac{p_{n+1}}{(\lfloor n)^2} - \frac{p_n}{(\lfloor n-1)^2} = (n+1)^2; \ldots \text{ and } \frac{p_2}{(\lfloor 1)^2} - \frac{p_1}{1} = 2^2.$$

Hence, by addition

$$\frac{p_{n+1}}{(\lfloor n)^2} - 1 = 2^2 + 3^2 + \ldots + (n+1)^2;$$

$$\therefore \frac{p_{n+1}}{(\lfloor n)^2} = 1^2 + 2^2 + 3^2 + \ldots + (n+1)^2; \text{ and } \frac{q_{n+1}}{(\lfloor n)^2} = 1;$$

$$\therefore \frac{p_{n+1}}{q_{n+1}} = 1^2 + 2^2 + \ldots + (n+1)^2 = \frac{(n+1)(n+2)(2n+3)}{6}.$$

§11.

Here

$$u_n = (2n+1)u_{n-1} - (n^2 - 1)u_{n-2};$$

or

$$u_n - nu_{n-1} = (n+1)\{u_{n-1} - (n-1)u_{n-2}\};$$

$$\dotfill$$

$$u_3 - 3u_2 = 4(u_2 - 2u_1).$$

Hence, by multiplication we obtain

$$u_n - nu_{n-1} = (n+1)n \ldots 4(u_2 - 2u_1).$$

Now

$$p_1 = 2, \; q_1 = 1; \; p_2 = 10, \; q_2 = 2;$$

$$\therefore p_n - np_{n-1} = \lfloor n+1, \; q_n - nq_{n-1} = 0.$$

Hence

$$q_n = nq_{n-1} = n(n-1)q_{n-2} = \ldots = \lfloor n.$$

Again

$$\frac{p_n}{\lfloor n} - \frac{p_{n-1}}{\lfloor n-1} = n+1; \ldots; \; \frac{p_2}{\lfloor 2} - \frac{p_1}{\lfloor 1} = 3;$$

whence, by addition

$$\frac{p_n}{\lfloor n} - 2 = 3 + 4 + \ldots (n+1);$$

that is,

$$\frac{p_n}{\lfloor n} = \frac{n(n+3)}{2}; \text{ and therefore } \frac{p_n}{q_n} = \frac{n(n+3)}{2}.$$

§12.
Here
$$u_{n+1} = (n+2)u_n - (n+2)u_{n-1};$$
$$\therefore \frac{u_{n+1}}{n+2} = u_n - u_{n-1} = \frac{n+1}{n+1}u_n - \frac{n}{n}u_{n-1};$$
$$\therefore \frac{u_{n+1}}{n+2} - \frac{u_n}{n+1} = n\left(\frac{u_n}{n+1} - \frac{u_{n-1}}{n}\right);$$

$$\cdots\cdots\cdots\cdots\cdots\cdots\cdots\cdots\cdots\cdots$$

$$\frac{u_3}{4} - \frac{u_2}{3} = 2\left(\frac{u_2}{3} - \frac{u_1}{2}\right).$$

Hence, by multiplication
$$\frac{u_{n+1}}{n+2} - \frac{u_n}{n+1} = \lfloor n \left(\frac{u_2}{3} - \frac{u_1}{2}\right).$$

Now
$$p_1 = 2,\ q_1 = 2;\ p_2 = 6,\ q_2 = 3;$$
$$\therefore \frac{q_{n+1}}{n+2} - \frac{q_n}{n+1} = 0;$$

so that
$$\frac{q_{n+1}}{n+2} = \frac{q_n}{n+1} = \frac{q_{n-1}}{n} = \ldots = \frac{q_1}{2} = 1.$$

Again
$$\frac{p_{n+1}}{n+2} - \frac{p_n}{n+1} = \lfloor n\ ;\ldots;\ \text{and}\ \frac{p_2}{3} - \frac{p_1}{2} = \lfloor 1\ ;$$

by addition,
$$\frac{p_{n+1}}{n+2} = 1 + \lfloor 1 + \lfloor 2 + \lfloor 3 + \ldots + \lfloor n\ ;$$

hence we have the required result.

§13.
Here
$$u_{n+1} = (n+2)u_n - nu_{n-1};$$
$$\therefore u_{n+1} - (n+1)u_n = u_n - nu_{n-1};$$

$$\cdots\cdots\cdots\cdots\cdots\cdots\cdots\cdots\cdots\cdots$$

$$u_3 - 3u_2 = u_2 - 2u_1.$$

Hence, by multiplication
$$u_{n+1} - (n+1)u_n = u_2 - 2u_1.$$

Now
$$p_1 = 1,\ q_1 = 1;\ p_2 = 3,\ q_2 = 2;$$
$$\therefore q_{n+1} - (n+1)q_n = 0;$$

$$\therefore q_{n+1} = (n+1)q_n = (n+1)nq_{n-1} = \ldots = \lfloor n+1\ .$$

Again
$$p_{n+1} - (n+1)p_n = 1;$$
$$\therefore \frac{p_{n+1}}{\lfloor n+1} - \frac{p_n}{\lfloor n} = \frac{1}{\lfloor n+1}\ ;\ldots;\ \text{and}\ \frac{p_2}{\lfloor 2} - \frac{p_1}{\lfloor 1} = \frac{1}{\lfloor 2}\ ;$$
$$\therefore \frac{p_{n+1}}{q_{n+1}} = 1 + \frac{1}{\lfloor 2} + \frac{1}{\lfloor 3} + \ldots + \frac{1}{\lfloor n+1}\ ;$$
$$\therefore \frac{p_{n+1}}{q_{n+1}} = 1 + \frac{1}{\lfloor 2} + \frac{1}{\lfloor 3} + \ldots + \frac{1}{\lfloor n+1}\ ;\ \text{so that}\ \frac{p_\infty}{q_\infty} = e - 1.$$

§14.
Here
$$u_n = nu_{n-1} + (2n+2)u_{n-2};$$
$$\therefore u_n - (n+2)u_{n-1} = -2\{u_{n-1} - (n+1)u_{n-2}\};$$

$$\cdots\cdots\cdots\cdots\cdots\cdots\cdots\cdots\cdots\cdots$$

$$u_3 - 5u_2 = -2(u_2 - 4u_1).$$

Hence, by multiplication
$$u_n - (n+2)u_{n-1} = (-1)^{n-2}2^{n-2}(u_2 - 4u_1).$$

also
$$p_1 = 4,\ q_1 = 1;\ p_2 = 8,\ q_2 = 8;$$
$$\therefore p_n - (n+2)p_{n-1} = (-1)^{n-1}2^{n+1},$$
$$q_n - (n+2)q_{n-1} = (-1)^{n-2}2^n;$$
$$\therefore \frac{p_n}{\lfloor n+2} - \frac{p_{n-1}}{\lfloor n+1} = \frac{(-1)^{n-1}2^{n+1}}{\lfloor n+2},$$

$$\frac{q_n}{\lfloor n+2} - \frac{q_{n-1}}{\lfloor n+1} = \frac{(-1)^{n-2}2^n}{\lfloor n+2};$$

. . . .

$$\frac{p_2}{\lfloor 4} - \frac{p_1}{\lfloor 3} = \frac{(-1)2^3}{\lfloor 4}, \quad \frac{q_2}{\lfloor 4} - \frac{q_1}{\lfloor 3} = \frac{2^2}{\lfloor 4}.$$

Hence, by addition

$$\frac{p_n}{\lfloor n+2} = \frac{2^2}{\lfloor 3} - \frac{2^3}{\lfloor 4} + \frac{2^4}{\lfloor 5} - \cdots$$

and

$$\frac{q_n}{\lfloor n+2} = \frac{1}{\lfloor 3} + \frac{2^2}{\lfloor 4} - \frac{2^3}{\lfloor 5} + \frac{2^4}{\lfloor 6} - \cdots$$

$$\therefore \frac{p_n}{q_n} = \frac{1}{2}\left(\frac{2^3}{\lfloor 3} - \frac{2^4}{\lfloor 4} + \frac{2^5}{\lfloor 5} - \cdots\right) \div \frac{1}{4}\left(\frac{4}{\lfloor 3} + \frac{2^4}{\lfloor 4} - \frac{2^5}{\lfloor 5} + \cdots\right).$$

Now

$$e^{-2} = 1 - 2 + \frac{2^2}{\lfloor 2} - \frac{2^3}{\lfloor 3} + \frac{2^4}{\lfloor 4} - \cdots$$

$$\therefore \frac{p_\infty}{q_\infty} = \frac{1}{2}\left(1 - 2 + \frac{2^2}{\lfloor 2} - e^{-2}\right)$$

$$\div \frac{1}{4}\left(\frac{4}{\lfloor 3} + e^{-2} - 1 + 2 - \frac{2^2}{\lfloor 2} + \frac{2^3}{\lfloor 3}\right).$$

$$= \frac{1}{2}(1 - e^{-2}) \div \frac{1}{4}(1 + e^{-2}) = \frac{2(1 - e^{-2})}{1 + e^{-2}} = \frac{2(e^2 - 1)}{e^2 + 1}.$$

§15.

$$u_n = nu_{n-1} + 3(n+2)u_{n-2};$$
$$u_n - (n+3)u_{n-1} = -3\{u_{n-1} - (n+2)u_{n-2}\};$$

. . . .

$$u_3 - 6u_2 = -3(u_2 - 5u_1).$$

Hence, by multiplication

$$u_n - (n+3)u_{n-1} = (-3)^{n-2}(u_2 - 5u_1).$$

Now

$$p_1 = 9, \quad q_1 = 1; \quad p_2 = 18, \quad q_2 = 14;$$

$$\therefore p_n - (n+3)p_{n-1} = (-1)^{n-1}3^{n+1}, \quad q_n - (n+3)q_{n-1} = (-1)^{n-2}3^n;$$

$$\frac{p_n}{\lfloor n+3} - \frac{p_{n-1}}{\lfloor n+2} = \frac{(-1)^{n-1}3^{n+1}}{\lfloor n+3}, \quad \frac{q_n}{\lfloor n+3} - \frac{q_{n-1}}{\lfloor n+2} = \frac{(-1)^{n-2}3^n}{\lfloor n+3};$$

. . . .

$$\frac{p_2}{\lfloor 5} - \frac{p_1}{\lfloor 4} = \frac{(-1)3^3}{\lfloor 5}, \quad \frac{q_2}{\lfloor 5} - \frac{q_1}{\lfloor 4} = \frac{3^2}{\lfloor 5};$$

whence, by addition

$$\frac{p_n}{\lfloor n+3} = \frac{3^2}{\lfloor 4} - \frac{3^3}{\lfloor 5} + \frac{3^4}{\lfloor 6} - \cdots = \frac{1}{9}\left(\frac{3^4}{\lfloor 4} - \frac{3^5}{\lfloor 5} + \frac{3^6}{\lfloor 6} - \cdots\right);$$

and

$$\frac{q_n}{\lfloor n+3} - \frac{1}{\lfloor 4} = \frac{3^2}{\lfloor 5} - \frac{3^3}{\lfloor 6} + \frac{3^4}{\lfloor 7} - \cdots = \frac{1}{27}\left(\frac{3^5}{\lfloor 5} - \frac{3^6}{\lfloor 6} + \frac{3^7}{\lfloor 7} - \cdots\right).$$

Now

$$e^{-3} = 1 - 3 + \frac{3^2}{\lfloor 2} - \frac{3^3}{\lfloor 3} + \frac{3^4}{\lfloor 4} - \cdots = -2 + \frac{3^4}{\lfloor 4} - \frac{3^5}{\lfloor 5} + \frac{3^6}{\lfloor 6} - \cdots;$$

$$\therefore \frac{p_\infty}{q_\infty} = \frac{1}{9}(e^{-3} + 2) \div \frac{1}{27}\left(\frac{27}{\underline{4}} - 2 + \frac{3^4}{\underline{4}} - e^{-3}\right)$$

$$= \frac{6(e^{-3} + 2)}{5 - 2e^{-3}} = \frac{6(2e^3 + 1)}{5e^3 - 2}.$$

§16.

Here
$$u_1 = \frac{p_1}{q_1}, \; u_2 = \frac{p_2}{q_2}, \; u_3 = \frac{p_3}{q_3}, \dots$$

where
$$p_n = q_{n-1}, \text{ and } q_n = q_{n-1} + p_{n-1} = q_{n-1} + q_{n-2}.$$

Hence $q_1 + q_2 x + q_3 x^2 + q_4 x^3 + \dots$ is a recurring series in which the scale of relation is $1 - x - x^2$.

Also
$$q_1 = b, \; q_2 = a + b,$$

$$\therefore q_1 + q_2 x + q_3 x^2 + q_4 x^3 + \dots = \frac{q_1 + (q_2 - q_1)x}{1 - x - x^2} = \frac{b + ax}{1 - x - x^2}.$$

Let
$$\frac{b + ax}{1 - x - x^2} = \frac{A}{1 - \alpha x} + \frac{B}{1 - \beta x};$$

then
$$q_n = A\alpha^{n-1} + B\beta^{n-1}; \; p_n = q_{n-1} = A\alpha^{n-2} + B\beta^{n-2};$$

$$\therefore \frac{p_n}{q_n} = \frac{A\alpha^{n-2} + B\beta^{n-2}}{A\alpha^{n-1} + B\beta^{n-1}}.$$

Now α and β are to be found from the equations $\alpha + \beta = 1, \alpha\beta = -1$; let α be the greater of the quantities, then
$$\alpha = \frac{1 + \sqrt{5}}{2}, \; \beta = \frac{1 - \sqrt{5}}{2},$$

so that $\alpha > 1$ and $\beta < 1$; hence the limit when n is infinite of α^n is ∞ and of β^n is 0;

$$\therefore \frac{p_\infty}{q_\infty} = \frac{A\alpha^{n-2}}{A\alpha^{n-1}} = \frac{1}{\alpha} = \frac{2}{1 + \sqrt{5}} = \frac{\sqrt{5} - 1}{2}.$$

§17.

We have
$$u_n = (r + 1)u_{n-1} - ru_n;$$

that is,
$$u_n - (r + 1)u_{n-1} + ru_n = 0.$$

Thus the series $u_1 + u_2 x + u_3 x^2 + \dots$ is a recurring series, whose scale of relation is $1 - (r + 1)x + rx^2$, and whose generating function is
$$\frac{u_1 + \{u_2 - (r + 1)u_1\}x}{1 - x - x^2}.$$

Now
$$p_1 = r, \; q_1 = r + 1, \; p_2 = r(r + 1), \; q_2 = r^2 + r + 1,$$

$$\therefore p_1 + p_2 x + p_3 x^2 + p_4 x^3 + \dots = \frac{r}{1 - (r + 1)x + rx^2}$$

$$= \frac{r}{r - 1}\left(\frac{r}{1 - rx} - \frac{1}{1 - x}\right);$$

$$\therefore p_n = \frac{r}{r - 1}(r^n - 1).$$

Similarly
$$q_1 + q_2 x + q_3 x^2 + q_4 x^3 + \dots = \frac{r + 1 - rx}{1 - (r + 1)x + rx^2}$$

$$= \frac{1}{r - 1}\left(\frac{r^2}{1 - rx} - \frac{1}{1 - x}\right);$$

$$\therefore q_n = \frac{1}{r - 1}(r^{n+1} - 1).$$

Thus we have
$$\frac{p_n}{q_n} = \frac{r(r^n - 1)}{r^{n+1} - 1}.$$

§18.

We have
$$u_n = (a_n + 1)u_{n-1} - a_n u_{n-2};$$

that is,
$$u_n - u_{n-1} = a_n(u_{n-1} - u_{n-2});$$
$$\dotfill$$
$$u_3 - u_2 = a_3(u_2 - u_1).$$

Hence, by multiplication
$$u_n - u_{n-1} = a_3 a_4 a_5 \ldots a_n(u_2 - u_1).$$

Now $\quad p_1 = a_1,\ q_1 = a_1 + 1,\ p_2 = a_1(a_2 + 1), q_2 = a_1 a_2 + a_1 + 1;$
$$\therefore p_n - p_{n-1} = a_1 a_2 \ldots a_n,\ q_n - q_{n-1} = a_1 a_2 \ldots a_n;$$
$$\dotfill$$
$$p_2 - p_1 = a_1 a_2,\ q_2 - q_1 = a_1 a_2;$$
$$p_1 = a_1,\ q_1 = 1 + a_1.$$

Hence, by addition
$$p_n = \quad a_1 + a_1 a_2 + a_1 a_2 a_3 + \ldots + a_1 a_2 a_3 \ldots a_n;$$
$$q_n = 1 + a_1 + a_1 a_2 + a_1 a_2 a_3 + \ldots + a_1 a_2 a_3 \ldots a_n;$$
$\therefore \ 1 + p_n = q_n;$ and $p_n,\ q_n$ are both infinite in the limit; hence the continued fraction tends to the limit 1.

§19. The convergents to $\dfrac{1}{1+}\dfrac{1}{2+}\dfrac{1}{1+}\dfrac{1}{2+} \ldots$ are
$$\frac{1}{1},\ \frac{2}{3},\ \frac{3}{4},\ \frac{8}{11},\ \frac{11}{15},\ \frac{30}{41},\ \frac{41}{56};$$
and the convergents to $1 - \dfrac{1}{4-}\dfrac{1}{4-}\dfrac{1}{4-} \ldots$ are
$$\frac{1}{1},\ \frac{3}{4},\ \frac{11}{15},\ \frac{41}{56},\ \frac{153}{209}, \ldots.$$

Let $\dfrac{p_1}{q_1},\ \dfrac{p_2}{q_2},\ \dfrac{p_3}{q_3}, \ldots;\ \dfrac{r_1}{s_1},\ \dfrac{r_2}{s_2},\ \dfrac{r_3}{s_3}, \ldots$ denote the two sets of convergents; then $p_1 = r_1,\ p_3 = r_2,\ p_5 = r_3,\ p_7 = r_4, \ldots$; and similarly for q and s.

Now $p_{2n-1} = p_{2n-2} + p_{2n-3},\ p_{2n-2} = 2p_{2n-3} + p_{2n-4},\ p_{2n-3} = p_{2n-4} + p_{2n-5};$
whence
$$p_{2n-1} - 4p_{2n-3} + p_{2n-5} = 0;$$
but
$$r_n - 4r_{n-1} + r_{n-2} = 0;$$
thus
$$p_9 = 4p_7 - p_5 = 4r_4 - r_3 = r_5;$$
$$p_{11} = 4p_9 - p_7 = 4r_5 - r_4 = r_6;$$
$$\dotfill$$
hence generally $p_{2n-1} = r_n$. Similarly $q_{2n-1} = s_n$.

§20.

We have
$$p_{3n} = p_{3n-1} - p_{3n-2},$$
$$p_{3n-1} = 2p_{3n-2} - p_{3n-3},$$
$$p_{3n-2} = 5p_{3n-3} - p_{3n-4},$$
$$p_{3n-3} = p_{3n-4} - p_{3n-5},$$
$$p_{3n-4} = 2p_{3n-5} - p_{3n-6}.$$

From the first three equations, $p_{3n} = 4p_{3n-3} - p_{3n-4}$; from the last two equations $2p_{3n-3} = p_{3n-4} - p_{3n-6}$. By combining these results we have $p_{3n} = 2p_{3n-3} - p_{3n-6}$; so that the scale of relation is $1 - 2x + x^2$.

Now $\quad p_3 = 1,\ q_3 = 4,\ p_6 = 2,\ q_6 = 7;$
$$\therefore p_3 + p_6 x + p_9 x^2 + \ldots + p_{3n} x^{n-1} + \ldots = \frac{p_3 + (p_6 - 2p_3)x}{1 - 2x + x^2}$$
$$= \frac{1}{1 - 2x + x^2}.$$

Similarly
$$q_3 + q_6 x + q_9 x^2 + \ldots + q_{3n} x^{n-1} + \ldots = \frac{4 - x}{1 - 2x + x^2}.$$
$$\therefore p_{3n} = n, \text{ and } q_{3n} = 4n - (n - 1) = 3n + 1.$$

§21. This may be proved ab initio, or it may be deduced from the Example in Art. 444 as follows.

$$\frac{1}{1+}\frac{2}{2+}\frac{3}{3+}\frac{4}{4+}\frac{5}{5+}\cdots = \frac{1}{1+}\frac{1}{1+}\frac{3}{2\cdot3+}\frac{2\cdot4}{4+}\frac{5}{5+}\cdots$$

$$= \frac{1}{1+}\frac{1}{1+}\frac{1}{2+}\frac{2\cdot4}{3\cdot4}\frac{3\cdot5}{5+}\cdots$$

$$= \frac{1}{1+}\frac{1}{1+}\frac{1}{2+}\frac{2}{3+}\frac{3\cdot5}{4\cdot5+}\frac{4\cdot6}{6+}\cdots$$

$$= \frac{1}{1+}\frac{1}{1+}\frac{1}{2+}\frac{2}{3+}\frac{3}{4+}\cdots$$

[Compare Art. 448.]

Thus
$$e - 1 = 1 + \frac{1}{1+}\frac{1}{2+}\frac{2}{3+}\cdots$$

If x denotes the value of the given expression, we have

$$e - 1 = 1 + \frac{1}{1+x}; \text{ whence } x = \frac{3-e}{e-2}.$$

Now
$$x < \frac{1}{2} \text{ and } > \frac{1}{2+}\frac{2}{3} \text{ or } \frac{3}{8}.$$

$$\therefore \frac{3-e}{e-2} < \frac{1}{2} \text{ and } > \frac{3}{8}; \text{ that is, } 8 < 3e \text{ and } 30 > 11e.$$

EXAMPLES XXXI b

§1.

Put
$$\frac{1}{u_r} - \frac{1}{u_{r+1}} = \frac{1}{u_r + x_r};$$

then
$$(u_{r+1} - u_r)(u_r + x_r) = u_r u_{r+1};$$

so that
$$x_r = \frac{u_r^2}{u_{r+1} - u_r}, \text{ and therefore } \frac{1}{u_0} - \frac{1}{u_1} = \frac{1}{u_0 + \dfrac{u_0^2}{u_1 - u_0}};$$

and so on as in Art. 447.

§2.

Put
$$\frac{1}{a_r} + \frac{x}{a_r a_{r+1}} = \frac{1}{a_r + y_r};$$

then
$$(a_r + y_r)(a_{r+1} + x) = a_r a_{r+1}; \text{ whence } y_r = -\frac{a_r x}{a_{r+1} + x};$$

and so on as in Ex. 1, Art. 447.

§3. Let $\dfrac{r-1}{r-2} = \dfrac{r}{r-x_1}$; then $x_1 = \dfrac{r}{r-1}$; replacing r by $r+1$, we have

$$\frac{r}{r-1} = \frac{r+1}{r+1-x_2}, \text{ where } x_2 = \frac{r+1}{r}.$$

Similarly

$$\frac{r+1}{r} = \frac{r+2}{r+2-x_3}, \text{ where } x_3 = \frac{r+2}{r+1}; \text{ and so on .}$$

§4.

We have
$$\frac{2n}{n+1} = \frac{1}{1-x_1}, \text{ where } x_1 = \frac{n-1}{2n};$$

$$\frac{n-1}{2n} = \frac{1}{4-y_1}, \text{ where } y_1 = \frac{2(n-2)}{n-1};$$

replacing n by $n-2$, we have

$$\frac{2(n-2)}{n-1} = \frac{1}{1 - \dfrac{n-3}{2(n-2)}}, \text{ and } \frac{n-3}{2(n-2)} = \frac{1}{4 - \dfrac{2(n-4)}{n-3}}.$$

Moreover since the numerators in the fractions
$$\frac{n-1}{2n}, \quad \frac{2(n-2)}{n-1}, \quad \frac{n-3}{2(n-2)}, \quad \frac{2(n-4)}{n-3}, \dots$$
diminish by unity, there will be n components on the right.

§5. We know that
$$\frac{1}{u_1} + \frac{1}{u_2} + \frac{1}{u_3} + \dots + \frac{1}{u_{n+1}}$$
$$= \frac{1}{u_1 -} \frac{u_1^2}{u_1 + u_2 -} \frac{u_2^2}{u_2 + u_3 -} \dots \frac{u_n^2}{u_n + u_{n+1}}.$$

On putting $u_1 = 1, u_2 = 2, u_3 = 3, \dots$ we obtain the result.

§6. In the equation of Ex. 5, on putting
$$u_1 = 1^2, \quad u_2 = 2^2, \quad u_3 = 3^2, \dots u_{n+1} = (n+1)^2,$$
we obtain the result.

§7. We have $\quad e^x = 1 + \dfrac{x}{1} + \dfrac{x^2}{\underline{2}} + \dfrac{x^3}{\underline{3}} + \dfrac{x^4}{\underline{4}} + \dots$

In Example 2 on putting $a_0 = 1$, $a_1 = 1$, $a_2 = 2$, $a_3 = 3, \dots$, the result follows at once.

§8. Let $\dfrac{1}{a} - \dfrac{1}{ab} = \dfrac{1}{a+\beta}$; thus $(b-1)(a+\beta) = ab$; and $\beta = \dfrac{a}{b-1}$;
$$\therefore \frac{1}{a} - \frac{1}{ab} = \frac{1}{a +} \frac{a}{b-1}.$$

Similarly
$$\frac{1}{a} - \frac{1}{ab} + \frac{1}{abc} = \frac{1}{a} - \frac{1}{a}\left(\frac{1}{b} - \frac{1}{bc}\right) = \frac{1}{a} - \frac{1}{a(b+\gamma)}, \text{ suppose,}$$
$$= \frac{1}{a +} \frac{a}{b - 1 + \gamma} = \frac{1}{a +} \frac{a}{b - 1 +} \frac{b}{c - 1}; \text{ and so on.}$$

§9. From Art. 447, we have
$$\frac{1}{u_1} + \frac{1}{u_2} + \frac{1}{u_3} + \frac{1}{u_4} + \dots = \frac{1}{u_1 -} \frac{u_1^2}{u_1 + u_2 -} \frac{u_2^2}{u_2 + u_3 -} \frac{u_3^2}{u_3 + u_4 -} \dots$$

$$\therefore \frac{1}{r} + \frac{1}{r^4} + \frac{1}{r^9} + \frac{1}{r^{16}} = \frac{1}{r -} \frac{r^2}{r + r^4 -} \frac{r^8}{r^4 + r^9 -} \frac{r^{18}}{r^9 + r^{16} -} \dots$$
$$= \frac{1}{r -} \frac{r}{r^3 + 1 -} \frac{r^7}{r^4 + r^9 -} \frac{r^{18}}{r^9 + r^{16} -} \dots;$$

on reducing as explained in Art. 448 we have the result required.

§10. This is an easy consequence from Art. 448. Thus
$$\frac{a_1}{a_1 +} \frac{a_2}{a_2 +} \frac{a_3}{a_3} = \frac{1}{1 +} \frac{a_2}{a_1 a_2 +} \frac{a_1 a_3}{a_3} = \frac{1}{1 +} \frac{1}{a_1 +} \frac{a_1 a_3}{a_2 a_3} = \frac{1}{1 +} \frac{1}{a_1 +} \frac{a_1}{a_2}.$$

§11.

We have $\qquad P = \dfrac{a}{a +} \dfrac{b}{b +} \dfrac{c}{c +} \dfrac{d}{d +} \dots$
$$= \frac{1}{1 +} \frac{b}{ab +} \frac{ac}{c +} \frac{d}{d +} \dots$$
$$= \frac{1}{1 +} \frac{1}{a +} \frac{ac}{bc +} \frac{bd}{d +} \dots$$
$$= \frac{1}{1 +} \frac{1}{a +} \frac{a}{b +} \frac{b}{c +} \dots \text{ [Compare Art. 448.]}$$

Thus $\qquad P = \dfrac{1}{1 +} \dfrac{1}{a + Q}$
$$= \frac{a + Q}{a + 1 + Q}.$$

§12. From Ex.2, Art. 447, we have

$$\frac{1}{a_1} - \frac{x}{a_2} + \frac{x^2}{a^3} - \frac{x^3}{a^4} + \ldots = \frac{1}{a_1+} \frac{a_1^2 x}{a_2 - a_1 x+} \frac{a_2^2 x}{a_3 - a_2 x+} \frac{a_3^2 x}{a_4 - a_3 x+} \ldots$$

Hence $\dfrac{1}{q_1} - \dfrac{x}{q_1 q_2} + \dfrac{x^2}{q_2 q_3} - \dfrac{x^3}{q_3 q_4} + \ldots$

$$= \frac{1}{q_1+} \frac{q_1^2 x}{q_1 q_2 - q_1 x+} \frac{q_1^2 q_2^2 x}{q_2 q_3 - q_1 q_2 x+} \frac{q_2^2 q_3^2 x}{q_3 q_4 - q_2 q_3 x + \ldots}$$

$$= \frac{1}{q_1+} \frac{x}{\dfrac{q_2 - x}{q_1}+} \frac{x}{\dfrac{q_3 - q_1 x}{q_2}+} \frac{x}{\dfrac{q_4 - q_2 x}{q_3}+} \ldots$$

$$= \frac{1}{a_1+} \frac{x}{a_2+} \frac{x}{a_3+} \frac{x}{a_4+} \ldots ,$$

$\because q_1 = a_1; q_2 = a_1 a_2 + x; q_3 = a_3 q_2 + x q_1; q_4 = a_4 q_3 + x q_2.$

EXAMPLES XXXII : Probability

§1. Two dice may be thrown in 36 ways, and five may be made up by 1, 4; 4, 1; 3, 2; 2, 3; that is, in 4 ways.

$$\therefore \text{ the chance of throwing five } = \frac{4}{36} = \frac{1}{9}.$$

Similarly, since six may be thrown in 5 ways, the chance of throwing six is $\frac{5}{36}$.

§2. A queen and a knave can be drawn together in 16 ways; any two cards may be drawn in $^{52}C_2$, or 1326 ways;

$$\therefore \text{ the chance required } = \frac{16}{1326} = \frac{8}{663}.$$

§3. Three balls can be drawn from 16 in $^{16}C_3$, or $16 \times 5 \times 7$ ways.

Three white balls can be drawn in $^5C_3$, or 10 ways.

$$\therefore \text{ the chance required } = \frac{10}{16 \times 5 \times 7} = \frac{1}{56}.$$

§4. The total number of ways of tossing the 4 coins is 2^4; of these $^4C_2$, or 6 ways are favorable to the event.

$$\therefore \text{ the chance required } = \frac{6}{2^4} = \frac{3}{8}.$$

§5. Let $\frac{2}{3}x$ and x be the respective probabilities of the first and second event; then, since one of the events must happen, $\frac{2}{3}x + x = 1$, or $x = \frac{3}{5}$.

$$\therefore \text{ the odds in favor of the second event are 3 to 2.}$$

§6. The total.number of draws is $^{52}C_4$, and of these 4 are favorable;

$$\therefore \text{ the required chance } = \frac{4 \times 1 \cdot 2 \cdot 3 \cdot 4}{52 \cdot 51 \cdot 50 \cdot 49} = \frac{4}{270725}.$$

§7. The number of ways in which thirteen persons can sit at a round table is $\lfloor 12$; and two particular persons can sit side by side in $2\lfloor 11$ ways.

$$\therefore \text{ the required chance } = \frac{2\lfloor 11}{\lfloor 12} = \frac{1}{6}.$$

$$\therefore \text{ the odds against the event are 5 to 1.}$$

Or thus: Call the specified persons A and B; then besides A's place, wherever, it may be, there are 12 places of which two are adjacent to $A's$ place and ten are not adjacent. Thus the odds are 5 to 1 against the event.

§8. The chance of A happening is $\frac{3}{11}$; the chance of B is $\frac{2}{7}$, and the chance of C is $1 - \frac{3}{11} - \frac{2}{7}$, or $\frac{34}{77}$. Thus the odds against C are 43 to 34.

§9. The chance of throwing 4 with one die is $\frac{1}{6}$.

With two dice 8 can be thrown as follows :

$$6, 2; \; 5, 3; \; 4, 4;$$

the first two of these can each occur in 2 ways ; therefore 8 can be made up in 5 ways, and the chance of throwing 8 is $\frac{5}{36}$.

With three dice 12 can be thrown as follows :

$$6, 5, 1; \; 6, 4, 2; \; 6, 3, 3; \; 5, 5, 2; \; 5, 4, 3; \; 4, 4, 4;$$

the first, second, and fifth of these can each occur in 6 ways, the third and fourth in 3 ways, and the last in only 1 way;
thus 12 can be thrown in $6 + 6 + 3 + 3 + 6 + 1$ ways, that is, in 25 ways.

$$\therefore \text{ the chance of throwing } 12 = \frac{25}{216}.$$

$$\therefore \text{ the three chances are } \frac{1}{6}, \frac{5}{36}, \frac{25}{216};$$

which are as $36 : 30 : 25$.

§10. One card from each suit may be dropped in 13^4 ways; and any four cards may be dropped in $^{52}C_4$ ways.

$$\therefore \text{ the required chance } = \frac{13^4}{^{52}C_4} = \frac{2197}{20825}.$$

§11. A can draw all blanks in $^9C_3$ ways, and he can draw three tickets $^{12}C_3$ ways. His chance of drawing all blanks is therefore $\frac{21}{55}$.

$$\therefore \text{ his chance of a prize is } \frac{34}{55}.$$

In the same way it will be found that $B's$ chance of a prize is $\frac{13}{28}$.

$$\therefore A's \text{ chance } : B's \text{ chance } :: \frac{34}{55} : \frac{13}{28}$$
$$:: 952 : 715.$$

§12. With two dice, 6 can be made up in the 5 following ways:
$$1, 5; \ 5, 1; \ 2, 4; \ 4, 2; \ 3, 3; .$$

$$\therefore \text{ the chance of throwing six with 2 dice } = \frac{5}{6^2}.$$

With three dice, 6 can be thrown in each of the following ways :
$$1, 1, 4; \ 1, 2, 3; \ 2, 2, 2;$$
the first of these may occur in 3 ways, the second in 6, and the last in 1 only.

$$\therefore \text{ the chance of throwing six with 3 dice } = \frac{10}{6^3}.$$

With four dice, 6 can be made up in the following ways:
$$1, 1, 1, 3; \ 1, 1, 2, 2;$$
the first of these may occur in 4 ways, and the second in 6;

$$\therefore \text{ the chance of throwing six with 4 dice } = \frac{10}{6^4}.$$

Thus these chances are as $18 : 6 : 1$.

§13. The 8 volumes can be placed on the shelf in $\lfloor 8$ ways; volumes of the same works will be altogether in $\lfloor 3 \times \lfloor 3 \times \lfloor 4$ ways; for the sets of volumes admit of $\lfloor 3$ permutations, and the volumes in two of the sets admit of $\lfloor 3$ and $\lfloor 4$ permutations respectively.

$$\text{Thus the required chance } = \frac{\lfloor 3 \times \lfloor 3 \times \lfloor 4}{\lfloor 8}.$$

§14. B will win if he throws more than 9.

He can throw 12 in 1 way, 11 in 2 ways, 10 in 3 ways.

That is out of the 36 ways in which he can throw the two dice, he can throw more than 9 in 6 ways.

$$\therefore \text{ the required chance } = \frac{1}{6}.$$

§15. There are 7 letters which can be placed altogether in $\lfloor 7$ ways.

As the two vowels are not to be separated we may consider them as a single letter; and we can then have $\lfloor 6$ different arrangements in which the two vowels come together in the same order. Therefore, since the two vowels may change places, the required chance $= \dfrac{2\lfloor 6}{\lfloor 7} = \dfrac{2}{7}$.

§16. The number of favorable ways is the same as the number of ways in which the 9 other cards forming the hand can be chosen, which is $^{48}C_9$. And the total number of ways in which the hand can be made up is $^{52}C_{13}$.

$$\therefore \text{ the required chance } = \frac{^{48}C_9}{^{52}C_{13}} = \frac{11}{4165}.$$

§17. The number of different ways in which the coins can be placed is $\dfrac{\lfloor 7}{\lfloor 4 \lfloor 3}$, or 35. And there are only 5 different ways of placing the coins so that the extreme places are occupied by half crowns.

$$\therefore \text{ the required chance } = \frac{5}{35} = \frac{1}{7}.$$

In the general case the total number of possible arrangements is $\dfrac{\lfloor m+n}{\lfloor m \lfloor n}$, only $\dfrac{\lfloor m+n-2}{\lfloor m \lfloor n-2}$ of which are favorable, and thus the chance is

$$\frac{n(n-1)}{(m+n)(m+n-1)}.$$

EXAMPLES XXXII b

§1. The chance of throwing an ace in the first throw is $\dfrac{1}{6}$, and in the second the chance is also $\dfrac{1}{6}$. The chance of not throwing ace in the second throw is $1 - \dfrac{1}{6} = \dfrac{5}{6}$.

$$\therefore \text{ the required chance } = \frac{1}{6} \times \frac{5}{6} = \frac{5}{36}.$$

§2. The knave, queen, king can each be drawn in 4 ways. Any 3 cards can be drawn in $^{52}C_3$ ways.

$$\therefore \text{ the required chance } = 4^3 \div \frac{52 \cdot 51 \cdot 50}{\lfloor 3} = \frac{16}{5525}.$$

§3. The chance that the first fails is $\dfrac{5}{7}$, and that the second fails $\dfrac{5}{11}$; the chance that both do not fail is $1 - \dfrac{5}{7} \times \dfrac{5}{11} = \dfrac{52}{77}$.

§4. $A's$ chance of failure is $\dfrac{4}{7}$; $B's$ chance of failure is $\dfrac{5}{12}$. The chance that both will not fail is $1 - \dfrac{4}{7} \times \dfrac{5}{12} = \dfrac{16}{21}$.

§5. The chance of selecting the first compartment is $\dfrac{1}{2}$, and then the chance of drawing a sovereign is $\dfrac{2}{5}$. Therefore the chance of a sovereign from the first compartment $\dfrac{1}{2} \times \dfrac{2}{5}$, or $\dfrac{1}{5}$. Similarly the chance of a sovereign from the other compartment is $\dfrac{1}{2} \times \dfrac{2}{3}$, or $\dfrac{1}{3}$. And since the two cases are mutually exclusive, the required chance $= \dfrac{1}{3} + \dfrac{1}{5} = \dfrac{8}{15}$.

§6. The chance of an even number the first time is $\frac{8}{17}$; the chance of an odd number the second time is $\frac{9}{17}$. Therefore the required chance

$$= \frac{8 \times 9}{17 \times 17} = \frac{72}{289}.$$

§7.
(1) The chance that the second card is of a different suit from the first is $\frac{39}{51}$; the chance of the third card being of a suit differing from the first and second is $\frac{26}{50}$; and the chance of the fourth being of a suit differing from all the preceding cards is $\frac{13}{49}$.

Hence the chance that all four are of different suits

$$= \frac{39 \cdot 26 \cdot 13}{51 \cdot 50 \cdot 49} = \frac{2197}{20825}.$$

(2) The chance that the second card is not of the same value as the first is $\frac{48}{51}$; that the third differs from the first and second is $\frac{44}{50}$; and the chance of the fourth not being of the value of the first, second, or third is $\frac{40}{49}$.

Therefore the chance that no two are of equal value $= \frac{48 \cdot 44 \cdot 40}{51 \cdot 50 \cdot 42} = \frac{2816}{4165}$.

§8. The chance of failing to throw an ace in each of the five trials is $\left(\frac{5}{6}\right)^5$; therefore the chance of succeeding once at least is $1 - \left(\frac{5}{6}\right)^5$, or $\frac{4651}{7776}$.

§9. In order that a majority may be favorable, the reviews must all be favorable, or the first, the second, or the third must be unfavorable. The chances for these four cases are respectively:

$$\frac{5 \cdot 4 \cdot 3}{7^3}, \quad \frac{2 \cdot 4 \cdot 3}{7^3}, \quad \frac{5 \cdot 3 \cdot 3}{7^3}, \quad \frac{5 \cdot 4 \cdot 4}{7^3};$$

and the sum of these is $\frac{209}{343}$.

§10. The chance that they are alternately of different colors beginning with white is $\frac{5}{8} \cdot \frac{3}{7} \cdot \frac{4}{6} \cdot \frac{2}{5} = \frac{1}{14}$.

The chance that they are alternately of different colors beginning with black is $\frac{3}{8} \cdot \frac{5}{7} \cdot \frac{2}{6} \cdot \frac{4}{5} = \frac{1}{14}$.

∴ the chance that they are alternately of different colors

$$= \frac{1}{14} + \frac{1}{14} = \frac{1}{7}.$$

§11. As in Example 8, the chance of not failing three times in succession is $1 - \left(\frac{5}{6}\right)^3 = \frac{91}{216}$.

§12. If the last digit in the product is not 1, 3, 7, or 9, it mast be 0, or 5, or even.

Therefore none of the four numbers must end in $0, 2, 4, 5, 6, 8$.

And the chance that each of the four numbers should not end in any of these is $\frac{2}{5}$. Thus the required chance $= \left(\frac{2}{5}\right)^4 = \frac{16}{625}$.

§13. The sovereign can only be in the second purse if both the following events have happened :

(1) the sovereign was among the 9 coins taken out of the first purse and put into the second;

(2) it was not among the 9 coins put from the second purse into the first.

The chance of 1. is $\frac{9}{10}$, and the chance of 2. when 1. has happened is $\frac{10}{19}$.

$$\therefore \text{ the chance of both } = \frac{9}{10} \times \frac{10}{19} = \frac{9}{19}.$$

Hence the chance that the sovereign is in the first purse

$$= 1 - \frac{9}{19} = \frac{10}{19}.$$

§14. The number of ways in which 10 things may be divided into two classes containing 5 of one and 5 of another is $\dfrac{\lfloor 10}{\lfloor 5 \, \lfloor 5}$. And the total number of ways in which the tossing of the two coins may occur is 2^{10}.

$$\therefore \text{ the chance required } = \frac{1}{2^{10}} \times \frac{\lfloor 10}{\lfloor 5 \, \lfloor 5} = \frac{63}{256}.$$

§15. The total number of ways in which the coins may fall is 2^8; and there are 8 ways in which one head can appear.

$$\therefore \text{ the required chance } = \frac{8}{2^8} = \frac{1}{32}.$$

§16. $B's$ chance in any round is $\frac{3}{4}$ of $A's$ chance, and $C's$ is $\frac{3}{4}$ of $B's$ chance. If $x = A's$ chance in the long run, we have $x + \frac{3}{4}x + \frac{9}{16}x = 1$.

Thus $x = \frac{16}{37}$, and the respective chances are $\frac{16}{37}, \frac{12}{37}, \frac{9}{37}$.

§17. The chance that A draws a sovereign $= \frac{3}{7}$;

the chance that A fails and B succeeds $= \frac{4}{7} \cdot \frac{1}{2}$;

the chance that A and B fail, and then A succeeds $= \frac{4}{7} \cdot \frac{1}{2} \cdot \frac{3}{5}$;

the chance that A fails twice, and then B succeeds $= \frac{4}{7} \cdot \frac{1}{2} \cdot \frac{2}{5} \cdot \frac{3}{4}$;

the chance that A and B fail twice and then B succeeds $= \frac{4}{7} \cdot \frac{1}{2} \cdot \frac{2}{5} \cdot \frac{1}{4} \cdot \frac{3}{3}$;

$$\therefore A's \text{ chance } = \frac{3}{7} + \frac{4}{7} \cdot \frac{1}{2} \cdot \frac{3}{5} + \frac{4}{7} \cdot \frac{1}{2} \cdot \frac{2}{5} \cdot \frac{1}{4} = \frac{22}{35};$$

$$\text{and } B's \text{ chance } = \frac{13}{35}.$$

§18. Call the specified persons A and B. Then besides $A's$ place, wherever it may be, there are $n-1$ places of which two are adjacent to $A's$ place and $n-3$ are not adjacent to it. Therefore the odds against A and B sitting together are $n-3$ to 2.

§19. The chance that B rides $A = \dfrac{2}{3}$; and the chance of $A's$ winning on this hypothesis $= \dfrac{2}{3} \times \dfrac{1}{6} = \dfrac{1}{9}$.

The chance that C rides $A = \dfrac{1}{3}$; and the chance of $A's$ winning on this hypothesis $= \dfrac{1}{3} \times \dfrac{3}{6} = \dfrac{1}{6}$. Therefore $A's$ chance $= \dfrac{1}{9} + \dfrac{1}{6} = \dfrac{5}{18}$; thus the odds against him are 13 to 5.

§20. Four at least will arrive safely if 5 are safe, or 4 safe.

$$\therefore \text{ the chance } = \left(\frac{9}{10}\right)^5 + 5\left(\frac{9}{10}\right)^4\left(\frac{1}{10}\right) = \frac{45927}{50000}.$$

EXAMPLES XXXII c

§1. As in Art. 463, the required chance is the sum of the first three terms in $\left(\dfrac{3}{5} + \dfrac{2}{5}\right)^5$; that is,

$$\frac{1}{5^5}\left\{3^5 + 5\cdot 3^4\cdot 2 + 10\cdot 3^3\cdot 2^2\right\};$$

which reduces to $\dfrac{2133}{3125}$.

§2. The number of ways of obtaining 12 is the coefficient of x^{12} in the expansion of $\left(x^2 + x^3\right)^5$; also the coins can be thrown in 2^5 ways.

Now the coefficient of x^{12} in the expansion of $x^{10}(1+x)^5 = 10$;

$$\therefore \text{ the required chance } = \frac{10}{32} = \frac{5}{16}.$$

§3. In order to win three at least, he must win all four, or lose the first, or the second or the third, or the fourth. The respective chances of these 5 events are

$$\left(\frac{2}{3}\right)^4, \ \frac{1}{3}\cdot\frac{1}{3}\cdot\frac{2}{3}\cdot\frac{2}{3}, \ \frac{2}{3}\cdot\frac{1}{3}\cdot\frac{1}{3}\cdot\frac{2}{3}, \ \frac{2}{3}\cdot\frac{2}{3}\cdot\frac{2}{3}\cdot\frac{1}{3},$$

and their sum is $\dfrac{4}{9}$, which is the required chance.

§4. Let x be the value in shillings of the unknown coins; then the chance of drawing a sovereign is $\dfrac{5}{9}$, and of drawing one of the others is $\dfrac{4}{9}$.

Therefore the probable value of a draw $= \dfrac{5}{9} \times 20 + \dfrac{4}{9}x$ shillings;

$$\therefore 4x + 5 \times 20 = 12 \times 9; \text{ whence } x = 2.$$

Thus the coins are florins.

§5. The chance is the sum of the second, fourth, sixth, ... terms in the expansion of $\left(\dfrac{1}{2} + \dfrac{1}{2}\right)^n$;

that is, the sum of the odd coefficients in $(1+1)^n$ divided by 2^n.

$$\therefore \text{ the required chance } = \frac{2^{n-1}}{2^n} = \frac{1}{2}.$$

§6. Two coins can be drawn in 10 ways, and two sovereigns in 1 way ;

$$\therefore \text{ expectation on this ground } = \frac{1}{10} \times 40 = 4 \text{ shillings.}$$

One sovereign and one shilling can be drawn in 6 ways, and $A's$ expectation on this ground $= \dfrac{3}{5} \times 21 = 12\frac{3}{5}$ shillings. Two shillings can be drawn in 3 ways, and $A's$ expectation $= \dfrac{3}{10} \times 2 = \dfrac{3}{5}$ shillings.

$\therefore$ on the whole $A's$ expectation $= 17\dfrac{1}{5}$ shillings.

Or more simply thus:

The probable value of a draw is $\dfrac{2}{5}$ of the amount in the bag.

$\therefore A's$ expectation $= \dfrac{2}{5}$ of 43 shillings $= 17\dfrac{1}{5}$ shillings.

§7. In his first throw the fourth man's chance is $\dfrac{1}{2^4}$, in his next throw it is $\dfrac{1}{2^{10}}$. and so on; therefore his chance is the sum of the infinite series

$$\dfrac{1}{2^4} + \dfrac{1}{2^{10}} + \dfrac{1}{2^{16}} + \cdots$$

Thus the chance $= \dfrac{1}{2^4} \div \left(1 - \dfrac{1}{2^6}\right) = \dfrac{4}{63}$.

§8. The required chance is obtained by dividing the coefficient of x^6 in $(x + x^2 + x^3)^3$ by 3^3.

Now $(x + x^2 + x^3)^3 = x^3(1 + x + x^2)^3 = x^3 \left(\dfrac{1 - x^3}{1 - x}\right)^3$; we have therefore to find the coefficient of x^3 in $(1 - x^3)^3(1 - x)^{-3}$, that is, in

$$(1 - 3x^3 + 3x^6 - x^9)(1 + 3x + 6x^2 + 10x^3 + \ldots).$$

This coefficient $= -3 + 10 = 7$.

$\therefore$ the required chance $= \dfrac{7}{27}$.

§9. If the sum of the numbers is less than 15, the numbers must be $3, 3, 3, 3$ or $3, 3, 3, 5$. And in this last combination of numbers the 5 may occur in any one of the four throws ; thus there are 5 cases favorable, and 16 cases in all.

$\therefore$ the chance required $= \dfrac{5}{16}$.

§10. The three dice can be thrown in 216 ways. The number of ways in which the dice can be thrown so as to have a total of 10 is the coefficient of x^{10} in $(x + x^2 + x^3 + \ldots + x^6)^3$; that is, in $x^3 \left(\dfrac{1 - x^6}{1 - x}\right)^3$;

now this is the same as the coefficient of x^7 in $(1 - x^6)^3(1 - x)^{-3}$, and is found to be 27;

$\therefore$ the required chance $= \dfrac{27}{216} = \dfrac{1}{8}$.

§11. In order to win the set, B must win 2 games before A wins 3.

Therefore by Art. 466,

$$B's \text{ chance} = \left(\dfrac{1}{2}\right)^2 \left\{1 + 2 \cdot \dfrac{1}{2} + \dfrac{2 \cdot 3}{1 \cdot 2} \left(\dfrac{1}{2}\right)^2\right\}, \text{ or } \dfrac{11}{16}.$$

$\therefore B's$ share $= £11$, and $A's$ share $= £5$.

§12. The number of ways in which the 3 dice may fall is 6^3, or 216.

In order to lose, B may throw anything from 3 to 8 inclusive; the number of ways in which this may be done is the sum of the coefficients of the powers of x from 3 to 8 inclusive in the expansion of

$$\left(x + x^2 + x^3 + \ldots + x^6\right)^3.$$

This expression $= x^3 \left(\dfrac{1 - x^6}{1 - x}\right)^3 = x^3(1 - x^6)^3(1 - x)^{-3}$

$$= x^3(1 - 3x^6 + 3x^{12} - x^{18})$$

$$(1 + 3x + 6x^2 + 10x^3 + 15x^4 + 21x^5 + \ldots).$$

Thus the number of ways $= 1 + 3 + 6 + 10 + 15 + 21 = 56$;

hence the chance that B loses is $\dfrac{56}{216} = \dfrac{7}{27}$.

§13. The chance of drawing a sovereign in the two coins is $\dfrac{4}{10}$, or $\dfrac{2}{5}$.

In this case $C's$ expectation $= \dfrac{1}{2} \times \dfrac{2}{5}$ of 21 shillings $= \dfrac{21}{5}$ shillings.

The chance that both coins drawn are shillings $= \dfrac{3}{5}$, and in this case

$C's$ expectation $= \dfrac{1}{2} \times \dfrac{3}{5}$ of 2 shillings $= \dfrac{3}{5}$ of a shilling.

Thus the whole expectation $= \dfrac{24}{5} = 4\frac{4}{5}$ shillings.

Or more simply thus:

The probable value of any two coins $= \dfrac{2}{5}$ of 24 shillings; and $C's$ expectation is half of this sum.

§14. With the notation of Art. 462, we have $p = \dfrac{1}{6}, q = \dfrac{5}{6}, n = 5$; hence

the chance of throwing exactly three aces is $^5C_3 \cdot \left(\dfrac{1}{6}\right)^3 \cdot \left(\dfrac{5}{6}\right)^2$, or $\dfrac{250}{7776}$.

The chance of throwing three aces at least is

$$\left(\frac{1}{6}\right)^5 + 5\left(\frac{1}{6}\right)^4 \cdot \frac{5}{6} + \frac{5 \cdot 4}{1 \cdot 2}\left(\frac{1}{6}\right)^3 \left(\frac{5}{6}\right)^2.$$

Thus the chance is $\dfrac{276}{7776}$.

§15. The chance of throwing 7 with two dice is $\dfrac{1}{6}$, and the chance of

throwing 4 is $\dfrac{1}{12}$. Thus $A's$ chance in each trial is double of $B's$.

Now we require $B's$ expectation in the long run, the throwing being continued until one or other of them wins.

Let $x = B's$ chance on this supposition, then clearly $2x = A's$ chance, and therefore $x + 2x = 1$.

Therefore $x = \dfrac{1}{3}$, and $B's$ expectation $= \dfrac{1}{3}$ of $5s$. $- \dfrac{2}{3}$ of $2s. = 4d.$

§16.. The two dice may be thrown in 4×6 or 24 ways.

The numbers of ways in which $2, 3, 4, \ldots 10$ may be thrown are given by the coefficients of those powers of x in the expansion of

$$(x + x^2 + x^3 + \ldots + x^6)(x + x^2 + x^3 + x^4).$$

In the question before us, the required event will happen if any of the numbers from 5 to 10 inclusive be thrown.

Thus the required chance

$$= \frac{4}{24} + \frac{4}{24} + \frac{4}{24} + \frac{3}{24} + \frac{2}{24} + \frac{1}{24} = \frac{18}{24} = \frac{3}{4}.$$

§17. Let the purse contain n coins in all. Then the expectation from the

first draw is $\dfrac{1}{n}(M + m)$.

Now the chance of a second draw is $\dfrac{n-1}{n}$, and here it is certain that

M remains, also $n - 2$ other coins, each of which has an average value

$\dfrac{m}{n-1}$; their total value is therefore $\dfrac{n-2}{n-1} \cdot m$;

Hence the expectation from 2nd draw

$$= \frac{n-1}{n} \cdot \frac{1}{n-1}\left\{M + \frac{n-2}{n-1}m\right\}$$

$$= \frac{1}{n}\left\{M + \frac{n-2}{n-1}m\right\}.$$

Similarly the chance of a 3rd draw is $\dfrac{n-1}{n}\cdot\dfrac{n-2}{n-1}$, in which it is certain

that M remains and $n-3$ other coins of average value $\dfrac{m}{n-1}$.

$\therefore$ the expectation from 3rd draw

$$= \frac{n-1}{n}\cdot\frac{n-2}{n-1}\cdot\frac{1}{n-2}\left\{M + \frac{n-3}{n-1}\cdot m\right\}$$

$$= \frac{1}{n}\left\{M + \frac{n-3}{n-1}\cdot m\right\};$$

and so on; the expectation from the last draw being $\dfrac{1}{n}\{M + 0\}$.

$\therefore$ the whole expectation

$$= M\left(\frac{1}{n} + \frac{1}{n} + \dots \text{ to } n \text{ terms }\right)$$

$$+ \frac{m}{n}\left\{1 + \frac{n-2}{n-1} + \frac{n-3}{n-1} + \dots \text{ to } \overline{n-1} \text{ terms }\right\}$$

$$= M + \frac{(n-1)n}{2}\cdot\frac{m}{n}\cdot\frac{1}{n-1} = M + \frac{1}{2}m.$$

[This problem and solution are due to the Rev. T. C. Simmons, M.A.]

§18. The total number of ways in which three tickets may be drawn is

$$\frac{6n(6n-1)(6n-2)}{1\cdot 2\cdot 3} = n(6n-1)(6n-2).$$

To find the number of ways in which the sum of the numbers drawn is $6n$ we may proceed as follows :

First suppose is 0 drawn, then we have to make up $6n$ in all possible ways from two of the numbers $1, 2, 3, \dots 6n-1$; this can be done in $3n-1$ ways. Then suppose 1 is drawn ; we have to make up $6n-1$ from two of the numbers $2, 3, \dots 6n-1$; this can be done in $3n-2$ ways.

If 2 is drawn, we have to make up $6n-2$ from two of the numbers $3, 4, \dots 6n-1$; this can be done in $3n-4$ ways; if 3 is drawn, there are $3n-5$ ways of making up the number.

Finally, if $2n-2$ is drawn, there are only two ways of making up the numbers, viz. $2n-2, 2n-1, 2n+3$, and $2n-2, 2n, 2n+2$; while if $2n-1$ is drawn, there is only one way, viz. $2n-1, 2n, 2n+1$.

Hence the number of ways of making up $6n$ is the sum of $2n$ terms, which may be arranged in n pairs as follows :

$$\{(3n-1) + (3n-2)\} + \{(3n-4) + (3n-5)\}+$$
$$\dots + (5+4) + (2+1)$$

$$= (6n-3) + (6n-9) + (6n-12) + \dots = 3n^2.$$

Thus the required chance $= 3n^2 \div n(6n-1)(6n-2)$.

EXAMPLES XXXII d

[Where the wording of a question admits of two interpretations, as in the Example on page 305, we have here adopted the first method of solution there explained.]

§1. There are four equally likely hypotheses, namely, the bag may have contained 4 white balls, or 3, or 2, or 1.

And $$p_1 = 1, \; p_2 = \frac{3}{4}, \; p_3 = \frac{2}{4}, \; p_4 = \frac{1}{4}.$$

Thus the required chance $= \dfrac{p_1}{\Sigma(p)} = \dfrac{4}{10} = \dfrac{2}{5}$.

§2. The four hypotheses here are 6 black balls, or 5, or 4, or 3, and these are all equally likely.

And $\qquad p_1 = 1, \quad p_2 = \dfrac{5}{6} \cdot \dfrac{4}{5} \cdot \dfrac{3}{4}, \quad p_3 = \dfrac{4}{6} \cdot \dfrac{3}{5} \cdot \dfrac{2}{4}, \quad p_4 = \dfrac{3}{6} \cdot \dfrac{2}{5} \cdot \dfrac{1}{4};$

$$\therefore \frac{p_1}{20} = \frac{p_2}{10} = \frac{p_3}{4} = \frac{p_4}{1} = \frac{\Sigma(p)}{35}.$$

$$\therefore \text{ the required chance } = \frac{p_4}{\Sigma(p)} = \frac{1}{35}.$$

§3. If the letter came from Clifton, there are 6 pairs of consecutive letters of which ON is one. Therefore the chance that this Was the legible couple on the Clifton hypothesis is $\dfrac{1}{6}$.

If the letter came from London, out of 5 pairs of consecutive letters 2 are ON. Therefore the chance that this was the legible couple on the London hypothesis is $\dfrac{2}{5}$.

Therefore the a posteriori chances that the letter was from Clifton or London are

$$\frac{\frac{1}{6}}{\frac{1}{6} + \frac{2}{5}}, \text{ and } \frac{\frac{2}{5}}{\frac{1}{6} + \frac{2}{5}} \text{ respectively.}$$

Thus the required chance $= \dfrac{12}{17}$.

§4. A could lose in two ways; either by B winning or by C winning. The probabilities of these two events are $\dfrac{3}{10}, \dfrac{2}{10}$ respectively. Therefore A's a priori chance of losing was $\dfrac{5}{10}$, or $\dfrac{1}{2}$. But after the accident his chance of losing becomes $\dfrac{2}{3}$; that is, his chance of losing is increased in the ratio of 4 to 3. Therefore, also, B's and C's chances of winning are increased in the same ratio. Thus B's chance of winning $= \dfrac{3}{10} \times \dfrac{4}{3} = \dfrac{2}{5}$; and C's chance of winning $= \dfrac{2}{10} \times \dfrac{4}{3} = \dfrac{4}{15}$.

§5. There are n equally likely hypotheses, for the purse may have contained any number of sovereigns from 1 to n.

Thus $\qquad p_1 = \dfrac{1}{n}, \quad p_2 = \dfrac{2}{n}, \quad p_3 = \dfrac{3}{n}, \dots p_n = \dfrac{n}{n}.$

$$\therefore \text{ the required chance } = \frac{p_1}{\Sigma(p)} = \frac{2}{n(n+1)}.$$

§6. There are two cases: either the coin had two heads, or it had a head and a tail.

Thus $\qquad\qquad P_1 = \dfrac{1}{10}, \quad P_2 = \dfrac{9}{10}.$

Also $\qquad\qquad p_1 = 1, \quad p_2 = \left(\dfrac{1}{2}\right)^5;$

Therefore $\qquad\qquad \dfrac{Q_1}{2^5} = \dfrac{Q_2}{9} = \dfrac{1}{41}.$

Thus $\qquad\qquad Q_1 = \dfrac{32}{41} = $ the required chance.

§7. We have five cases to consider, for the bag may contain 1, 2, 3, 4, or

5 red balls, and we suppose these to be all equally likely. Hence

$$p_1 = \left(\frac{1}{5}\right)^2, \quad p_2 = \left(\frac{2}{5}\right)^2, \quad p_3 = \left(\frac{3}{5}\right)^2, \quad p_4 = \left(\frac{4}{5}\right)^2, \quad p_5 = \left(\frac{5}{5}\right)^2.$$

$$\therefore \frac{Q_1}{1^2} = \frac{Q_2}{2^2} = \frac{Q_3}{3^2} = \frac{Q_4}{4^2} = \frac{Q_5}{5^2} = \frac{1}{55}.$$

The chance of now drawing two red balls

$$= (Q_1 \times 0) + \left(Q_2 \times \frac{2}{5} \cdot \frac{1}{4}\right) + \left(Q_3 \times \frac{3}{5} \cdot \frac{2}{4}\right) + \left(Q_4 \times \frac{4}{5} \cdot \frac{3}{4}\right)$$

$$+ (Q_5 \times 1)$$

$$= \frac{1}{55} \left\{ \frac{2}{5} + \frac{27}{10} + \frac{48}{5} + 25 \right\} = \frac{377}{550}.$$

§8. See Case II. in the Example to Art. 473, whence it appears that the chance of 5 shillings is $\left(\frac{1}{2}\right)^5$, of 4 shillings $\frac{5}{2^5}$, of 3 shillings $\frac{10}{2^5}$, of 2 shillings $\frac{10}{2^5}$; thus

$$P_1 = \frac{1}{32}, \quad P_2 = \frac{5}{32}, \quad P_3 = \frac{10}{32}, \quad P_4 = \frac{10}{32}.$$

Also $\quad p_1 = 1, \quad p_2 = \frac{4}{5} \times \frac{3}{4}, \quad p_3 = \frac{3}{5} \times \frac{2}{4}, \quad p_4 = \frac{2}{5} \times \frac{1}{4}.$

$$\therefore p_1 P_1 = \frac{1}{32}, \quad p_2 P_2 = \frac{3}{32}, \quad p_3 P_3 = \frac{3}{32}, \quad p_4 P_4 = \frac{1}{32}.$$

$$\therefore \frac{Q_1}{1} = \frac{Q_2}{3} = \frac{Q_3}{3} = \frac{Q_4}{1} = \frac{1}{8}.$$

Hence the probable value in shillings of the remaining coins

$$= \frac{1}{8} \times 3 + \frac{3}{8} \times \frac{5}{2} + \frac{3}{8} \times 2 + \frac{1}{8} \times \frac{3}{2} = \frac{9}{4} = 2\frac{1}{4} \text{ shillings.}$$

§9. Reckon the result of the last two throws in one total. Then the whole throw of 15 can be made up as follows: $3 + 12$, $4 + 11$, $5 + 10$, $6 + 9$; and these four cases can occur in 1, 2, 3, 4 ways respectively, all of which are equally likely;

$$\therefore \frac{p_1}{1} = \frac{p_2}{2} = \frac{p_3}{3} = \frac{p_4}{4};$$

$$\therefore Q_2 = \frac{p_2}{\Sigma(p)} = \frac{2}{10} = \frac{1}{5}.$$

§10. Denote A's and B's veracities by p and p', then the required probability is

$$p(1 - p') + p'(1 - p);$$

that is, $\quad \frac{3}{4} \cdot \frac{1}{6} + \frac{5}{6} \cdot \frac{1}{4} = \frac{8}{24} = \frac{1}{3}.$

§11. There are two hypotheses; (i) their coincident testimony is true, (ii) it is false.

With the notation of Art. 478, we have

$$P_1 = \frac{1}{6}, \qquad\qquad P_2 = \frac{5}{6};$$

$$p_1 = \frac{2}{3} \times \frac{4}{5}, \qquad\qquad p_2 = \frac{1}{25} \times \frac{1}{3} \times \frac{1}{5};$$

for in estimating p_2 we must take into account the chance that A and B will both select the red ball when it has not been drawn. Thus

$$P_1 p_1 : P_2 p_2 = 8 \cdot \frac{1}{5} = 40 : 1;$$

hence the probability that the statement is true is $\frac{40}{41}.$

§12. The antecedent chance that the lost card is a spade is $\frac{1}{4}$, because there are 4 suits; and the chance that it is not a spade is $\frac{3}{4}$.

Thus
$$P_1 = \frac{1}{4}, \quad P_2 = \frac{3}{4};$$

also
$$p_1 = \frac{12 \cdot 11}{51 \cdot 50}, \quad p_2 = \frac{13 \cdot 12}{51 \cdot 50}.$$

$$\therefore \frac{Q_1}{11} = \frac{Q_2}{3 \times 13} = \frac{1}{50}.$$

$\therefore Q_1 = \frac{11}{50} =$ the chance that the missing card was a spade.

§13. There are three hypotheses; A may have won £5, £1, or nothing, for B and C may both have been mistaken.

Thus
$$P_1 = \frac{1}{10}, \quad P_2 = \frac{1}{10}, \quad P_3 = \frac{8}{10}.$$

Since B's veracity is represented by $\frac{2}{3}$, and C's by $\frac{3}{4}$, we have

$$p_1 = \frac{2}{3} \times \frac{1}{4}, \quad p_2 = \frac{1}{3} \times \frac{3}{4}, \quad p_3 = \frac{1}{4} \times \frac{1}{3}.$$

$$\therefore \frac{Q_1}{2} = \frac{Q_2}{3} = \frac{Q_3}{8} = \frac{1}{13}.$$

Thus A's expectation $= \frac{2}{13}$ of £5 $+ \frac{3}{13}$ of £1 $=$ £1.

§14. There are three equally likely hypotheses; for the purse may contain 2, or 3, or 4 sovereigns.

Now
$$p_1 = \frac{1}{6}, \quad p_2 = \frac{1}{2}, \quad p_3 = 1;$$

$\therefore$ the chance that all are sovereigns $= \frac{p_3}{\Sigma(p)} = \frac{3}{5}.$

Again
$$\frac{p_1}{1} = \frac{p_2}{3} = \frac{p_3}{6} = \frac{\Sigma(p)}{10}.$$

$$\therefore Q_1 = \frac{1}{10}, \quad Q_2 = \frac{3}{10}, \quad Q_3 = \frac{6}{10}.$$

$\therefore$ the chance that another drawing will give a sovereign

$$= \left(Q_1 \times \frac{1}{2}\right) + \left(Q_2 \times \frac{3}{4}\right) + (Q_3 \times 1) = \frac{1}{20} + \frac{9}{40} + \frac{6}{10} = \frac{7}{8}.$$

§15. At first, B's chance of winning his race is $\frac{1}{5}$; similarly C's chance is $\frac{1}{3}$, and D's chance is $\frac{1}{3}$.

Therefore after the 2nd race is known to have been won by B or D,

$$\text{B's chance : certainty} :: \frac{1}{5} : \frac{1}{3};$$

that is, B's chance $= \frac{3}{8}.$

Therefore, the chance of P winning his bet $= 1 \times \frac{3}{8} \times \frac{1}{3} = \frac{1}{8}$; and the chance of P losing it is $\frac{7}{8}.$

Thus P's expectation $= \frac{1}{8}$ or £120 $- \frac{7}{8}$ of £8 $=$ £8.

§16. We have n cases to consider, for there may be 1, 2, 3, ... n white balls; and all these cases are equally likely, so that $P_1 = P_2 = P_3 = \ldots = P_n$.

If there were r white balls, the chance of drawing two white balls in this case would be $\left(\dfrac{r}{n}\right)^2$.

$$\therefore \frac{Q_1}{\left(\dfrac{1}{n}\right)^2} = \frac{Q_2}{\left(\dfrac{2}{n}\right)^2} = \ldots = \frac{Q_r}{\left(\dfrac{r}{n}\right)^2} = \ldots = \frac{1}{\Sigma\left(\dfrac{r^2}{n^2}\right)}.$$

Thus $\dfrac{Q_r}{\left(\dfrac{r}{n}\right)^2} = \dfrac{6n}{(n+1)(2n+1)}$, and $Q_r = \dfrac{6r^2}{n(n+1)(2n+1)}$.

And the chance of another drawing giving a black ball

$$= \sum_{r=1}^{r=n} \frac{n-r}{n} \cdot \frac{6r^2}{n(n+1)(2n+1)}$$

$$= \frac{6\Sigma r^2}{n(n+1)(2n+1)} - \frac{6\Sigma r^3}{n^2(n+1)(2n+1)}$$

$$= 1 - \frac{6n^2(n+1)^2}{4n^2(n+1)(2n+1)} = 1 - \frac{3(n+1)}{2(2n+1)} = \frac{1}{2}(n-1)(2n+1)^{-1}.$$

§17. Represent the two coins by A and B. Then the a priori chance that B is with A is $\dfrac{n-1}{mn-1}$, for wherever A is placed, there remain $mn-1$ possible positions for B, $n-1$ of which are favourable. Hence the a priori chance that B and A are not together is $\dfrac{n(m-1)}{mn-1}$.

Now consider the $m-r$ purses which have not been examined. If A and B are together, the chance that they occur in these purses is $\dfrac{m-r}{m}$. If A and B are apart the chance that they both occur in these purses is

$$\frac{(m-r)(m-r-1)}{m(m-1)};$$

for $m(m-1)$ is the total number of ways in which they can occur separately in any two purses whatever, and $(m-r)(m-r-1)$ is the number of ways in which they can occur separately in any two of the purses we are considering.

Hence the required chance

$$= \frac{n-1}{mn-1} \cdot \frac{m-r}{m}$$

$$\div \left\{ \frac{n-1}{mn-1} \cdot \frac{m-r}{m} + \frac{n(m-1)}{mn-1} \cdot \frac{(m-r)(m-r-1)}{m(m-1)} \right\}$$

$$= \frac{n-1}{mn-nr-1}.$$

§18. The chance that A and B both get the correct result is $\dfrac{1}{8} \cdot \dfrac{1}{12}$; the chance that they both get an incorrect result is $\dfrac{7}{8} \cdot \dfrac{11}{12}$; and therefore the chance that they get the same incorrect result is $\dfrac{1}{1001} \cdot \dfrac{7}{8} \cdot \dfrac{11}{12} = \dfrac{1}{13 \cdot 8 \cdot 12}$.

Thus the chance that their solution is correct is to the chance that it is incorrect as 1 to $\dfrac{1}{13}$, or as 13 to 1.

§19. Let p be the a priori probability of the event; then the probability that their statement is true is to the probability that it is false as

$$\left(\frac{5}{6}\right)^{10} p \text{ is to } (1-p)\left(\frac{1}{6}\right)^{10}.$$

Therefore $\dfrac{5^{10}p}{1-p}$ represents the odds in favour of the event. Now in order that the odds in favour of the event may be at least five to one, we must have $\dfrac{5^{10}p}{1-p}$ not less than 5; that is, 5^9p must he not less than $1-p$, or $(5^9+1)p$ must be not less than 1. Hence p must be not less than $\dfrac{1}{5^9+1}$.

EXAMPLES XXXII e

§1. By writing down the different combinations it is easy to see that 12 can be thrown in 1 way, 11 in 2 ways, 10 in 3 ways, 9 in 4 ways, 8 in 5 ways, 7 in 6 ways. Therefore out of the 36 possible ways of throwing the dice there are $1+2+3+4+6+6$, or 21 ways favourable to throwing 7 or more. Thus the chance of throwing at least 7 is $\dfrac{7}{12}$.

§2. The nine coins can be arranged in $\lfloor 9$ ways; but the five sovereigns can be arranged in the odd places and the four shillings in the even places in $\lfloor 5 \times \lfloor 4$ ways. Hence the chance that they will be drawn alternately beginning with a sovereign is $\dfrac{\lfloor 5 \times \lfloor 4}{\lfloor 9} = \dfrac{1}{126}$.

Or thus: The number of ways in which nine things can be arranged, when five are alike of one sort, and four are alike of another sort, is $\dfrac{\lfloor 9}{\lfloor 5 \times \lfloor 4}$ or 126, and all these ways are equally likely.

$$\therefore \text{ the required chance} = \dfrac{1}{126}.$$

§3. See XXXII. b. Example 20.

§4. The first person's chance is $\dfrac{1}{n}$; if he fails, since there are $n-1$ tickets left, the second person's chance is $\dfrac{n-1}{n} \cdot \dfrac{1}{n-1} = \dfrac{1}{n}$. If the first two fail, the third person draws from $n-2$ tickets, and his chance is

$$\dfrac{n-1}{n} \cdot \dfrac{n-2}{n-1} \cdot \dfrac{1}{n-2} = \dfrac{1}{n};$$

and so on. Thus each person's chance is $\dfrac{1}{n}$.

§5. The chance that the first bag is chosen is $\dfrac{1}{2}$; and the chance of choosing one white and one red is $5 \times 3 \div {}^8C_2$. Again the chance that the second bag is chosen is $\dfrac{1}{2}$, and the chance of choosing one of each colour is now $4 \times 5 \div {}^9C_2$.

$$\therefore \text{ the required chance} = \dfrac{1}{2} \cdot \dfrac{15}{28} + \dfrac{1}{2} \cdot \dfrac{20}{36} = \dfrac{275}{504}.$$

§6 .

A's chance

$$= \dfrac{1}{6}\left\{1 + \left(\dfrac{5}{6}\right)^5 + \left(\dfrac{5}{6}\right)^{10} + \ldots\right\} = S \text{ suppose;}$$

B's chance

$$= \dfrac{5}{6} \cdot \dfrac{1}{6}\left\{1 + \left(\dfrac{5}{6}\right)^5 + \left(\dfrac{5}{6}\right)^{10} + \ldots\right\} = \dfrac{5}{6}S.$$

Similarly C's chance is $\left(\dfrac{5}{6}\right)^2 S$, while D's chance is $\left(\dfrac{5}{6}\right)^3 S$, and E's chance

is $\left(\dfrac{5}{6}\right)^4 S$. Thus their respective chances are as

$$1 : \frac{5}{6} : \left(\frac{5}{6}\right)^2 : \left(\frac{5}{6}\right)^3 : \left(\frac{5}{6}\right)^4.$$

§7. Three squares may be chosen in $^{64}C_3$ ways; two white and one black, or two black and one white may be chosen in $^{32}C_2 \times 32$ ways. Thus the required chance

$$= \frac{2 \times {}^{32}C_2 \times 32}{{}^{64}C_3} = \frac{16}{21}.$$

§8. The two dice may be thrown in 24 ways. The number of ways in which $2, 3, 4, \ldots 10$ may be thrown respectively are given by the coefficients of those powers of x in the expansion of

$$\left(x + x^2 + x^3 + \ldots + x^6\right)\left(x + x^2 + x^3 + x^4\right).$$

Multiplying out, we get

$$x^2 + 2x^3 + 3x^4 + 4x^5 + 4x^6 + 4x^7 + 3x^8 + 2x^9 + x^{10}.$$

Thus the chances of throwing 5, 6, 7 are equal. Also the average value of the throw is

$$\frac{1}{24} \cdot 2 + \frac{2}{24} \cdot 3 + \frac{3}{24} \cdot 4 + \frac{4}{24} \cdot 5 + \frac{4}{24} \cdot 6 + \frac{4}{24} \cdot 7$$
$$+ \frac{3}{24} \cdot 8 + \frac{2}{24} \cdot 9 + \frac{1}{24} \cdot 10 = \frac{144}{24} = 6.$$

The average value of the throw may also be obtained as follows :

The chances of throwing 10 and 2 are equal; as also the chances of throwing 9 and 3, 8 and 4, 7 and 5; and in each case the average value is 6. Therefore on the whole the average value is 6.

§9. When A tries with B, his chance is $\dfrac{1}{4}$, and B's is $\dfrac{3}{4}$; when A tries with C, his chance is $\dfrac{3}{5}$, and C's is $\dfrac{2}{5}$; when A tries with D, his chance is $\dfrac{4}{7}$, and D's is $\dfrac{3}{7}$.

A may either win with all three, or fail with B, or fail with C, or fail with D. The chances of these four cases are

$$\frac{1}{4} \cdot \frac{3}{5} \cdot \frac{4}{7}, \quad \frac{3}{4} \cdot \frac{3}{5} \cdot \frac{4}{7}, \quad \frac{1}{4} \cdot \frac{2}{5} \cdot \frac{4}{7}, \quad \frac{1}{4} \cdot \frac{3}{5} \cdot \frac{3}{7}.$$

The sum of these four chances gives the required chance.

§10. In order that the 4th person may have a throw the preceding 3 persons must all fail : thus $\left(\dfrac{7}{8}\right)^3 \cdot \dfrac{1}{8}$ is the chance that he will win the stake at his first throw. If he fails and all the other 3 persons fail he gets a second throw; so that at his second trial his chance of winning is $\left(\dfrac{7}{8}\right)^7 \cdot \dfrac{1}{8}$; and so on. Thus his whole chance is the sum of an infinite G. P. of which the first term is $\left(\dfrac{7}{8}\right)^3 \cdot \dfrac{1}{8}$, and the common ratio is $\left(\dfrac{7}{8}\right)^4$.

§11. In order to win, A must win 2 games before B wins 3. Thus by Art. 466, A's chance

$$= \left(\frac{1}{2}\right)^2 \left\{1 + 2 \cdot \frac{1}{2} + \frac{2 \cdot 3}{1 \cdot 2}\left(\frac{1}{2}\right)^2\right\} = \frac{11}{16}.$$

Therefore A's chance is to B's as 11 to 5.

§12. There are two hypotheses : either he has drawn two sovereigns, or one sovereign and one shilling.

Therefore $\qquad P_1 = \dfrac{3}{10}, \qquad\qquad\qquad\qquad P_2 = \dfrac{6}{10}.$

Also $\qquad\qquad p_1 = 1, \qquad\qquad\qquad\qquad p_2 = \dfrac{1}{2}.$

$$Q_1 = Q_2 = \frac{1}{2}.$$

§13. Consider six players, A, B, C, D, E, F, then A's chanee $= \dfrac{1}{6}.$

$B's$ chance $= \dfrac{5}{6} \cdot \dfrac{1}{3}$;	the chance that B throws and fails	$= \dfrac{5}{6} \cdot \dfrac{2}{3}.$
$C's$ ...... $= \dfrac{5}{9} \cdot \dfrac{1}{2}$;	................C............	$= \dfrac{5}{9} \cdot \dfrac{1}{2}.$
$D's$ ...... $= \dfrac{5}{18} \cdot \dfrac{2}{3}$;	................D............	$= \dfrac{5}{18} \cdot \dfrac{1}{3}.$
$E's$ ...... $= \dfrac{5}{54} \cdot \dfrac{5}{6}$;	................E............	$= \dfrac{5}{54} \cdot \dfrac{1}{6}.$
$F's$ ...... $= \dfrac{5}{54} \cdot \dfrac{1}{6} \cdot 1.$		

Now A, C, E are identical, so are B, D, F.

$$\therefore\ A's \text{ chance} = \frac{1}{6} + \frac{5}{18} + \frac{25}{324} = \frac{169}{324}.$$

$$\therefore\ B's \text{ chance} = \frac{5}{18} + \frac{5}{27} + \frac{5}{324} = \frac{155}{324}.$$

§14. Denote the persons by A and B.

(1) The chance that A obtains entrance $= \dfrac{6}{7}$, in which case there are 6 equally likely places for B, namely 5 inside and 1 outside. Wherever A may be seated there is only 1 case favourable to B's gaining an opposite seat.

$$\therefore\ \text{the required chance} = \frac{6}{7} \times \frac{1}{6} = \frac{1}{7}.$$

(2) The chance that A obtains a middle seat is $\dfrac{2}{7}$, in which case there are 2 favourable, 4 unfavourable positions for B, all equally likely. Therefore the chance that A and B are adjacent, B being at the end of the carriage is $\dfrac{2}{7} \times \dfrac{2}{6} = \dfrac{2}{21}$. Similarly the chance that A and B are adjacent, A being at the end of the carriage is $\dfrac{2}{21}$. These events are mutually exclusive; hence the whole chance of A and B being adjacent $= \dfrac{4}{21}.$

(2) may also be solved as follows :

The total number of pairs of positions in which A and B can be adjacent $= 4$. The total number of pairs of positions they can occupy without restriction, inside or outside, is $^7C_2 = 21.$

$$\therefore\ \text{the required chance} = \frac{4}{21}.$$

§15. In order that a number may be divisible by 11, the difference of the sum of the digits in the odd and even places must be either zero, or a multiple of 11. [See Art. 84.]

Here the difference cannot be zero, so that we have to divide 59 into two parts whose difference is 11; these parts are 35 and 24.

But the sum of three digits cannot be equal to 35; hence the seven digits must be such that the sum of the four odd ones is 35, and the sum of the three even ones is 24.

Now the number of ways in which 7 digits may be arranged so as to make 59 is equal to the coefficient of x^{59} in the expansion of

$$\left(x^0 + x^1 + x^2 + \ldots + x^9\right)^7$$

and since the coefficients of terms equidistant from the beginning and end are equal, this is equal to the coefficient of x^4 in the above expansion; that is, is equal to the coefficient of x^4 in the expansion of $\left(1 - x^{10}\right)^7 \left(1 - x\right)^{-7}$. This coefficient is 210.

Again, the number of ways in which 4 digits can be arranged so as to make 35 is equal to the coefficient of x^{35} in the expansion of

$$\left(x^0 + x^1 + x^2 + \ldots + x^9\right)^4 ;$$

and is therefore equal to 4.

Similarly, the number of ways in which 3 digits can be arranged to make 24 is equal to the coefficient of x^{24} in the expansion of

$$\left(x^0 + x^1 + x^2 + \ldots + x^9\right)^3$$

and is therefore equal to 10.

Each way of arranging the odd digits may be associated with each way of arranging the even digits;

$$\text{hence the required chance} = \frac{4 \times 10}{210} = \frac{4}{21}.$$

§16. The number of favourable cases is the coefficient of x^{12} in the expansion of $x^3(1 - x^6)^3(1 - x)^3$.   [See Ex. 2, Art. 466.]

Putting this in the form

$$x^3(1 - 3x^6 + \ldots)(1 + 3x + 6x^2 + 10x^3 + \ldots + 55x^9 + \ldots),$$

we easily see that the number of favourable cases is $55 - 30$, or 25.

$$\text{Thus the required chance} = \frac{25}{6^3} = \frac{25}{216}.$$

§17. The total number of drawings is 7^4. The number of ways in which the sum of the drawings will amount to 8 is the coefficient of x^8 in the expansion of

$$\left(x^0 + x^1 + x^2 + \ldots x^6\right)^4 .$$

This expression $= (1 - x^7)^4(1 - x)^{-4} = (1 - 4x^7 + \ldots)(1 + 4x + \ldots + 165x^8 + \ldots).$

Thus the coefficient of x^8 is $165 - 16$, or 149, and the required chance is

$$\frac{149}{2401}.$$

§18.

(1) We must find the coefficient of x^{10} in the expansion of

$$\left(x^0 + x^0 + x^0 + x^0 + x^0 + x^1 + x^2 + x^3 + x^4 + x^5\right)^3$$

and divide it by 10^3.

Put y for $x + x^2 + x^3 + x^4 + x^5$; then $(5 + y)^3 = 5^3 + 3 \cdot 5^2 y + 3 \cdot 5y^2 + y^3$. The coefficient of x^{10} comes from the last two terms only, and is equal to $15 + 18$. Thus the required chance $= \dfrac{33}{1000}$.

(2) There are now two favourable cases, namely those in which the tickets 1, 4, 5, or the tickets 2, 3, 5 are drawn. And the whole number of cases is $^{10}C_3$, since the chance is just the same as if

the three tickets were drawn simultaneously. Thus the required chance $= \frac{2}{120} = \frac{1}{60}$.

§19.

(1) If the last digit be 1, 3, 7, or 9, none of the numbers can be even or end in 0 or 5 ; that is, we have a choice of 4 digits with which to end each of our n numbers. Thus the required chance $= \frac{4^n}{10^n} = \left(\frac{2}{5}\right)^n$.

(2) If the last digit be 2, 4, 6, or 8, none of the numbers can end in 0 or 5 and one of the last digits must be even. Now 8^n is the number of ways in which we can exclude 0 and 5; and of these we have further to exclude the 4^n cases in which the last digit can be selected solely from 1, 3, 7, or 9. Thus the required chance $= \frac{8^n - 4^n}{10^n} = \frac{4^n - 2^n}{5^n}$.

(3) If the last digit is 5, one of the numbers must end in 5 and all the rest must be odd. Now 5^n is the number of ways in which an odd digit can be chosen to end the number, but to ensure 5 being one of them we must exclude the 4^n ways in which an odd digit can be chosen solely from 1, 3, 7 or 9.

$$\therefore \text{ the required chance} = \frac{5^n - 4^n}{10^n}.$$

(4) We have now to subtract the sum of the previous chances from unity.

§20. This is a particular case of Ex. 17, XXXII. c. and may be solved in the same way. Or we may proceed as follows.

If the dummy is drawn first the value of the draw is nothing.
If it is drawn second, the value of the two draws

$$= \frac{1}{4}(£1. + £1. + 1s. + 1s.) = 10s.\ 6d.$$

If it is drawn third, the value of the three draws

$$= \frac{1}{6}(£2. + £1.\ 1s. + £1.\ 1s. + £1.\ 1s. + 2s.) = £1.\ 1s.$$

If it is drawn fourth, the value of the four draws

$$= \frac{1}{4}(£2.\ 1s. + £2.\ 1s. + £1.\ 2s. + £1.\ 2s.) = £1.\ 11s.\ 6d.$$

If it is drawn fifth, the proceeds of the five draws $= £2.\ 2s.$
All these cases are equally likely ; hence the whole expectation

$$= \frac{1}{5}(0 + 10s.\ 6d. + £1.\ 1s. + £1.\ 11s.\ 6d + £2.\ 2s.) = £1.\ 1s.$$

§21. The chance of throwing 10 with 3 dice is $\frac{1}{8}$.　　　[See XXXII. c. Example 10.]

A throws first, and the chance that B has a throw is $\frac{7}{8}$. So that if x be A's chance of winning, B's chance is $\frac{7}{8}x$, and C's chance is $\left(\frac{7}{8}\right)^2 x$. And the sum of these three chances is 1, since they continue throwing until the event happens.

$$\therefore x\left\{1 + \frac{7}{8} + \left(\frac{7}{8}\right)^2\right\} = 1, \text{ whence } x = \left(\frac{8}{13}\right)^2.$$

§22. The solution of this Example is exactly similar to that of XXXII. d. Example 11.

§23. The chance of drawing the single counter marked 1 is $\dfrac{2}{n(n+1)}$; the chance of drawing one of the two counters marked 4 is $\dfrac{4}{n(n+1)}$; the chance of drawing one of the three counters marked 9 is $\dfrac{6}{n(n+1)}$; and so on.

∴ the required expectation in shillings

$$= \frac{2}{n(n+1)} \left\{ 1^3 + 2^3 + 3^3 + \ldots + n^3 \right\} = \frac{n(n+1)}{2}.$$

§24. The number of ways in which a man may have all 10 things is 1; the number of ways in which he may have 9 things is 10×2, for $^{10}C_9 = 10$, and in each case the remaining thing may be given in 2 ways. Similarly he may have 8 things in $\dfrac{10 \cdot 9}{1 \cdot 2} \cdot 2^2$ ways, for after taking away a combination of 8 things the remaining 2 may be given in 2^2 ways.

Similarly a man may have 7 things in $\dfrac{10 \cdot 9 \cdot 8}{1 \cdot 2 \cdot 3} \cdot 2^3$ ways, and he may have 6 things in $\dfrac{10 \cdot 9 \cdot 8 \cdot 7}{1 \cdot 2 \cdot 3 \cdot 4} \cdot 2^4$ ways. And the total number of ways, in which 10 things can be given among 3 persons is 3^{10}.

∴ the chance of a man having more than 5 things

$$= \frac{1 + 20 + 180 + 960 + 3360}{3^{10}} = \frac{4521}{3^{10}} = \frac{1507}{19683}.$$

§25. Let the rod be divided into n equal divisions A_1A_2, A_2A_3, A_3A_4, ... and let the random points of division be denoted by P_1, P_2, P_3, ... Then first it is necessary that one of the random points falls in each division; the chance of this is $\dfrac{\lfloor n}{n^n}$, for the total number of cases is the number of ways in which n places can be occupied by n things when repetitions are allowed, and the number of favourable cases is the number of ways in which n places can be occupied by n things when repetitions are not allowed.

Again A_1P_1 must be greater than A_2P_2, or P_1P_2 would exceed $\left(\dfrac{1}{n} \right)^{th}$ of the rod; therefore A_1P_1, A_2P_2, ... are in descending order of magnitude. The chance that this particular order will occur is $\dfrac{1}{\lfloor n}$, for the number of orders in which they can occur is $\lfloor n$, and all are equally likely.

Thus the required chance $= \dfrac{1}{n^n}$.

§26. Denote the two purses by B_1 and B_2. Four cases are a priori, namely,

(1) a sovereign may be transferred from B_1 to B_2,

(2) a shilling . . . . B_1 to B_2,

(3) a sovereign . . . . B_2 to B_1,

(4) a shilling . . . . B_2 to B_1.

Then since the chance of drawing from either purse is $\dfrac{1}{2}$, we have

$$P_1 = \frac{1}{2} \times \frac{3}{4}; \quad P_2 = \frac{1}{2} \times \frac{1}{4}; \quad P_3 = \frac{1}{2} \times \frac{1}{4}; \quad P_4 = \frac{1}{2} \times \frac{3}{4}.$$

In (1), B_1 has 2 sovereigns, 1 shilling, B_2 has 2 sovereigns, 3 shillings; so that $p_1 = \dfrac{1}{3} \times \dfrac{3}{5} = \dfrac{1}{5}$.

In (2), B_1 has 3 sovereigns, B_2 has 1 sovereign, 4 shillings; so that $p_2 = 0$.

In (3), B_1 has 4 sovereigns, 1 shilling, B_2 has 3 shillings; so that

$$p_1 = \frac{1}{5} \times 1 = \frac{1}{5}.$$

In (4), B_1 has 3 sovereigns, 2 shillings, B_2 has 1 sovereign, 2 shillings; so that

$$p_4 = \frac{2}{5} \times \frac{2}{3} = \frac{4}{15}.$$

$$\therefore \; P_1 p_1 = \frac{3}{40}; \; P_2 p_2 = 0; \; P_3 p_3 = \frac{1}{40}; \; P_4 p_4 = \frac{4}{40}.$$

$$\therefore \; \frac{Q_1}{3} = \frac{Q_3}{1} = \frac{Q_4}{4}; \; \text{whence } Q_4 = \frac{1}{2}.$$

Now for the second trial we have only to consider the case which corresponds to Q_4, for in none of the other cases could a shilling be drawn from each purse.

$$\therefore \; \text{the required chance} = Q_4 \times \frac{1}{4} \times \frac{1}{2} = \frac{1}{16}.$$

§27. Draw tangents to the circle at the three random points, thus forming a second triangle. Then if the first triangle is acute angled, the circle is inscribed in the second; and if the first triangle is obtuse angled, the circle is escribed to the second. Hence the required result follows as in Ex. 3 of Art. 481.

[This problem and solution are due to the Rev. T. C. Simmons.]

§28. Let A, B, C he the three points; then in favourable eases the sum of any two of the angles of the triangle ABC must be greater than the third. That is, the triangle must be acute angled, and by Ex. 27 the chance of this is $\dfrac{1}{4}$.   [Rev. T. C. Simmons.]

§29. Let AB be the straight line divided at P and Q; let $AB = a$, $AP = x$, $BQ = y$.

A ————————— P   Q   B

Then the favourable cases require

$$x < \frac{a}{2}, \; y < \frac{a}{2}, \; PQ < \frac{a}{2};$$

$$\therefore \; a - (x + y) < \frac{a}{2}, \text{ or } x + y > \frac{a}{2}.$$

And the possible cases require $x + y < a$.

Tate a pair of rectangular axes OC, OD; let OC, OD be each equal to a, so that CD is the line $x + y = a$. Bisect CD, OC, OD in E, F, G respectively.

Then GF is the line $x + y = \dfrac{a}{2}$; and GE, EF are the lines $y = \dfrac{a}{2}$, $x = \dfrac{a}{2}$.

Now the favourable cases are restricted to points in the triangle EGF, and the possible cases include all points in the triangle OCD.

Thus the required chance $= \dfrac{1}{4}$.

Or thus: If the 3 parts of the line are x, y, z we must have $x + y + z = a$, while $x + y > z$, $y + z > x$, $z + x > y$. Therefore x, y, z must each be $< \dfrac{a}{2}$.

Therefore if we take three rectangular axes OA, OB, OC, and make OA, OB, OC each equal to a, the plane $x + y + z = a$ includes the points which give the possible cases, while the favourable cases are restricted to the triangle DEF, where D, E, F are the middle points of BC, CA, AB respectively.

Thus the required chance

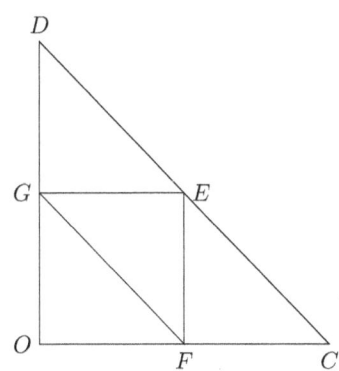

$$= \frac{1}{4}.$$

§30. Let p_1, p_2 be the a priori probabilities of drawing 4 sovereigns from the 1st and 2nd purses respectively.

Then $p_1 = 1$, and $p_2 =$

$${}^{10}C_4 \div {}^{25}C_4 = \frac{21}{1265}.$$

$$\therefore \frac{Q_1}{1265} = \frac{Q_2}{21} = \frac{1}{1286}.$$

$$\therefore Q_1 = \frac{1265}{1286}; \quad Q_2 = \frac{21}{1286}.$$

Again the probable value of the next draw in pounds is

$$Q_1 \times 1 + Q_2 \left(\frac{6}{21} + \frac{15}{21} \times \frac{1}{20} \right) = \pounds \frac{5087}{5144}.$$

§31. Let AB be the straight line of length a, and let the random points P, Q be at distances x, y from one end of the line.

A ——————— P   Q   B

Now in favourable cases we must have

$$x > b + y, \text{ or } y > b + x.$$

Again in possible cases we must have $x > 0$ and $< a$; $y > 0$ and $< a$.

Take a pair of rectangular axes and make OC, OD each equal to a. Draw the line $y = b + x$ represented by GH in the figure; and the line $x = b + y$, represented by EF.

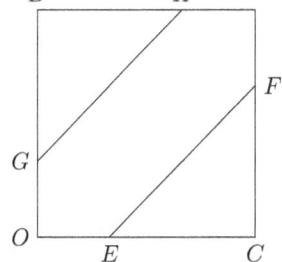

Then $OE = OG = b$; $CE = DG = a - b$.

Now the favourable cases are restricted to points within the triangles CEF, GDH, while for possible cases we may have all points in the figure CD.

Thus the required chance $= \left(\dfrac{a - b}{a} \right)^2$.

§32. In the line AB let points P, Q be taken in the order $APQB$ so that $AP = x$, $BQ = y$, $PQ = a - x - y$. Then in favourable cases we must have $x < b$, $y < b$, $a - x - y < b$; and in possible cases $x + y < a$.

Take a pair of rectangular axes OC, OD, and make OC, OD each equal to a.

A ——————— P   Q   B

Let $OE = OF = CG = DH = b$.

Then GH is the line $x + y = a - b$; and parallels to the axes through E and F are the lines $y = b$, $x = b$.

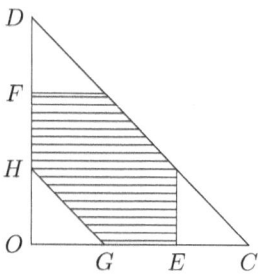

Figure 0.1: Fig. 1

(1) When $b > \dfrac{a}{2}$, the favourable cases will be restricted to the shaded area in Fig. 1, and the required chance $= 1 - 3\left(\dfrac{a-b}{a}\right)^2$.

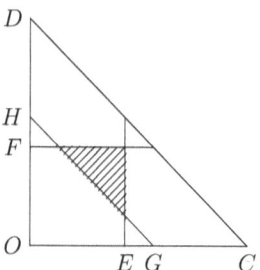

Figure 0.2: Fig. 2

(2) When $b < \dfrac{a}{2}$, the favourable cases will be restricted to the shaded area in Fig. 2. This consists of a right-angled isosceles triangle each of whose sides $= OF - EG = b - (a - 2b) = 3b - a$.

$$\therefore \text{ the required chance} = \left(\dfrac{3b-a}{a}\right)^2.$$

§33. Let AB be the line of length $a+b$, and on it measure $AP = x$, $PQ = a$;

A   P   Q   B

also let $AR = y$, $RS = b$.

A   R   S   B

Then for possible cases we must have $x > 0$, and $< b$; $y > 0$, and $< a$. If Q be beyond S the common part is $y + b - x$; but if S be beyond Q the common part is $x + a - y$; and one of these two must happen. Hence for favourable cases we must have

$$y + b - x < c, \text{ or } x + a - y < c.$$

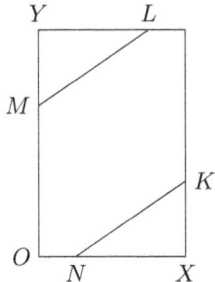

Take a, pair of rectangular axes and make $OY = a$, and $OX = b$.

Draw the line $x = y + b - c$, represented by NK in the figure; and the line $y = x + a - c$, represented by ML.

Thus the favourable cases are restricted to the two triangles XNK, YML; while for possible cases we have any points in the rectangle XY.

Thus the required chance $= \dfrac{c^2}{ab}$.

If RS is to lie within PQ we must have two conditions satisfied simultaneously; namely, $x < y$ and $x + a > y + b$.

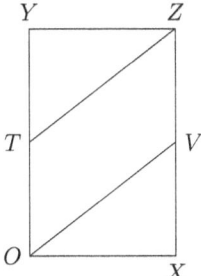

Draw the line $y = x$, represented by OV in the figure; and the line $y = x + a - b$, represented by TZ.

Then the favourable cases are restricted to points in the area $TOVZ$.

$$\therefore \text{ the required chance} = \frac{ab - b^2}{ab} = \frac{a - b}{a}.$$

[This solution is due to Professor R. S. Heath, Sc.D.]

§34. Let AB be the straight line, and let $AP = x$, $PQ = a$; also let $AP' = y$, $P'Q' = b$, and let all the measurements be made from left to right.

Then in possible cases we must have $x > 0$ and $< b + c$, and $y > 0$ and $< a + c$. Also in favourable cases we must have

$$\left.\begin{array}{l} AQ - AP' < d \\ (1)\ x + a - y < d \\ y - x > a - d \end{array}\right\}, \text{ or } (2)\ \left.\begin{array}{l} AQ' - AP < d \\ b + y - x < d \\ x - y > b - d \end{array}\right\}.$$

Take a pair of rectangular axes OC, OD. Let $OC = b + c$, $OD = a + c$, $OE = b$, $EF = d, CE = c$. Also let $OE = a$, $E'F' = d$, $E'D = c$.

Then $OF' = a - d$, $OF = b - d$.

A   P   Q   B

A   P'   Q'   B

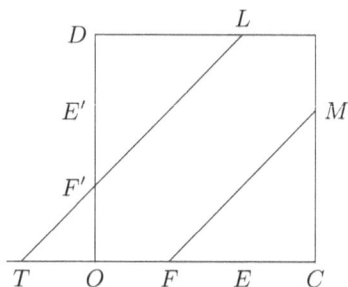

Draw the line $y - x = a - d$ represented by $LF'T$ in the figure, and the line $x - y = b - d$, represented by MF.

Then $DF' = DL = CM = CF = c + d$. And the favourable eases are restricted to points in the triangles CFM, $DF'L$, while the possible cases include all points in the rectangle OC, OD.

$$\text{Thus the required chance} = \frac{(c + d)^2}{(c + a)(c + b)}.$$

§35. The chance that C travels first class $= \dfrac{l}{l + m + n}$; the chance that A travels in any particular first class compartment $= \dfrac{1}{l} \cdot \dfrac{\lambda}{\lambda + \mu + \nu}$; therefore the chance that C and A travel together, both in the same first class compartment

$$= \frac{\lambda}{(l + m + n)(\lambda + \mu + \nu)}.$$

The chance that A and C travel together both in the same second class compartment

$$= \frac{\mu}{(l + m + n)(\lambda + \mu + \nu)}.$$

Similarly for companionship in any the same third class compartment. Thus the chance of A and C being companions in some compartment

$$= \frac{\lambda + \mu + \nu}{(l + m + n)(\lambda + \mu + \nu)} = \frac{1}{l + m + n}.$$

Hence the chance that A is with C, and B with $D = \dfrac{1}{(l + m + n)^2}$; and the chance that A is with one lady and B with the other is $\dfrac{2}{(l + m + n)^2}$.

Again, the chance that A and B both travel first class $= \dfrac{\lambda^2}{(\lambda + \mu + \nu)^2}$, and the chance that they both travel first class in the same compartment

$$= \frac{1}{l} \cdot \frac{\lambda^2}{(\lambda + \mu + \nu)^2}.$$

Thus the whole chance of their traveling together in some compartment

$$= \left(\frac{\lambda^2}{l} + \frac{\mu^2}{m} + \frac{\nu^2}{n} \right) \cdot \frac{1}{(\lambda + \mu + \nu)^2} = \frac{\lambda^2 mn + \mu^2 nl + \nu^2 ln}{lmn(\lambda + \mu + \nu)^2}.$$

Now $C's$ chance is the same for every compartment; therefore the chance that A and B are together and C also in their company

$$= \frac{1}{l + m + n} \cdot \frac{\lambda^2 mn + \mu^2 nl + \nu^2 ln}{lmn(\lambda + \mu \nu)^2};$$

and the chance that A and B are together and one or other of the ladies with them is the double of this, or

$$\frac{2}{l+m+n} \cdot \frac{\lambda^2 mn + \mu^2 nl + \nu^2 lm}{lmn(\lambda + \mu + \nu)^2}.$$

We have to prove that this is greater than $\dfrac{2}{(l+m+n)^2}$.

This will be the case if

$$(\lambda^2 mn + \mu^2 nl + \nu^2 lm)(l+m+n) > lmn(\lambda + \mu + \nu)^2,$$

that is, if $l(\mu^2 n^2 + \nu^2 m^2) + m(\nu^2 l^2 + \lambda^2 n^2) + n(\lambda^2 m^2 + \mu^2 l^2) > 2lmn(\mu\nu + \nu\lambda + \lambda\mu)$, an inequality which always holds except when $l : m : n = \lambda : \mu : \nu$.

[This problem and solution are due to the Rev. T. C. Simmons.]

EXAMPLES XXXIII a

§1. Subtracting the first column from the second, and also from the third, we have

$$\begin{vmatrix} 1 & 1 & 1 \\ 35 & 37 & 34 \\ 23 & 26 & 25 \end{vmatrix} = \begin{vmatrix} 1 & 0 & 0 \\ 35 & 2 & -1 \\ 23 & 3 & 2 \end{vmatrix} = \begin{vmatrix} 2 & -1 \\ 3 & 2 \end{vmatrix} = 7.$$

§2. Adding together the first and last columns, we obtain a column in which each of the constituents is double of the corresponding constituent in the middle column; hence the result is zero.

§3. Keeping the second column unaltered, first multiply it by 4 and subtract from the first column; then multiply the second column by 7 and subtract from the third; thus

$$\begin{vmatrix} 13 & 3 & 23 \\ 30 & 7 & 53 \\ 39 & 9 & 70 \end{vmatrix} = \begin{vmatrix} 1 & 3 & 2 \\ 2 & 7 & 4 \\ 3 & 9 & 7 \end{vmatrix} = \begin{vmatrix} 1 & 3 & 0 \\ 2 & 7 & 0 \\ 3 & 9 & 1 \end{vmatrix} = \begin{vmatrix} 1 & 3 \\ 2 & 7 \end{vmatrix} = 1.$$

§4. Here

$$\begin{vmatrix} a & h & g \\ h & b & f \\ g & f & c \end{vmatrix} = a(bc - f^2) - h(ch - fg) + g(fh - bg).$$

§5. Here

$$\begin{vmatrix} 1 & z & -y \\ -z & 1 & x \\ y & -x & 1 \end{vmatrix} = 1(1 + x^2) - z(-z - xy) - y(xz - y).$$

§6. Here

$$\begin{vmatrix} 1 & 1 & 1 \\ 1 & 1+x & 1 \\ 1 & 1 & 1+y \end{vmatrix} = \begin{vmatrix} 1 & 0 & 0 \\ 1 & x & 0 \\ 1 & 0 & y \end{vmatrix} = xy.$$

§7. Adding together all the columns we obtain a new determinant in which all the constituents of one column are zero; hence the value of the determinant is zero.

§8. Add together the second and third row and subtract the sum from the first row; thus

$$\begin{vmatrix} b+c & a & a \\ b & c+a & b \\ c & c & a+b \end{vmatrix} = \begin{vmatrix} 0 & -2c & -2b \\ b & c+a & b \\ c & c & a+b \end{vmatrix}$$

$$= 2cb(a+b-c) - 2bc(b-c-a) = 4abc.$$

§9. Adding together all the columns we obtain a new determinant in which all the constituents of one column are $1 + \omega + \omega^2$, that is, equal to zero; hence the value of the determinant is zero.

§10. Since ω^3 is equal to 1, we have

$$\begin{vmatrix} 1 & \omega^3 & \omega^2 \\ \omega^3 & 1 & \omega \\ \omega^2 & \omega & 1 \end{vmatrix} = \begin{vmatrix} 1 & 1 & \omega^2 \\ 1 & 1 & \omega \\ \omega^2 & \omega & 1 \end{vmatrix} = \begin{vmatrix} 1 & 0 & \omega^2 \\ 1 & 0 & \omega \\ \omega^2 & \omega - \omega^2 & 1 \end{vmatrix}$$

$$= -(\omega - \omega^2)(\omega - \omega^2) = -\omega^2 + 2\omega^3 - \omega^4 = 2 - (\omega^2 + \omega) = 3.$$

§11. The result of the elimination is

$$\begin{vmatrix} a & c & b \\ c & b & a \\ b & a & c \end{vmatrix} = 0; \text{ that is }, a(bc - a^2) - c(c^2 - ab) + b(ac - b^2) = 0.$$

§12. We have

$$\begin{vmatrix} a & b & c \\ x & y & z \\ p & q & r \end{vmatrix} = - \begin{vmatrix} x & y & z \\ a & b & c \\ p & q & r \end{vmatrix} = + \begin{vmatrix} x & y & z \\ p & q & r \\ a & b & c \end{vmatrix}.$$

Again,

$$\begin{vmatrix} a & b & c \\ x & y & z \\ p & q & r \end{vmatrix} = \begin{vmatrix} a & x & p \\ b & y & q \\ c & z & r \end{vmatrix} = - \begin{vmatrix} b & y & q \\ a & x & p \\ c & z & r \end{vmatrix} = + \begin{vmatrix} y & b & q \\ x & a & p \\ z & c & r \end{vmatrix}.$$

§13. (1) The given equation is a quadratic, and clearly vanishes when $x = a$, or when $x = b$; hence the solution is $x = a$ or b.

(2) Add together the first and second rows, and subtract twice the third row from the sum; thus

$$\begin{vmatrix} 15 - 2x & 11 & 10 \\ 11 - 3x & 17 & 16 \\ 12 - 3x & 0 & 0 \end{vmatrix} = 0;$$

hence

$$(12 - 3x) \begin{vmatrix} 11 & 10 \\ 17 & 16 \end{vmatrix} = 0;$$

therefore $12 - 3x = 0$, and $x = 4$.

§14. From Art. 495, this determinant can be expressed as the sum of eight determinants, all of which vanish with the exception of

$$\begin{vmatrix} b & c & a \\ q & r & p \\ y & z & x \end{vmatrix} \text{ and } \begin{vmatrix} c & a & b \\ r & p & q \\ z & x & y \end{vmatrix}; \text{ each of which is equal to } \begin{vmatrix} a & b & c \\ p & q & r \\ x & y & z \end{vmatrix}.$$

§15. This determinant vanishes if $a = b$, and therefore must contain $a - b$ as a factor; similarly it contains $b - c$ and $c - a$ as factors; and therefore being of the third degree must be equal to $k(b - c)(c - a)(a - b)$. By comparing the coefficients of bc^2, we see that $k = 1$.

§16. As in Ex. 15, the determinant is divisible by $(b - c)(c - a)(a - b)$, and being of the fourth degree the remaining factor must be $k(a + b + c)$. By comparing the coefficients of bc^3, we see that $k = 1$.   [See Ex. 2, Art. 522.]

§17. As in Ex. 15, the determinant is divisible by $(y - z)(z - x)(x - y)$; the remaining factor must be of the form $A(x^2 + y^2 + z^2) + B(yz + zx + xy)$. Since the highest power of x in the determinant is x^3, a comparison of the coefficients of $x^4 y$ shews that A must be zero, and a comparison of the coefficients of $x^2 y^3$ shews that $B = 1$.   [See Ex. 3, Art. 522.]

§18. On expansion, the determinant

$$= -2a\{4bc - (b + c)^2\} - (a + b)\{-2c(a + b) - (b + c)(c + a)\} +$$
$$+ (c + a)\{(b + c)(a + b) + 2b(c + a)\}$$

$$= 2(b + c)(c + a)(a + b) + 2a(b + c)^2$$
$$+ 2b(c + a)^2 + 2c(a + b)^2 - 8abc$$

$$= 2(b + c)(c + a)(a + b) + 2(b^2 c + bc^2 + c^2 a + a^2 b + ab^2 + 2abc)$$
$$= 2(b + c)(c + a)(a + b) + 2(b + c)(c + a)(a + b).$$

§Alternative Solution. If $a + b = 0$, so that $b = -a$, the determinant

$$= \begin{vmatrix} -2a & 0 & c + a \\ 0 & 2a & c - a \\ c + a & c - a & -2c \end{vmatrix} = 2a\{4ac + (c - a)^2\} - 2a(c + a)^2 = 0;$$

hence the determinant is divisible by $a + b$; similarly it is divisible by $b + c$ and $c + a$; and therefore must be equal to $k(b + c)(c + a)(a + b)$. To find

k, put $a = 0$, $b = 1$, $c = 1$; thus

$$2k = \begin{vmatrix} 0 & 1 & 1 \\ 1 & -2 & 2 \\ 1 & 2 & -2 \end{vmatrix} = 8.$$

§19. It is easy to shew that the determinant vanishes when $a = 0$, $b = 0$, $c = 0$; hence it is divisible by abc.

Again, the determinant

$$= \begin{vmatrix} (b+c)^2 - a^2 & 0 & a^2 \\ b^2 - (c+a)^2 & (c+a)^2 - b^2 & b^2 \\ 0 & c^2 - (a+b)^2 & (a+b)^2 \end{vmatrix}.$$

Here both the first and second columns contain $a + b + c$ as a factor; hence the given determinant must be divisible by $(a+b+c)^2$, and since it is of six dimensions the remaining factor must be of the form $k(a+b+c)$. Thus the given determinant must be equal to $kabc(a+b+c)^3$.

Hence k must be equal to the coefficient of a^4bc in the expanded determinant.

Now the term a^4bc can only arise from the product $(b+c)^2(c+a)^2(a+b)^2$, and its coefficient in this product is 2; hence $k = 2$.

§20. As in Art. 498, the product of the determinants

$$\begin{vmatrix} a_1 & b_1 & c_1 \\ a_2 & b_2 & c_2 \\ a_3 & b_3 & c_3 \end{vmatrix} \text{ and } \begin{vmatrix} x_1 & y_1 & z_1 \\ x_2 & y_2 & y_2 \\ x_3 & y_3 & y_3 \end{vmatrix}$$

is

$$\begin{vmatrix} a_1x_1 + b_1y_1 + c_1z_1 & a_2x_1 + b_2y_1 + c_2z_1 & a_3x_1 + b_3y_1 + c_3z_1 \\ a_1x_2 + b_1y_2 + c_1z_2 & a_2x_2 + b_2y_2 + c_2z_2 & a_3x_2 + b_3y_2 + c_3z_2 \\ a_1x_3 + b_1y_3 + c_1z_3 & a_2x_3 + b_2y_3 + c_2z_3 & a_3x_3 + b_3y_3 + c_3z_3 \end{vmatrix}.$$

By the above formula

$$\begin{vmatrix} 0 & c & b \\ c & 0 & a \\ b & a & 0 \end{vmatrix} \times \begin{vmatrix} 0 & c & b \\ c & 0 & a \\ b & a & 0 \end{vmatrix} = \begin{vmatrix} c^2 + b^2 & ab & ac \\ ab & c^2 + a^2 & bc \\ ac & bc & b^2 + a^2 \end{vmatrix}.$$

§21. Substituting the three sets of values for x, y, z, we have the equations :

$$la_1 + mb_1 + nc_1 = 0;$$
$$la_2 + mb_2 + nc_2 = 0;$$
$$la_3 + mb_3 + nc_3 = 0;$$

whence by eliminating l, m, n, we have

$$\begin{vmatrix} a_1 & b_1 & c_1 \\ a_2 & b_2 & c_2 \\ a_3 & b_3 & c_3 \end{vmatrix} = 0.$$

§22. From the general result given in Ex. 20, the constituents of the first column of the determinant-product are

$$\lambda(a^2 + \lambda^2) + c(ab + c\lambda) - b(ca - b\lambda);$$
$$-c(a^2 + \lambda^2) + \lambda(ab + c\lambda) + a(ca - b\lambda);$$
$$b(a^2 + \lambda^2) - a(ab + c\lambda) + \lambda(ca - b\lambda);$$

these expressions reduce to $\lambda^3 + \lambda(a^2 + b^2 + c^2)$; 0; 0 respectively. Similarly for the constituents of the second and third columns. Thus the product is a determinant whose constitue-nts are zero except in the leading diagonal, where each constituent is $\lambda(\lambda^2 + a^2 + b^2 + c^2)$.

§23. The constituents of the first column of the determinant-product are

$$(a + ib)(\alpha - i\beta) + (c + id)(\gamma - i\delta);$$
$$(a + ib)(-\gamma - i\delta) + (c + id)(\alpha + i\beta).$$

On reduction these become
$$a\alpha + b\beta + c\gamma + d\delta - i(a\beta - b\alpha + c\delta - d\gamma),$$
and $\qquad -a\gamma + b\delta + c\alpha - d\beta - i(a\delta + b\gamma - c\beta - d\alpha)$ respectively,
that is, $A - iB$ and $-C - iD$ suppose.

Similarly the constituents of the second column are
$$(-c + id)(\alpha - i\beta) + (a - ib)(\gamma - i\delta),$$
and $\qquad (-c + id)(-\gamma - i\delta) + (a - ib)(\alpha + i\beta);$
which reduce to $C - iD$ and $A + iB$ respectively.

The three determinants in the question when expanded become
$$a^2 + b^2 + c^2 + d^2; \quad \alpha^2 + \beta^2 + \gamma^2 + \delta^2;$$
$$A^2 + B^2 + C^2 + D^2 \text{ respectively.}$$

Hence $\qquad (a^2 + b^2 + c^2 + d^2)(\alpha^2 + \beta^2 + \gamma^2 + \delta^2)$
$$= (a\alpha + b\beta + c\gamma + d\delta)^2 + (a\beta - b\alpha + c\delta - d\gamma)^2$$
$$+ (a\gamma - b\delta - c\alpha + d\beta)^2 + (a\delta + b\gamma - c\beta - d\alpha)^2.$$

§24. The given determinant vanishes when $b = c$, hence it contains $b - c$ as a factor. Similarly it contains $c - a$ and $a - b$ as factors. Hence the determinant must be divisible by $(b - c)(c - a)(a - b)$. But by Ex. 15 the expression $(b - c)(c - a)(a - b)$ can be expressed as a determinant, and therefore the given determinant is equal to the product of two determinants.

Assume then that the given determinant is equal to the product of
$$\begin{vmatrix} 1 & a & a^2 \\ 1 & b & b^2 \\ 1 & c & c^2 \end{vmatrix} \text{ and } \begin{vmatrix} p & q & r \\ s & t & u \\ x & y & z \end{vmatrix}.$$

To find the unknown quantities we have (as in Ex. 20) the equations
$$p + aq + a^2 r = 1, s + at + a^2 u = bc + ad, x + ay + a^2 z = b^2 c^2 + a^2 d^2,$$
$$p + bq + b^2 r = 1, s + bt + b^2 u = ca + bd, x + by + b^2 z = c^2 a^2 + b^2 d^2,$$
$$p + cq + c^2 r = 1, s + ct + c^2 u = ab + cd, x + cy + c^2 z = a^2 b^2 + c^2 d^2.$$

From these equations we find,
$$p = 1, \qquad q = 0, \qquad r = 0;$$
$$s = bc + ca + ab, \qquad t = d - a - b - c, \qquad u = 1;$$
$$x = b^2 c^2 + c^2 a^2 + a^2 b^2 + abc(a + b + c),$$
$$y = -(b + c)(c + a)(a + b), \qquad z = d^2 + bc + ca + ab.$$

[Since $q = 0$, $r = 0$, it is unnecessary to find the values of s and x.]

The second determinant thus becomes
$$\begin{vmatrix} 1 & 0 & 0 \\ bc + ca + ab & d - a - b - c & 1 \\ b^2 c^2 + c^2 a^2 + a^2 b^2 + abc(a + b + c) & -(b + c)(c + a)(a + b) & d^2 + bc + ca + ab \end{vmatrix},$$

and therefore is equal to
$$d^3 - (a + b + c)d^2 + (bc + ca + ab)d$$
$$- (a + b + c)(bc + ca + ab) + (b + c)(c + a)(a + b);$$
that is, equal to $\qquad d^3 - (a + b + c)d^2 + (bc + ca + ab)d - abc,$
or $\qquad (d - a)(d - b)(d - c).$

Note. The preceding equations may be easily solved by the ordinary rules : thus taking the equations in x, y, z, on subtracting the second equation from the first and dividing by $a - b$, we obtain
$$y + (a + b)z = -c^2(a + b) + d^2(a + b);$$
similarly $\qquad y + (a + c)z = -b^2(a + c) + d^2(a + c);$
whence $\qquad z = a(b + c) + bc + d^2;$

The values of x and y are easily found by substitution.

The solution of these and similar equations may however be some-times more easily obtained by trial from a consideration of the fact that p, q, r, s, t, u, x, y, z must be symmetrical functions of a, b, c, or constant.

Thus take the equation $s + at + a^2u = bc + ad$; here the term bc must belong to s; hence $bc + ca + ab$ is part of the value of s; if there be any other part, denote it by s'; then $(bc + ca + ab) + s' + at + a^2u = bc + ad$; that is, $s' + at + a^2u = -ab - ac + ad$. Here the terms $-ab - ac + ad$ arise from part of the value of at, so that $t = -a - b - c + d$. On substituting this value of t, we have $s' + a^2u = a^2$, which is satisfied by $s' = 0$, $u = 1$. Thus $s = bc + ca + ab$, $t = -a - b - c + d$, $u = 1$.

§24. Alternative Solution. If $d = a$, the second and third rows are identical; thus the determinant is divisible by $a - d$; similarly it is divisible by $b - d$ and $c - d$; hence the determinant being a cubic in d must be equal to $(a - d)(b - d)(c - d) f(a, b, c)$.

To find the value of $f(a, b, c)$, put $d = 0$;
thus

$$abc\, f(a, b, c) = \begin{vmatrix} 1 & bc & b^2c^2 \\ 1 & ca & c^2a^2 \\ 1 & ab & a^2b^2 \end{vmatrix}.$$

This determinant vanishes when $a = 0$, for then the second and third rows are identical; also when $b = c$ for the same reason; hence the de-terminant must be equal to $kabc(b - c)(c - a)(a - b)$. It is easy to see that $k = -1$;

hence $\qquad\qquad f(a, b, c) = -(b - c)(c - a)(a - b)$.

§25. As in Ex. 24, the given determinant

$$= \begin{vmatrix} 1 & a & a^2 \\ 1 & b & b^2 \\ 1 & c & c^2 \end{vmatrix} \times \begin{vmatrix} p & q & r \\ s & t & u \\ x & y & z \end{vmatrix};$$

where

$$p + aq + a^2r = bc - a^2, \qquad\qquad s + at + a^2u = -bc + ca + ab,$$
$$p + bq + b^2r = ca - b^2, \qquad\qquad s + bt + b^2u = bc - ca + ab,$$
$$p + cq + c^2r = ab - c^2, \qquad\qquad s + ct + c^2u = bc + ca - ab,$$

$$x + ay + a^2z = a^2 + ab + ac + bc,$$
$$x + by + b^2z = b^2 + ab + ac + bc,$$
$$x + cy + c^2z = c^2 + ab + ac + bc.$$

From these equations, we obtain

$p = bc + ca + ab$,	$q = -(a + b + c)$,	$r = 0$;
$s = -(bc + ca + ab)$,	$t = 2(a + b + c)$,	$u = -2$;
$x = bc + ca + ab$,	$y = 0$,	$z = 1$.

Thus the second determinant

$$= \begin{vmatrix} bc + ca + ab & -(a + b + c) & 0 \\ -(bc + ca + ab) & 2(a + b + c) & -2 \\ bc + ca + ab & 0 & 1 \end{vmatrix}$$

$$= (bc + ca + ab)(a + b + c) \begin{vmatrix} 1 & -1 & 0 \\ -1 & 2 & -2 \\ 1 & 0 & 1 \end{vmatrix} = 3(a + b + c)(bc + ca + ab).$$

§25. Alternative Solution. Multiply the second row by -1 and add the result to the sum of the first and third rows : thus the determinant

$$= \begin{vmatrix} 3bc & 3ca & 3ab \\ -bc + ca + ab & bc - ca + ab & bc + ca - ab \\ (a+b)(a+c) & (b+c)(b+a) & (c+a)(c+b) \end{vmatrix}.$$

Remove the factor 3, and multiply the new first row by 2 and add to the second; the new second row has now $bc + ca + ab$ for each of its constituents; thus the determinant

$$= 3(bc + ca + ab) \begin{vmatrix} bc & ca & ab \\ 1 & 1 & 1 \\ (a+b)(a+c) & (b+c)(b+a) & (c+a)(c+b) \end{vmatrix}.$$

This last determinant is clearly $= k(b - c)(c - a)(a - b)(a + b + c)$.
To find k put $a = 2$, $b = 1$, $c = 0$; thus

$$-6k = \begin{vmatrix} 0 & 0 & 2 \\ 1 & 1 & 1 \\ 6 & 3 & 2 \end{vmatrix} = 2 \begin{vmatrix} 1 & 1 \\ 6 & 3 \end{vmatrix} = -6;$$

whence $k = 1$, and the last determinant $= (b - c)(c - a)(a - b)(a + b + c)$.

§26. Changing the columns into rows, we may write

$$\begin{vmatrix} (a-x)^2 & (b-x)^2 & (c-x)^2 \\ (a-y)^2 & (b-y)^2 & (c-y)^2 \\ (a-z)^2 & (b-z)^2 & (c-z)^2 \end{vmatrix} = \begin{vmatrix} 1 & a & a^2 \\ 1 & b & b^2 \\ 1 & c & c^2 \end{vmatrix} \times \begin{vmatrix} f & g & h \\ l & m & n \\ p & q & r \end{vmatrix}.$$

Hence $(a-x)^2 = f + ag + a^2 h$, $(b-x)^2 = f + bg + b^2 h$, $(c-x)^2 = f + cg + c^2 h$; and similarly for l, m, n; p, q, r.
From these equations, we have by inspection

$$f = x^2, \ g = -2x, \ h = 1, \& c.$$

Thus the second determinant

$$= \begin{vmatrix} x^2 & -2x & 1 \\ y^2 & -2y & 1 \\ z^2 & -2z & 1 \end{vmatrix} = 2 \begin{vmatrix} 1 & x & x^2 \\ 1 & y & y^2 \\ 1 & z & z^2 \end{vmatrix} = 2(y-z)(z-x)(x-y).$$

§27. The given expression may be written

$$(u\alpha + w'\beta + v'\gamma)\alpha + (w'\alpha + v\beta + u'\gamma)\beta + (v'\alpha + u'\beta + w\gamma)\gamma.$$

Now suppose that, for all values of α, β, γ,

$$\frac{u\alpha + w'\beta + v'\gamma}{l} = \frac{w'\alpha + v\beta + u'\gamma}{m} = \frac{v'\alpha + u'\beta + w\gamma}{n} \tag{55}$$

where l, m, n are constants; then the given expression will be the product of two linear factors proportional to $(u\alpha + w'\beta + v'\gamma)(l\alpha + m\beta + n\gamma)$.

The necessary condition is that (55) should hold for all values of α, β, γ, and therefore for such values as simultaneously satisfy

$$u\alpha + w'\beta + v'\gamma = 0,$$
$$w'\alpha + v\beta + u'\gamma = 0,$$
$$v'\alpha + u'\beta + w\gamma = 0.$$

Eliminating α, β, γ, we obtain

$$\begin{vmatrix} u & w' & v' \\ w' & v & u' \\ v' & u' & w \end{vmatrix} = 0.$$

§28. By Art. 495 the determinant can be expressed as the sum of eight determinants.
The terms containing x^3 will be obtained from

$$\begin{vmatrix} a^2 x & abx & acx \\ abx & b^2 x & bcx \\ acx & bcx & c^2 x \end{vmatrix} \text{ or } abcx^3 \begin{vmatrix} a & a & a \\ b & b & b \\ c & c & c \end{vmatrix}.$$

Thus the coefficient of x^3 is zero.

The terms containing x^2 will be obtained from

$$\begin{vmatrix} u & abx & acx \\ w' & b^2x & bcx \\ v' & bcx & c^2x \end{vmatrix} + \text{two similar determinants.}$$

Thus the coefficient of x^2 is also zero.
The coefficient of x

$$= a\begin{vmatrix} a & w' & v' \\ b & v & u' \\ c & u' & w \end{vmatrix} + b\begin{vmatrix} u & a & v' \\ w' & b & u' \\ v' & c & w \end{vmatrix} + c\begin{vmatrix} u & w' & a \\ w' & v & b \\ v' & u' & c \end{vmatrix}$$

$$= -\begin{vmatrix} 0 & a & b & c \\ a & u & w' & v' \\ b & w' & v & u' \\ c & v' & u' & w \end{vmatrix} = -\begin{vmatrix} u & w' & v' & a \\ w' & v & u' & b \\ v' & u' & w & c \\ a & b & c & 0 \end{vmatrix}$$

Lastly the term independent of x is

$$\begin{vmatrix} u & w' & v' \\ w' & v & u' \\ v' & u' & w \end{vmatrix}$$

$$\therefore x = \begin{vmatrix} u & w' & v' \\ w' & v & u' \\ v' & u' & w \end{vmatrix} \div \begin{vmatrix} u & w' & v' & a \\ w' & v & u' & b \\ v' & u' & w & c \\ a & b & c & 0 \end{vmatrix}.$$

EXAMPLES XXXIII b

§1. Subtract the first column from the second, the second from the third, and the third from the fourth : thus we obtain

$$\begin{vmatrix} 1 & 1 & 1 & 1 \\ 1 & 2 & 3 & 4 \\ 1 & 3 & 6 & 10 \\ 1 & 4 & 10 & 20 \end{vmatrix} = \begin{vmatrix} 1 & 0 & 0 & 0 \\ 1 & 1 & 1 & 1 \\ 1 & 2 & 3 & 4 \\ 1 & 3 & 6 & 10 \end{vmatrix} = \begin{vmatrix} 1 & 1 & 1 \\ 2 & 3 & 4 \\ 3 & 6 & 10 \end{vmatrix}$$

$$= \begin{vmatrix} 1 & 0 & 0 \\ 2 & 1 & 1 \\ 3 & 3 & 4 \end{vmatrix} = \begin{vmatrix} 1 & 1 \\ 3 & 4 \end{vmatrix} = 1.$$

[This is a particular case of Ex. 18. The solution of Ex. 9 is similar.]
§2. Form a new determinant by adding together the first and second rows for one row, and the third and fourth rows for another row; it will be found that the new determinant has two rows identical; hence the result is zero.
§3. Subtract the second column from the first, the third from the second, and the fourth from the third; thus the given determinant

$$= \begin{vmatrix} a-1 & 0 & 0 & 1 \\ 1-a & a-1 & 0 & 1 \\ 0 & 1-a & a-1 & 1 \\ 0 & 0 & 1-a & a \end{vmatrix} = (a-1)^3\begin{vmatrix} 1 & 0 & 0 & 1 \\ -1 & 1 & 0 & 1 \\ 0 & -1 & 1 & 1 \\ 0 & 0 & -1 & a \end{vmatrix}$$

On subtracting the first column from the last, we obtain

$$(a-1)^3\begin{vmatrix} 1 & 0 & 0 & 0 \\ -1 & 1 & 0 & 2 \\ 0 & -1 & 1 & 1 \\ 0 & 0 & -1 & a \end{vmatrix} = (a-1)^3\begin{vmatrix} 1 & 0 & 2 \\ -1 & 1 & 1 \\ 0 & -1 & a \end{vmatrix}$$

$$= (a-1)^3\begin{vmatrix} 1 & 0 & 0 \\ -1 & 1 & 3 \\ 0 & -1 & a \end{vmatrix} = (a-1)^3(a+3).$$

§4. By subtracting the third column from the second, and the fourth from the third, we have

$$
\begin{vmatrix} 0 & 1 & 1 & 1 \\ 1 & b+c & a & a \\ 1 & b & c+a & b \\ 1 & c & c & a+b \end{vmatrix} = \begin{vmatrix} 0 & 0 & 0 & 1 \\ 1 & b+c-a & 0 & a \\ 1 & b-c-a & c+a-b & b \\ 1 & 0 & c-a-b & a+b \end{vmatrix}
$$

$$
= - \begin{vmatrix} 1 & b+c-a & 0 \\ 1 & b-c-a & c+a-b \\ 1 & 0 & c-a-b \end{vmatrix} = - \begin{vmatrix} 1 & b+c-a & 0 \\ 0 & -2c & c+a-b \\ 0 & -b-c+a & c-a-b \end{vmatrix}
$$

$$
= 2c(c-a-b) - (b+c-a)(c+a-b) = a^2 + b^2 + c^2 - 2bc - 2ca - 2ab.
$$

§5. From the first column subtract three times the third, from the second subtract twice the third, and from the fourth subtract four times the third; thus

$$
\begin{vmatrix} 3 & 2 & 1 & 4 \\ 15 & 29 & 2 & 14 \\ 16 & 19 & 3 & 17 \\ 33 & 39 & 8 & 38 \end{vmatrix} = \begin{vmatrix} 0 & 0 & 1 & 0 \\ 9 & 25 & 2 & 6 \\ 7 & 13 & 3 & 5 \\ 9 & 23 & 8 & 6 \end{vmatrix}
$$

$$
= \begin{vmatrix} 9 & 25 & 6 \\ 7 & 13 & 5 \\ 9 & 23 & 6 \end{vmatrix} = \begin{vmatrix} 3 & 1 & 6 \\ 2 & -7 & 5 \\ 3 & -1 & 6 \end{vmatrix} = \begin{vmatrix} 3 & 1 & 6 \\ 2 & -7 & 5 \\ 0 & -2 & 0 \end{vmatrix} = 2 \begin{vmatrix} 3 & 6 \\ 2 & 5 \end{vmatrix} = 2 \times 3 = 6.
$$

§6. By subtracting the second column from the first, the third from the second, the fourth from the third, we have

$$
\begin{vmatrix} 1+a & 1 & 1 & 1 \\ 1 & 1+b & 1 & 1 \\ 1 & 1 & 1+c & 1 \\ 1 & 1 & 1 & 1+d \end{vmatrix} = \begin{vmatrix} a & 0 & 0 & 1 \\ -b & b & 0 & 1 \\ 0 & -c & c & 1 \\ 0 & 0 & -d & 1+d \end{vmatrix}.
$$

This last determinant

$$
= a \begin{vmatrix} b & 0 & 1 \\ -c & c & 1 \\ 0 & -d & 1+d \end{vmatrix} - \begin{vmatrix} -b & b & 0 \\ 0 & -c & c \\ 0 & 0 & -d \end{vmatrix}
$$

$$
= ab(c+cd+d) + acd + bcd = abcd + abc + abd + acd + bcd.
$$

§7. Adding together all the columns we obtain a column in which each constituent is $x+y+z$; hence the given determinant is divisible by $x+y+z$.

Multiply the columns by 1, -1, 1, -1 respectively and add the products; we obtain a column in which each constituent contains $z + x - y$ as a factor; thus the given determinant is divisible by $z + x - y$; similarly it is divisible by $x + y - z$, and $y + z - x$; and therefore must be equal to

$$
k(x + y + z)(y + z - x)(z + x - y)(x + y - z).
$$

By inspection of the coefficient of x^4 we find that $k = -1$.

§8. The given determinant

$$
= \frac{1}{a} \begin{vmatrix} 0 & ax & y & z \\ -x & 0 & c & b \\ -y & -ac & 0 & a \\ -z & -ab & -a & 0 \end{vmatrix}
$$

$$
= \frac{1}{a} \begin{vmatrix} 0 & ax-by+cz & y & z \\ -x & 0 & c & b \\ -y & 0 & 0 & a \\ -z & 0 & -a & 0 \end{vmatrix} = -\frac{1}{a}(ax-by+cz) \begin{vmatrix} -x & c & b \\ -y & 0 & a \\ -z & -a & 0 \end{vmatrix}
$$

$$
= -\frac{1}{a}(ax-by+cz) \times (-a^2 x - acz + aby) = (ax - by + cz)^2.
$$

This determinant belongs to the class of skew determinants, in which every constituent in the leading diagonal is zero, and the corresponding constituents on each side of the diagonal are equal in magnitude but opposite in sign. It is shewn in works on Determinants that a skew determinant of an odd order vanishes, whilst a skew determinant of an even order is a perfect square. See Muir's Determinants, Arts. 157, 159.

§9. Proceeding as in Example 1 (but operating on the rows) we see that the determinant

$$= \begin{vmatrix} a & b & c & d \\ 0 & a & a+b & a+b+c \\ 0 & a & 2a+b & 3a+2b+c \\ 0 & a & 3a+b & 6a+3b+c \end{vmatrix} = a \begin{vmatrix} a & a+b & a+b+c \\ a & 2a+b & 3a+2b+c \\ a & 3a+b & 6a+3b+c \end{vmatrix}.$$

In like manner this last determinant

$$= a \begin{vmatrix} a & a+b & a+b+c \\ 0 & a & 2a+b \\ 0 & a & 3a+b \end{vmatrix} = a^2 \begin{vmatrix} a & 2a+b \\ a & 3a+b \end{vmatrix} = a^4.$$

§10. Replace ω^3 by 1; then by Art. 497 it is easy to prove that

$$\begin{vmatrix} 1 & \omega & \omega^2 & 1 \\ \omega & \omega^2 & 1 & 1 \\ \omega^2 & 1 & 1 & \omega \\ 1 & 1 & \omega & \omega^2 \end{vmatrix} \times \begin{vmatrix} 1 & \omega & \omega^2 & 1 \\ \omega & \omega^2 & 1 & 1 \\ \omega^2 & 1 & 1 & \omega \\ 1 & 1 & \omega & \omega^2 \end{vmatrix} = \begin{vmatrix} 1 & 1 & -2 & 1 \\ 1 & 1 & 1 & -2 \\ -2 & 1 & 1 & 1 \\ 1 & -2 & 1 & 1 \end{vmatrix}.$$

For the constituents of the first row of the determinant-product are

$$1 + \omega^2 + \omega^4 + 1; \quad \omega + \omega^3 + \omega^2 + 1;$$
$$\omega^2 + \omega + \omega^2 + \omega; \quad 1 + \omega + \omega^3 + \omega^2;$$

but since $\omega^3 = 1$ and $1 + \omega + \omega^2 = 0$, these reduce to 1, 1, -2, 1 respectively. Similarly for the other rows.

The numerical value of the determinant on the right

$$= \begin{vmatrix} 1 & 0 & 0 & 0 \\ 1 & 0 & 3 & -3 \\ -2 & 3 & -3 & 3 \\ 1 & -3 & 3 & 0 \end{vmatrix} = 27 \begin{vmatrix} 0 & 1 & -1 \\ 1 & -1 & 1 \\ -1 & 1 & 0 \end{vmatrix} = -27.$$

Thus the square of the given determinant is equal to -27.

§11. The determinant formed by the minors of the determinant

$$\begin{vmatrix} a & h & g \\ h & b & f \\ g & f & c \end{vmatrix} \text{ is } \begin{vmatrix} bc - f^2 & fg - ch & hf - bg \\ fg - ch & ca - g^2 & gh - af \\ hf - bg & gh - af & ab - h^2 \end{vmatrix}.$$

This second determinant is therefore the square of the first. [Art. 498.]

Hence the second determinant $= (abc + 2fgh - af^2 - bg^2 - ch^2)^2$.

But in order that the three given equations may hold, the second determinant must vanish; hence $abc + 2fgh - af^2 - bg^2 - ch^2 = 0$.

Alternative Solution. From the first two equations, the value of x is proportional to $(ch - fg)(af - gh) - (bg - hf)(g^2 - ca)$, which is equal to

$$g(abc + 2fgh - af^2 - bg^2 - ch^2).$$

Similarly the value of y is proportional to $f(abc + 2fgh - af^2 - bg^2 - ch^2)$; and the value of z is proportional to $c(abc + 2fgh - af^2 - bg^2 - ch^2)$.

Substituting in the third equation, we have

$$(abc + 2fgh - af^2 - bg^2 - ch^2)$$
$$\{g(bg - hf) + f(af - gh) + c(h^2 - ab)\} = 0;$$

that is, $\qquad (abc + 2fgh - af^2 - bg^2 - ch^2)^2 = 0$.

§12. The solution is similar to that of the next example.

§13. We may express the value of any one of the unknown quantities as the quotient of one determinant by another; thus the value of y is given by the equation

$$\begin{vmatrix} a & b & c \\ a^2 & b^2 & c^2 \\ a^3 & b^3 & c^3 \end{vmatrix} y = \begin{vmatrix} a & k & c \\ a^2 & k^2 & c^2 \\ a^3 & k^3 & c^3 \end{vmatrix};$$

that is, $abc(b-c)(c-a)(a-b)y = akc(k-c)(c-a)(a-k);$

whence $$y = \frac{k(k-c)(k-a)}{b(b-c)(b-a)}.$$

§14. Here

$$\begin{vmatrix} 1 & 1 & 1 & 1 \\ a & b & c & d \\ a^2 & b^2 & c^2 & d^2 \\ a^3 & b^3 & c^3 & d^3 \end{vmatrix} u = \begin{vmatrix} 1 & 1 & 1 & 1 \\ a & b & c & k \\ a^2 & b^2 & c^2 & k^2 \\ a^3 & b^3 & c^3 & k^3 \end{vmatrix} :$$

that is, $(b-c)(c-a)(a-b)(a-d)(b-d)(c-d)u$
$= (b-c)(c-a)(a-b)(a-k)(b-k)(c-k);$ [Art. 505]

whence $$u = \frac{(k-a)(k-b)(k-c)}{(d-a)(d-b)(d-c)}.$$

Examples 15 and 16 may be solved after the manner of the first solution of XXXIII. a. Ex. 24. Here however we shall give another method.

§15. When $d = a$, the 3rd row $= (-1) \times 2^{nd}$ row;

When $d = b$, the 3rd row $= (-1) \times 1^{st}$ row;

When $d = c$, the 2nd row $= (-1) \times 1^{st}$ row.

Hence the determinant vanishes in all these cases, and since it is a cubic in d it must be equal to $f(a, b, c) \times (a - d)(b - d)(c - d)$.

To find the value of $f(a, b, c)$, put $d = 0$;

then $$abc\, f(a, b, c) = \begin{vmatrix} b+c-a & bc & abc \\ c+a-b & ca & abc \\ a+b-c & ab & abc \end{vmatrix};$$

hence $$f(a, b, c) = \begin{vmatrix} 1 & b+c-a & bc \\ 1 & c+a-b & ca \\ 1 & a+b-c & ab \end{vmatrix};$$

$\therefore f(a, b, c) = bc\{(a+b-c) - (c+a-b)\} + \dots + \dots$
$= 2\{bc(b-c) + ca(c-a) + ab(a-b)\}$
$= -2(b-c)(c-a)(a-b).$

§16. For the new second column, multiply the first column by -2, the last by -2, and add the results to the second; thus the determinant

$$= \begin{vmatrix} a^2 & -a^2-b^2-c^2 & bc \\ b^2 & -a^2-b^2-c^2 & ca \\ c^2 & -a^2-b^2-c^2 & ab \end{vmatrix} = (a^2+b^2+c^2)\begin{vmatrix} 1 & a^2 & bc \\ 1 & b^2 & ca \\ 1 & c^2 & ab \end{vmatrix}$$

$= (a^2+b^2+c^2)(b-c)(c-a)(a-b)\, f(a, b, c);$

where $f(a, b, c)$ being of one dimension must be equal to $k(a+b+c)$. It is easy to see that $k = 1$.

§17. Adding together all the columns we see that the determinant is divisible by $a+b+c+d+e+f$.

Multiplying the columns by $1, -1, 1, -1, 1, -1$ respectively, and adding the results we see that $a-b+c-d+e-f$ is a factor of the determinant.

Multiplying the columns by $1, \omega, \omega^2, 1, \omega, \omega^2$ respectively, and adding the results, it follows that $a + \omega b + \omega^2 c + d + \omega e + \omega^2 f$ is a factor of the determinant.

Similarly we may shew that

$$a + \omega^2 b + \omega c + d + \omega^2 e + \omega f;$$
$$a - \omega b + \omega^2 c - d + \omega e - \omega^2 f;$$

$$a - \omega^2 b + \omega c - d + \omega^2 e - \omega f;$$

are factors of the determinant.

Hence the determinant is the product of these six factors and some constant, which is obviously unity.

Taking these factors in pairs, it follows that the determinant is the product of the three expressions

$$(a + c + e)^2 - (b + d + f)^2;$$
$$(a + \omega^2 c + \omega e)^2 - (d + \omega^2 f + \omega b)^2;$$
$$(a + \omega c + \omega^2 e)^2 - (d + \omega f + \omega^2 b)^2.$$

The last of these factors

$$= (a^2 - d^2 + 2ce - 2bf) + \omega(e^2 - b^2 + 2ac - 2df)$$
$$+ \omega^2(c^2 - f^2 + 2ae - 2bd)$$
$$= A + \omega B + \omega^2 C,$$

where A, B, C have the values given in the question.

Similarly the second factor $= A + \omega^2 B + \omega C$.

and the first factor $= A + B + C$.

Hence the determinant

$$= (A + B + C)(A + \omega B + \omega^2 C)(A + \omega^2 B + \omega C)$$
$$= A^3 + B^3 + C^3 - 3ABC,$$

which is the expanded form of the determinant on the right side.

§18. The determinant in question is

$$\begin{vmatrix} 1 & 1 & 1 & 1 & 1 & 1 & \cdots \\ 1 & 2 & 3 & 4 & 5 & 6 & \cdots \\ 1 & 3 & 6 & 10 & 15 & 21 & \cdots \\ 1 & 4 & 10 & 20 & 35 & 56 & \cdots \\ 1 & 5 & 15 & 35 & 70 & 126 & \cdots \\ \cdots & \cdots & \cdots & \cdots & \cdots & \cdots & \cdots \\ \cdots & \cdots & \cdots & \cdots & \cdots & \cdots & \cdots \end{vmatrix} \quad \text{[Art. 393.]}$$

If we form a new determinant by subtracting each row from the row immediately beneath it, we obtain a determinant in which each constituent of the first column vanishes except the first; thus the determinant

$$= \begin{vmatrix} 1 & 2 & 3 & 4 & 5 & \cdots \\ 1 & 3 & 6 & 10 & 15 & \cdots \\ 1 & 4 & 10 & 20 & 35 & \cdots \\ 1 & 5 & 15 & 35 & 70 & \cdots \\ \cdots & \cdots & \cdots & \cdots & \cdots & \cdots \\ \cdots & \cdots & \cdots & \cdots & \cdots & \cdots \end{vmatrix}.$$

This determinant consists of $n - 1$ rows, and the constituents of the successive rows are easily seen to be the first $n - 1$ terms of the figurate numbers of the 2nd, 3rd, 4th, ..., n^{th} orders.   [Art. 393.]

In like manner the last determinant

$$= \begin{vmatrix} 1 & 3 & 6 & 10 & \cdots \\ 1 & 4 & 10 & 20 & \cdots \\ 1 & 5 & 15 & 35 & \cdots \\ \cdots & \cdots & \cdots & \cdots & \cdots \\ \cdots & \cdots & \cdots & \cdots & \cdots \end{vmatrix}$$

The constituents of the successive rows of this determinant are the first $n - 2$ terms of the figurate numbers of the 3rd, 4th, ..., n^{th} orders.

Proceeding in this manner, the determinant will at length reduce to

$$\begin{vmatrix} 1 & n-1 \\ 1 & n \end{vmatrix};$$

and therefore its value is unity.

EXAMPLES XXXIV : Miscellaneous Theorems and Examples

EXAMPLES XXXIV a

§1. Here the multiplier is -5; hence as in Art. 515,

$$
\begin{array}{rrrrrr}
3 & 11 & 0 & 90 & -19 & 53 \\
 & -15 & 20 & -100 & 50 & -155 \\
\hline
 & -4 & 20 & -10 & 31 & -102 \\
\end{array}
$$

§2. The given expression vanishes when $x = 3$; hence
$$162 - 189 + 3a + b = 0;\ \text{that is,}\ 3a + b = 27.$$

§3. As in Art. 517, we have

$$
\begin{array}{r|rrrr|rr}
1 & 1 & - & 5 & + & 9 & - & 6 & - & 16 & + & 13 \\
+\ 3 & & + & 3 & - & 2 & & & & & & \\
-\ 2 & & & & - & 6 & + & 4 & & & & \\
 & & & & + & 3 & - & 2 & & & & \\
 & & & & & & + & 3 & - & 2 & & \\
\hline
 & 1 & - & 2 & + & 1 & + & 1 & - & 15 & + & 11. \\
\end{array}
$$

Therefore the quotient is $x^3 - 2x^2 + x + 1$; and the remainder is $-15x + 11$.

§4. As in Art. 517, we have

$$
\begin{array}{r|rrrrrrr}
1 & 1 & - & 2 & - & 4 & + & 19 & - & 31 & + & (12 & + & a) \\
0 & & & 0 & + & 7 & - & 5 & & & & & & \\
7 & & & & & 0 & - & 14 & + & 10 & & & & \\
-\ 5 & & & & & & & 0 & + & 21 & - & 15 & & \\
\hline
 & 1 & - & 2 & + & 3 & + & 0 & + & 0 & + & (a & - & 3). \\
\end{array}
$$

Thus the remainder is $a - 3$, and will therefore vanish when $a = 3$.

§5. As in Art. 517, we have

$$
\begin{array}{r|l}
1 & 1 \\
\\
+\ 5 & +5 - 7 - 1 \mid + 8 \\
\\
-\ 7 & \quad\ +25 - 35 \mid - 5 + 40 \\
\\
-\ 1 & \qquad\quad +90 \mid - 126 - 18 + 144 \\
\\
+\ 8 & \qquad\qquad\ \mid + 270 - 378 - 54 + 432 \\
\hline
 & 1 + 5 + 18 + 54 \mid + 147 - 356 + 90 + 432. \\
\end{array}
$$

Therefore the quotient is $x^{-4} + 5x^{-5} + 18x^{-6} + 54x^{-7}$; and the remainder is $147x^{-4} - 356x^{-5} + 90x^{-6} + 432x^{-7}$.

§6. $a(b - c)^3 + b(c - a)^3 + c(a - b)^3 = k(b - c)(c - a)(a - b)(a + b + c)$.

To find k, put $a = 2$, $b = 1$, $c = 0$; thus $-6k = -6$; that is, $k = 1$.

§8. It is clear that $(a + b + c)^3 - (b + c - a)^3 - (c + a - b)^3 - (a + b - c)^3 = kabc$; and on putting $a = b = c = 1$, we find $k = 24$.

§9. $a(b - c)^2 + b(c - a)^2 + c(a - b)^2 + 8abc$ vanishes when $b = -c$ and is therefore divisible by $b + c$. Similarly it is divisible by $c + a$ and $a + b$; it must therefore be equal to $k(b + c)(c + a)(a + b)$. On putting $a = 1$, $b = 1$, $c = 0$, we find that $k = 1$.

§10. This expression is equal to
$$(b - c)(c - a)(a - b)\{k(a^2 + b^2 + c^2) + l(bc + ca + ab)\}.$$
If $\qquad a = 2$, $\qquad b = 1$, $\qquad c = 0$, then $\qquad -14 = -2(5k + 2l)$;

and if
$$a = 1, \qquad b = -1, \qquad c = 0, \text{ then} \qquad 2 = 2(2k - l);$$
whence we find $k = 1$, $l = 1$.

§11. This expression vanishes when $a = 0$, $b = 0$, $c = 0$; and also when $b + c = 0$, $c + a = 0$, $a + b = 0$. Thus the expression is equal to $kabc(b + c)(c + a)(a + b)$. By putting $a = b = c = 1$, we find $k = 1$.

§12. This expression vanishes when $a = 0$, $b = 0$, $c = 0$. Moreover it is symmetrical and of four dimensions in a, b, c; therefore it must be equal to $kabc(a + b + c)$ where k is a constant. By putting $a = b = c = 1$, we find $k = 12$.

§13. This expression is equal to
$$abc\{k(a^2 + b^2 + c^2) + l(bc + ca + ab)\}.$$

If $\qquad a = b = c = 1$, then $\qquad\qquad 240 = 3(k + l);$

and if $\quad a = 1, b = 1, c = 1$, then $\qquad -240 = -(3k + l);$

whence $k = 80$, $l = 0$.

§14. This expression is equal to $k(b - c)(c - a)(a - b)(x - a)(x - b)(x - c)$. The coefficient of x^3 in the given expression
$$= (b - c)^3 + (c - a)^3 + (a - b)^3 = 3(b - c)(c - a)(a - b).$$
Hence $k = 3$.

§15. If $x + y + z = 0$, then $x^3 + y^3 + z^3 = 3xyz$. The condition $x + y + z = 0$ is satisfied if $x = b + c - 2a$, $y = c + a - 2b$, $z = a + b - 2c$.

§16. See Solution to Ex. 6.

§17. The expression
$$2a(b + c)(c + a) + 2b(c + a)(a + b) + 2c(a + b)(b + c)$$
$$+ (b - c)(c - a)(a - b)$$
vanishes when $b + c = 0$, and therefore is equal to
$$k(b + c)(c + a)(a + b).$$
By putting $a = b = c = 1$, we find $k = 3$.

§18. When $a = b + c$, the expression
$$a^2(b + c) + b^2(c + a) + c^2(a + b) - a^3 - b^3 - c^3 - 2abc$$
$$= a^2(b + c - a) + a(b^2 - 2bc + c^2) + bc(b + c) - b^3 - c^3$$
$$= (b + c)(b - c)^2 + bc(b + c) - b^3 - c^3$$
$$= 0.$$
Hence the given expression may be written $k(b + c - a)(c + a - b)(a + b - c)$, and a comparison of the terms involving a^3 shews that $k = 1$.

§19. The expression $a^3(b^2 - c^2) + b^3(c^2 - a^2) + c^3(a^2 - b^2)$
$$= (b - c)(c - a)(a - b)\{k(a^2 + b^2 + c^2) + l(bc + ca + ab)\}.$$

If $\qquad a = 2, \qquad b = 1, \qquad c = 0, \qquad$ then $4 = -2(5k + 2l);$

and if

$\qquad a = 1, \qquad b = -1, \qquad c = 0, \qquad$ then $2 = 2(2k - l);$

whence $k = 0$, and $l = -1$. Thus
$$a^3(b^2 - c^2) + b^3(c^2 - a^2)$$
$$+ c^3(a^2 - b^2) = -(b - c)(c - a)(a - b)(bc + ca + ab).$$

§20. It is easy to show that
$$(b - c)(b + c - 2a)^2 + (c - a)(c + a - 2b)^2 + (a - b)(a + b - 2c)^2$$
$$= -9(b - c)(c - a)(a - b);$$
and
$$(b - c)^2(b + c - 2a) + (c - a)^2(c + a - 2b) + (a - b)^2(a + b - 2c)$$
$$= -4(b - c)(c - a)(a - b);$$

§21. Since $(y+z)(z+x)(x+y) = x^2(y+z) + y^2(z+x) + z^2(x+y) + 2xyz$,

we have $(y+z)^2(z+x)^2(x+y)^2$

$$= \Sigma x^4(y+z)^2 + 2x^2y^2(x+z)(y+z) + 2x^2z^2(x+y)(z+y)$$
$$+ 2y^2z^2(z+x)(y+x) + 4x^3yz(y+z) + 4xy^3z(z+x)$$
$$+ 4xyz^3(x+y) + 4x^2y^2z^2$$
$$= \Sigma x^4(y+z)^2 + 2\Sigma x^3y^3 + 6\Sigma x^3y^2z + 10x^2y^2z^2$$
$$= \Sigma x^4(y+z)^2 + 2(\Sigma x^3y^3 + 3\Sigma x^3y^2z + 6x^2y^2z^2) - 2x^2y^2z^2$$
$$= \Sigma x^4(y+z)^2 + 2\Sigma(xy)^3 - 2x^2y^2z^2. \qquad \text{[Art. 522.]}$$

§22. On multiplication, we have $\Sigma(ab - c^2)(ac - b^2)$

$$= a^2bc + ab^2c + abc^2 - ab^3 - ac^3 - a^3b - bc^3 - a^3c - b^3c$$
$$+ b^2c^2 + a^2c^2 + a^2b^2$$
$$= (bc + ca + ab)^2$$
$$- (a^2bc + ab^2c + abc^2 + ab^3 + ac^3 + a^3b + bc^3 + a^3c + b^3c)$$
$$= (bc + ca + ab)^2 - (a^2 + b^2 + c^2)(bc + ca + ab)$$
$$= (bc + ca + ab)(bc + ca + ab - a^2 - b^2 - c^2).$$

§23. $abc(\Sigma a)^3 - (\Sigma bc)^3$

$$= abc(\Sigma a^3 + 3\Sigma a^2b + 6abc) - (\Sigma b^3c^3 + 3\Sigma a^3b^2c + 6a^2b^2c^2)$$
$$= abc\Sigma a^3 + 3\Sigma a^3b^2c - \Sigma b^3c^3 - 3\Sigma a^3b^2c$$
$$= abc\Sigma a^3 - \Sigma b^3c^3.$$

This last expression

$$= abc(a^3 + b^3 + c^3) - (b^3c^3 + c^3a^3 + a^3b^3)$$
$$= bc(a^4 - b^2c^2) + ab^3(bc - a^2) + ac^3(bc - a^2)$$
$$= (a^2 - bc)\{bc(a^2 + bc) - ab^3 - ac^3\}$$
$$= (a^2 - bc)(b^2 - ca)(c^2 - ab).$$

§24. Let $b - c = x$, $c - a = y$, $a - b = z$, then we have to show that

$$\Sigma x^3(y - z) = 0 \text{ when } x + y + z = 0.$$

Now $x^3(y - z) + y^3(z - x) + z^3(x - y) = k(y - z)(z - x)(x - y)(x + y + z)$;
which vanishes because of the zero factor $x + y + z$.

§25. The solution is similar to that of the next Example.

§26. From a formula given in Art. 523, we have

$$x^3 + y^3 + z^3 - 3xyz$$
$$= \frac{1}{2}(x + y + z)\{(y - z)^2 + (z - x)^2 + (x - y)^2\}$$
$$= \frac{1}{2}(a + b + c)\{4(b - c)^2 + 4(c - a)^2 + 4(a - b)^2\}$$
$$= 4(a^3 + b^3 + c^3 - 3abc).$$

§27. Let $x = s - a$, $y = s - b$, $z = s - c$.   By Art. 523,

$$x^3 + y^3 + z^3 - 3xyz$$
$$= \frac{1}{2}(x + y + z)\{(y - z)^2 + (z - x)^2 + (x - y)^2\}$$
$$= \frac{1}{2}(3s - a - b - c)\{(b - c)^2 + (c - a)^2 + (a - b)^2\}$$
$$= \frac{1}{2}(a + b + c)\{(b - c)^2 + (c - a)^2 + (a - b)^2\}$$
$$= a^3 + b^3 + c^3 - 3abc.$$

§28. The common denominator is $(b - c)(c - a)(a - b)(x - a)(x - b)(x - c)$,
and the numerator

$$= -\Sigma a(b - c)(x - b)(x - c)$$

$$= -x^2 \Sigma a(b-c) + x\Sigma a(b-c)(b+c) - abc\Sigma(b-c)$$
$$= x\Sigma a(b^2 - c^2)$$
$$= (b-c)(c-a)(a-b)x.$$

Note. It is easy to prove the converse result by resolving
$$\frac{x}{(x-a)(x-b)(x-c)}$$ into partial fractions.

§29. The common denominator is $(b-c)(c-a)(a-b)$,

and the numerator $= -\Sigma a^2(b-c) + \Sigma(b^2 + c^2)(b-c)$
$$= -\Sigma a^2(b-c) + \Sigma\{(b^3 - c^3) - bc(b-c)\}$$
$$= -\Sigma a^2(b-c) - \Sigma bc(b-c)$$
$$= (b-c)(c-a)(a-b) + (b-c)(c-a)(a-b)$$
$$= 2(b-c)(c-a)(a-b).$$

§30. The common denominator is $(b-c)(c-a)(a-b)(x+a)(x+b)(x+c)$;

the numerator $= -\Sigma(a+p)(a+q)(b-c)(x+b)(x+c)$
$$= -\Sigma(a+p)(a+q)$$
$$\{x^2(b-c) + x(b^2 - c^2) + bc(b-c)\}.$$

The coefficient of $x^2 = -\Sigma a^2(b-c) - (p+q)\Sigma a(b-c)$
$$- pq\Sigma(b-c)$$
$$= (b-c)(c-a)(a-b).$$

The coefficient of $x = -\Sigma a^2(b^2 - c^2) - (p+q)\Sigma a(b^2 - c^2)$
$$- pq\Sigma(b^2 - c^2)$$
$$= (p+q)(b-c)(c-a)(a-b).$$

The term independent of x
$$= -abc\Sigma a(b-c) - abc(p+q)\Sigma(b-c)$$
$$- pq\Sigma bc(b-c)$$
$$= pq(b-c)(c-a)(a-b).$$

Thus the numerator $= (b-c)(c-a)(a-b)\{x^2 + (p+q)x + pq\}.$

Note. The converse of this result is easily proved by Partial Fractions.

§31. The numerator is $-\Sigma bcd(b-c)(b-d)(c-d).$

This expression is of six dimensions, and vanishes when $b = c$, or $c = a$, or $a = b$, or $a = d$, or $b = d$, or $c = d$; hence it must be equal to
$$k(b-c)(c-a)(a-b)(a-d)(b-d)(c-d).$$

A comparison of the coefficients of $b^3 c^2 d$ shews that $k = -1$; hence the numerator becomes
$$-(b-c)(c-a)(a-b)(a-d)(b-d)(c-d).$$

§32. The numerator $= -\Sigma a^4(b-c)(b-d)(c-d).$

This expression consists of four terms and vanishes when $b = c$, or $c = a$, or $a = b$, or $a = d$, or $b = d$, or $c = d$, and therefore is divisible by
$$(b-c)(c-a)(a-b)(a-d)(b-d)(c-d).$$

Further the given expression is of seven dimensions; hence the remaining factor is $k(a+b+c+d)$. A comparison of the coefficients of $a^4 b^2 c$ shews that $k = 1$. Hence the numerator is equal to
$$(b-c)(c-a)(a-b)(a-d)(b-d)(c-d)(a+b+c+d).$$

§33. Taking $xyzp^2 s^2$ for common denominator, we find the numerator
$$= \Sigma(p^2 - sy^2)(p^2 - sz^2)x = p^4\Sigma x - p^2 s\Sigma x(y^2 + z^2) + s^2 xyz\Sigma yz.$$

Now $\Sigma x(y^2 + z^2) = (x+y+z)(yz+zx+xy) - 3xyz$; hence the above expression
$$= p^4 s - p^2 s\{s(yz + zx + xy) - 3p^2\} + p^2 s^2\Sigma yz = 4p^4 s.$$

Thus the given expression $= \dfrac{4p^4 s}{xyzp^2 s^2} = \dfrac{4p^4 s}{p^4 s^2} = \dfrac{4}{s}.$

EXAMPLES XXXIV b

§1. From the given condition, we have
$$3(b+c)(c+a)(a+b) = 0;$$
hence $\qquad b+c = 0,$ or $c+a = 0,$ or $a+b = 0.$

If $b+c = 0,$ then $a+b+c = a;$ also $b = -c,$ so that $b^{2n+1} = -c^{2n+1}.$ From this it is easy to see that each side of the identity is equal to $a^{2n+1}.$

§2. We have $X^3 + Y^3 = (X+Y)(X+\omega Y)(X+\omega^2 Y).$
If $X = a + \omega b + \omega^2 c,$ $Y = a + \omega^2 b + \omega c,$ then
$$X + Y = 2a - b - c;$$
$$X + \omega Y = (1+\omega)(a+b) + 2\omega^2 c = \omega^2(2c - a - b);$$
$$X + \omega^2 Y = (1+\omega^2)(a+c) + 2\omega b = \omega(2b - c - a).$$

§3. The expression $(x+y)^n - x^n - y^n$ vanishes when $x = 0,$ and when $y = 0,$ and is therefore divisible by $xy.$

If $x = \omega y,$ the expression $= y^n\{(1+\omega)^n - 1 - \omega^n\} = -y^n\{1 + \omega^n + (\omega^2)^n\},$ n being odd; and this vanishes since n is not a multiple of 3.

Similarly it vanishes when $x = \omega^2 y.$ Thus the expression is divisible by $xy(x - \omega y)(x - \omega^2 y);$ that is, by $xy(x^2 + xy + y^2).$

§4. The expression $X^3 + Y^3 + Z^3 - 3XYZ$ is divisible by $X + Y + Z,$ and therefore vanishes when $X + Y + Z = 0.$ This condition is satisfied if
$$X = a(bz - cy), \quad Y = b(cx - az), \quad Z = c(ay - bx).$$

§5. By multiplication, $(b - cx)(c - ax)(a - bx)$
$$= abc - x(a^2 b + b^2 c + c^2 a) + x^2(ab^2 + bc^2 + ca^2) - abcx^3.$$

Substituting for x in succession the values $1, \omega, \omega^2,$ the terms involving abc destroy each other, and by adding the results the other terms are zero, since $1 + \omega + \omega^2 = 0.$

§6. This theorem is involved in that of Art. 525.

§7. By writing $-b$ for $b,$ $-y$ for y and putting $c = 0,$ $z = 0,$ the theorem follows from that of Example 6.

Or we may prove it directly as follows:
$$(a^2 + ab + b^2)(x^2 + xy + y^2) = (a - \omega b)(a - \omega^2 b)(x - \omega y)(x - \omega^2 y).$$

Now $(a - \omega b)(x - \omega^2 y) = ax + by - \omega bx + (1+\omega)ay$
$$= ax + by + ay - \omega(bx - ay);$$
and $(a - \omega^2 b)(x - \omega y) = ax + by - \omega^2 bx + (1+\omega^2)ay$
$$= ax + by + ay - \omega^2(bx - ay);$$

Thus the product
$$= (A - \omega B)(A - \omega^2 B) = A^2 + AB + B^2,$$
where $\qquad A = ax + by + ay, \quad B = bx - ay.$

§8. Let $X = a^2 + 2bc,$ $Y = b^2 + 2ca,$ $Z = c^2 + 2ab;$ then
$$X + Y + Z = (a+b+c)^2,$$
$$X + \omega Y + \omega^2 Z = (a + \omega^2 b + \omega c)^2,$$
$$X + \omega^2 Y + \omega Z = (a + \omega b + \omega^2 c)^2.$$

§9. Let $X = a^2 - bc,$ $Y = b^2 - ca,$ $Z = c^2 - ab,$ then
$$X + Y + Z = (a + \omega b + \omega^2 c)(a + \omega^2 b + \omega c),$$
$$X + \omega Y + \omega^2 Z = (a + b + c)(a + \omega b + \omega^2 c),$$
$$X + \omega^2 Y + \omega Z = (a + b + c)(a + \omega^2 b + \omega c).$$

§10. Let $X = a^2 + b^2 + c^2,$ $Y = bc + ca + ab,$ then
$$X^3 + 2Y^3 - 3XY^2 = (X - Y)^2(X + 2Y)$$

$$= (X - Y)^2(a + b + c)^2$$
$$= \{(a + b + c)(X - Y)\}^2$$
$$= \{(a + b + c)(a^2 + b^2 + c^2 - bc - ca - ab)\}^2.$$

Note. Proceeding as in the Example of Art. 526, we have

$$a^2 + b^2 + c^2 = -2q, \qquad a^3 + b^3 + c^3 = 3r, \qquad a^4 + b^4 + c^4 = 2q^2,$$
$$a^5 + b^5 + c^5 = -5qr, \qquad a^6 + b^6 + c^6 = 3r^2 - 2q^3, \qquad a^7 + b^7 + c^7 = 7q^2 r,$$

where $q = bc + ca + ab$ and $r = abc$.

From these relations it is easy to prove Ex. $11 - 15$.

§16. Since $\dfrac{b-c}{a} + \dfrac{c-a}{b} + \dfrac{a-b}{c} = \dfrac{bc(b-c) + ca(c-a) + ab(a-b)}{abc}$

$$= -\frac{(b-c)(c-a)(a-b)}{abc},$$

the given expression becomes

$$\frac{1}{abc}\{a(a-b)(a-c) + b(b-c)(b-a) + c(c-a)(c-b)\}.$$

Now $a(a-b)(a-c) = a^2(a - b - c) + abc = 2a^3 + abc$, since $b + c = -a$.

Thus the expression

$$= \frac{1}{abc}(2a^3 + 2b^3 + 2c^3 + 3abc)$$
$$= \frac{1}{abc}(6r + 3abc) = 9. \qquad \text{[See the Note above.]}$$

§17.
$$(b^2c + c^2a + a^2b - 3abc)(bc^2 + ca^2 + ab^2 - 3abc) \qquad (56)$$
$$= (b^2c + c^2a + a^2b)(bc^2 + ca^2 + ab^2) - 3abc\Sigma a^2 b + 9a^2b^2c^2$$

Now $(b^2c + c^2a + a^2b)(bc^2 + ca^2 + ab^2)$

$$= b^3c^3 + c^3a^3 + a^3b^3 + abc(a^3 + b^3 + c^3) + 3a^2b^2c^2$$
$$= \{(bc + ca + ab)^3 - 3abc\Sigma a^2 b - 6a^2b^2c^2\}$$
$$\quad + abc(a^3 + b^3 + c^3) + 3a^2b^2c^2.$$

Hence from (56), the expression

$$= (bc + ca + ab)^3 - 6abc\Sigma a^2 b + 6a^2b^2c^2 + abc(a^3 + b^3 + c^3).$$

But $\Sigma a^2 b = (a + b + c)(a^2 + b^2 + c^2) - 3abc = -3abc$, since $a + b + c = 0$.

Also $a^3 + b^3 + c^3 = 3abc$;

hence the expression $= (bc + ca + ab)^3 + 18a^2b^2c^2 + 6a^2b^2c^2 + 3a^2b^2c^2.$

§18. Put $y - z = a$, $z - x = b$, $x - y = c$; then $a + b + c = 0$, and we have to prove that

$$25(a^7 + b^7 + c^7)(a^3 + b^3 + c^3) = 21(a^5 + b^5 + c^5)^2.$$

This easily follows from the Note preceding the solution of Ex. 16; for

$$a^7 + b^7 + c^7 = 7q^2 r, \quad a^3 + b^3 + c^3 = 3r, \quad a^5 + b^5 + c^5 = -5qr.$$

§19. Put $y - z = a$, $z - x = b$, $x - y = c$, so that $a + b + c = 0$; then we have to prove that

$$(a^2 + b^2 + c^2)^3 - 54a^2b^2c^2 = 2(b - c)^2(c - a)^2(a - b)^2.$$

Since $c = -(a + b)$, the left side of this expression becomes

$$8(a^2 + ab + b^2)^3 - 54a^2b^2(a + b)^2$$
$$= 8\{(a - b)^2 + 3ab\}^3 - 54a^2b^2\{(a - b)^2 + 4ab\}$$
$$= 8(a - b)^6 + 72ab(a - b)^4 + 162a^2b^2(a - b)^2$$
$$= 2(a - b)^2\{2(a - b)^2 + 9ab\}^2$$
$$= 2(a - b)^2(2a^2 + 5ab + 2b^2)^2$$
$$= 2(a - b)^2(2a + b)^2(a + 2b)^2$$

$$= 2(a-b)^2(a-c)^2(b-c)^2, \text{ since } a+b = -c.$$

The theorem may also be deduced from Art. 574, Ex. 2.

§20. Put $b-c = \alpha$, $c-a = \beta$, $a-b = \gamma$, so that $\alpha+\beta+\gamma = 0$; then we have to shew that

$$\alpha^6 + \beta^6 + \gamma^6 - 3\alpha^2\beta^2\gamma^2 = 2\left(\frac{\alpha^2+\beta^2+\gamma^2}{2}\right)^3.$$

This is easily proved from the Note preceding the solution of Ex. 16, for

$$\alpha^6 + \beta^6 + \gamma^6 = 3r^2 - 2q^3, \quad \alpha\beta\gamma = r, \quad \alpha^2+\beta^2+\gamma^2 = -2q.$$

§21. Proceeding as in Ex. 20, we have to prove that

$$\alpha^7 + \beta^7 + \gamma^7 = 7\alpha\beta\gamma\left(\frac{\alpha^2+\beta^2+\gamma^2}{2}\right)^2.$$

[See the Note preceding Ex. 16.]

§22. Suppose that $a = 0$, in which case $c = -b$; then the left-hand side becomes

$$4b^3(y-z)^3 - 6b^3(y-z)(x^2+y^2+z^2) - 4b^3(y-z)(z-x)(x-y)$$
$$= 2b^3(y-z)\{2(y-z)^2 - 3(x^2+y^2+z^2) + 2(x-y)(x-z)\}$$
$$= 2b^3(y-z)\{-x^2 - y^2 - z^2 - 2xy - 2xz - 2yz\}$$
$$= -2b^3(y-z)(x+y+z)^2$$
$$= 0, \text{ since } x+y+z = 0.$$

Hence the left-hand side vanishes when $a = 0$; and similarly it vanishes when $b = 0$, $c = 0$, $x = 0$, $y = 0$, $z = 0$, and thus may be put equal to $kabcxyz$.

To find k, put $a = 1$, $b = 1$, $c = -2$, $x = 1$, $y = 1$, $z = -2$; thus

$$4k = 4 \times 6^3 - 3 \times 6 \times 6 \times 6 = 6^3; \text{ whence } k = 54.$$

§23. Assume $(1+ax)(1+bx)(1+cx)(1+dx) = 1 + qx^2 + rx^3 + sx^4$, where $q = \Sigma ab$, $r = \Sigma abc$, $s = abcd$. By proceeding as in Art. 526, we have

$$\Sigma a^2 = -2q, \quad \Sigma a^3 = 3r, \quad \Sigma a^5 = -5qr;$$

whence
$$\frac{\Sigma a^5}{5} = \frac{\Sigma a^3}{3} \cdot \frac{\Sigma a^2}{2}.$$

§24. With the notation of the preceding Example, we have

$$\left(\Sigma a^3\right)^2 = 9r^2 = 9\left(\Sigma abc\right)^2.$$

Since $d = -(a+b+c)$, we have

$$bcd + cda + dab + abc = -(a+b+c)(bc+ca+ab) + abc$$
$$= -(b+c)(c+a)(a+b).$$

$$\therefore (bcd + cda + dab + abc)^2 = (b+c)^2(c+a)^2(a+b)^2.$$

But
$$(c+a)(a+b) = bc + a(a+b+c) = bc - ad;$$

and similarly

$$(a+b)(b+c) = ca - bd, \text{ and } (b+c)(c+a) = ab - cd.$$

§25.

We have

$$8(s-b)(s-c)(\sigma^2 - a^2)$$
$$= (a-b+c)(a+b-c)(b^2+c^2-a^2)$$
$$= (a^2 - b^2 - c^2 + 2bc)(b^2+c^2-a^2)$$
$$= 2bc(b^2+c^2-a^2) - (b^2+c^2-a^2)^2$$
$$= 2bc(b^2+c^2+a^2) - 4a^2bc - (b^2+c^2+a^2)^2 + 4a^2(b^2+c^2).$$

Hence $8\Sigma(s-b)(s-c)(\sigma^2 - a^2) + 40abcs$

$$= 2(bc + ca + ab)(a^2+b^2+c^2) - 4(a^2bc + b^2ca + c^2ab)$$
$$- 3(a^2+b^2+c^2)^2 + 8(b^2c^2 + c^2a^2 + a^2b^2)$$

$$+ 20(a^2bc + b^2ca + c^2ab)$$
$$= -3(a^2 + b^2 + c^2)^2 + 2(bc + ca + ab)(a^2 + b^2 + c^2)$$
$$+ 8(bc + ca + ab)^2$$
$$= \{-(a^2 + b^2 + c^2) + 2(bc + ca + ab)\}$$
$$\{3(a^2 + b^2 + c^2) + 4(bc + ca + ab)\}$$
$$= \{-2\sigma^2 + (4s^2 - 2\sigma^2)\}\{(6\sigma^2 + 2(4s^2 - 2\sigma^2)\}$$
$$= 8(s^2 - \sigma^2)(4s^2 + \sigma^2).$$

§26. We have $A^3 + B^3 = (A + B)(A + \omega B)(A + \omega^2 B)$.
On putting
$$A = x^3 + 6x^2y + 3xy^2 - y^3, \quad B = y^3 + 6xy^2 + 3x^2y - x^3,$$
we obtain $\qquad A + B = 9x^2y + 9xy^2 = 9xy(x + y).$
Also
$$A + \omega B = x^3(1 - \omega) + 3x^2y(2 + \omega) + 3xy^2(1 + 2\omega) - y^3(1 - \omega)$$
$$= (1 - \omega)(x^3 - 3\omega^2 x^2y + 3\omega^4 xy^2 - \omega^6 y^3),$$
for $\qquad 2 + \omega = 1 + \omega + \omega^3 = -\omega^2 + \omega^3 = -\omega^2(1 - \omega);$
and $\qquad 1 + 2\omega = -(\omega + \omega^2) + 2\omega = \omega(1 - \omega) = \omega^4(1 - \omega).$
Thus $\qquad A + \omega B = (1 - \omega)(x - \omega^2 y)^3.$
By writing ω^2 for ω and ω for ω^2, we have
$$A + \omega^2 B = (1 - \omega^2)(x - \omega y)^3.$$
$$\therefore A^3 + B^3 = 9xy(x + y)(1 - \omega)(1 - \omega^2)(x - \omega y)^3(x - \omega^2 y)^3$$
$$= 27xy(x + y)(x^2 + xy + y^2)^3,$$

§27. The numerator $= -\Sigma a^5(b - c)(b - d)(c - d)$.
This expression is divisible by $(b - c)(c - a)(a - b)(a - d)(b - d)(c - d)$; the remaining factor must be a symmetrical function of two dimensions; hence the numerator
$$= (b - c)(c - a)(a - b)(a - d)(b - d)(c - d)(k\Sigma a^2 + l\Sigma ab).$$
If $a = 2$, $b = 1$, $c = -1$, $d = 0$, then $60 = 12(6k - l)$ or $6k - l = 5$.
If $a = 3$, $b = 2$, $c = -1$, $d = 0$, then
$$-300 = -12(14k + 11l) \text{ or } 14k + 11l = 25.$$
From these equations we obtain $k = 1$, $l = 1$. Thus the remaining factor is
$$a^2 + b^2 + c^2 + d^2 + ab + ac + ad + bc + bd + cd.$$
§28. Since all the signs are positive, the factors consists of positive terms only. By writing $2a^2b^2c^2$ in the form of $a^2b^2c^2 + a^2b^2c^2$, we see that the expression consists of eight terms; also it is of the sixth degree; hence the factors will be binomial expressions of the second degree, and by trial it is easy to verify that they are $a^2 + bc$, $b^2 + ca$, $c^2 + ab$.

EXAMPLES XXXIV c

§2. Here m and n are the roots of the quadratic equation $xt^2 - yt + a = 0$; hence $mn =$ the product of the roots $= \dfrac{a}{x}$; thus $\dfrac{a}{x} + 1 = 0$; that is, $x + a = 0$.
§3. Squaring and adding the first two equations, we have
$$(x^2 + y^2)(m^2 + n^2) = a^2(m^2 + n^2)^2; \text{ that is, } x^2 + y^2 = a^2.$$
§4. From the second and fourth equations, we have
$$-a + ap(q + r) = 2a - x; \text{ that is, } ap(q + r) = 3a - x.$$
By substituting from the first equation, $ap^2 = x - 3a$;
and from the third and fourth equations, we find $-ap = y$;

whence by eliminating p, we obtain, $y^2 = a(x - 3a)$.

§5. From the given equations, we have by cross multiplication
$$\frac{x^2}{2a^4 - 3a} = \frac{x}{a^3 - 1} = \frac{1}{a^2};$$
whence by eliminating x, we obtain $(a^3 - 1)^2 = a^2(2a^4 - 3a)$.

§6. Square and add; then we have $(x^2 + y^2)(1 + m^2) = 2a^2(1 + m^2)$.

§7. We have $b^2c^2 = x^2yz = a^2x^2$; thus $x^2 = \dfrac{b^2c^2}{a^2}$, $y^2 = \dfrac{c^2a^2}{b^2}$, $z^2 = \dfrac{a^2b^2}{c^2}$.

§8. From the first and second equations, we have
$$4pq = (p + q)^2 - (p - q)^2 = \frac{y^2}{x^2} - k^2(1 + pq)^2.$$
Substitute fro pq from the third equation.

§9. We have $x + y = \dfrac{b^2}{a}$, and $x^2 + xy + y^2 = \dfrac{c^3}{a}$; whence $xy = \dfrac{b^4}{a^2} - \dfrac{c^3}{a}$.

Thus $\dfrac{4b^4}{a^2} - \dfrac{4c^3}{a} = 4xy = (x + y)^2 - (x - y)^2 = \dfrac{b^4}{a^2} - a^2.$

§10. From the last two equations, we have $2x^2y^2 = b^4 - c^4$; and from the first two equations $2xy = a^2 - b^2$; hence $2(b^4 - c^4) = (a^2 - b^2)^2$.

§11. We have $x + ax = ax + by + cz + du = y + by = z + cz = u + du = k$ say.

Thus $x = \dfrac{k}{1 + a}$, $y = \dfrac{k}{1 + b}$, $z = \dfrac{k}{1 + c}$, $u = \dfrac{k}{1 + d}$.

Substitute these values in the equation $ax + by + cz + du = k$.

§12. Putting $(1 + xk)(1 + yk)(1 + zk) = 1 + qk^2 + rk^3$, and proceeding as in the example of Art. 526, we have
$$x^2 + y^2 + z^2 = -2q, \quad x^3 + y^3 + z^3 = 3r, \quad x^5 + y^5 + z^5 = -5qr.$$
Thus $a^2 = -2q$, $b^3 = 3r$, $c^5 = -5qr$; whence $6c^5 = 5a^2b^3$,

§13. It is easy to see that $ab = 3 + \Sigma\dfrac{x^2}{yz} + \Sigma\dfrac{yz}{x^2}$, and $c = 2 + \Sigma\dfrac{x^2}{yz} + \Sigma\dfrac{yz}{x^2}$;

hence $ab = c + 1$.

§14. We have $x^2(y + z) = a^3$, $y^2(z + x) = b^3$, $z^2(x + y) = c^3$, $xyz = abc$;

hence $a^3b^3c^3 = x^2y^2z^2(y + z)(z + x)(x + y)$;

that is, $abc = (y + z)(z + x)(x + y)$

$$= x^2(y + z) + y^2(z + x) + z^2(x + y) + 2xyz$$
$$= a^3 + b^3 + c^3 + 2abc.$$

§15. From the first two equations, we have
$$ax = 3x^2 + y^2, \quad by = x^2 + 3y^2.$$
Multiplying the first of these equations by y, we have $ac^2 = 3x^2y + y^3$.

Similarly $bc^2 = x^3 + 3xy^2$.

Thus $(a + b)c^2 = (y + x)^3$, and $(a - b)c^2 = (y - x)^3$.

$\therefore \left\{(a + b)^{\frac{2}{3}} - (a - b)^{\frac{2}{3}}\right\}c^{\frac{4}{3}} = (y + x)^2 - (y - x)^2 = 4xy = 4c^2.$

§16. By multiplication, $64a^2b^2c^2x^2y^2z^2 = (y + z)^2(z + x)^2(x + y)^2$;

that is, $(y + z)(z + x)(x + y) = \pm 8abcxyz$;

or $\dfrac{y}{z} + \dfrac{z}{y} + \dfrac{z}{x} + \dfrac{x}{z} + \dfrac{x}{y} + \dfrac{y}{x} + 2 = \pm 8abc.$

But from the given equations, we have
$$\frac{y}{z} + 2 + \frac{z}{y} = 4a^2; \quad \frac{z}{x} + 2 + \frac{x}{z} = 4b^2; \quad \frac{x}{y} + 2 + \frac{y}{x} = 4c^2.$$
Hence $\pm 8abc = (4a^2 - 2) + (4b^2 - 2) + (4c^2 - 2) + 2.$

§17. By multiplication, $abcx^2y^2z^2 = (y + z - x)^2(z + x - y)^2(x + y - z)^2$
$$= (-x^3 - y^3 - z^3 + y^2z + yz^2 + z^2x + zx^2 + x^2y + xy^2 - 2xyz)^2;$$

$$\therefore abc = \left(\frac{y}{z} + \frac{z}{y} + \frac{z}{x} + \frac{x}{z} + \frac{x}{y} + \frac{y}{x} - \frac{x^2}{yz} - \frac{y^2}{zx} - \frac{z^2}{xy} - 2\right)^2.$$

$$\because a = \frac{x^2 - y^2 - z^2 + 2yz}{yz} = \frac{x^2}{yz} - \frac{y}{z} - \frac{z}{y} + 2;$$

hence the expression within brackets
$$= (2 - a) + (2 - b) + (2 - c) - 2 = 4 - a - b - c;$$
$$\therefore abc = (4 - a - b - c)^2.$$

§18. Substitute $y = c - 2x$ in the second and first equations; thus

$$x^2 - cx + b = 0 \tag{57}$$
$$2x^3 - cx^2 + a = 0 \tag{58}$$

From (57) and (58),
$$cx^2 - 2bx + a = 0 \tag{59}$$

From (57) and (59),
$$\frac{x^2}{ac - 2b^2} = \frac{x}{a - bc} = \frac{1}{2b - c^2}.$$

By eliminating x, we have $(ac - 2b^2)(2b - c^2) = (a - bc)^2$.

§19. We have $ax^2 + by^2 + cz^2 = ax + by + cz = yz + zx + xy = 0.$

From the first two equations, we have
$$c(ax^2 + by^2) = -c^2z^2 = -(ax + by)^2;$$
$$\therefore a(a + c)x^2 + 2abxy + b(b + c)y^2 = 0 \tag{60}$$

From the third equation, $z = -\dfrac{xy}{x + y}$. Also $cz = -(ax + by);$

hence
$$ax + by = \frac{cxy}{x + y}.$$
$$\therefore ax^2 + (a + b - c)xy + by^2 = 0 \tag{61}$$

From (60) and (61), we obtain by cross multiplication
$$\frac{x^2}{b(b - c)(a - b - c)} = \frac{xy}{-ab(a - b)} = \frac{y^2}{a(a - c)(a - b + c)}.$$

Hence the eliminant is
$$a^2b^2(a - b)^2 = ab(a - c)(b - c)(a - b - c)(a - b + c),$$
or
$$ab(a - b)^2 = (a - c)(b - c)(a - b - c)(a - b + c);$$
$$\therefore ab(a - b)^2 = \{ab - (a + b)c + c^2\}\{(a - b)^2 - c^2\}.$$

Erasing the term $ab(a - b)^2$ on each side and dividing by c, we have
$$(a + b)(a - b)^2 - (a - b)^2c + abc - (a + b)c^2 + c^3 = 0;$$
that is,
$$a^3 + b^3 + c^3 - b^2c - bc^2 - ca^2 - a^2b - ab^2 + 3abc = 0;$$
or
$$\Sigma a^3 - \Sigma a^2 b + 3abc = 0.$$
But
$$\Sigma a^3 = (a + b + c)^3 - 3\Sigma a^2 b - 6abc;$$
and
$$\Sigma a^2 b = (b + c)(c + a)(a + b) - 2abc.$$

Hence the eliminant is
$$\{(a + b + c)^3 - 3(b + c)(c + a)(a + b)\}$$
$$- \{(b + c)(c + a)(a + b) - 2abc\} + 3abc = 0;$$
or
$$(a + b + c)^3 - 4(b + c)(c + a)(a + b) + 5abc = 0.$$

§20. We have $c(ax^2 + by^2) = c^2 = (ax + by)^2,$
$$\therefore a(a - c)x^2 + 2abxy + b(b - c)y^2 = 0 \tag{62}$$
Again,
$$(ax + by)(x + y) = xy;$$
$$\therefore ax^2 + (a + b - 1)xy + by^2 = 0 \tag{63}$$

From (62) and (63), we obtain by cross multiplication
$$\frac{x^2}{b\{2ab - (b - c)(a + b - 1)\}} = \frac{xy}{-ab(a - b)}$$

$$= \frac{y^2}{a\{(a-c)(a+b-1)-2ab\}};$$

$$\therefore \{2ab-(b-c)(a+b-1)\}\{(a-c)(a+b-1)-2ab\}=ab(a-b)^2;$$

$$\therefore 2ab(a+b-1)(a+b-2c)-4a^2b^2-(a-c)(b-c)(a+b-1)^2$$
$$= ab(a-b)^2;$$

$$\therefore 2ab(a+b)(a+b-1)-4abc(a+b-1)-4a^2b^2-ab(a+b-1)^2$$
$$+ c(a+b)(a+b-1)^2-c^2(a+b-1)^2=ab(a-b)^2;$$

whence by arranging according to powers of c, we have

$$ab(a+b-1)(a+b+1)+c(a+b-1)\{(a-b)^2-(a+b)\}$$
$$-c^2(a+b-1)^2=ab(a-b)^2;$$

$$\therefore c^2(a+b-1)^2-c(a+b-1)\{(a-b)^2-(a+b)\}+ab=0.$$

§21. Substituting $z = \dfrac{abc}{xy}$ in the first three equations, we have

$$ax^2-bcx+abc=0, \quad by^2-cay+abc=0, \quad x^2y^2-abxy+abc^2=0.$$

From these last two equations, we have

$$\frac{y^2}{a^2bc(c^2-bx)}=\frac{-y}{abc(x^2-bc)}=\frac{1}{ax(b^2-cx)}.$$

Hence
$$a^3bcx(c^2-bx)(b^2-cx)=a^2b^2c^2(x^2-bc)^2;$$

$$\therefore bcx^4-abcx^3+x^2(ab^3+ac^3-2b^2c^2)-ab^2c^2x+b^3c^3=0.$$

It remains to eliminate x between this equation and $ax^2-bcx+abc=0$.
Multiply the first equation by a, and the second by bcx^2 and subtract;
thus
$$bc(bc-a^2)x^3+(a^2b^3+a^2c^3-3ab^2c^2)x^2-a^2b^2c^2x+ab^3c^3=0.$$

Multiply this equation by a, and the equation $ax^2-bcx+abc=0$ by
$bc(bc-a^2)x$, and subtract; then
$$(a^3b^3+a^3c^3+b^3c^3-4a^2b^2c^2)x^2-ab^3c^3x+a^2b^3c^3=0.$$

Multiply $ax^2-bcx+abc=0$ by ab^2c^2 and subtract from this last equation,
then
$$(a^3b^3+a^3c^3+b^3c^3-5a^2b^2c^2)x^2=0;$$

hence the eliminant is $b^3c^3+c^3a^3+a^3b^3=5a^2b^2c^2.$

§22.

Put
$$\frac{a}{x}(x-p)=\frac{b}{y}(y-q)=\frac{c}{z}(z-r)=k;$$

then
$$x=\frac{ap}{a-k}, \quad y=\frac{bq}{b-k}, \quad z=\frac{cr}{c-k}.$$

Also
$$1=(x+y+z)^2=x^2+y^2+z^2+2(yz+zx+xy)$$
$$=1+2(yz+zx+xy);$$

therefore
$$yz+zx+xy=0.$$

Thus
$$bcqr(a-k)+carp(b-k)+abpq(c-k)=0;$$

whence

$$k=\frac{abc(qr+rp+pq)}{bcqr+carp+abpq}; \text{ so that } a-k=\frac{ap(car+abp-bcr-bcq)}{bcqr+carp+abpq};$$

$$\therefore x=\frac{ap}{a-k}=\frac{bcqr+carp+abpq}{a(bq+cr)-bc(q+r)};$$

and similar values for y and z.

By substituting in $x+y+z=1$, we obtain as the eliminant

$$\frac{1}{a(bq+cr)-bc(q+r)}+\frac{1}{b(cr+ap)-ca(r+p)}$$

$$+ \frac{1}{c(ap+bq)-ab(p+q)} = \frac{1}{bcqr+carp+abpq}.$$

§23. Divide by y^3 and put $\dfrac{x}{y} = z$, then

$$az^3 + bz^2 + cz + d = 0, \quad a'z^3 + b'z^2 + c'z + d' = 0;$$

we have

$$\frac{a}{a'} = \frac{bz^2+cz+d}{b'z^2+c'z+d'}; \quad \frac{az+b}{a'z+b'} = \frac{cz+d}{c'z+d'}; \quad \frac{az^2+bz+c}{a'z^2+b'z+c'} = \frac{d}{d'}.$$

Multiply up and eliminate z^2 and z from the three equations so obtained.

EXAMPLES XXXV : Theory of Equations

EXAMPLES XXXV a

§4. Corresponding to the first pair of roots there will be a quadratic factor $x^2 - 2ax + (a^2 - b^2)$; and corresponding to the second pair a quadratic factor $x^2 + 2ax + (a^2 - b^2)$.

Thus the required equation is
$$\{x^2 + (a^2 - b^2)\}^2 - 4a^2x^2 = 0.$$

§5. Corresponding to the two given roots there is a factor $x^2 - 8x + 7$.

By writing the equation in the form
$$x^2(x^2 - 8x + 7) - 8x(x^2 - 8x + 7) + 5(x^2 - 8x + 7) = 0,$$
we see that the other two roots are obtained from the quadratic equation
$$x^2 - 8x + 5 = 0.$$

§6. Let the roots be $a, -a, b$;
$$\text{then the sum of the roots } = b = -4;$$
also the sum of the products two at a time $= -a^2 = -\dfrac{9}{4}$.

§7. Let a, a, b be the roots; then
$$2a + b = -5; \quad a^2 + 2ab = \frac{23}{4}; \quad a^2b = -\frac{3}{2}.$$

Eliminating b from the first two equations, we get $12a^2 + 40a - 23 = 0$; whence $a = \dfrac{1}{2}$ or $-\dfrac{23}{6}$; and from the first equation we find $b = -6$ or $-\dfrac{8}{3}$.

It will be found on trial that $a = \dfrac{23}{6}$, $b = -\dfrac{8}{3}$ do not satisfy the third equation $a^2b = -\dfrac{3}{2}$; hence the roots are $\dfrac{1}{2}, \dfrac{1}{2}, -6$.

§8. Let $\dfrac{a}{r}$, a, ar be the roots; then $a^3 = 8$, and $a\left(r + 1 + \dfrac{1}{r}\right) = \dfrac{26}{3}$; whence $a = 2$, $r = 3$ or $\dfrac{1}{3}$; thus the roots are $\dfrac{2}{3}$, 2, 6.

§9. Let $3a$, $4a$, b be the roots; then
$$7a + b = \frac{1}{2}, \quad 7ab + 12a^2 = -11, \quad a^2b = 1.$$

From the first two, by eliminating b, we have $74a^2 - 7a - 22 = 0$; whence $a = -\dfrac{1}{2}$ or $\dfrac{22}{37}$. Taking $a = -\dfrac{1}{2}$, we get $b = 4$; thus the roots are $-\dfrac{3}{2}$, -2, 4. It will be found that the other value of a is inadmissible.

§10. Let $2a$, a, b be the roots; then
$$3a + b = -\frac{23}{12}, \quad 2a^2 + 3ab = \frac{3}{8}, \quad 2a^2b = \frac{3}{8}.$$

By eliminating b from the first two of these, we get $56a^2 + 46a + 3 = 0$; whence $a = -\dfrac{3}{4}$ or $-\dfrac{1}{14}$. The first of these values gives $b = \dfrac{1}{3}$, the other being inadmissible.

§11. Let a, $-a$, b, c be the roots; then
$$b + c = \frac{1}{4}, \quad (b+c)a^2 = \frac{3}{4}, \quad -a^2bc = \frac{9}{8}.$$

Thus $a = \pm\sqrt{3}$, $bc = -\dfrac{3}{8}$, $b + c = \dfrac{1}{4}$; whence $b = \dfrac{3}{4}$, $c = -\dfrac{1}{2}$.

§12. Let $\dfrac{a}{r}$, a, ar be the roots; then $a^3 = -\dfrac{8}{27}$, $a\left(\dfrac{1}{r} + 1 + r\right) = \dfrac{39}{54}$.

Thus $a = -\dfrac{2}{3}$, and $12r^2 + 25r + 12 = 0$; whence $r = -\dfrac{4}{3}$, or $-\dfrac{3}{4}$.

§13. Let $a - d$, a, $a + d$ be the roots; then, as in Art. 541, we have

$$a = \frac{1}{2}, \quad 3a^2 - d^2 = \frac{11}{16}; \text{ whence } d = \pm\frac{1}{4}.$$

§14. Let a, b, c, d be the roots, and suppose that $cd = 2$; then

$$a + b + c + d = \frac{29}{6}, \quad ab + ac + ad + bc + bd + cd = \frac{40}{6}.$$

$$abc + acd + abd + bcd = \frac{7}{6}, \quad abcd = -2.$$

Thus $ab = -1$. By substituting $ab = -1$ and $cd = 2$, we have

$$-c + 2a - d + 2b = \frac{7}{6}, \quad c + a + d + b = \frac{29}{6};$$

whence by addition, $a + b = 2$; and since $ab = -1$, we easily obtain $1 \pm \sqrt{2}$ for two of the roots.

We now have $c + d = \frac{17}{6}$, $cd = 2$; whence $c = \frac{4}{3}$, $d = \frac{3}{2}$.

§15. Let $a - 3d, a - d, a + d, a + 3d$ be the roots; then $4a = 2$, and $a = \frac{1}{2}$.

Also $(a^2 - 9d^2)(a^2 - d^2) = 40$; hence $(1 - 36d^2)(1 - 4d^2) = 640$; that is, $144d^4 - 40d^2 - 639 = 0$; or $(4d^2 - 9)(36d^2 + 71) = 0$;

thus $d = \pm\frac{3}{2}$; and the roots are $-4, -1, 2, 5$.

§16. Denote the roots by $\dfrac{a}{r^3}, \dfrac{a}{r}, ar, ar^3$; then

$$\text{the product of the roots} = a^4 = \frac{192}{27} = \frac{64}{9}; \text{ whence } a^2 = \frac{8}{3}.$$

The sum of the products of the roots two at a time

$$= a^2 \left(\frac{1}{r^4} + \frac{1}{r^2} + 2 + r^2 + r^4 \right) = \frac{494}{27};$$

thus

$$\left(r^2 + \frac{1}{r^2} \right)^2 + \left(r^2 + \frac{1}{r^2} \right) = \frac{247}{36};$$

whence

$$r^2 + \frac{1}{r^2} = \frac{13}{6}, \text{ and therefore } r^2 = \frac{3}{2}.$$

Thus $a^2 r^2 = \dfrac{8}{3} \times \dfrac{3}{2} = 4$, or $ar = 2$; and therefore the roots are $3, 2, \dfrac{4}{3}, \dfrac{8}{9}$.

§17. Let $a, b, \dfrac{a + b}{2}$ be the roots; then

$$\frac{3}{2}(a + b) = -\frac{81}{18}, \quad \frac{ab(a + b)}{2} = -\frac{10}{3};$$

whence, $a + b = -3$, $ab = \dfrac{20}{9}$; and therefore $a = -\dfrac{5}{3}, b = -\dfrac{4}{3}$.

§18. (1) Here we have

$$a + b + c = p, \quad ab + bc + ca = q, \quad abc = r.$$

$$\therefore \Sigma \frac{1}{a^2} = \frac{\Sigma(a^2 b^2)}{a^2 b^2 c^2} = \frac{(ab + bc + ca)^2 - 2abc(a + b + c)}{a^2 b^2 c^2} = \frac{q^2 - 2rp}{r^2}.$$

(2) $\Sigma \dfrac{1}{a^2 b^2} = \dfrac{a^2 + b^2 + c^2}{a^2 b^2 c^2} = \dfrac{p^2 - 2q}{r^2}.$

§19. (1) Here $a + b + c = 0$, $ab + bc + ca = q$, $abc = -r$.

$$\therefore \Sigma(b - c)^2 = 2(a^2 + b^2 + c^2) - 2(bc + ca + ab)$$
$$= 2(a + b + c)^2 - 6(bc + ca + ab)$$
$$= -6q.$$

(2) Since $a + b + c = 0$, we have

$$\Sigma(b + c)^{-1} = \Sigma \left(-\frac{1}{a} \right) = -\frac{bc + ca + ab}{abc} = \frac{q}{r}.$$

§20. (1) Here we have $\Sigma a = 0$, $\quad \Sigma ab = q$, $\quad \Sigma abc = -r$, $\quad abcd = s$.
$$\therefore \Sigma a^2 = (\Sigma a)^2 - 2\Sigma ab = -2q.$$

(2) Again, $\qquad\qquad \Sigma a^3 = 3\Sigma abc = -3r.$ [See XXXIV. b. Ex. 23.]

§21. Here $\Sigma a = 0, \Sigma ab = q, abc = -r$. Multiply the equation through by x, then substitute a, b, c for x successively and add the results; thus we obtain $\Sigma a^4 + q\Sigma a^2 + r\Sigma a = 0$;
$$\therefore \Sigma a^4 = -q\Sigma a^2 = -q\{(\Sigma a)^2 - 2\Sigma ab\} = 2q^2.$$

EXAMPLES XXXV b

Examples $1 - 12$ do not require full solution, as they all depend on Arts. $543 - 545$, and the method of procedure is explained in Art. 545. As further illustrations the following solutions will be sufficient.

§1. Corresponding to the two roots $\dfrac{1 + \sqrt{-3}}{2}, \dfrac{1 - \sqrt{-3}}{2}$ we have the quadratic factor $x^2 - x + 1$. The equation is now easily put in the form $(3x^2 - 7x - 6)(x^2 - x + 1) = 0$. Thus the other roots are obtained from
$$3x^2 - 7x - 6 = 0.$$

§5. Here four of the roots are $\pm\sqrt{3}, 1 \pm 2\sqrt{-1}$. Corresponding to these pairs of roots we have the factors $x^2 - 3$ and $x^2 - 2x + 5$. Also the equation may be written $(x + 1)(x^2 - 3)(x^2 - 2x + 5) = 0$; hence the remaining root is -1.

§6. The equation required has the following pairs of roots:
$$+\sqrt{3} + \sqrt{-2}, \quad +\sqrt{3} - \sqrt{-2}; \quad -\sqrt{3} + \sqrt{-2}, \quad -\sqrt{3} - \sqrt{-2}.$$
Corresponding to these we have the quadratic factors
$$x^2 - 2\sqrt{3}.x + 5 \text{ and } x^2 + 2\sqrt{3}.x + 5.$$
Thus the equation is
$$(x^2 + 2\sqrt{3}.x + 5)(x^2 - 2\sqrt{3}.x + 5) = 0, \text{ or } x^4 - 2x^2 + 25 = 0.$$

§10. Here we have the quadratic factors $x^2 - 48$ and $x^2 - 10x + 29$ corresponding to the two pairs of roots; hence the equation is
$$(x^2 - 48)(x^2 - 10x + 29) = 0.$$

§12. The equation whose roots are $\sqrt{2} + \sqrt{3} \pm \sqrt{-1}$
is $\qquad (x - \sqrt{2} - \sqrt{3})^2 + 1 = 0$, or $x^2 + 6 - 2\sqrt{2}.x - 2\sqrt{3}.x - 2\sqrt{6} = 0$.

Similarly the equation whose roots are $\sqrt{2} - \sqrt{3} \pm \sqrt{-1}$
is $\qquad\qquad x^2 + 6 - 2\sqrt{2}.x + 2\sqrt{3}.x + 2\sqrt{6} = 0$;
these two equations are equivalent to
$$(x^2 + 6 - 2\sqrt{2}.x)^2 - (2\sqrt{3}.x + 2\sqrt{6})^2 = 0,$$
$$\text{or } x^4 + 8x^2 + 12 - 4\sqrt{2}.x^3 = 0.$$

Hence the equation whose roots are $-\sqrt{2} \pm \sqrt{3} \pm \sqrt{-1}$
is $\qquad\qquad x^4 + 8x^2 + 12 + 4\sqrt{2} \cdot x^3 = 0$;
thus the required equation is
$$(x^4 + 8x^2 + 12)^2 - (4\sqrt{2}.x^3)^2 = 0,$$
$$\text{or } x^8 - 16x^6 + 88x^4 + 192x^2 + 144 = 0.$$

§13. Denote the equation by $f(x) = 0$; then in $f(x)$ there is one change of sign, so that there cannot be more than one positive root. Again, $f(-x)$ has only one change of sign; therefore there cannot be more than one negative root. Hence there must be at least two imaginary roots.

[By Art. 554, we know that the equation has one positive and one negative root.]

§14. Here $f(x)$ has three changes of sign, and $f(-x)$ has no change of sign. Therefore the equation has no negative roots and at most three positive roots. Hence it has at least four imaginary roots since it is of the seventh degree.

§15.

Here $$f(x) = x^{10} - 4x^6 + x^4 - 2x - 3,$$

and $$f(-x) = x^{10} - 4x^6 + x^4 + 2x - 3;$$

thus there are three changes of sign in $f(x)$; and three changes of sign in $f(-x)$; hence there cannot be more than three positive roots nor more than three negative roots; hence at least four of the roots must be imaginary.

[By Art. 554, we see that the equation has one positive root and also one negative root.]

§16. Since $f(x)$ has two changes of sign, the equation has at most two positive roots. And since $f(-x)$ has only one change of sign, there cannot be more than one negative root. Therefore it must have at least six imaginary roots.

§17. (1) Let $a, b, -b$ be the roots; then $a = p, -b^2 = q, -ab^2 = r$; by eliminating a, b we obtain $pq = r$, which is the required relation.

(2) Let $\dfrac{a}{k}, a, ak$ be the roots; then

$$a^3 = r, \quad \frac{a}{k} + a + ak = p, \quad \frac{a^2}{k} + a^2 + a^2k = q;$$

thus $\dfrac{p}{q} = \dfrac{1}{a}$, or $p^3a^3 = q^3$; that is, $p^3r = q^3$.

§18. Let $a - 3d, a - d, a + d, a + 3d$ be the roots. Then by Art.539 we have after easy reduction the relations

$$4a = -p, \quad 6a^2 - 10d^2 = q, \quad 4a^3 - 20ad^2 = -r.$$

From the last two of these equations we have, on eliminating d,

$$12a^3 - 2aq = 4a^3 + r, \text{ or } 8a^3 - 2aq = r.$$

Multiply by 8, transpose, and put $4a = -p$; thus we obtain $p^3 - 4pq + 8r = 0$.

In the second case assume for the roots, a, ak, ak^2, ak^3; then we have

$$a(1 + k + k^2 + k^3) = -p, \quad a^3k^3(1 + k + k^2 + k^3) = -r, \quad a^4k^6 = s;$$

whence it is easily seen that $p^2s = r^2$.

§19. Put $1 - x = y$, then we have $(1 - y)^n - 1 = 0$.

Expand and divide by y; thus

$$y^{n-1} - ny^{n-2} + \frac{1}{2}n(n-1)y^{n-3} - \ldots + (-1)^{n-1}n = 0.$$

If $y_1, y_2, y_3, \ldots, y_{n-1}$ denote the roots of this equation, we have

$$y_1y_2y_3 \ldots y_{n-1} = n;$$

that is, $$(1 - \alpha)(1 - \beta)(1 - \gamma) \ldots = n.$$

§20. Here $\Sigma a = p, \quad \Sigma ab = q, \quad abc = r.$

$$\therefore \Sigma a^2b^2 = (\Sigma ab)^2 - 2abc\Sigma a = q^2 - 2rp.$$

§21. Here $(b + c)(c + a)(a + b) = (ab + bc + ca)(a + b + c) - abc = pq - r.$

§22.

Here $\Sigma\left(\dfrac{b}{c} + \dfrac{c}{b}\right) = \Sigma\dfrac{b^2 + c^2}{bc}$

$$= \frac{1}{r}(ab^2 + ac^2 + bc^2 + ba^2 + ca^2 + cb^2)$$

$$= \frac{1}{r}\{(a + b + c)(ab + bc + ca) - 3abc\}$$

$$= \frac{1}{r}(pq - 3r).$$

§23.

Here $\Sigma a^2b = a^2b + a^2c + b^2c + b^2a + c^2a + c^2b$

$$= pq - 3r, \text{ as in Example 22.}$$

§24. Here we have $\Sigma a = -p,\quad \Sigma ab = q,\quad \Sigma abc = -r,\quad abcd = s.$

Now
$$\Sigma a^2 bc = a^2(bc + bd + cd) + \dots + \dots + \dots$$
$$= a(abc + abd + acd) + \dots + \dots + \dots$$
$$= a(-r - bcd) + \dots + \dots + \dots$$
$$= -r(a + b + c + d) - 4abcd$$
$$= pr - 4s.$$

§25. Substitute a, b, c, d for x successively and add the results; thus we obtain
$$\Sigma a^4 + p\Sigma a^3 + q\Sigma a^2 + r\Sigma a + 4s = 0.$$

Now
$$\Sigma a = -p, \text{ and } \Sigma a^2 = p^2 - 2q.$$

Also
$$\Sigma a^3 = (\Sigma a)^3 - 3\Sigma a^2 b - 6\Sigma abc \qquad\qquad \text{[Art. 522.]}$$
$$= -p^3 - 3(3r - pq) + 6r. \qquad\qquad \text{[Art. 542, Ex. 2.]}$$

$$\therefore \Sigma a^4 = p^4 + 3p(3r - pq) - 6pr - q(p^2 - 2q) + pr - 4s$$
$$= p^4 - 4p^2 q + 2q^2 + 4pr - 4s.$$

EXAMPLES XXXV c

§1. Proceeding as in Art. 549, we have

$$
\begin{array}{ccccc}
1 & 10 & 39 & 76 & 65 \\
1 & 6 & 15 & 16 \mid & 1 \\
1 & 2 & 7 \mid & -12 & \\
1 & -2 \mid & 15 & & \\
1 \mid & -6 & & &
\end{array}
$$

$$\therefore f(x - 4) = x^4 - 6x^3 + 15x^2 - 12x + 1.$$

§2. We have

$$
\begin{array}{ccccc}
1 & -12 & 17 & -9 & 7 \\
1 & -9 & -10 & -39 \mid & -110 \\
1 & -6 & -28 \mid & -123 & \\
1 & -3 \mid & -37 & & \\
1 \mid & 0 & & &
\end{array}
$$

$$\therefore f(x + 3) = x^4 - 37x^2 - 123x - 110.$$

§3. We have

$$
\begin{array}{ccccc}
2 & 0 & -13 & 10 & -19 \\
2 & 2 & -11 & -1 \mid & -20 \\
2 & 4 & -7 \mid & -8 & \\
2 & 6 \mid & -1 & & \\
2 \mid & 8 & & &
\end{array}
$$

$$\therefore f(x + 1) = 2x^4 + 8x^3 - x^2 - 8x - 20.$$

§4. We have

$$
\begin{array}{ccccc}
1 & 16 & 72 & 64 & -120 \\
1 & 12 & 24 & -32 \mid & -1 \\
1 & 8 & -8 \mid & 0 & \\
1 & 4 \mid & -24 & & \\
1 \mid & 0 & & &
\end{array}
$$

$$\therefore f(x - 4) = x^4 - 24x^2 - 1.$$

§5. By Art. 548, we have
$$f(x + h) - f(x - h)$$

$$= 2\left\{ hf'(x) + \frac{h^3}{\underline{|3}}f'''(x) + \frac{h^5}{\underline{|5}}f^v(x) + \frac{h^7}{\underline{|7}}f^{vii}(x) \right\}.$$

Now $f'(x) = 8ax^7 + 5bx^4 + c$; $f'''(x) = 8 \cdot 7 \cdot 6ax^5 + 5 \cdot 4 \cdot 3bx^2$;

and therefore $\qquad \dfrac{f'''(x)}{\underline{|(3)}} = 56ax^5 + 10bx^2$;

also $\qquad \dfrac{f^v(x)}{\underline{|5}} = 56ax^3 + b$; $\dfrac{f^{vii}(x)}{\underline{|7}} = 8ax$.

$\therefore f(x + h) - f(x - h)$
$= 2\{h(8ax^7 + 5bx^4 + c) + h^3(56ax^5 + 10bx^2) + h^5(56ax^3 + b) + h^7 \cdot 8ax\}$,
which easily reduces to the form given in the answer.

§6. Here $f(0) = 6$, and $f(-1) = -22$; thus $f(0)$ and $f(-1)$ have different signs, and therefore there is a root between 0 and -1.

§7.

Here $\qquad\qquad f(2) = 16 - 40 + 12 + 70 - 70 = -12$;
$\qquad\qquad\qquad f(3) = 81 - 135 + 27 + 105 - 70 = 8$;
$\qquad\qquad\qquad \therefore f(x) = 0$ has a root between 2 and 3.

Again, $f(-2)$ is negative, and $f(-3)$ is positive; therefore $f(x) = 0$ has a root between -2 and -3.

Examples 8 and 9 may be solved in the same way.

§10.

Here $\qquad\qquad f(x) = x^4 - 9x^2 + 4x + 12$,
and $\qquad\qquad f'(x) = 4x^3 - 18x + 4$.

The H.C.F. of these two expressions will be found to be $x - 2$. Thus $(x - 2)^2$ is a factor of $f(x)$.

Now $\quad f(x) = (x - 2)^2(x^2 + 4x + 3) = (x - 2)^2(x + 3)(x + 1)$;
thus the roots are $2, 2, -3, -1$.

§11. Proceeding as in Ex. 10, we find that the H.C.F. of $f(x)$ and $f'(x)$ is $x^2 - 2x + 1$; hence $(x - 1)^3$ is a factor of $f(x)$.

Now $f(x) = (x - 1)^3(x - 3)$; thus the roots are $1, 1, 1, 3$.

§12.

Here $\qquad\qquad f(x) = x^5 - 13x^4 + 67x^3 - 171x^2 + 216x - 108$,
and $\qquad\qquad f'(x) = 5x^4 - 52x^3 + 201x^2 - 342x + 216$.

$\therefore 2f(x) + f'(x) = x(2x^4 - 21x^3 + 82x^2 - 141x + 90) = x\phi(x)$ say.

$\therefore 2f'(x) - 5\phi(x) = x^3 - 8x^2 + 21x - 18$;

and since this expression divides $\phi(x)$, it is the H.C.F. of $f(x)$ and $f'(x)$.

Now $\qquad\qquad x^3 - 8x^2 + 21x - 18 = (x - 3)^2(x - 2)$;
and $f(x) = (x - 3)^3(x - 2)^2 = 0$; and therefore the roots are $3, 3, 3, 2, 2$.

§13.

Here $\qquad\qquad f(x) = x^5 - x^3 + 4x^2 - 3x + 2$,
and $\qquad\qquad f'(x) = 5x^4 - 3x^2 + 8x - 3$.

The H.C.F. of these expressions is $x^2 - x + 1$.
Hence $f(x) = (x^2 - x + 1)^2(x + 2)$, and the roots are

$$-2, \frac{1 \pm \sqrt{-3}}{2}, \frac{1 \pm \sqrt{-3}}{2}.$$

§14. Here it will be found that $f(x) = (2x - 1)^3(x + 2)$;

thus the roots are $\quad \dfrac{1}{2}, \dfrac{1}{2}, \dfrac{1}{2}, -2$.

§15. Here it will be found that
$$f(x) = (x - 1)^3(x^3 - 3x - 2) = (x - 1)^3(x + 1)^2(x - 2).$$

Thus the roots are $1, 1, 1, -1, -1, 2$.

§16.. Here $\qquad f(x) = (x^2 - 3)^2(x^2 - 2x + 2)$.

Therefore the roots are $\pm\sqrt{3}, \pm\sqrt{3}, 1 \pm \sqrt{-1}$.

§17. Here $\qquad f(x) = (x - a)^2\{x^2 + (a - b)x - ab\}$.

Therefore the roots are $a, a, b, -a$.

§18. Donate the two equations by $f(x) = 0$ and $F(x) = 0$. Then it will be found that the H.C.F. of $f(x)$ and $F(x)$ is $2x^2 - 3$.

Also $\qquad\qquad f(x) = (2x^2 - 3)(x^2 - x + 2)$;

and $\qquad\qquad F(x) = (2x^2 - 3)(2x^2 - x + 3)$.

Thus the roots are $\pm\sqrt{\dfrac{3}{2}}, \dfrac{1 \pm \sqrt{-7}}{2}$; and $\pm\sqrt{\dfrac{3}{2}}, \dfrac{1 \pm \sqrt{-23}}{4}$.

Example 19 may be solved in the same way.

§20. If $f(x) = 0$ has equal roots, $f(x) = 0, f'(x) = 0$ have a common root.

Thus $\qquad\qquad\qquad x^n - px^2 + r = 0 \qquad\qquad\qquad (64)$

and $\qquad\qquad\qquad nx^{n-1} - 2px = 0 \qquad\qquad\qquad (65)$

have a common root, and the required condition will be obtained by eliminating x between them.

From (65) we have $x^{n-2} = \dfrac{2p}{n}$.

Multiply (64) by n and (65) by x; then by subtraction, $p(n - 2)x^2 = nr$, that is, $x^2 = \dfrac{nr}{p(n - 2)}$.

$$\therefore \left\{\frac{nr}{p(n-2)}\right\}^{n-2} = \left(\frac{2p}{n}\right)^2.$$

§21. If $f(x) = 0$ has three equal roots, $f(x)$ and $f'(x)$ must have a common quadratic factor, and $f'(x) = 0$ must have two equal roots.

Now $f'(x) = 2x(2x^2 + q)$, so that one root of $f'(x) = 0$ is zero, and the other two are the roots of $2x^2 + q = 0$ which are equal in magnitude but not in eign. Thus $f(x) = 0$ cannot have three equal roots.

§22. We may write the two equations in the following forms:

$$x^2 + \frac{b}{a}x + 1 = 0 \qquad\qquad\qquad (66)$$

and $\qquad\qquad\qquad (x - 1)(x^2 - x + 1) = 0 \qquad\qquad\qquad (67)$

Now in (67) we have one real root and two imaginary roots. Therefore by Art. 543 the two equations may have one common real root, or two common imaginary roots. In the first case $x = 1$ satisfies (66) and thus $b = -2a$. In the second case $x^2 + \dfrac{b}{a}x + 1$ must be identical with the quadratic factor $x^2 - x + 1$, and thus $b = -a$.

§23. Here we have

$$f(x) = x^n + nx^{n-1} + n(n - 1)x^{n-2} + \ldots + \lfloor n = 0,$$

and $\qquad\quad f'(x) = \qquad nx^{n-1} + n(n - 1)x^{n-2} + \ldots + \lfloor n = 0.$

Now if $f(x)$ has a pair of equal roots, $f(x)$ and $f'(x)$ must have a factor of the form $x - a$. Therefore also $f(x) - f'(x)$ must have a factor of this form. But $f(x) - f'(x) = x^n$, and it follows that $f(x) = 0$ cannot have equal roots.

§24.

Here $\qquad\qquad\qquad f'(x) = 5x^4 - 20a^3x + b^4 \qquad\qquad\qquad (68)$

and $\qquad\qquad\qquad f''(x) = 20x^3 - 20a^3 \qquad\qquad\qquad (69)$

If $f(x) = 0$ has three equal roots $f'(x)$ and $f''(x)$ must have a common linear factor; from (69) it is evident that this factor must be $x - a$. Thus

$x = a$ must satisfy the given equation. On substituting for x we get the required relation.

§25.

Here
$$f'(x) = 4x^3 + 3ax^2 + 2bx + c,$$
$$f''(x) = 12x^2 + 6ax + 2b;$$

and if $f(x) = 0$ has three equal roots, $f'(x)$ and $f''(x)$ must have a common factor.

Hence
$$4x^3 + 3ax^2 + 2bx + c = 0 \tag{70}$$
and
$$12x^2 + 6ax + 2b = 0 \tag{71}$$

must have a common root.

Multiply (70) by 3 and (71) by x; thus by subtraction
$$3ax^2 + 4bx + 3c = 0 \tag{72}$$

Eliminating x^2 between (71) and (72), we get $(6a^2 - 16b)x = 12c - 2ab;$

whence $x = \dfrac{6c - ab}{3a^2 - 8b}$, which is the common root.

§26.

Here
$$x^5 + qx^3 + rx^2 + t = 0 \tag{73}$$
and
$$5x^4 + 3qx^2 + 2rx = 0 \tag{74}$$

must have a common root.

Multiply (74) by x and (73) by 5 and subtract; thus
$$2qx^3 + 3rx^2 + 5t = 0 \tag{75}$$

Multiply (2) by q and divide by x; thus
$$5qx^3 + 3q^2x + 2rq = 0 \tag{76}$$

Eliminating qx^3 between (75) and (76) thus we find x is one of the roots of
$$15rx^2 - 6q^2x + 25t - 4qr = 0.$$

§27. By the method of Art. 563 we have
$$\frac{f'(x)}{f(x)} = \frac{1}{x-a} + \frac{1}{x-b} + \frac{1}{x-c} = \Sigma \left(\frac{1}{x} + \frac{a}{x^2} + \frac{a^2}{x^3} + \ldots \right)$$
$$= \frac{3}{x} + \frac{s_1}{x^2} + \frac{s_2}{x^3} + \ldots + \frac{s_6}{x^7} + \ldots.$$

Now $f(x) = x^3 - x - 1$, and $f'(x) = 3x^2 - 1$.

```
1|3 + 0 − 1
0|    0 + 3 + 3
1|        0 + 0 + 0
1|            0 + 2 + 2
 |                0 + 3 + 3
 |                    0 + 2 + 2
 |                        0 + 5 + 5
 |3 + 0 + 2 + 3 + 2 + 5 + 5 +
```

Thus $s_6 = 5$.

§28. Proceeding as in the last Example, we have to find the coefficients of $\dfrac{1}{x^5}$ and $\dfrac{1}{x^7}$ in the quotient of $4x^3 - 3x^2 - 14x + 1$ by $x^4 - x^3 - 7x^2 + x + 6$.

```
 1|4 − 3 − 14 + 1
 1|    4 + 28 − 4 − 24
 7|        1 + 7 − 1 − 6
−1|            15 + 105 − 15 − 90
−6|                19 + 133 − 19
  |                    99 + 693
  |                        211 + ...
  |4 + 1 + 15 + 19 + 99 + 112 + 795 + ...
```

Thus $s_4 = 99, s_6 = 795$.

EXAMPLES XXXV d

§1. Put $x = \dfrac{y}{q}$ and multiply each term by q^3; thus

$$y^3 - 4y^2 q + \frac{1}{4} y q^2 - \frac{q^3}{9} = 0.$$

By putting $q = 6$ all the terms become integral, and we obtain
$$y^3 - 24y^2 + 9y - 24 = 0.$$
Example 2 may be solved in the same way.

Examples 3 and 4 are reciprocal equations which present no difficulty; they may be solved like the Example in Art. 133.

§5. Here $x = 1$ is evidently a root. On removing the factor $x - 1$, we have
$$x^4 - 4x^3 + 5x^2 - 4x + 1 = 0;$$
$$\therefore \left(x^2 + \frac{1}{x^2} \right) - 4 \left(x + \frac{1}{x} \right) + 5 = 0,$$
$$\text{or } \left(x + \frac{1}{x} \right)^2 - 4 \left(x + \frac{1}{x} \right) + 3 = 0;$$

whence $x + \dfrac{1}{x} = 3$ or 1. By solving these two quadratics we obtain
$$x = 1, \frac{3 \pm \sqrt{5}}{2}, \frac{1 \pm \sqrt{-3}}{2}.$$

§6. Divide all through by x^3 and rearrange; thus
$$4 \left(x^3 + \frac{1}{x^3} \right) - 24 \left(x^2 + \frac{1}{x^2} \right) + 57 \left(x + \frac{1}{x} \right) - 73 = 0;$$
$$4 \left(x + \frac{1}{x} \right)^3 - 24 \left(x + \frac{1}{x} \right)^2 + 57 \left(x + \frac{1}{x} \right) - 25 = 0$$

By inspection $x + \dfrac{1}{x} = 1$ satisfies the equation. On removing the factor corresponding to this root we have the equation
$$4 \left(x + \frac{1}{x} \right)^2 - 20 \left(x + \frac{1}{x} \right) + 25 = 0;$$

The roots of this equation in $x + \dfrac{1}{x}$ are $\dfrac{5}{2}, \dfrac{5}{2}$. Thus finally we have to solve the quadratics
$$x + \frac{1}{x} = 1, \ x + \frac{1}{x} = \frac{5}{2}, \ x + \frac{1}{x} = \frac{5}{2}.$$

§7. Put $x = \dfrac{1}{y}$, then the resulting equation $32y^3 - 48y^2 + 22y - 3 = 0$ has its roots in A.P., and may be solved like Example 1 in Art. 541.

Example 8 may be treated similarly.

§9. The equation $by^3 - y^2 + ay - 1 = 0$ has its roots in A.P. Let them be denoted by $a - d, a, a + d$; then $3a = \dfrac{1}{b}$; that is, $a = \dfrac{1}{3b}$.

Thus the mean root of the original equation is $3b$.

§10. The equation $y^4 + 2y^3 - 21y^2 - 22y + 40 = 0$ has its roots in A.P.

Assume $a - 3d, a - d, a + d, a + 3d$ for the roots, and proceed as in Ex. 15. XXXV.a.

§11. Here, since the sum of the roots of the equation is 6, we must decrease root by 2. We have therefore to substitute $x + 2$ for x, which is effected by Horner's process using $x - 2$ as divisor.

1	-6	10	-3
1	-4	2	$\mid$ 1
1	-2	$\mid$ -2	
1	$\mid$ 0		
1			

Thus the transformed equation is $x^3 - 2x + 1 = 0$.

§12. Here we have to increase each root by 1; therefore using $x + 1$ as divisor in Horner's process, we have

1	4	2	-4	-2
1	3	-1	-3	$\mid$ 1
1	2	-3	$\mid$ 0	
1	1	$\mid$ -4		
1	$\mid$ 0			
1				

Thus the transformed equation is $x^4 - 4x^2 + 1 = 0$.

§13. Here we have to increase each root by 1. Therefore using $x + 1$ as divisor in Horner's process, we have

1	5	3	1	1	-1
1	4	-1	2	-1	$\mid$ 0
1	3	-4	6	$\mid$ -7	
1	2	-6	$\mid$ 12		
1	1	$\mid$ -7			
1	$\mid$ 0				

Thus the transformed equation is $x^5 - 7x^3 + 12x^2 - 7x = 0$.

§14. Here we have to decrease each root by 2. Therefore using $x - 2$ as divisor in Horner's process, we have

1	-12	0	0	3	-17	300
1	-10	-20	-40	-77	-171	$\mid$ -42
1	-8	-36	-112	-301	$\mid$ -773	
1	-6	-48	-208	$\mid$ -717		
1	-4	-56	$\mid$ -320			
1	-2	$\mid$ -60				
1	$\mid$ 0					

Thus the transformed equation is $x^6 - 60x^4 - 320x^3 - 717x^2 - 773x - 42 = 0$.

§15. Here we have to use $x + \dfrac{3}{2}$ as divisor.

1	0	$-\dfrac{1}{4}$	$-\dfrac{3}{4}$
1	$-\dfrac{3}{2}$	2	$\mid$ $-\dfrac{15}{4}$
1	-3	$\mid$ $\dfrac{13}{2}$	
1	$\mid$ $-\dfrac{9}{2}$		
1			

Thus the transformed equation is $x^3 - \dfrac{9}{2}x^2 + \dfrac{13}{2}x - \dfrac{15}{4} = 0$.

Examples 16 and 17 may be solved in the same way.

§18. Put $y = x^2$, so that $x = \sqrt{y}$; then after transposing we have

$$y^2 + 2y + 1 = (y + 1)\sqrt{y};$$

whence $\qquad y^4 + 4y^3 + 6y^2 + 4y + 1 = y(y^2 + 2y + 1).$

§19. Put $y = x^3$, so that $x = \sqrt[3]{y}$;

then $\qquad y + 3y^{\frac{2}{3}} + 2 = 0$; or $(y + 2)^3 = (-3y^{\frac{2}{3}})^3$;

which reduces to $y^3 + 33y^2 + 12y + 8 = 0$.

§20. When $x = a$ in the given equation, $y = \dfrac{k}{a}$ in the transformed equation. Thus the transformed equation will be obtained by substituting $y = \dfrac{k}{x}$, or $x = \dfrac{k}{y}$ in the given equation.

§21. If $x = a$, then $y = b^2 c^2 = \dfrac{b^2 c^2 a^2}{a^2} = \dfrac{r^2}{\dfrac{x^2}{r^2}}$.

We have therefore to substitute $x^2 = \dfrac{r^2}{y}$ in $x(x^2 + q) = -r$.

Hence
$$\frac{r}{\sqrt{y}}\left(\frac{r^2}{y} + q\right) = -r;$$

that is,
$$\frac{r^2}{y}\left(\frac{r^4}{y^2} + \frac{2qr^2}{y} + q^2\right) = r^2,$$

or
$$y^3 - q^2 y^2 - 2qr^2 y - r^4 = 0.$$

§22. If $x = a$, then $y = \dfrac{b+c}{a^2} = -\dfrac{a}{a^2} = -\dfrac{1}{x}$;

thus we have only to substitute $x = -\dfrac{1}{y}$ in the given equation.

§23. If $x = a$, then $y = \dfrac{abc+1}{a} = \dfrac{1-r}{x}$;

thus we have to substitute $x = \dfrac{1-r}{y}$ in the given equation.

§24. If $x = a$, then $y = a(b+c) = a(-a) = -a^2 = -x^2$.
We have now to substitute $x = \sqrt{-y}$ in the given equation.
Thus $(\sqrt{-y})(-y+q) = -r$; that is, $y^3 - 2qy^2 + q^2 y + r^2 = 0$.
§25. Here as in Ex.19 we have only to put $x = \sqrt[3]{y}$.

Thus
$$y + qy^{\frac{1}{3}} + r = 0, \text{ or } (y+r)^3 = -q^3 y;$$

that is,
$$y^3 + 3ry^2 + (3r^2 + q^3)y + r^3 = 0.$$

§26. If $x = a$, then $y = \dfrac{b^2 + c^2}{bc} = \dfrac{(b+c)^2}{bc} - 2 = \dfrac{a^2}{bc} - 2 = -\dfrac{a^3}{r} - 2$.

Therefore the transformed equation will be obtained by putting
$$y = -\frac{(x^3 + 2r)}{r}, \text{ or } x^3 = -r(2+y).$$

Now $x^3 + r = -qx$; hence $qx + r = r(2+y)$, or $qx = r(y+1)$; and therefore the new equation is
$$\{(y+1)r\}^3 = -q^3 r(2+y),$$
$$\text{or } r^3 y^3 + 3r^3 y^2 + (3r^2 + q^3)ry + r(r^2 + 2q^3) = 0.$$

§27. Let $y = x^3$, so that $x = \sqrt[3]{y}$.
From the given equation we have $y + ab = -y^{\frac{1}{3}}(ay^{\frac{1}{3}} + b)$.
Cube each side; thus

$$(y+ab)^3 = -y\{a^3 y + b^3 + 3aby^{\frac{1}{3}}(ay^{\frac{1}{3}} + b)\};$$
$$\therefore y^3 + a^3 b^3 + 3aby(y+ab) = -y\{a^3 y + b^3 - 3ab(y+ab)\};$$
$$\therefore y^3 + a^3 y^2 + b^3 y + a^3 b^3 = 0.$$

§28. The sum of the roots $= c = 5$; hence one of the roots is 5.
The equation may now be written
$$(x-5)(x^4 - 5x^2 + 4) = 0, \text{ or } (x-5)(x^2 - 4)(x^2 - 1) = 0.$$
Thus the roots are $\pm 2, \pm 1, 5$.
§29. Write $\dfrac{1}{y}$ for x; then the equation

$$y^3 + \frac{3q}{r}y^2 + \frac{3p}{r}y + \frac{1}{r} = 0$$

has its roots in A.P.

Denote the roots by $a - d, a, a + d$; then

$$3a = -\frac{3q}{r}; \text{ that is, } a = -\frac{q}{r}.$$

Now

$$a(a^2 - d^2) = -\frac{1}{r}; \text{ whence } q\left(\frac{q^2}{r^2} - d^2\right) = 1, \text{ or } d^2 = \frac{q^2}{r^2} - \frac{1}{q}.$$

Also

$$3a^2 - d^2 = \frac{3p}{r}.$$

Thus

$$\frac{3q^2}{r^2} - \left(\frac{q^2}{r^2} - \frac{1}{q}\right) = \frac{3p}{r}.$$

EXAMPLES XXXV e

As usual, we shall denote the imaginary cube roots of 1 by ω and ω^2, so that $2\omega = -1 + \sqrt{-3}$, and $2\omega^2 = -1 - \sqrt{-3}$; also $1 + \omega + \omega^2 = 0$.

§1. Putting $x = y + z$, we find $3yz - 18 = 0$, or $y^3 z^3 = 216$.
Also $y^3 + z^3 = 216$; thus $y^3 = 27, z^3 = 8$.
Thus the real root is $3 + 2$ or 5; and the imaginary roots are

$$3\omega + 2\omega^2 = \omega + 2(\omega + \omega^2) = \omega - 2 = \frac{-5 + \sqrt{-3}}{2};$$

and

$$3\omega^2 + 2\omega = \omega^2 - 2 = \frac{-5 - \sqrt{-3}}{2};$$

§2. Here $3yz + 72 = 0$, that is $y^3 z^3 = (-24)^3 = -12^3 \cdot 2^3$.
Also $y^3 + z^3 = 1720 = 12^3 - 2^3$; whence $y^3 = 12^3$, and $z^3 = -2^3$.
Thus the real root is $12 - 2$ or 10; and one of the imaginary roots is
$$12\omega - 2\omega^2 = 12\omega + 2(1 + \omega) = 2 + 7(-1 + \sqrt{-3}) = -5 + 7\sqrt{-3}.$$
The other root is got by changing the sign of $\sqrt{-3}$.

§3. Here $3yz = -63$, or $y^3 z^3 = -21^3 = -7^3 \cdot 3^3$.
Also $y^3 + z^3 = 316$; whence $y^3 = 7^3$, and $z^3 = -3^3$.
The real roots is $7 - 3 = 4$; and one of the imaginary roots is
$$7\omega - 3\omega^2 + 10\omega = -2 + 5\sqrt{-3}.$$

§4. Here $3yz = -21$, or $y^3 z^3 = -7^3$.
Also $y^3 + z^3 = -342$; thus $y = -7, z = 1$.
The real root is $-7 + 1 = -6$; and one of the imaginary roots is
$$-7\omega + \omega^2 = -1 - 8\omega = 3 - 4\sqrt{-3}.$$

§5. Let $x = \frac{1}{t}$; then $t^3 - 9t + 28 = 0$. Putting $t = y + z$, we have $3yz = 9$, or $y^3 z^3 = 27$. Also $y^3 + z^3 = -28$; whence $y^3 = -27$, and $z^3 = -1$.
Thus the real value of t is $-3 - 1$ or -4; and one of the imaginary values is $-3\omega - \omega^2 = 1 - 2\omega = 2 - \sqrt{-3}$.

Thus the values of x are $-\frac{1}{4}$, and $\frac{1}{2 \pm \sqrt{-3}}$ or $\frac{2 \mp \sqrt{-3}}{7}$.

§6. This equation may be written $(x - 5)^3 - 108(x - 5) + 432 = 0$; that is $t^3 - 108t + 432$, where $t = x - 5$.
Putting $t = y + z$, we have $3yz = 108$, or $y^3 z^3 = 36^3 = 6^3 \cdot 6^3$.
Also $y^3 + z^3 = -432$; whence $y^3 = -6^3$ and $z^3 = -6^3$.
Thus the real value of t is $-6 - 6$ or 12; and the other roots are $-6\omega - 6\omega^2$ and $-6\omega^2 - 6\omega$, which are both equal to 6.
Thus the values of x are $-7, 11, 11$.

§7. Multiply the given equation by 4, then
$$8x^3 + 12x^2 + 12x + 4 = 0, \text{ or } (2x+1)^3 + 3(2x+1) = 0;$$
whence $2x + 1 = 0$, and $(2x+1)^2 + 3 = 0$.

§8. Here $3yz = -12$, or $y^3z^3 = -4^3$.
Also $y^3 + z^3 = 12$; whence $y^3 = 16, z^3 = -4$.
Thus the real root is $\sqrt[3]{16} - \sqrt[3]{4} = 2\sqrt[3]{2} - \sqrt[3]{4}$.
Examples 9 – 17 may be solved by the methods given in Arts. 582, 583; but usually shorter solutions may be easily found.

§9. Here $x^4 = 3x^2 + 42x + 40$; on adding $6x^2 + 9$ to each side, we have
$$x^4 + 6x^2 + 9 = 9x^2 + 42x + 49; \text{ that is, } x^2 + 3 = \pm(3x+7);$$
thus $\qquad x^2 - 3x - 4 = 0$, and $x^2 + 3x + 10 = 0$.

§10. Here $x^4 = 10x^2 + 20x + 16$, and therefore $x^4 - 6x^2 + 9 = 4x^2 + 20x + 25$;
that is, $x^2 - 3 = \pm(2x+5)$.
   Thus $x^2 - 2x - 8 = 0$, and $x^2 + 2x + 2 = 0$.

§11. Here $\qquad\qquad\qquad x^4 + 9x^2 - 10 + 8x(x^2 - 1) = 0;$
that is,
$$(x^2 - 1)(x^2 + 10) + 8x(x^2 - 1) = 0, or (x^2 - 1)(x^2 + 8x + 10) = 0;$$
thus $\qquad\qquad\qquad x^2 - 1 = 0$, and $x^2 + 8x + 10 = 0$.

§12. Here $\qquad\qquad\qquad x^4 - 7x^2 + 12 + 2x(x^2 - 4) = 0;$
that is,
$$(x^2 - 4)(x^2 - 3) + 2x(x^2 - 4) = 0, \text{ or } (x^2 - 4)(x^2 + 2x - 3) = 0;$$
thus $\qquad\qquad\qquad x^2 - 4 = 0$, and $x^2 + 2x - 3 = 0$.

§13. Here $x^4 = 3x^2 + 6x + 2$, and therefore $x^4 + x^2 + \dfrac{1}{4} = 4x^2 + 6x + \dfrac{9}{4}$;

that is, $\qquad\qquad\left(x^2 + \dfrac{1}{2}\right)^2 = \pm\left(2x + \dfrac{3}{2}\right)^2.$

   Thus $x^2 - 2x - 1 = 0$, and $x^2 + 2x + 2 = 0$.

§14. Here $x^4 - 2x^3 = 12x^2 - 10x - 3$, and therefore by adding $-3x^2 + 4x + 4$ to each side, we have $(x^2 - x - 2)^2 = 9x^2 - 6x + 1$; thus
$$x^2 - x - 2 = \pm(3x - 1); \text{ that is, } x^2 - 4x - 1 = 0,$$
$$\text{and } x^2 + 2x - 3 = 0.$$

§15. This is a reciprocal equation and may be put in the form
$$4\left(x + \dfrac{1}{x}\right)^2 - 20\left(x + \dfrac{1}{x}\right) + 25 = 0;$$
whence $x + \dfrac{1}{x} = \dfrac{5}{2}$. Thus the roots are $2, 2, \dfrac{1}{2}, \dfrac{1}{2}$.

§16. By inspection $x = 1$ is a root, and on removing the corresponding factor $x - 1$, we have $x^4 - 5x^3 - 22x^2 - 5x + 1 = 0$;
that is,
$$\left(x + \dfrac{1}{x}\right)^2 - 5\left(x + \dfrac{1}{x}\right) - 24 = 0; \text{ whence } x + \dfrac{1}{x} = 8 \text{ or } -3.$$

§17. The first derived equation is $4x^3 + 27x^2 + 24x - 80 = 0$; hence by Art. 559, this equation and the given equation must have a common root. Now the highest common factor of
$$x^4 + 9x^3 + 12x^2 - 80x - 192 \text{ and } 4x^3 + 27x^2 + 24x - 80$$
is easily found to be $x^2 + 8x + 16$ or $(x+4)^2$.
   Thus $x^4 + 9x^3 + 12x^2 - 80x - 192$ contains the factor $x + 4$ repeated three times; the remaining factor is $x - 3$; hence the roots are $-4, -4, -4, 3$.

§18. If $x^4 = (x^2 + ax + b)^2$, we have $2ax^3 + (a^2 + 2b)x^2 + 2abx + b^2$. By supposition this reduces to the form $x^3 + qx + r = 0$; hence

$$a^2 + 2b = 0, \; q = b, \; r = \frac{b^2}{2a}.$$

From these equations, we have $q = -\dfrac{a^2}{2}$, and $r = \dfrac{a^3}{8}$;

thus $\qquad\qquad r^2 = \dfrac{a^6}{64} = -\dfrac{q^3}{8}$, or $q^3 + 8r^2 = 0$.

Suppose that $8x^3 - 36x + 27 = 0$ can be thrown into the form

$$x^4 = (x^2 + ax - b)^2;$$

then we have $a^2 - 2b = 0$, and $\dfrac{2a}{8} = \dfrac{2ab}{36} = \dfrac{b^2}{27}$;

hence $\qquad\qquad b = \dfrac{36}{8} = \dfrac{9}{2}$, and $a = \dfrac{4b^2}{27} = 3$;

these values satisfy the equation $a^2 - 2b = 0$.

Thus $\qquad x^4 = \left(x^2 + 3x - \dfrac{9}{2}\right)^2$, that is, $x^2 = \pm\left(x^2 + 3x - \dfrac{9}{2}\right)$;

hence $\qquad\qquad 3x - \dfrac{9}{2} = 0$, or $2x^2 + 3x - \dfrac{9}{2} = 0$.

§19. The required condition may be obtained by eliminating x between the two equations. [See Art. 528.]

We have $px^2 + 2qx + r = 0$, and $x^2 + 2px + q = 0$, whence by cross multiplication, $x^2 : x : 1 = 2(q^2 - pr) : r - pq : 2(p^2 - q)$; hence

$$4(p^2 - q)(q^2 - pr) = (pq - r)^2.$$

According to the second supposition, the first expression is divisible by the second without remainder. Now

$$(x^3 + 3px^2 + 3qx + r) = (x + p)(x^2 + 2px + q) + (2q - 2p^2)x + r - pq;$$

hence $2(q - p^2)x + (r - pq) = 0$ for all values of x; and therefore $p^2 - q = 0, \, pq - r = 0$. Thus $pr = p^2 q = q^2$.

§20. By the conditions of the question, $ax^3 + 3bx^2 + 3cx + d = 0$, and its first derived equation $ax^2 + 2bx + c = 0$ must have a common root; hence $bx^2 + 2cx + d = 0$, and $ax^2 + 2bx + c = 0$ must have a common root. Eliminating x^2, we have $2(ac - b^2)x + ad - bc = 0$; whence $x = \dfrac{bc - ad}{2(ac - b^2)}$.

§21. We have $x^4 + px^3 + qx^2 + rx + s = x^4 + px^3 + qx^2 + rx + \dfrac{r^2}{p^2}$

$$= \left(x^2 + \frac{p}{2}x + \frac{r}{p}\right)^2 - \left(\frac{p^2}{4} + \frac{2r}{p} - q\right)x^2.$$

Hence $\qquad \left(x^2 + \dfrac{p}{2}x + \dfrac{r}{p}\right)^2 = \left(\dfrac{p^2}{4} + \dfrac{2r}{p} - q\right)x^2 = a^2x^2$, say;

thus $\qquad\qquad x^2 + \dfrac{p}{2}x + \dfrac{r}{p} = \pm ax.$

§22. The equation whose roots are $\pm\sqrt{6} - 2$ is $x^2 + 4x - 2 = 0$; hence the other roots are given by $x^4 - 4x^3 + 8x - 4 = 0$.

Thus $\qquad x^4 - 4x^3 + 4x^2 = 4x^2 - 8x + 4$, or $x^2 - 2x = \pm(2x - 2)$;

that is, $\qquad\qquad x^2 - 4x + 2 = 0$, and $x^2 = 2.$

§23. Here $y = \beta + \gamma + \delta + \dfrac{1}{\beta\gamma\delta} = (\alpha + \beta + \gamma + \delta) - \alpha + \dfrac{\alpha}{\alpha\beta\gamma\delta}$;

that is, $\qquad\qquad y = 0 - \alpha + \dfrac{\alpha}{s} = \alpha\left(\dfrac{1}{s} - 1\right)$;

thus if x has the value α, then y has the value $\alpha\left(\dfrac{1}{s} - 1\right)$; and we have to substitute $y = x\left(\dfrac{1}{s} - 1\right)$, or $x(1 - s) = sy$ in the original equation.

Now
$$x^4(1 - s)^4 + qx^2(1 - s)^4 + rx(1 - s)^4 + s(1 - s)^4 = 0;$$
$$\therefore s^3 y^4 + qs(1 - s)^2 y^2 + r(1 - s)^3 y + (1 - s)^4 = 0.$$

§24.
We have
$$\alpha + \beta + \gamma + \delta = p,$$
$$\alpha\beta + \alpha\gamma + \alpha\delta + \beta\gamma + \beta\delta + \gamma\delta = q,$$
$$\alpha\beta\gamma + \alpha\beta\delta + \alpha\gamma\delta + \beta\gamma\delta = r,$$
$$\alpha\beta\gamma\delta = s.$$

(1) Suppose that $\qquad \alpha + \beta = \gamma + \delta.$
In this case $p = 2(\alpha + \beta)$;
$$q = \alpha\beta + (\alpha + \beta)(\gamma + \delta) + \gamma\delta = \alpha\beta + (\alpha + \beta)^2 + \gamma\delta;$$
$$r = \alpha\beta(\gamma + \delta) + \gamma\delta(\alpha + \beta)$$
$$= \alpha\beta(\alpha + \beta) + \gamma\delta(\alpha + \beta) = (\alpha + \beta)(\alpha\beta + \gamma\delta).$$
Thus $4q = 4(\alpha\beta + \gamma\delta) + p^2$, and $2r = p(\alpha\beta + \gamma\delta)$;
whence we obtain
$$4pq = 8r + p^3.$$

(2) Suppose that $\alpha\beta = \gamma\delta$.
In this case $s = \alpha^2\beta^2$,
$$r = \alpha\beta(\gamma + \delta) + \gamma\delta(\alpha + \beta) = \alpha\beta(\alpha + \beta + \gamma + \delta) = p\alpha\beta;$$
hence
$$r^2 = p^2\alpha^2\beta^2 = p^2 s.$$

§25. Denote the roots by a and $\dfrac{1}{a}$; then we have $a^5 - 209a + 56 = 0$, and $56a^5 - 209a^4 + 1 = 0$. Eliminating a^5, we have
$$209a^4 - 56 \times 209a + (56)^2 - 1 = 0;$$
but $\qquad 56^2 - 1 = 57 \cdot 55 = 19 \cdot 11 \cdot 3 \cdot 5 = 209 \times 15;$
hence $\qquad a^4 - 56a + 15 = 0.$

Similarly by eliminating the constant from the two above equations and dividing by a, we have $15a^4 - 56a^3 + 1 = 0$.

From these last two equations, we find
$$a^3 - 15a + 4 = 0, \text{ and } 4a^3 - 15a^2 + 1 = 0;$$
finally eliminating a^3, we have $a^2 - 4a + 1 = 0$; whence $a = 2 \pm \sqrt{-3}$.

§26. Denote the product of the two roots by y; then these two roots are given by the quadratic equation $x^2 - 5x + y = 0$; hence $x^5 - 409x + 285$ must be divisible by $x^2 - 5x + y$. It will be found that the quotient is
$$x^3 + 5x^2 + (25 - y)x + (125 - 10y),$$
and the remainder
$$(y^2 - 75y + 216)x + 5(2y^2 - 25 + 57),$$
or $\qquad (y - 3)(y - 72)x + 5(y - 3)(2y - 19).$
Thus the remainder vanishes when $y = 3$, and therefore the two roots are given by $x^2 - 5x + 3 = 0$.

§27. If $i = \sqrt{-1}$, then $(1 + a^2)(1 + b^2)(1 + c^2)\ldots$
$$= (1 + ia)(1 + ib)(1 + ic)\ldots \times (1 - ia)(1 - ib)(1 - ic)\ldots$$
$$= (1 - ip_1 + i^2 p_2 - i^3 p_3 + \ldots) \times (1 + ip_1 + i^2 p_2 + i^3 p_3 + \ldots)$$
$$= \{(1 - p_2 + p_4 - \ldots) - i(p_1 - p_3 + p_5 - \ldots)\}$$
$$\times \{(1 - p_2 + p_4 - \ldots) + i(p_1 - p_3 + p_5 - \ldots)\}$$
$$= (1 - p_2 + p_4 - \ldots)^2 + (p_1 - p_3 + p_5 - \ldots)^2.$$

§28. The given equation may be written
$$(x^2 - 4x + 3)^2 = x^2 - 4x + 4 = (x - 2)^2;$$
hence $\qquad x^2 - 4x + 3 = \pm(x - 2);$
that is, $\qquad x^2 - 5x + 5 = 0, \text{ or } x^2 - 3x + 1 = 0.$

If we put $x = 4 - y$, the above equations become $y^2 - 3y + 1 = 0$, and $y^2 - 5y + 5 = 0$ respectively, and we merely reproduce the original equation.

MISCELLANEOUS EXAMPLES

§1. If a is the first term and d the common difference, we have
$$2s_1 = n\{2a + (n-1)d\}, \; 2s_2 = 2n\{2a + (2n-1)d\}, \; 2s_3 = 3n\{2a + (3n-1)d\};$$
hence $\dfrac{2s_1}{n} + \dfrac{2s_3}{3n} = 2 \cdot \dfrac{2S_2}{2n}$; that is, $3s_1 + s_3 = 3s_2$.

§2. We have $\dfrac{x-y}{1} = \dfrac{x+y}{7} = \dfrac{xy}{24}$; that is, $3x = 4y$, and $xy = 24(x - y)$.
Hence $3x^2 = 24(4x - 3x)$; therefore (excluding zero solutions) $x = 8, y = 6$.

§3. If r be the radix, $5r + 2 = 2(2r + 5)$; whence $r = 8$.

§4. (1) By rearranging, we have $(x+2)(x-4)(x+3)(x-5) = 44$;
that is $(y-8)(y-15) = 44$; where $y = x^2 - 2x$. We easily obtain $y = 4$
or 10; hence $x^2 - 2x - 4 = 0$, or $x^2 - 2x - 19 = 0$. Thus the solutions are
$1 \pm \lfloor 5, 1 \pm 2\lfloor 5$.

(2)We have $xy + xz = -2, \; -2xy + yz = -21, \; 2xz - yz = 5$.
Solving these as equations in xy, xz, yz we obtain
$$xy = 3, xz = -5, yz = -15; \text{ whence } xyz = \pm 15.$$

§5. We have $\quad 2a + (p-1)d = 0$.
the sum of the next q terms = sum of $(p+q)$ terms - sum of p terms
$$= \frac{p+q}{2}\{2a + (p+q-1)d\} - 0.$$

Thus the sum is
$$(p+q)\left\{a - \frac{(p+q-1)a}{p-1}\right\} = -\frac{(p+q)qa}{p-1}.$$

§6. (1)One solution is obviously $x = 1$. On reduction the equation becomes
$$(a+b)\{ab + (a^2 - b^2)x - abx^2\} = a^3x - ab^2 + a^2bx^2 - b^3x.$$

The product of the roots $= -\dfrac{ab(a+b) + ab^2}{a^2b + ab(a+b)} = -\dfrac{a + 2b}{2a + b}$;

which is therefore the value of the second root.

(2)If $c = a + b$; then
$$c^3 = (a+b)^3 = a^3 + b^3 + 3ab(a+b) = a^3 + b^3 + 3abc;$$
that is, $\quad 3abc = c^3 - a^3 - b^3$.
Hence the given equation is equivalent to
$$3\sqrt[3]{12x(2x-3)(x-1)} = 12(x-1) - x - (2x-3) = 9(x-1),$$
or $\quad 12x(2x-3)(x-1) = 27(x-1)^3$;
whence $\quad x - 1 = 0$, or $4x(2x-3) = 9(x-1)^2$.

§7. we have $(1+d)(1+33d) = (1+9d)^2$; that is, $48d^2 - 16d = 0$; thus $d = 0$
or $d = \dfrac{1}{3}$.

§8. Here $\alpha + \beta = -p, \; \alpha\beta = q$; hence
$$\alpha^2 + \alpha\beta + \beta^2 = p^2 - q; \quad \alpha^2 - \alpha\beta + \beta^2 = p^2 - 3q;$$
$$\alpha^3 + \beta^3 = p(p^2 - 3q);$$
and
$$\alpha^4 + \alpha^2\beta^2 + \beta^4 = (\alpha^2 + \alpha\beta + \beta^2)(\alpha^2 - \alpha\beta + \beta^2) = (p^2 - q)(p^2 - 3q).$$

§9. If $2x = a + a^{-1}$, then $4x^2 - 4 = a^2 + 2 + a^{-2} - 4 = (a - a^{-1})^2$. Denoting
the given expression by E. we have
$$4E = (a + a^{-1})(b + b^{-1}) + (a - a^{-1})(b - b^{-1}) = 2(ab + a^{-1}b^{-1}).$$

§10. Without altering the value of the whole expression, we may double each of the expressions under the radical signs. Now

$$8 + 2\sqrt{15} = (\sqrt{5} + \sqrt{3})^2 \text{ and } 12 + 2\sqrt{35} = (\sqrt{7} + \sqrt{5})^2;$$

hence the required value $= \dfrac{(\sqrt{5} + \sqrt{3})^3 + (\sqrt{5} - \sqrt{3})^3}{(\sqrt{7} + \sqrt{5})^3 - (\sqrt{7} - \sqrt{5})^3}$

$$= \frac{5\sqrt{5} + 3 \cdot \sqrt{5} \cdot (\sqrt{3})^2}{3(\sqrt{7})^2\sqrt{5} + 5\sqrt{5}} = \frac{5+9}{21+5} = \frac{7}{13}.$$

§11. Replacing α and β by the more usual forms ω and ω^2, we have

$$\alpha^4 + \beta^4 + \alpha^{-1}\beta^{-1} = \omega^4 + \omega^3 + \omega^{-3} = \omega + \omega^2 + 1 = 0.$$

§12. This follows from the fact that

$$r^4 + 2r^3 + 4r^2 + 3r + 2 = (r^2 + r + 1)(r^2 + r + 2).$$

§13. Let x and y denote the number of yards that A and B run in a second; then $\dfrac{1760 - 11}{y} - \dfrac{1760}{x} = 57.$

Again $\dfrac{1760}{y} - 81 = \dfrac{1760 - 88}{x}.$

To eliminate x, multiply the second equation by 20, and the first by 19, and subtract; thus

$$\frac{1}{y}(20 \times 1760 - 19 \times 1749) = 20 \times 81 - 19 \times 57;$$

$$\text{or } \frac{1}{y}(1760 + 209) = 81 + 456;$$

hence $y = \dfrac{11}{3}$, and therefore $x = \dfrac{88}{21}.$

Thus A takes 420 seconds, and B 480 seconds.

§14.. See Ex. 4, Art. 137. Thus from the first three equations we have

$$\frac{x}{a^4 - b^2c^2} = \frac{y}{b^4 - c^2a^2} = \frac{z}{c^4 - a^2b^2} = k.$$

Substituting for x, y, z in $x + y + z = 0$, we get

$$a^4 + b^4 + c^4 = b^2c^2 + c^2a^2 + a^2b^2.$$

§15.

We have $\qquad (a - b)x^2 + (b - c)xy + (c - a)y^2 = 0,$

or $\qquad (x - y)\{(a - b)x - (c - a)y\} = 0.$

Taking $x = y$, we have $x^2 = y^2 = \dfrac{d}{a + b + c}.$

Taking $(a - b)x = (c - a)y$, we find $\dfrac{x}{c - a} = \dfrac{y}{a - b} = k$, where

$$ak^2(a^2 + b^2 + c^2 - bc - ca - ab) = d.$$

§16. Suppose that the waterman can row x miles per hour in still water, and that the stream flows y miles per hour; then he can row $x + y$ miles per hour with the stream, and $x - y$ miles against the stream. Thus

$$\frac{48}{x + y} + \frac{48}{x - y} = 14, \text{ and } \frac{x + y}{4} = \frac{x - y}{3};$$

whence $x = 7y$, and $y = 1, x = 7.$

§17.

(1) The expression $= (a + b)(a + c) \times (b + c)(b + a) \times (c + a)(c + b)$

$$= (b + c)^2(c + a)^2(a + b)^2.$$

(2) The expression $= \dfrac{1}{2}\{2 - 2x + 2\sqrt{(5 - 4x)(2x - 3)}\}$

$$= \frac{1}{2}(\sqrt{5 - 4x} + \sqrt{2x - 3})^2.$$

§18.

(1) The coefficient $= \dfrac{1}{\underline{6}} \cdot \dfrac{10}{3} \cdot \dfrac{7}{3} \cdot \dfrac{4}{3} \cdot \dfrac{1}{3} \cdot \dfrac{2}{3} \cdot \dfrac{5}{3} \cdot 3^6 = \dfrac{35}{9}.$

(2) We have

$$\left(\frac{4}{3}x^2 - \frac{3}{2x}\right)^9 = x^{18}\left(\frac{4}{3} - \frac{3}{2x^3}\right)^9.$$

Hence the term required is the coefficient of $\dfrac{1}{x^{18}}$ in the expansion of the last binomial, and is therefore equal to

$$\frac{\underline{9}}{\underline{6}\,\underline{3}}\left(\frac{4}{3}\right)^3\left(\frac{3}{2}\right), \text{ or } 2268.$$

§19. (1) We have $\left(2 - \dfrac{1}{x-1}\right) - \left(3 - \dfrac{2}{x-2}\right) + \left(1 + \dfrac{6}{x-3}\right) = 0;$

that is, $\qquad \dfrac{-1}{x-1} + \dfrac{2}{x-2} + \dfrac{6}{x-3} = 0;$

whence $\qquad 7x^2 - 21x + 12 = 0, \text{ and } x = \dfrac{21 \pm \sqrt{105}}{14}.$

(2) From the given equations, we have

$$\frac{x^2 - xy - y^2}{(x+y)(ax+by)} = \frac{-ab}{2ab(a+b)} = \frac{-1}{2(a+b)};$$

that is, $\qquad (3a+2b)x^2 - (a+b)xy - (2a+b)y^2 = 0.$

Thus $\qquad x - y = 0, \text{ or } (3a+2b)x + (2a+b)y = 0.$

If $x - y = 0$, then from $x^2 - y^2 = xy - ab$, we find $x^2 = y^2 = ab.$

If $\dfrac{x}{2a+b} = \dfrac{y}{-(3a+2b)} = k$, then from $(x+y)(ax+by) = 2ab(a+b),$

we have $\qquad -(a+b)(2a^2 - 2ab - b^2)k^2 = 2ab(a+b;)$

that is, $\qquad k^2(b^2 + ab - a^2) = ab.$

§20. When the expression is a perfect square,

$$4ac(b-c)(a-b) = b^2(c-a)^2;$$

and therefore arranging according to powers of b, we have

$$b^2(c+a)^2 - 4ac(a+c)b + 4a^2c^2 = 0;$$

that is, $b(c+a) - 2ac = 0$; which proves the proposition.

§21. Since $(y+z-2x)^2 - (y-z)^2 = (2y-2x)(2z-2x) = 4(x-y)(x-z),$

we have

$$(x-y)(x-z) + (y-z)(y-x) + (z-x)(z-y) = 0.$$

Put $y - z = a, z - x = b, x - y = c$; then $bc + ca + ab = 0$, while $a+b+c = 0.$

$\therefore (a+b+c)^2 - 2(bc+ca+ab) = 0$; that is, $a^2 + b^2 + c^2 = 0;$

thus $a = 0, b = 0, c = 0.$

§22.

$$\dot{3}e\dot{5}8\dot{2}6i(1et5$$
$$\underline{1}$$
$$2e)2e5$$
$$\underline{281}$$
$$3tt)3482$$
$$\underline{3304}$$
$$3e85)17t61$$
$$\underline{17t61}$$

Let r denote the radix of the scale; then

$$\frac{1}{5} = \left(\frac{1}{r} + \frac{7}{r^2}\right) + \left(\frac{1}{r^3} + \frac{7}{r^4}\right) + \ldots;$$

that is,
$$\frac{1}{5} = \left(\frac{1}{r} + \frac{7}{r^2}\right) \div \left(1 - \frac{1}{r^2}\right) = \frac{r+7}{r^2-1};$$
or $r^2 - 5r - 36 = 0$; whence $r = 9$.

§23. We know that
$$2(ab + ac + ad + \ldots + bc + bd + \ldots) = (a + b + c + d + \ldots)^2$$
$$- (a^2 + b^2 + c^2 + d^2 + \ldots).$$
From this the required result at once follows, since
$$(1 + 2 + 3 + \ldots + n)^2 = 1^3 + 2^3 + 3^3 + \ldots + n^3.$$

§24. Denote his weekly wages by x pence, and the price of a loaf by y pence; then we have
$$\frac{x}{20} - \frac{20y}{40} = 6, \text{ and } \frac{7\frac{1}{2}x}{100} - \frac{20y}{10} = 1\frac{1}{2};$$
whence $x = 180$, $y = 6$.

§25. Denote the numbers by $a - 3d, a - d, a + d, a + 3d$; thus
$$4a = 48, \text{ or } a = 12.$$
Hence $\quad (12 - 3d)(12 + 3d) : (12 - d)(12 + d) = 27 : 35$;
or $35(16 - d^2) = 3(144 - d^2)$; that is $d^2 = 4$.

§26.

(1)By inspection, one root is unity; also the product of the roots is $\dfrac{c(a - b)}{a(b - c)}$;

thus the second root is $\dfrac{c(a - b)}{a(b - c)}$.

(2) By an easy reduction we see that $x + \dfrac{ab}{x - a - b} = x + \dfrac{cd}{x - c - d}$;

that is $\quad ab(x - c - d) = cd(x - a - b)$.

§27.

(1)By transposing and squaring, we have
$$a - x + b - x + 2\sqrt{(a - x)(b - x)} = c - x.$$
Repeating the process, we obtain $(a + b - c - x)^2 = 4(a - x)(b - x)$;
that is,
$$a^2 + b^2 + c^2 - 2ab - 2ab - 2ac - 2bc + 2(a + b + c)x - 3x^2 = 0,$$
or $\quad (a + b + c)^2 + 2(a + b + c)x - 3x^2 = 4(bc + ca + ab)$.

(2)Since $x^3 + y^3 + z^3 = 3xyz$ when $x + y + z = 0$, we have in the present case $a + b + c = 3\sqrt[3]{abc}$; therefore $(a + b + c) = 27abc$.

§28. Suppose that the length of the journey is x miles, and the velocity of the train y miles per hour; then
$$1 + 1 + \frac{x - y}{\frac{3}{5}y} = \frac{x}{y} + 3;$$

that is, $\quad \dfrac{5(x - y)}{3y} - \dfrac{x}{y} = 1$, or $x = 4y$.

Again, the train takes $1\frac{1}{2}$ hours more in traveling 50 miles at the reduced speed than it does in traveling 50 miles at the original speed; thus
$$\frac{50}{\frac{3}{5}y} - \frac{50}{y} = 1\frac{1}{2}; \text{ whence } y = \frac{200}{9}. \text{ Therefore } x = \frac{800}{9}.$$

§29. From the first two equations by cross multiplication, we have $\dfrac{x}{3} = \dfrac{y}{4} = \dfrac{z}{5} = k$ say; hence $k^3(27 + 64 + 125) = 216$; that is, $k^3 = 1$.

§30. Suppose the two mathematical papers A and B were fastened together and considered as one. We should thus obtain $2\lfloor 5$ permutations among the five papers, since the mathematical papers themselves admit of two arrangements, and these cases are all ineligible. Also the whole number of permutations without restriction is $\lfloor 6$; therefore the required number of arrangements is $\lfloor 6 - 2\lfloor 5$, or 480.

§31. Let x, y, z denote the number of half-crowns, shillings and four-penny-pieces respectively; then $x + y + z = 60$. Also $30x + 12y + 4z = 1250$; that is, $15x + 6y + 2z = 625$. Eliminating z we have $13x + 4y = 505$; of which the general solution is $x = 1 + 4t$, and $y = 123 - 13t$; hence $z = 9t - 64$. Thus t must be greater than $\dfrac{64}{9}$ and less than $\dfrac{123}{13}$; that is, t may have the values 8 and 9. Thus $x = 33$, $y = 19$, $z = 8$; or $x = 37$, $y = 6$, $z = 17$.

§32. Subtracting the first expression from the second we have
$$(b - a)x^2 + 3x + 2.$$
Multiplying the first expression by 8, and the second by 6, and subtracting we have $2x\{x^2 + (4a - 3b)x + 2\}$. Thus both $(b - a)x^2 + 3x + 2$ and $x^2 + (4a - 3b)x + 2$ must divide each of the given expressions, multiplied if necessary by some positive integer.

In these two quadratic expressions the term independent of x is the same; hence the coefficients of x^2 and x must be the same; thus $b - a = 1$, and $4a - 3b = 3$; whence $a = 6$, $b = 7$.

§33. Suppose that A, B, C together do the work in x hours; then A alone can do the work in $x + 6$ hours, B alone $x + 1$ hours, and C alone in $2x$ hours. Hence working together they can do $\dfrac{1}{x + 6} + \dfrac{1}{x + 1} + \dfrac{1}{2x}$ of the work in one hour; but they also do $\dfrac{1}{x}$ of the work in one hour;

hence $$\frac{1}{x + 6} + \frac{1}{x + 1} + \frac{1}{2x} = \frac{1}{x};$$

that is, $\quad 2x(2x + 7) = (x + 6)(x + 1)$, or $3x^2 + 7x - 6 = 0$.

Thus $(3x - 2)(x + 3) = 0$; whence $x = \dfrac{2}{3}$.

§34. Eliminating y, we have $b^2cx^2 + d(1 - ax)^2 = b^2$,

or $$(b^2c + a^2d)x^2 - 2adx + d - b^2 = 0.$$

By Hypothesis, this equation must have equal roots; hence
$$(b^2c + a^2d)(d - b^2) = a^2d^2;$$

that is, $\quad b^2(b^2c + a^2d) = b^2cd$, or $b^2c + a^2d = cd$.

Also the sum of the roots $= \dfrac{2ad}{b^2c + a^2d} = 2x$;

therefore $\quad x = \dfrac{ad}{b^2c + a^2d} = \dfrac{ad}{cd} = \dfrac{a}{c}$. By symmetry $y = \dfrac{b}{d}$.

§35.
Here
$$(1 - 2x + 2x^2)^{-\frac{1}{2}} = 1 + \frac{1}{2}(2x - 2x^2) + \frac{1}{2} \cdot \frac{3}{4}(2x - 2x^2)^2$$
$$+ \frac{1}{2} \cdot \frac{3}{4} \cdot \frac{5}{6}(2x - 2x^2)^3 + \frac{1}{2} \cdot \frac{3}{4} \cdot \frac{5}{6} \cdot \frac{7}{8}(2x)^4 + \dots$$
$$= 1 + (x - x^2) + \frac{3}{2}(x^2 - 2x^3 + x^4) + \frac{5}{2}(x^3 - 3x^4) + \frac{35}{8}x^4 + \dots$$
$$= 1 + x + \frac{x^2}{2} - \frac{x^3}{2} - \frac{13x^4}{8} + \dots$$

§36. Denote the roots by a and a^2; then $a + a^2 = -p$, and $a^3 = q$.
Hence $-p^3 = a^6 + 3a^5 + 3a^4 + a^3 = q^2 + q + 3a^3(a^2 + a) = q^2 + q - 3pq$.

§37. Arranging the equation in the form $x(x^3 - 1) - 5(x^2 + x + 1) = 0$,
we have $(x^2 + x + 1)(x^2 - x - 5) = 0$.

§38. Subtracting numerator from denominator, we have $x^2 - 4x + 3$, that
is $(x - 1)(x - 3)$.

Hence numerator and denominator must be divisible by $x - 1$, or by
$x - 3$, and must therefore vanish when $x = 1$, or when $x = 3$.

If $x = 1$, we have $a = 8$, and in this case
$$\frac{x^3 - 8x^2 + 19x - 12}{x^3 - 9x^2 + 23x - 15} = \frac{x^2 - 7x + 12}{x^2 - 8x + 15} = \frac{x - 4}{x - 5}.$$
If $x = 3$, we find also that $a = 8$.

§39. This equation is equivalent to
$$a^2 + b^2 + c^2 - bc - ca - ab + 3x^2 + 3y^2 + 3z^2 = 0,$$
or $\qquad \frac{1}{2}\{(b - c)^2 + (c - a)^2 + (a - b)^2\} + 3x^2 + 3y^2 + 3z^2 = 0;$
and therefore $b - c = 0, c - a = 0, a - b = 0, x = 0, y = 0, z = 0$.

§40. With the notation of Art. 187,
$$T_{r+1} = \frac{\frac{3}{2} + r - 1}{r}\left(\frac{2x}{3}\right)T_r = \frac{\frac{3}{2} + r - 1}{r} \cdot \frac{4}{7} \cdot T_r;$$
$$\therefore T_{r+1} > T_r \text{ so long as } \frac{6 + 4r - 4}{7r} > 1, \text{ or } 2 > 3r.$$
Therefore the first term is the greatest.

§41. Denote the numbers by x and y;
then $\qquad (x + y)(x^2 + y^2) = 5500, \text{ and } (x - y)(x^2 - y^2) = 352;$
hence $\qquad \frac{(x + y)(x^2 + y^2)}{(x - y)(x^2 - y^2)} = \frac{5500}{352}, \text{ that is } \frac{x^2 + y^2}{(x - y)^2} = \frac{125}{8};$
whence
$$117x^2 - 250xy + 117y^2 = 0, \text{ or } (13x - 9y)(9x - 13y) = 0.$$
Thus $\dfrac{x}{13} = \dfrac{y}{9} = k$ say; and therefore $352 = 4k \times 88k^2$; whence $k = 1$.

§42. From the data, $x^2 + y^2 + z^2 = \lambda^2(a^2 + b^2 + c^2) - 2\lambda(b^2 + 3c^2) + b^2 + 9c^2$
$$= \frac{(1 + b^2 + 3c^2)^2}{a^2 + b^2 + c^2} - \frac{2(b^2 + 3c^2)(1 + b^2 + 3c^2)}{a^2 + b^2 + c^2} + b^2 + 9c^2$$
$$= \frac{(1 + b^2 + 3c^2)(1 - b^2 - 3c^2)}{a^2 + b^2 + c^2} + b^2 + 9c^2$$
$$= \frac{1 - b^4 - 6b^2c^2 - 9c^4 + a^2(b^2 + 9c^2) + (b^2 + c^2)(b^2 + 9c^2)}{a^2 + b^2 + c^2}$$
$$= \frac{1 + 4b^2c^2 + 9c^2a^2 + a^2b^2}{a^2 + b^2 + c^2}.$$

§43.
(1) Add $x^2 + 4$ to each side; then $x^4 + 4x^2 + 4 = x^2 + 16x + 64$;
whence $x^2 + 2 = \pm(x + 8)$; that is, $x^2 - x - 6 = 0$, or $x^2 + x + 10 = 0$.
(2) From the given squations, we have
$$x^2 - y^2 + x - y = 0, \text{ or } (x - y)(x + y + 1) = 0.$$
Thus $\qquad\qquad x = y, \text{ or } x + y + 1 = 0.$
Similarly $\qquad\qquad x = z, \text{ or } x + z + 1 = 0.$

If $x = y = z$, we have $2x^2 - x - 1 = 0$; whence $x = 1$ or $-\dfrac{1}{2}$.

If $x = y$ and $x + z + 1 = 0$, we have $2x^2 + x = 0$; whence $x = 0$ or $-\dfrac{1}{2}$.

If $x + y + 1 = 0$ and $x = z$, we also obtain $2x^2 + x = 0$.

If $x + y + 1 = 0$ and $x + z + 1 = 0$, we obtain $2x^2 + 3x + 1$;

whence $\qquad\qquad\qquad x = -1$ or $-\dfrac{1}{2}$.

§44.

$$\begin{aligned}
\log(x + z) + \log(x - 2y + z) &= \log\{(x + z)^2 - 2y(x + z)\} \\
&= \log\{(x + z)^2 - 4xz\} \\
&= \log(x - z)^2 = 2\log(x - z).
\end{aligned}$$

§45. $1 + \dfrac{1}{2} \cdot \dfrac{1}{4} + \dfrac{1 \cdot 3}{2 \cdot 4}\left(\dfrac{1}{4}\right)^2 + \dfrac{1 \cdot 3 \cdot 5}{2 \cdot 4 \cdot 6}\left(\dfrac{1}{4}\right)^3 + \ldots$

$$= \left(1 - \dfrac{1}{4}\right)^{-\frac{1}{2}} = \left(\dfrac{3}{4}\right)^{-\frac{1}{2}} = \left(\dfrac{4}{3}\right)^{\frac{1}{2}} = \dfrac{2\sqrt{3}}{3};$$

that is, $1 + \dfrac{1}{4}S = \dfrac{2\sqrt{3}}{3}$; whence $S = \dfrac{4}{3}(2\sqrt{3} - 3)$.

§46.

$$\text{Each fraction} = \frac{\text{sum of numerators}}{\text{sum of denominators}} = \frac{5(x + y + z)}{a + b + c}.$$

$$\begin{aligned}
\text{Again, each fraction} &= \frac{(3x + 2y) + 2(3y + 2z) + 3(3z + 2x)}{(3a - 2b) + 2(3b - 2c) + 3(3c - 2a)} \\
&= \frac{9x + 8y + 13z}{-3a + 4b + 5c};
\end{aligned}$$

$$\text{thus} \quad \frac{5(x + y + z)}{a + b + c} = \frac{9x + 8y + 13z}{5c + 4b - 3a}.$$

§47. The first place can be filled in 17 ways, and the last place also in 17 ways, since the consonants may be repeated. The vowels can be placed in 5×4 ways; hence the number of ways $= 17 \times 17 \times 20 = 5780$.

§48. Suppose that at first x persons voted for the motion, then $600 - x$ voted against the motion, and it was therefore lost by $600 - 2x$ votes.

Suppose that y persons changed their minds, then in the second case $x + y$ voted for the motion, and $600 - x - y$ against it; thus the motion was carried by $2(x + y) - 600$ votes.

Hence $2(x + y) - 600 = 2(600 - 2x)$, and $\dfrac{x + y}{600 - x} = \dfrac{8}{7}$;

whence $x = 250$, $y = 150$.

§49. The expression on the left $= \dfrac{1 - x}{2}\log(1 + x) - \dfrac{1 + x}{2}\log(1 - x)$

$$= \frac{1}{2}\{\log(1 + x)\} - \log(1 - x)\} - \frac{x}{2}\{\log(1 + x) + \log(1 - x)\}$$

$$= \left(x + \frac{x^3}{3} + \frac{x^5}{5} + \frac{x^7}{7} + \ldots\right) + x\left(\frac{x^2}{2} + \frac{x^4}{4} + \frac{x^6}{6} + \ldots\right)$$

$$= x + x^3\left(\frac{1}{2} + \frac{1}{3}\right) + x^5\left(\frac{1}{4} + \frac{1}{5}\right) + x^7\left(\frac{1}{6} + \frac{1}{7}\right) + \ldots$$

§50. Let x denote the number of men in the side of the hollow square; then the number of men in the hollow square $x = x^2 - (x - 6)^2 = 12x - 36$.

Hence $(12x - 36) + 25 = (\sqrt{x} + 22)^2$; from which we obtain $x - 4\sqrt{x} - 45 = 0$, and $x = 81$.

§51. (1) Divide throughout by $\sqrt[m]{a^2 - x^2}$;

thus $\qquad\qquad \sqrt[m]{\dfrac{a + x}{a - x}} + 2\sqrt[m]{\dfrac{a - x}{a + x}} + 3$;

whence $\qquad\qquad \sqrt[m]{\dfrac{a + x}{a - x}} = 1$ or 2, and $\dfrac{a + x}{a - x} = 1$ or 2^m.

(2) We have $(x-a)^{\frac{1}{2}}(x-b)^{\frac{1}{2}} - (x-c)^{\frac{1}{2}}(x-d)^{\frac{1}{2}}$

$$= \overline{(x-c} - \overline{x-a})^{\frac{1}{2}}(\overline{x-d} - \overline{x-b})^{\frac{1}{2}}$$

$$= \{(\overline{x-c} - \overline{x-a})(\overline{x-d} - \overline{x-b})\}^{\frac{1}{2}}.$$

Square both sides; then

$(x-a)(x-b) + (x-c)(x-d) - 2\{(x-a)(x-b)(x-c)(x-d)\}^{\frac{1}{2}} = (x-a)(x-b) + (x-c)(x-d) - (x-a)(x-d) - (x-b)(x-c);$

hence

$$(x-a)(x-d) + (x-b)(x-c)$$
$$-2\{(x-a)(x-b)(x-c)(x-d)\}^{\frac{1}{2}} = 0;$$

that is, $\qquad (x-a)^{\frac{1}{2}}(x-d)^{\frac{1}{2}} - (x-b)^{\frac{1}{2}}(x-c)^{\frac{1}{2}} = 0;$

whence, by transposing and squaring, $(x-a)(x-d) = (x-b)(x-c).$

§52. We have $\qquad \sqrt[8]{4} = 2^{\frac{2}{3}} = \left(\dfrac{1}{2}\right)^{-\frac{2}{3}} = \left(1 - \dfrac{1}{2}\right)^{-\frac{2}{3}};$

expanding by the Binomial Theorem we obtain the series on the right.

§53. Put $\qquad u = \sqrt[3]{6(5x+6)}$ and $v = \sqrt[3]{5(6x-11)};$

then $\qquad\qquad u - v = 1,$ and $u^3 - v^3 = 91.$

But $\qquad u^3 - 3uv(u - v) - v^3 = 1;$ and therefore $uv = 30.$

From these equations we easily obtain $u = 6$ or $-5, v = 5$ or -6. Thus we have finally $6(5x+6) = 216$ or -125; that is, $x = 6$ or $-\dfrac{161}{30}.$

§54. After the first operation the first vessel contains $a - c$ gallons of wine, the second contains c gallons of wine.

At the second operation $\dfrac{(a-c)}{a} \times c$ gallons of wine are removed from the first vessel, and $\dfrac{c}{b} \times c$ gallons of wine removed from the second vessel; these quantities are equal if $\dfrac{a-c}{a} = \dfrac{c}{b}$, or $c(a + b) = ab$; that is, after the first operation equal quantities of wine are removed from the two vessels, and therefore the amount of wine in each will always remain the same after any number of operations.

§55. From the data, we have $\dfrac{m+n}{2} = \sqrt{ab} = \dfrac{ma+nb}{m+n};$

hence $ma + nb = (m+n)\sqrt{ab} = 2\sqrt{ab} \times \sqrt{ab} = 2ab$; and $m + n = 2\sqrt{ab}$. From these equations we easily find m and n.

§56. Let $x + y + z = c$, a constant. By hypothesis $(c - 3y)(c - 3z) = yz$; that is, $c^2 - 3c(y + z) + 9yz = myz;$

hence $\qquad (9 - m)yz = c(3y + 3z - c) = c(2y + 2z - x);$

thus $2y + 2z - x$ varies as yz.

§57. We have $(1 + x)^n = 1 + {}^nC_1 x + {}^nC_2 x^2 + \ldots + {}^nC_{r-2}x^{r-2} + {}^nC_{r-1}x^{r-1} + {}^nC_r x^r + \ldots;$

and $(1 + x)^{-3} = 1 - 3x + \dfrac{3 \cdot 4}{1 \cdot 2}x^2 - \dfrac{4 \cdot 5}{1 \cdot 2}x^3 + \ldots + (-1)^r \dfrac{(r+1)(r+2)}{\lfloor 2}x^r + \ldots$

The given series is twice the coefficient of x^r in the product of the two series on the right; thus $\frac{1}{2}S$ = the coefficient of x^r in $(1 + x)^{n-3} = {}^{n-3}C_r;$

that is, $\qquad\qquad\qquad S = 2 \times {}^{n-3}C_r.$

§58. (1) We have identically, $(2x - 1) - (3x - 2) = 1 - x = (4x - 3) - (5x - 4);$

dividing each side of this equation by the corresponding side of the given equation, we have

$$\sqrt{2x-1} - \sqrt{3x-2} = \sqrt{4x-3} - \sqrt{5x-4}.$$

By addition, $\sqrt{2x-1} = \sqrt{4x-3}$; whence we obtain $x = 1$.

(2) Put $x^2 - 16 = y^4$, so that $x^2 = y^4 + 16$;

then $\quad 4(y^3 + 8) = y^4 + 16 + 16y$, or $y^4 - 4y^3 + 16y - 16 = 0$.

Thus $y^4 - 16 = 4y(y^2 - 4)$; whence $y^2 - 4 = 0$, and $y^2 - 4y + 4 = 0$; so that the values of y are 2 and -2; and therefore $x^2 = 32$, and $x = \pm 4\sqrt{2}$.

§59. Clearing of fractions we have

$$\{(y-z) + x(y^2 - z^2) + x^2yz(y-z)\} + \ldots + \ldots = 0;$$

that is, $\quad x(y^2 - z^2) + y(z^2 - x^2) + z(x^2 - y^2) = 0$;

hence $(y-z)(z-x)(x-y) = 0$, and two of the quantities x, y, z must be equal.

§60. Denote the number of males and females by m and f respectively; then $m + f = p$.

Again $\dfrac{bm}{1000} + \dfrac{cf}{100} = \dfrac{ap}{100}$; that is $bm + cf = ap$. From these equations we have

$$(b-c)m = (a-c)p, \text{ and } (b-c)f = (b-a)p.$$

§61. If $x^{\frac{a}{b}} = \left(\dfrac{a}{b}\right)^{\frac{2a^2}{a^2-b^2}}$, then $x^{\frac{b}{a}} = (x^{\frac{a}{b}})^{\frac{b^2}{a^2}} = \left(\dfrac{a}{b}\right)^{\frac{2b^2}{a^2-b^2}}$;

hence

$$x^{\frac{a}{b}} + x^{\frac{b}{a}} = \left(\dfrac{a}{b}\right)^{\frac{a^2+b^2}{a^2-b^2}} \left\{ \left(\dfrac{a}{b}\right)^{\frac{a^2-b^2}{a^2-b^2}} + \left(\dfrac{a}{b}\right)^{\frac{b^2-a^2}{a^2-b^2}} \right\}$$

$$= \left(\dfrac{a}{b}\right)^{\frac{a^2+b^2}{a^2-b^2}} \left\{ \dfrac{a}{b} + \dfrac{b}{a} \right\};$$

whence the required result at once follows.

§62.

$$(1 - x + x^2 - x^3)^{-1} = \dfrac{1}{1 - x + x^2 - x^3}$$

$$= \dfrac{1+x}{1-x^4} = (1+x)(1-x^4)^{-1}$$

$$= (1+x)(1 + x^4 + \ldots + x^{4n-4} + x^{4n} + x^{4n+4} + \ldots).$$

Thus the coefficient of x^{4n} is unity.

§63. By simplifying each side separately, we have

$$\dfrac{a(x-a) + b(x-b)}{ab} = \dfrac{b(x-b) + a(x-a)}{(x-a)(x-b)};$$

hence the numerators being equal, the denominator must be equal; thus

$$a(x-a) + b(x-b) = 0, \text{ or } (x-a)(x-b) = ab.$$

§64. If x is the common difference of the A.P., we have $b = a + (n-1)x$. Similarly if y is the common difference of the reciprocal A.P.,

$$\dfrac{1}{b} = \dfrac{1}{a} + (n-1)y; \text{ whence } y = \dfrac{a-b}{ab(n-1)}.$$

Hence the rth term of the A.P. $=$

$$a + \dfrac{(r-1)(b-a)}{n-1} = \dfrac{a(n-r) + b(r-1)}{n-1};$$

and the $(n-r+1)^{th}$ term of the reciprocal A.P.

$$= \dfrac{1}{a} + \dfrac{(n-r)(a-b)}{ab(n-1)} = \dfrac{a(n-r) + b(r-1)}{n-1}.$$

Hence the product required

$$= \dfrac{a(n-r) + b(r-1)}{n-1} \times \dfrac{ab(n-1)}{a(n-r) + b(r-1)} = ab.$$

§65. Applying the condition for equal roots, we have
$$p^2(1+q)^2 = \{p^2 - 2(q-1)\}\{p^2 + 2q(q-1)\};$$
that is,
$$p^2(1+q)^2 = p^4 + p^2(2q^2 - 4q + 2) - 4q(q-1)^2;$$
or
$$p^4 + p^2(q^2 - 6q + 1) - 4q(q-1)^2 = 0;$$
thus
$$(p^2 - 4q)\{p^2 + (q-1)^2\} = 0;$$
and as the last factor is positive, we must have $p^2 - 4q = 0$.

§66. We have
$$(a+b)^2 = 9ab;$$
that is,
$$a + b = 3\sqrt{ab}, \text{ or } \frac{1}{3}(a+b) = \sqrt{ab};$$

$$\therefore \log\left\{\frac{1}{3}(a+b)\right\} = \log(\sqrt{ab}) = \frac{1}{2}\log(ab) = \frac{1}{2}(\log a + \log b).$$

§67. Let d be the common difference of the reciprocal A.P.; then
$$\frac{1}{c} = \frac{1}{a} + (n+1)d; \text{ whence } d = \frac{a-c}{ac(n+1)}.$$
Hence the first and last means of the reciprocall A.P. are
$$\frac{1}{a} + \frac{a-c}{ac()n+1}, \text{ and } \frac{1}{a} + \frac{n(a-c)}{ac(n+1)}.$$
Thus the difference between the first and last mean of the H.P.
$$= ac(n+1)\left\{\frac{1}{a+nc} - \frac{1}{c+na}\right\}$$
$$= \frac{ac(n+1)(a-c)(n-1)}{n^2ac + n(a^2 + c^2) + ac}$$
$$= ac(a-c),$$
provided that
$$n^2ac + n(a^2 + c^2) + ac = n^2 - 1;$$
that is, if
$$n^2(1 - ac) - n(a^2 + c^2) - (1 + ac) = 0.$$

§68. We have
$$\frac{(n+2)(n+1)n(n-1)}{\underline{|8}} : 1 = 57 : 16;$$

that is,
$$(n+2)(n+1)n(n-1) = \frac{57\underline{|8}}{16} = 57 \cdot 7 \cdot 6 \cdot 5 \cdot 4 \cdot 3;$$
hence the product of the four consecutive integers
$$= 19 \cdot 7 \cdot 6 \cdot 5 \cdot 4 \cdot 3 \cdot 3 = 21 \cdot 20 \cdot 19 \cdot 18.$$
Hence $n + 2 = 21$, and $n = 19$.

§69. Suppose that £100 stock was issued at £x, then the actual rate of interest would be $\dfrac{100}{x} \times 6\frac{1}{2}$.

If the loan had been issued at £$(x-3)$, the rate of interest would have been $\dfrac{100}{x-3} \times 6\frac{1}{2}$.

Hence
$$\frac{650}{x-3} - \frac{650}{x} = \frac{1}{3};$$
that is,
$$9 \times 650 = x(x-3); \text{ whence } x = 78.$$

§70. From the identities
$$(a+b)^3 - a^3 - b^3 = 3ab(a+b), \text{ and } (a-b)^3 - a^3 + b^3 = -3ab(a-b),$$
we have
$$(x^2 + x + 1)^3 - (x^2 + 1)^3 - x^3 = 3x(x^2 + 1)(x^2 + x + 1);$$
$$(x^2 - x + 1)^3 - (x^2 + 1)^3 + x^3 = 3x(x^2 + 1)(x^2 - x + 1);$$
$$(x^4 + x^2 + 1)^3 - (x^4 + 1)^3 - x^6 = 3x^2(x^4 + 1)(x^4 + x^2 + 1);$$
$$\therefore x^2(x^2 + 1)^2(x^2 + x + 1)(x^2 - x + 1) = x^2(x^4 + 1)(x^4 + x^2 + 1);$$
but
$$x^4 + x^2 + 1 = (x^2 + x + 1)(x^2 - x + 1);$$

$$\therefore x = 0, \ x^2 + x + 1 = 0, \ x^2 - x + 1, \text{ and } (x^2 + 1)^2 = x^4 + 1;$$

whence the solution is easily obtained.

§71. From the second equation, we have $y(x + l) = -(lx + m)$; hence by substituting in the first equation,

$$(lx + m)^2 - a(x + l)(lx + m) + b(x + l)^2 = 0,$$

or

$$x^2(l^2 - al + b) + (2lm - al^2 - am + 2bl) + (m^2 - alm + bl^2) = 0.$$

This equation is equivalent to $x^2 + ax + b = 0$,

if $\qquad \dfrac{l^2 - al + b}{1} = \dfrac{2lm - al^2 - am + 2bl}{a} = \dfrac{m^2 - alm + bl^2}{b}.$

From these equations we have $b(l^2 - al + b) = m^2 - alm + bl^2$, that is,

$$al(b - m) - (b^2 - m^2) = 0, \text{ or } (b - m)(al - b - m) = 0.$$

Therefore either $b = m$, or $b + m = al$.

If we put $b = m$, by equating the first two fractions we obtain

$$a(l^2 - al + m) = 4lm - al^2 - am,$$

or

$$a^2 l - 2a(l^2 + m) + 4lm = 0;$$

that is,

$$(a - 2l)(al - 2m) = 0,$$

or

$$a = 2l, \text{ or } al = 2m.$$

Thus

$$\text{either } b = m \text{ and } a = 2l,$$

$$\text{or } b = m \text{ and } al = 2m;$$

and these last two conditions are equivalent to the single condition $b + m = al$ which was obtained before.

§72.

(1) On reduction we have $3 \cdot 6^{2x} - 10 \cdot 6^x + 3 = 0$; whence $6^x = 3$ or $\dfrac{1}{3}$;

thus $\qquad x = \pm \dfrac{\log 3}{\log 6}, \text{ or } x = \pm \dfrac{47712}{77815} = \pm \cdot 614 \text{ nearly.}$

(2) On reduction, we have $10 \cdot 5^x - 29 \cdot 5^{\frac{x}{2}} + 10 = 0$;

whence $\qquad 5^{\frac{x}{2}} = \dfrac{5}{2} \text{ or } \dfrac{2}{5};$

thus $\qquad \dfrac{x}{2} = \pm \dfrac{\log 5 - \log 2}{\log 5} = \pm \dfrac{1 - 2\log 2}{1 - \log 2};$

whence $\qquad x = \pm \dfrac{79588}{69897} = \pm 1 \cdot 139 \text{ nearly.}$

§73. We have $\qquad x + y = 9 \text{ and } x^4 + y^4 = 2417;$

hence $\qquad 4x^3 y + 6x^2 y^2 + 4xy^3 = 9^4 - 2417 = 4144;$

or $\qquad xy(2x^2 + 3xy + 2y^2) = 2072;$

but $\qquad 2x^2 + 3xy + 2y^2 = 2(x + y)^2 - xy = 162 - xy;$

hence $\qquad xy(162 - xy) = 2072, \text{ or } (xy - 14)(xy - 148) = 0.$

The only admissible solution is obtained from $xy = 14$ and $x + y = 9$, which give $x = 7, y = 2$.

§74. Suppose that n is the number of hours; then A has walked $11 + 4n$ miles, while B has walked $\dfrac{n}{2}\left\{9 + (n - 1)\dfrac{1}{4}\right\}$ or $\dfrac{n(n + 35)}{8}$ miles.

Thus $\dfrac{n(n + 35)}{8} = 11 + 4n$; that is, $n^2 + 3n - 88 = 0$; whence $n = 8$.

§75. The expression $(\sqrt{3}+1)^{2m} + (\sqrt{3}-1)^{2m}$ is an integer and is therefore greater by 1 than the greatest integer in $(\sqrt{3}+1)^{2m}$, since $(\sqrt{3}-1)^{2m} < 1$. Hence the integer in question

$$= (\sqrt{3} + 1)^{2m} + (\sqrt{3} - 1)^{2m}$$

$$= (4 + 2\sqrt{3})^m + (4 - 2\sqrt{3})^m$$
$$= 2^m [(2 + \sqrt{3})^m + (2 - \sqrt{3})^m]$$
$$= 2^{m+1} \left[2^m + 2^{m-2} \cdot \frac{m(m-1)}{2} \cdot 3 + \ldots \right];$$

and is therefore a multiple of 2^{m+1}.

§76. The sum of the series $1, 3, 5, 7, \ldots$ to x terms is x^2, hence in the n groups there are n^2 terms. It will be observed that the last terms of the first, second, third, $\ldots$ groups are $1^2, 2^2, 3^2, \ldots$; hence the last term of the $(n-1)^{th}$ group is $(n-1)^2$; thus the first term of the nth group is $(n-1)^2 + 1$, and the number of terms in this group is $n^2 - (n-1)^2 = 2n - 1$.

Therefore the sum $= \dfrac{(2n-1)}{2} \{2(n-1)^2 + 2 + (2n-2)(1)\}$

$$= (2n-1)\{(n-1)^2 + n\} = 2n^3 - 3n^2 + 3n - 1.$$

§77. We have $(1 - x)^{\frac{1}{2}} = 1 - \dfrac{1}{2}x - \dfrac{\frac{1}{2} \cdot \frac{1}{2}}{\lfloor 2}x^2 - \dfrac{\frac{1}{2} \cdot \frac{1}{2} \cdot \frac{3}{2}}{\lfloor 3}x^3 - \ldots;$

also $\qquad\qquad (1 - x)^{-1} = 1 + x + x^2 + x^3 + \ldots;$

By multiplying together the two series on the right, we see that the coefficient of x^n in the product is $1 - S$; hence

$1 - S =$ the coefficient of x^n in $(1 - x)^{\frac{1}{2}} \times (1 - x)^{-1}$

$\qquad\quad =$ the coefficient of x^n in $(1 - x)^{-\frac{1}{2}}$

$$= \frac{1 \cdot 3 \cdot 5 \cdot 7 \ldots (2n - 1)}{2^n \lfloor n}.$$

§78. We have

$$\frac{1 + 2x}{1 - x + x^2} = \frac{(1 + 2x)(1 + x)}{1 + x^3}$$

$$= \frac{1 + 3x + 2x^2}{1 + x^3} = (1 + 3x + 2x^2)(1 + x^3)^{-1}$$

$$= (1 + 3x + 2x^2)\{1 - x^3 + x^6 + \ldots + (-1)^m x^{3m} + \ldots\}.$$

If $n = 3m$, the coefficient of $x^n = (-1)^m = (-1)^{\frac{n}{3}}$.

If $n = 3m + 1$, the coefficient of $x^n = 3(-1)^m = 3(-1)^{\frac{n-1}{3}}$.

If $n = 3m + 2$, the coefficient of $x^n = 2(-1)^m = 2(-1)^{\frac{n-2}{3}}$.

§79.

(1) Putting $x = ak, y = bk, z = ck$, we have $\dfrac{abck^3}{(a + b + c)k} = k$;

whence $\qquad\qquad\qquad k = 0$, or $\dfrac{a + b + c}{abc}$.

Thus $\qquad\qquad \dfrac{x}{a} = \dfrac{y}{b} = \dfrac{z}{c} = 0$, or $\dfrac{a + b + c}{abc}$.

(2) Equating the first two fractions, we have

$$x^2(y - z) + y^2(z - x) + z^2(x - y) = 0;$$

that is $\qquad\qquad (y - z)(z - x)(x - y) = 0.$

Putting $y - z = 0$, or $y = z$, we obtain

$$\frac{x}{y} + 1 + \frac{y}{x} = x + 2y = 3;$$

thus $x^2 - 2xy + y^2 = 0$, and $x + 2y = 3$; whence $x = y = 1$.

§80. The three arithmetic means between a and b are

$$\frac{3a + b}{4} \cdot \frac{a + b}{2} \cdot \frac{a + 3b}{4}.$$

Similarly the three arithmetic means between $\dfrac{1}{a}$ and $\dfrac{1}{b}$ are

$$\frac{a+3b}{4ab}, \frac{a+b}{2ab}, \frac{3a+b}{4ab}.$$

Hence we have

$$\frac{(3a+b)(a+b)(a+3b)}{32} = 7\frac{1}{2};$$

and

$$\frac{32a^3b^3}{(a+3b)(a+b)(3a+b)} = 3\frac{3}{5}.$$

Multiplying these equations together, we find that $a^3b^3 = 27$, or $ab = 3$.

Also $$(3a+b)(a+b)(a+3b) = 240;$$

that is $$(a+b)(3a^2+10ab+3b^2) = 240;$$

or $$(a+b)\{3(a+b)^2+4ab\} = 240.$$

Thus $(a+b)^3 + 4(a+b) - 80 = 0$; whence $a+b = 4$. Also $ab = 3$.

§81. Putting $x - a = u$ and $y - b = v$, we have

$$av - bu = c\sqrt{u^2+v^2}; \text{ that is, } (av-bu)^2 = c^2(u^2+v^2),$$

or $$(c^2-b^2)u^2 + 2abuv + (c^2-a^2)v^2 = 0.$$

For real roots we must have $a^2b^2 > (c^2-a^2)(c^2-b^2)$;

that is $$0 > c^2(c^2-a^2-b^2); \text{ hence } c^2 < a^2+b^2.$$

§82. If $(x+1)^2 > 5x-1$, then x^2-3x+2 or $(x-2)(x-1)$ is positive, so that x cannot lie between 1 and 2.

If $(x+1)^2 < 7x-3$, then x^2-5x+4 or $(x-4)(x-1)$ is negative, so that x must lie between 1 and 4. Thus $x = 3$.

§83. Since the logarithms of all numbers between 10^p and 10^{p+1} have characteristics p, we have

$$P = 10^{p+1} - 10^p = 10^p(10-1) = 9 \times 10^p.$$

Again since the logarithms of all fractions between $\dfrac{1}{10^{q-1}}$ and $\dfrac{1}{10^q}$ have characteristic $-q$, we see that $Q = 10^q - 10^{q-1} = 9 \times 10^{q-1}$.

Hence $\dfrac{P}{Q} = 10^{p-q+1}$, and therefore $\log P - \log Q = p-q+1$.

§84. The number of ways is equal to the coefficient of x^{20} in the expansion of $(x^3+x^4+x^5+\ldots)^5$; that is, to the coefficient of x^5 in $(1+x+x^2+\ldots)^5$.

This last expression is equal to $\dfrac{1}{(1-x)^5}$ or $(1-x)^{-5}$.

Hence the number of ways $= \dfrac{5 \cdot 6 \cdot 7 \cdot 8 \cdot 9}{1 \cdot 2 \cdot 3 \cdot 4 \cdot 5} = 126$.

§85. Denote the sums invested by £x and £$(x-3500)$.

The elder daughter receives the accumulated simple interest on £x for 4 years, the rate of interest being £4 on every £88; hence she receives

$$£x \times 4 \times \frac{4}{88}.$$

Similarly the younger daughter receives £$(x-3500) \times 7 \times \dfrac{3}{63}$;

thus $$\frac{2x}{11} = \frac{x-3500}{3}; \text{ whence } x = 7700.$$

§86. In the scale of 7 let the digits beginning from the left be x, y, z;

then $$49x + 7y + z = 81z + 9y + x;$$

that is $$24x - y - 40z = 0; \text{ or } y = 8(3x-5z).$$

Now y must be less than 7, and $3x - 5z$ is an integer; hence $3x - 5z$ must be equal to zero, and therefore $y = 0$. Again $\dfrac{x}{5} = \dfrac{z}{3} = k$ say; and thus $x = 5k, z = 3k$. But x and z are less than 7; hence $k = 1$; that is, $x = 5$ and $y = 3$.

§87. The sum of $m+n$ terms, and the sum of $m+p$ terms are each double of the sum of m terms; thus

$$\frac{m+n}{2}\{2a + (m+n-1)d\} = \frac{m+p}{2}\{2a + (m+p-1)d\}$$
$$= m\{2a + (m-1)d\} = \text{ s suppose.}$$

Therefore

$$2a + (m+n-1)d = \frac{2s}{m+n}, \text{ and } 2a + (m-1)d = \frac{s}{m};$$

whence $\qquad nd = s\left(\frac{2}{m+n} - \frac{1}{m}\right) = \frac{(m-n)s}{m(m+n)}.$

Similarly $\qquad pd = \frac{(m-p)s}{m(m+p)}.$

Hence $\qquad \dfrac{n}{p} = \dfrac{m(m-n)(m+p)}{m(m+n)(m-p)}$; or

$$\frac{(m+n)(m-p)}{mp} = \frac{(m+p)(m-n)}{mn}.$$

§88. Put $y - z = u, z - x = v, x - y = w$, so that $u + v + w = 0$;

then $\qquad \dfrac{1}{vw} + \dfrac{1}{wu} + \dfrac{1}{uv} = 0.$

Thus $\qquad \dfrac{1}{u^2} + \dfrac{1}{v^2} + \dfrac{1}{w^2} = \dfrac{1}{u^2} + \dfrac{1}{v^2} + \dfrac{1}{w^2} + \dfrac{2}{vw} + \dfrac{2}{wu} + \dfrac{2}{uv};$

or $\qquad \dfrac{1}{u^2} + \dfrac{1}{v^2} + \dfrac{1}{w^2} = \left(\dfrac{1}{u} + \dfrac{1}{v} + \dfrac{1}{w}\right)^2.$

§89. $\dfrac{1^m + 3^m + \ldots + (2n-1)^m}{n} > \left\{\dfrac{1 + 3 + 5 + \ldots + (2n-1)}{n}\right\}^m$; that is, $>$ n^m.

§90. Suppose that the three equations are equivalent to

$$(x - \beta)(x - \gamma) = 0, \ (x - \gamma)(x - \alpha) = 0, \ (x - \alpha)(x - \beta) = 0;$$

then $\qquad \beta + \gamma = p_1, \ \gamma + \alpha = p_2, \alpha + \beta = p_3;$

$$\beta\gamma = q_1, \ \gamma\alpha = q_2, \ \alpha\beta = q_3.$$

Thus $\qquad p_1^2 - 4q_1 = (\beta - \gamma)^2 = (p_2 - p_3)^2;$

that is, $4q_1 = p_1^2 - p_2^2 - p_3^2 + 2p_2p_3.$

Hence

$$4(q_1 + q_2 + q_3) = 2(p_2p_3 + p_3p_1 + p_1p_2) - p_1^2 - p_2^2 - p_3^2.$$

§91. Let $x = $ the common rate of A and B in miles per hour, and suppose that B starts y hours after A. Then when A is at L or at any previous instant B is xy miles behind A.

Now the rate of approach of B and the geese is $x - \dfrac{3}{2}$ miles per hour; therefore we may say that at this rate xy miles are covered while the geese go 5 miles at $\dfrac{3}{2}$ miles per hour.

Hence $\qquad \dfrac{xy}{x - \dfrac{3}{2}} = \dfrac{10}{3}$ $\qquad\qquad (77)$

Again when A meets the wagon, B is xy miles behind, and A and the wagon are $50 - 2x$ miles from L. When B meets the waggon, he is $31 + \dfrac{2}{3}x$ miles from L. Therefore the wagon has travelled in the interval $\left(31 + \dfrac{2}{3}x\right) - (50 - 2x)$ miles. And since the rate of approach of B and the wagon is

$x + \dfrac{9}{4}$ miles per hour we may say that xy miles are covered at this rate while $\dfrac{8}{3}x - 10$ miles are covered at $\dfrac{9}{4}$ miles per hour.

$$\text{Hence} \qquad \frac{xy}{x + \dfrac{9}{4}} = \frac{\dfrac{8}{3}x - 19}{\dfrac{9}{4}} \qquad (78)$$

By equating the values of xy from (77) and (78), we get a simple equation in x which gives $x = 9$; whence $y = \dfrac{25}{9}$ and $xy = 25$.

§92. Since $d = -(a + b + c)$, we have

$$abc + bcd + cda + dab = abc - (bc + ca + ab)(a + b + c)$$
$$= -(b + c)(c + a)(a + b)$$
$$= \sqrt{(a+b)(a+c)(b+c)(b+a)(c+a)(c+b)}.$$

Now $(a+b)(a+c) = a(a+b+c) + bc = bc - ad$; hence the required result follows at once.

§93. For the A.P. the common difference is $b - a$; hence the $(n+2)^{th}$ term is $a + (n+1)(b - a) = -na + (n+1)b$.

For the G.P. the common ratio is $\dfrac{b}{a}$; hence the $(n + 2)^{th}$ term is

$$a\left(\frac{b}{a}\right)^{n+1} = \frac{b^{n+1}}{a^n}.$$

For the reciprocal A.P., the $(n+2)^{th}$ term is $-\dfrac{n}{a} + \dfrac{n+1}{b}$; and therefore the $(n+2)^{th}$ term of the H.P. is $\dfrac{ab}{(n+1)a - nb}$.

When the three means are in G.P. we have

$$\frac{\{-na + (n+1)b\}ab}{(n+1)a - nb} = \frac{b^{2n+2}}{a^{2n}};$$

that is,

$$\frac{-na + (n+1)b}{(n+1)a - nb} = \frac{b^{2n+1}}{a^{2n+1}};$$

or

$$(n+1)\{ab^{2n+1} - a^{2n+1}b\} = n(b^{2n+2} - a^{2n+2}).$$

§94.

We have

$$\frac{x}{(x-a)(x-b)} = \frac{1}{a-b}\left(\frac{a}{x-a} - \frac{b}{x-b}\right)$$
$$= \frac{1}{a-b}\left\{-\left(1 - \frac{x}{a}\right)^{-1} + \left(1 - \frac{x}{b}\right)^{-1}\right\}.$$

Thus the coefficient of x^n is $\dfrac{1}{a-b}\left(-\dfrac{1}{a^n} + \dfrac{1}{b^n}\right) = \dfrac{a^n - b^n}{a^n b^n (a - b)}.$

We have

$$\frac{(1 + x^2)^n}{(1 - x)^3} = \frac{\{(1 - x)^2 + 2x\}^n}{(1 - x)^3}.$$

Expanding the numerator by the Binomial Theorem, and dividing each term of the expansion by $(1 - x)^3$, we obtain

$$(1 - x)^{2n-3} + 2nx(1 - x)^{2n-5} + \ldots + \frac{n(n-1)}{2}(1 - x)(2x)^{n-2} + \frac{n}{1 - x}(2x)^{n-1} + \frac{(2x)^n}{(1 - x)^3}.$$

Hence the coefficient of x^{2n} must come from the last two terms, and therefore is equal to $n2^{n-1} + \dfrac{2^n(n+1)(n+2)}{2}$ or $2^{n-1}(n^2 + 4n + 2)$.

§95. We have $15x^2 - 34xy + 15y^2 = 0$; whence $(5x - 3y)(3x - 5y) = 0$.

On reduction, the first equation gives $2(x - y) + \sqrt{x^2 - y^2} = 2(x - 1)$; that is, $\sqrt{x^2 - y^2} = 2(y - 1)$; whence $x^2 = 5y^2 - 8y + 4$.

Putting $5x = 3y$, we have $9y^2 = 25(5y^2 - 8y + 4)$, or $29y^2 - 50y + 25 = 0$; whence $\qquad\qquad\qquad 29y = 25 \pm 10\sqrt{-1}$.

Putting $3x = 5y$, we have $25y^2 = 9(5y^2 - 8y + 4)$, or $5y^2 - 18y + 9 = 0$; whence $y = 3$ or $\dfrac{3}{5}$.

§96. Let x denote the value of the continued fraction; then

$$x - 1 = \frac{1}{3+}\frac{1}{2+}\frac{1}{3+}\frac{1}{2+}\cdots = \frac{1}{3+}\frac{1}{2 + x - 1};$$

that is, $\qquad\qquad x - 1 = \dfrac{1}{3+}\dfrac{1}{x + 1} = \dfrac{x + 1}{3x + 4};$

or $3x^2 + x - 4 = x + 1$, and $3x^2 - 5 = 0$.

§97. The first part easily follows from Art. 69; but may be proved directly as follows:

$$n^3 = n^2 \cdot n = n^2 \left\{ \left(\frac{n+1}{2}\right)^2 - \left(\frac{n-1}{2}\right)^2 \right\}$$

$$= \frac{n^2(n+1)^2}{4} - \frac{n^2(n-1)^2}{4}.$$

This holds whether n is odd or even: but if n is odd, we also have

$$n^3 = \left(\frac{n^3+1}{2}\right)^2 - \left(\frac{n^3-1}{2}\right)^2,$$

which shows that there is a second way.

Finally $(n+1)^3 - n^3 = 3n^2 + 3n + 1 = k$ say; but $k = \left(\dfrac{k+1}{2}\right)^2 - \left(\dfrac{k-1}{2}\right)^2$; and $k = 3n(n+1) + 1$, and is therefore an odd integer, since $n(n+1)$ is even; hence both $\dfrac{k+1}{2}$ and $\dfrac{k-1}{2}$ are integers.

§98.

Here
$$2S = \frac{2}{\lfloor 3} + \frac{4}{\lfloor 5} + \frac{6}{\lfloor 7} + \frac{8}{\lfloor 9} + \dots$$

$$= \frac{3-1}{\lfloor 3} + \frac{5-1}{\lfloor 5} + \frac{7-1}{\lfloor 7} + \frac{9-1}{\lfloor 9} + \dots$$

$$= \frac{1}{\lfloor 2} - \frac{1}{\lfloor 3} + \frac{1}{\lfloor 4} - \frac{1}{\lfloor 5} + \frac{1}{\lfloor 6} - \frac{1}{\lfloor 7} + \dots$$

$$= \frac{1}{e}.$$

§99. We have $x = \dfrac{a}{b + y}$, and $y = \dfrac{c}{d + x}$; hence $xy + bx = a$, and $xy + dy = c$.

By subtraction, $bx - dy = a - c$.

§100. Assume for the scale of relation $1 - px - qx^2$.

[See Art. 324.]

Let $\qquad\qquad S = 1 + 5x + 7x^2 + 17x^3 + 31x^4 + \dots;$

then $\qquad -pxS = \qquad -px - 5px^2 - 7px^3 - 17px^4 - \dots;$

$\qquad\qquad -qx^2S = \qquad\qquad -qx^2 - 5qx^3 - 7qx^4 - \dots.$

$\therefore S(1 - px - qx^2) = 1 + (5 - p)x$;

the quantities p and q being given by $5p + q = 7$, and $7p + 5q = 17$; whence $p = 1, q = 2$.

Hence $S = \dfrac{1 + 4x}{1 - x - 2x^2} = \dfrac{2}{1 - 2x} - \dfrac{1}{1 + x}$; and the nth term $= \{2^n + (-1)^n\}x^{n-1}$.

The sum of n terms $= \dfrac{2(1 - 2^n x^n)}{1 - 2x} - \dfrac{1 - (-1)^n x^n}{1 + x}$.

§101. (1) Since $b = \dfrac{2ac}{a + c}$, by substituting we have

$$\frac{a + b}{2a - b} = \frac{a(a + c) + 2ac}{2a(a + c) - 2ac} = \frac{a + 3c}{2a}.$$

Therefore $\dfrac{a + b}{2a - b} + \dfrac{c + b}{2c - b} = \dfrac{a + 3c}{2a} + \dfrac{c + 3a}{2c} = 1 + \dfrac{3}{2}\left(\dfrac{a}{c} + \dfrac{c}{a}\right).$

And this last expression is greater than 4, since $\dfrac{a}{c} + \dfrac{c}{a} > 2.$

(2) We have

$$\begin{aligned}
b^2(a - c)^2 &= b^2(a^2 - 2ac + c^2) \\
&= 2b^2(a^2 + c^2) - b^2(a + c)^2 \\
&= 2b^2(a^2 + c^2) - 4a^2c^2 \\
&= 2\{b^2(a^2 + c^2) + 2a^2c^2 - 2acb(a + c)\},
\end{aligned}$$

since $2ac = b(a + c)$.

Thus $\qquad b^2(a - c)^2 = 2\{c^2(b - a)^2 + a^2(c - b)^2\}.$

§102. The given expression vanishes both when $x = a$, and $x = b$; hence

$$a^3 - 3ab^2 + 2c^3 = 0, \text{ and } 2c^3 - 2b^3 = 0.$$

This latter equation gives $b = c$, since by hypothesis b and c are real; and therefore $a^3 - 3ab^2 + 2b^3 = 0$, that is, $(a - b)^2(a + 2b) = 0$. Hence $a = b$, or $a = -2b$.

§103. Denote the number by $2n - 1, 2n + 1, 2n + 3$.

Now $1 + (2n - 1)^2 + (2n + 1)^2 + (2n + 3)^2 = 12(n^2 + n + 1)$; but $n^2 + n$ is even; hence $n^2 + n + 1$ is odd, and the sum is an odd multiple of 12.

§104. We have $\qquad ax^2 + 2bx + c = a\left(x + \dfrac{b}{a}\right)^2 + \dfrac{ac - b^2}{a}$;

if therefore a is positive, $\dfrac{ac - b^2}{a}$ is the least value of the expression; if a is negative, it is the greatest value.

From the given equation, we have

$$(x^2 - yz)^2 + (y^2 - zx)^2 + (z^2 - xy)^2 = 0;$$

hence $\qquad x^2 - yz = 0, \quad y^2 - zx = 0, \quad z^2 - xy = 0;$

and therefore $\qquad x^2 + y^2 + z^2 - yz - zx - xy = 0;$

that is, $\qquad (y - z)^2 + (z - x)^2 + (x - y)^2 = 0;$

whence the required result at once follows.

§105. By inspection the value of the expression $= \dfrac{\sqrt{1 + x}}{2} - \dfrac{\sqrt{1 - x}}{2}$; and the required result follows at once from the Binomial Theorem, since

$$\begin{aligned}
(1 + x)^{\frac{1}{2}} = 1 &+ \frac{1}{2}x - \frac{1}{2}\cdot\frac{x^2}{4} + \frac{1\cdot 3}{2\cdot 4}\cdot\frac{x^3}{6} \\
&- \frac{1\cdot 3\cdot 5}{2\cdot 4\cdot 6}\cdot\frac{x^4}{8} + \frac{1\cdot 3\cdot 5\cdot 7}{2\cdot 4\cdot 6\cdot 8}\cdot\frac{x^5}{10} - \cdots
\end{aligned}$$

§106.

We have $\qquad\qquad \alpha + \beta = -p \text{ and } \alpha\beta = q.$

also $\qquad\qquad \alpha^{2n} + p^n\alpha^n + q = 0, \text{ and } \beta^{2n} + p^n\beta^n + q = 0;$

whence

$$\alpha^{2n} - \beta^{2n} + p^n(\alpha^n - \beta^n) = 0, \text{ or } \alpha^n + \beta^n + p^n = 0;$$

thus $\qquad \alpha^n + \beta^n + (\alpha + \beta)^n = 0$, since n is an even integer;

and therefore $x^n + 1 + (x+1)^n = 0$, where $x = \dfrac{\alpha}{\beta}$ or $\dfrac{\beta}{\alpha}$.

§107. Denote the values of the continued fractions by x and y; then

$$x - a = \cfrac{b}{2a + (x - a)}; \text{ whence } x^2 = a^2 + b.$$

Similarly $y^2 = c^2 + d^2$; thus $x^2 - y^2 = a^2 + b - c^2 - d$.

§108. Let n be the number of persons; then the number of shillings that the last person receives is

$$1 + 1 + 2 + 3 + \ldots + (n - 1), \text{ or } 1 + \frac{n(n-1)}{2};$$

therefore $\qquad 1 + \dfrac{n(n-1)}{2} = 67$; whence $n = 12$.

The number of shillings distributed $= n + \dfrac{1}{2}\Sigma n(n-1)$

$$= n + \frac{1}{6}(n+1)n(n-1)$$

$$= 298, \text{ since } n = 12.$$

§109. (1) The equation is obviously satisfied by $x = a, y = b, z = c$, and being of the first degree there is only one solution.

or thus, $\qquad (b + c)x + ay + az = 2a(b + c),$

$$bx + (c + a)y + bz = 2b(c + a), \qquad cx + cy + (a + b)z = 2c(a + b).$$

Adding the first two equations together and subtracting the third from their sum, we have $bx + ay = 2ab$;

that is $\qquad \dfrac{x}{a} + \dfrac{y}{b} = 2.$

Similarly we may obtain the equations

$$\frac{y}{b} + \frac{z}{c} = 2, \text{ and } \frac{x}{a} + \frac{z}{c} = 2;$$

whence $\qquad \dfrac{x}{a} = \dfrac{y}{b} = \dfrac{z}{c} = 1.$

(2) Clearing of fractions, we have

$$3(x^2 + y^2)(1 + xy) = 40xy, \text{ and } 10xy(1 + x^2 + y^2) = 33(x^2 + y^2).$$

From these two equations,

$$x^2 + y^2 = \frac{40xy}{3(1 + xy)}, \text{ and } x^2 + y^2 = \frac{10xy}{33 - 10xy};$$

thus $\qquad \dfrac{40xy}{3(1 + xy)} = \dfrac{10xy}{33 - 10xy};$ whence $xy = 0$, or $= 3$.

The case $xy = 0$ may be excluded as the equations are not then satisfied. If $xy = 3$, then $x^2 + y^2 = 10$.

§110. Divide by $a - b$; then it will be sufficient to show that

$$a^{n-1} + a^{n-2}b + a^{n-3}b^2 + \ldots + ab^{n-2} + b^{n-1} > n(ab)^{\frac{n-1}{2}}.$$

This readily follows from the inequalities,

$a^{n-1} + b^{n-1} > 2(ab)^{\frac{n-1}{2}}$; $a^{n-2}b + ab^{n-2} > 2(ab)^{\frac{n-1}{2}}$; $a^{n-3}b^2 + a^2b^{n-3} > 2(ab)^{\frac{n-1}{2}}$;

and so on. [Compare Ex. 89.]

§111. Performing the operation of finding the greatest common measure of 396 and 763, we have

1	396	763	1	;
1	29	367	12	
1	10	19	1	
	1	9	9	

thus $\qquad \dfrac{763}{396} = 1 + \dfrac{1}{1+} \dfrac{1}{12+} \dfrac{1}{1+} \dfrac{1}{1+} \dfrac{1}{1+} \dfrac{1}{9}.$

The successive convergents are $\dfrac{1}{1}, \dfrac{2}{1}, \dfrac{25}{13}, \dfrac{27}{14}, \dfrac{52}{27}, \dfrac{79}{41}$.

Hence $79 \cdot 396 - 41 \cdot 763 = 1$, and therefore

$$948 \cdot 396 - 492 \cdot 763 = 12 = 396x - 763y.$$

Thus $\qquad \dfrac{x - 948}{763} = \dfrac{y - 492}{396} = t$ say;

whence $\qquad x = 948 + 763t; \quad y = 492 + 396t.$

§112. Suppose that A, B, C working alone would do the work in x, y, z days respectively. Then, since $B's$ and $C's$ joint daily work is m times $A's$ daily work,

we have $\qquad \dfrac{m}{x} = \dfrac{1}{y} + \dfrac{1}{z}.$

Therefore $\qquad \dfrac{m+1}{x} = \dfrac{1}{x} + \dfrac{1}{y} + \dfrac{1}{z} = \dfrac{n+1}{y} = \dfrac{p+1}{z}$ similarly;

which proves the first part of the question.

Again $\qquad \dfrac{1}{m+1} = \dfrac{1}{x} \div \left(\dfrac{1}{x} + \dfrac{1}{y} + \dfrac{1}{z} \right);$

$$\therefore \dfrac{1}{m+1} + \dfrac{1}{n+1} + \dfrac{1}{p+1} = 1.$$

And $\qquad \dfrac{m}{m+1} = 1 - \dfrac{1}{m+1}.$

$$\therefore \dfrac{m}{m+1} + \dfrac{n}{n+1} + \dfrac{p}{p+1} = 3 - \dfrac{1}{m+1} + \dfrac{1}{n+1} + \dfrac{1}{p+1} = 2.$$

§113. Let £C denote the constant expenses, B the number of boarders, and £P the profits on each boarder; then since each boarder pays £65 we have, $B(65 - P) = C + mB$, where m is some constant.

If $B = 50$, then $P = 9$; hence $2800 = C + 50m$;

If $B = 60$, then $P = 10\frac{2}{3}$; hence $3260 = C + 60m$;

whence $m = 46, C = 500$.

Putting $B = 80$, we have $80(65 - P) = 500 + (80 \times 46)$; whence $P = 12\frac{3}{4}$.

§114. We have $y = \dfrac{2x}{1 + x^2}$; hence $1 - y^2 = \left(\dfrac{1 - x^2}{1 + x^2} \right)^2.$

Taking logarithms, we have

$$-\log(1 - y^2) = 2\{\log(1 + x^2) - \log(1 - x^2)\};$$

that is, $\qquad y^2 + \dfrac{y^4}{2} + \dfrac{y^6}{3} + \ldots = 4 \left\{ x^2 + \dfrac{x^6}{3} + \dfrac{x^{10}}{5} + \ldots \right\}.$

§115.

We have $\qquad x(a^2 - x^2) = y(a^2 - y^2);$

that is, $\qquad (x - y)(x^2 + xy + y^2 - a^2) = 0.$

(1) Taking $x - y = 0$, we have $x = y = \pm c$; hence from the equation $bx = a^2 - y^2$, or $b^2 x^2 = (a^2 - y^2)^2$, we have $b^2 c^2 = (a^2 - c^2)^2$.

(2) Taking $x^2 + xy + y^2 = a^2$, and combining it with $xy = c^2$, we have

$$x^2 + y^2 = a^2 - c^2.$$

Again we have $\dfrac{x - y}{x^2 - y^2} = \dfrac{1}{b}$; or $x + y = b$, since $x - y$ is not zero.

$$\therefore x^2 + y^2 = b^2 - 2xy = b^2 - 2c^2.$$

By equating the two values of $x^2 + y^2$ we obtain $a^2 + c^2 - b^2 = 0$.

§116. The first result follows at once by putting $x = -1$ in the first of the given relations.

Multiplying together the two given expansions, we see that series (2) is the coefficient of x^{3r} in the product and is therefore equal to the coefficient of x^{3r} in $\{(x^2 + x + 1)(x - 1)\}^{3r}$, that is in $(x^3 - 1)^{3r}$; this is equal to the coefficient of y^r in $(y - 1)^{3r}$, and therefore to $(-1)^r \dfrac{\lfloor 3r}{\lfloor 2r \ \lfloor r}$.

§117. (1) From the second equation we have $(x - a)(y - b) = 0$; hence
$$x = a \text{ or } y = b.$$
Substitute $x = a$ in the first equation; then
$$(a - y)^2 + 2ab = a^2 + by, \text{ or } y^2 - 2ay - by + 2ab = 0;$$
whence
$$y = b, \text{ or } 2a.$$
Similarly if $y = b$, then $x = a$, or $2b$.

(2) From the second and third equations, we have
$$2y^2 - zx = 13 - 4y, \text{ or } zx = 2y^2 + 4y - 13;$$
multiply this by 2 and add to the first equation; thus
$$(x + z)^2 - y^2 = 4y^2 + 8y - 20,$$
and therefore substituting from the third equation, we obtain
$$(2 + y)^2 - y^2 = 4y^2 + 8y - 20, \text{ or } y^2 + y - 6 = 0;$$
whence $y = 2$ or -3.

Substituting these values of y, we find
$$x^2 + z^2 = 10, \text{ and } x + z = 4; \text{ or } x^2 + z^2 = 15, \text{ and } x + z = -1.$$

§118. By taking the n letters in pairs we can form $\dfrac{n(n-1)}{2}$ inequalities of the form $2\sqrt{a_1 a_2} < a_1 + a_2$.

By adding these together we obtain the required result, since in the sum each of the n letters will occur on the right-hand side $n - 1$ times.

Thus $2\sqrt{a_1 a_2} + 2\sqrt{a_1 a_3} + \ldots < (n - 1)(a_1 + a_2 + \ldots + a_n)$.

Divide both sides by $n(n - 1)$; then
$$\frac{\sqrt{a_1 a_2} + \sqrt{a_1 a_3} + \ldots \text{ to } \dfrac{n(n-1)}{2} \text{ terms}}{\dfrac{n(n-1)}{2}} < \frac{a_1 + a_2 + \ldots + a_n}{n},$$
which proves the second part of the question.

§119. We have
$$b^2 x^4 + a^2 y^4 = a^2 b^2 (x^2 + y^2);$$
that is,
$$b^2 x^2 (x^2 - a^2) = a^2 y^2 (b^2 - y^2).$$
But $x^2 - a^2 = b^2 - y^2$; hence $b^2 x^2 = a^2 y^2$,
and
$$(b^2 x^2 - a^2 y^2) = 0, \text{ or } b^4 x^4 + a^4 y^4 = 2a^2 b^2 x^2 y^2.$$
Now
$$b^4 x^6 + a^4 y^6 = (b^4 x^6 + a^4 y^6)(x^2 + y^2)$$
$$= b^4 x^8 + a^4 y^8 + x^2 y^2 (a^4 y^4 + b^4 x^4)$$
$$= b^4 x^8 + a^4 y^8 + 2a^2 b^2 x^4 y^4$$
$$= (b^2 x^4 + a^2 y^4)^2.$$

§120. (1) Here
$$\frac{2r + 1}{r^2 (r + 1)^2} = \frac{1}{r^2} - \frac{1}{(r + 1)^2};$$
hence the sum $= 1 - \dfrac{1}{(n + 1)^2}$.

(2) The series is the sum of the two series,
$$a(x^{n-1} + x^{n-2} + \ldots + x + 1);$$
and
$$b(x^{n-1} + 4x^{n-2} + 9x^{n-3} + \ldots + n^2).$$
The second series is a recurring series whose scale of relation is $\left(1 - \dfrac{1}{x}\right)^3$.

If we multiply the expression in brackets by $1 - \dfrac{3}{x} + \dfrac{3}{x^2} - \dfrac{1}{x^3}$, we shall find that the first terms of the product are $x^{n-1}+x^{n-2}$ and that the other terms are zero with the exception of some at the end. Also

the coefficient of $\dfrac{1}{x} = -3n^2 + 3(n-1)^2 - (n-2)^2 = -(n+1)^2$;

the coefficient of $\dfrac{1}{x^2} = 3n^2 - (n-1)^2 = 2n^2 + 2n - 1$;

the coefficient of $\dfrac{1}{x^3} = -n^2$.

$$\therefore S = a\frac{x^n - 1}{x-1}$$

$$+ \frac{b}{(x-1)^3}\left\{x^{n+2} + x^{n+1} - (n+1)^2 x^2 + (2n^2 + 2n - 1)x - n^2\right\}.$$

§121. Put $\dfrac{x+2}{2x^2 + 3x + 6} = y$; then $2yx^2 + (3y-1)x + 6y - 2 = 0$.

If x is real, $(3y-1)^2 > 8y(6y - 2)$;

that is, $1 + 10y - 39y^2$, or $(1+13y)(1-3y)$ must be positive; hence y cannot

be greater than $\dfrac{1}{3}$.

§122. (1) On reduction the given equation becomes
$$3x^4 + 14x^3 + 21x^2 + 14x + 3 = 0;$$

which is a reciprocal equation. Putting $x + \dfrac{1}{x} = z$,

we have $3z^2 + 14z + 15 = 0$, or $(3z+5)(z+3) = 0$.

Thus $3x^2 + 5x + 3 = 0$, or $x^2 + 3x + 1 = 0$.

(2) We have $3xy = -2z$, $xz = -6y$, $2yz = -3x$; hence by multiplication, $x^2 y^2 z^2 = -6xyz$; and therefore $xyz = 0$ or $xyz = -6$.

The given equations are clearly satisfied when $x = 0$, $y = 0$, $z = 0$.

If $xyz = -6$, we have $3x^2 = -2xyz = 12$; also $6y^2 = -xyz = 6$; and $2z^2 = -3xyz = 18$.

§123. Suppose that a_1, a_2, a_3, a_4 are the coefficients of x^r, x^{r+1}, x^{r+2}, x^{r+3} in the expansion of $(1+x)^n$; then

$$\frac{a_1}{a_1 + a_2} = \frac{1}{1 + \dfrac{a_2}{a_1}} = \frac{1}{1 + \dfrac{{}^nC_{r+1}}{{}^nC_r}} = \frac{1}{1 + \dfrac{n-r}{r+1}} = \frac{r+1}{n+1}.$$

Similarly $\dfrac{a_3}{a_3 + a_4} = \dfrac{r+3}{n+1}$; and $\dfrac{a_2}{a_2 + a_3} = \dfrac{r+2}{n+1}$;

whence the required result follows at once.

§124. (1) Let $\dfrac{x^3 + 7x^2 - x - 8}{(x^2 + x + 1)(x^2 - 3x - 1)} = \dfrac{Ax + B}{x^2 + x + 1} + \dfrac{Cx + D}{x^2 - 3x - 1}$;

then

$$x^3 + 7x^2 - x - 8 = (Ax + B)(x^2 - 3x - 1) + (Cx + D)(x^2 + x + 1) \qquad (79)$$

Put $x^2 = 3x + 1$;

then $x^3 + 7x^2 - x - 8 = x(3x+1) + 7x^2 - x - 8$

$$= 10x^2 - 8 = 30x + 2;$$

and $(Cx + D)(x^2 + x + 1) = (Cx + D)(4x + 2)$

$$= 4C(3x + 1) + 4Dx + 2Cx + 2D$$

$$= (14C + 4D)x + (4C + 2D).$$

Therefore $30x + 2 = (14C + 4D)x + (4C + 2D)$;

that is, $7C + 2D = 15$, and $2C + D = 1$; whence

$$C = \frac{13}{3}, \; D = -\frac{23}{3}.$$

Also by equating coefficients (1),

$$A + C = 1, \text{ and } -B + D = -8; \text{ whence } A = -\frac{10}{3}, \text{ and } B = \frac{1}{2}.$$

Thus

$$\frac{x^3 + 7x^2 - x - 8}{(x^2 + x + 1)(x^2 - 3x - 1)} = \frac{1}{3} \cdot \frac{13x - 23}{x^2 - 3x - 1} - \frac{1}{3} \cdot \frac{10x - 1}{x^2 + x + 1}.$$

(2) Here

$$\frac{3x - 8}{4 - 4x + x^2} = \frac{1}{4}(3x - 8)\left(1 - \frac{x}{2}\right)^{-2}$$

$$= \frac{1}{4}(3x - 8)\left\{1 + \ldots + \frac{rx^{r-1}}{2^{r-1}} + \frac{(r+1)x^r}{2^r}\right\} + \ldots\right\}.$$

Hence the coefficient of $x^r =$

$$\frac{1}{4}\left\{\frac{3r}{2^{r-1}} - \frac{8(r+1)}{2^r}\right\} = \frac{6r - 8(r+1)}{2^{r+2}} = -\frac{r+4}{2^{r+1}}$$

§125. If the scale of relation is $1 - px - qx^2$, we have

$$2 = -\frac{1}{2}p + \frac{5}{4}q; \quad l = 2p - \frac{1}{2}q; \quad 5 = pl + 2q; \quad 7 = 5p + ql.$$

From the first two equations, $9p = 5l + 4$, and $9q = 21 + 16$; hence

$$45 = l(5l + 4) + 2(2l + 16), \text{ or } 5l^2 + 8l - 13 = 0;$$

$$\text{whence } l = 1 \text{ or } -\frac{13}{5}.$$

The value $l = 1$ is the only one which satisfies the fourth equation $7 = 5p + ql$; thus $l = 1, p = 1, q = 2$, and the scale of relation is $1 - x - 2x^2$. Hence the generating function

$$= \frac{\frac{5}{4} - \frac{7}{4}x}{1 - x - 2x^2} = \frac{1}{4}\left(\frac{4}{1+x} + \frac{1}{1-2x}\right);$$

and the general term

$$= \frac{1}{4}\{2^{n-1} + 4(-1)^{n-1}\}x^{n-1} = \{2^{n-3} + (-1)^{n-1}\}x^{n-1}.$$

§126. From the first equations, we have

$$2a(y - z) - 3(y^2 - z^2) = (z - x)^2 - (x - y)^2;$$

or

$$2a - 3(y + z) = 2x - y - z;$$

that is

$$x + y + z = a.$$

Hence

$$2a - 3(z + x) = 2y - z - x;$$

$$\therefore 2a(z - x) - 3(z^2 - x^2) = 2y(z - x) - (z^2 - x^2).$$

But $2az - 3z^2 = (x - y)^2$; hence

$$2ax - 3x^2 = (x - y)^2 - 2y(z - x) + z^2 - x^2 = (y - z)^2.$$

§127. (1) We have $x^2 + xy - 2x + 6 = 0$, and $xy + y^2 - 2y - 9 = 0$; hence by addition, $(x + y)^2 - 2(x + y) - 3 = 0$; from which we find $x + y = 3$ or -1.

By subtraction, we have $(x - y)(x + y) - 2(x - y) + 15 = 0$.

If $x + y = 3$, we have $3(x - y) - 2(x - y) + 15 = 0$, or $x - y = -15$; whence $x = -6, y = 9$.

If $x + y = -1$, we have $-(x - y) - 2(x - y) + 15 = 0$, or $x - y = 5$; whence $x = 2, y = -3$.

(2) Taking logarithms we have

$$\log a(\log a + \log x) = \log b(\log b + \log y),$$

and

$$\log b \log x = \log a \log y.$$

For shortness put $\log a = A$, &c.; then we have

$$AX - BY = B^2 - A^2, \text{ and } BX = AY;$$

whence $X = -A$, and $Y = -B$; or $\log x = -\log a$, and $\log y = -\log b$;

thus

$$x = \frac{1}{a}, y = \frac{1}{b}.$$

§128.

(1)
$$x\sqrt{x^2+a^2} - \sqrt{x^4+a^4} = \frac{x^2(x^2+a^2)-(x^4+a^4)}{x\sqrt{x^2+a^2}+\sqrt{x^4+a^4}}$$

$$= \frac{a^2x^2 - a^4}{x\sqrt{x^2+a^2}+\sqrt{x^4+a^4}};$$

and when $x = \infty$ this becomes $\dfrac{a^2x^2}{x^2+x^2} = \dfrac{a^2}{2}$.

(2)
$$\frac{\sqrt{a+2x}-\sqrt{3x}}{\sqrt{3a+x}-2\sqrt{x}} = \frac{(a+2x)-3x}{\sqrt{a+2x}+\sqrt{3}}\cdot\frac{\sqrt{3a+x}+2\sqrt{x}}{(3a+x)-4x}$$

$$= \frac{1}{3}\cdot\frac{\sqrt{3a+x}+2\sqrt{x}}{\sqrt{a+2x}+\sqrt{3x}} = \frac{1}{3}\cdot\frac{4}{2\sqrt{3}} = \frac{2\sqrt{3}}{9}, \text{ when } x = a.$$

§129. Let x and y denote the two numbers; then $xy = 192$.

Let g denote their greatest common measure and l their least common multiple; then
$$3\tfrac{2}{4}\tfrac{5}{8} = \frac{g+l}{2} \div \frac{2gl}{g+l} = \frac{(g+l)^2}{4gl}.$$

Now $gl = xy = 192$. [See Elementary Algebra, Art. 163.]

Thus
$$(g+l)^2 = 169 \times 16;$$
so that $g + l = 52$. Also $gl = 192$; hence $g = 4$ and $l = 48$; that is, greatest common measure is 4, and the least common multiple is 48.

The numbers may therefore be denoted by $4p$ and $4q$ where p and q have no common factor. The least common multiple is $4pq$; hence $pq = 12$, and therefore $p = 3, q = 4$; or $p = 1, q = 12$.

§130. (1) If $a - b = c$, then $c^3 = a^3 - b^3 - 3ab(a-b) = a^3 - b^3 - 3abc$;

that is,
$$3abc = a^3 - b^3 - c^3.$$

Thus $3\sqrt[3]{2}\cdot\sqrt[3]{13x+37}\cdot\sqrt[3]{13x-37} = (13x+37)-(13x-37)-2 = 72$.

By cubing each side, we have
$$169x^2 - 1369 = 6912;$$

that is,
$$x^2 = 49, \text{ or } x = \pm 7.$$

(2) Multiply the first equation by $-a$, the second by b, and the third by c and add; then $2bc\sqrt{1-x^2} = b^2 + c^2 - a^2$;

that is,
$$4b^2c^2(1-x^2) = (b^2+c^2-a^2)^2,$$

or
$$4b^2c^2x^2 = 2b^2c^2 + 2c^2a^2 + 2a^2b^2 - a^4 - b^4 - c^4.$$

§131. We have $2^{\frac{3}{2}} = (1+1)^{\frac{3}{2}}$

$$= 1 + \frac{3}{2} + \frac{\frac{3}{2}\cdot\frac{1}{2}}{\underline{|2}} - \frac{\frac{3}{2}\cdot\frac{1}{2}\cdot\frac{1}{2}}{\underline{|3}} + \frac{\frac{3}{2}\cdot\frac{1}{2}\cdot\frac{1}{2}\cdot\frac{3}{2}}{\underline{|4}} - \frac{\frac{3}{2}\cdot\frac{1}{2}\cdot\frac{1}{2}\cdot\frac{3}{2}\cdot\frac{5}{2}}{\underline{|5}} + \dots$$

Therefore
$$2\sqrt{2} = 1 + \frac{3}{2} + \frac{3}{8} - 3S = \frac{23}{8} - 3S.$$

§132. Let r be the radix of the scale and suppose that when the number $ar^2 + br + c$ is multiplied by 2 the result is $cr^2 + br + a$. Then remembering that a, b, c must all be less than r and that c must be greater than a, it easily follows that $2c = r + a, 2b + 1 = r + b, 2a + 1 = c$. From the first and third of these equations we see that
$$(ar+c) \times 2 = r(c-1) + r + a = cr + a.$$
Again, $r = 2c - a = 2(2a+1) - a = 3a+1$, and only one out of every three consecutive numbers can be of this form.

§133.

The product
$$= \frac{(1+x^3)(1-x+x^2)}{(1-x)(1+x)(1-x)}$$

$$= \left(\frac{1 - x + x^2}{1 - x}\right)^2 = \left(1 + \frac{x^2}{1 - x}\right)^2$$

$$= 1 + 2x^2(1 - x)^{-1} + x^4(1 - x)^{-2}$$

Hence the coefficient of x^r

$$= 2 + (r - 3) = r - 1.$$

§134. Let x, y be the number of yards in the frontage and depth of the rectangle; then $3x + 2y = 96$, and we have to find the maximum value of xy subject to this restriction.

Now $96 \times 96 = (3x + 2y)^2 = 24xy + (3x - 2y)^2$, and therefore xy is a maximum when $3x - 2y = 0$, value of xy is then $96 \times 96 \div 24$, that is 384.

§135. The expression is of four dimensions, and obviously vanishes when $a = 0, b = 0, c = 0, d = 0$, and therefore must be equal to $kabcd$.

Putting $a = b = c = d = 1$, we have $k = 4^4 - 4 \cdot 2^4 = 192$.

§136. Assume $\qquad x^4 + ax^3 + bx^2 + cx + 1 = \left(x^2 + \frac{a}{2}x + 1\right)^2$,

and $\qquad x^4 + 2ax^3 + 2bx^2 + 2cx + 1 = (x^2 + ax + 1)^2$;

then by equating coefficients we must have

$$b = \frac{a^2}{4} + 2, c = a, 2b = a^2 + 2, 2c = 2a.$$

Thus $\qquad \dfrac{a^2}{2} + 4 = a^2 + 2$, that is, $a^2 = 4$;

hence $\qquad\qquad a = \pm 2, c = \pm 2, b = 3.$

§137. (1) After multiplying up and transposing, we have

$$\sqrt[3]{x + y} = -2\sqrt[3]{x - y}; \text{ that is, } (x + y) = -8(x - y), \text{ or } 9x = 7y.$$

Hence $\dfrac{x}{7} = \dfrac{y}{9} = k$, where $130k^2 = 65$, or $k^2 = \dfrac{1}{2}$.

(2) We have identically $(2x^2 + 1) - (2x^2 - 1) = 2$;

hence by division $\quad \sqrt{2x^2 + 1} - \sqrt{2x^2 - 1} = \sqrt{3 - 2x^2}$;

$$\therefore 4x^2 - 2\sqrt{4x^4 - 1} = 3 - 2x^2, \text{ or } 3(2x^2 - 1) = \sqrt{4x^4 + 1};$$

$$\therefore \sqrt{2x^2 - 1} = 0, \text{ or } 3\sqrt{2x^2 - 1} = \sqrt{2x^2 + 1}.$$

§138. Suppose the number of pounds received for the first lot is expressed by the digits x, y; then the price of each sheep $= \dfrac{10x + y}{10}$ pounds.

The number of pounds received for the second lot is expressed by the digits y, x; and the price of each sheep is $\dfrac{10y + x}{5}$ pounds.

Thus $\dfrac{10x + y}{10} - \dfrac{10y + x}{5} = \dfrac{1}{2}$; that is, $8x - 19y = 5$; whence $x = 3, y = 1$, since x and y are each less than ten.

§139. (1) The sum $= 2n(1 + 2 + 3 + 4 + \ldots) - (1 \cdot 1 + 2 \cdot 3 + 3 \cdot 5 + 4 \cdot 7 + \ldots)$.

Now

$$1 \cdot 1 + 2 \cdot 3 + 3 \cdot 5 + 4 \cdot 7 + \ldots = \Sigma n(2n - 1)$$

$$= 2\Sigma n^2 - \Sigma n$$

$$= \frac{n(n + 1)(2n + 1)}{3} - \frac{n(n + 1)}{2}.$$

Hence $S = \dfrac{2n \cdot n(n + 1)}{2} - \dfrac{n(n + 1)(2n + 1)}{3} + \dfrac{n(n + 1)}{2}$

$$= \frac{n(n + 1)}{2}\left\{2n - \frac{4n + 2}{3} + 1\right\} = \frac{n(n + 1)(2n + 1)}{6}.$$

(2) The general term of the series is $\dfrac{n(n + 1)}{2}$, and we have to find the

value of $\Sigma \dfrac{n^2(n+1)^2}{4}$.

Now $n^2(n+1)^2 = n(n+1)\{(n+2)(n+3) - 4(n+2) + 2\}$;

hence $4S = \dfrac{1}{5}n(n+1)(n+2)(n+3)(n+4) - n(n+1)(n+2)(n+3) + \dfrac{2}{3}n(n+1)(n+2)$

$= \dfrac{1}{15}n(n+1)(n+2)(3n^2 + 6n + 1)$.

(3) The general term of the series is $\dfrac{1}{2}(2n-1)2n$, or $n(2n-1)$.

Hence $\qquad S = \dfrac{n(n+1)(2n+1)}{3} - \dfrac{n(n+1)}{2} = \dfrac{1}{6}n(n+1)(4n-1)$.

§140. Proceeding as in Art. 526, we have identically

$$1 + qy^2 + ry^3 = (1 - \alpha y)(1 - \beta y)(1 - \gamma y).$$

Take logarithms and equate the coefficients of powers of y; then

$$\dfrac{\alpha^2 + \beta^2 + \gamma^2}{2} = -q, \quad \dfrac{\alpha^3 + \beta^3 + \gamma^3}{3} = r,$$

$$\dfrac{\alpha^4 + \beta^4 + \gamma^4}{4} = \dfrac{q^2}{2}, \quad \dfrac{\alpha^5 + \beta^5 + \gamma^5}{5} = -qr;$$

from these equations the required result at once follows.

§141. (1) Substituting for x from the first equation, we have

$$\dfrac{27}{y} = \dfrac{8}{3y-5} + 7;$$

that is, $21y^2 - 108y + 135 = 0$, or $7y^2 - 36y + 45 = 0$; whence $y = 3$ or $\dfrac{15}{7}$.

(2) From the first and third equations, we have

$$x^3 + y^3 + z^3 - 3xyz = 180;$$

dividing this equation by the second, $x^2 + y^2 + z^2 - yz - zx - xy = 12$.

Subtracting this equation from the square of the second, we obtain

$$yz + zx + xy = 71.$$

Thus $x + y + z = 15, yz + zx + xy = 71, xyz = 105$; hence x, y, z are the roots of the cubic equation $t^3 - 15t^2 + 71t - 105 = 0$, and are therefore equal to $3, 5, 7$.

§142. When $x = a$ in the given equation, $y = b + c - a$ in the transformed equation.

Now $b + c - a = a + b + c - 2a = -q - 2a$. If therefore we put $y = -(q + 2x)$ we have only to eliminate x between this and the given equation.

Now $8x^3 + 8qx^2 + 8r = 0$, and $-2x = y + q$. Hence we have

$$(q+y)^3 - 2q(q+y)^2 + 8r = 0.$$

§143. (1) We have

$$xS = \qquad nx + (n-1)x^2 + (n-2)x^3 + \ldots + 2x^{n-1} + x^n;$$
$$S = n + (n-1)x + (n-2)x^2 + (n-3)x^3 + \ldots + x^{n-1};$$

hence $(x-1)S = -n + (x + x^2 + x^3 + \ldots + x^{n-1} + x^n)$.

(2) Let the scale of relation be $1 - px - qx^2 - rx^3$; then

$$S = 3 - x - 2x^2 - 16x^3 - 28x^4 - 676x^5 - \ldots$$
$$-pxS = \quad -3px + px^2 + 2px^3 + 16px^4 + 28px^5 + \ldots$$
$$-qx^2S = \qquad\qquad -3qx^2 + qx^3 + 2qx^4 + 16qx^5 + \ldots$$
$$-rx^3S = \qquad\qquad\qquad -3rx^3 + rx^4 + 2rx^5 + \ldots$$

Thus $\quad S(1 - px - qx^2 - rx^3) = 3 - (3p+1)x - (3q - p + 2)x^2$;

where $\qquad\qquad 2p + q - 3r = 16$;

$$16p + 2q + r = 28;$$
$$28p + 16q + 2r = 676;$$

whence $\qquad\qquad p = -5, q = 50, r = 8$.

Hence
$$S = \frac{3 + 14x - 157x^2}{1 + 5x - 50x^2 - 8x^3}.$$

(3) Applying the method of differences, we have

	6		9		14		23		40
		3		5		9		17	
			2		4		8		

Hence as in Art. 401 we may assume $u_n = a \cdot 2^{n-1} + bn + c$; and we have
$$a + b + c = 6,\ 2a + 2b + c = 9,\ 4a + 3b + c = 14;$$

whence
$$a = 2,\ b = 1,\ c = 3.$$

Thus $\quad u_n = 2^n + n + 3$; and $S_n = (2^{n+1} - 2) + \dfrac{1}{2}n(n+1) + 3n.$

§144. We have from the first equation $a(yz + zx + xy) = xyz$; also from the second and third equations,
$$2(yz + zx + xy) = (x + y + z)^2 - (x^2 + y^2 + z^2) = b^2 - c^2.$$

Now $\quad d^3 = x^3 + y^3 + z^3$
$$= (x + y + z)^3 - 3(x + y + z)(yz + zx + xy) + 3xyz$$
$$= b^3 - 3(yz + zx + xy)(b - a);$$
$$\therefore 2d^3 = 2b^3 - 3(b^2 - c^2)(b - a).$$

Again, $\quad b^3 - d^3 = 3\{(x + y + z)(yz + zx + xy) - xyz\}$
$$= 3(y + z)(z + x)(x + y)$$
which by hypothesis is not zero. Hence b cannot be equal to d.

§145. The first derived function of $3x^4 + 16x^3 + 24x^2 - 16$ is
$$12x(x^2 + 4x + 4);$$
and the H.C.F. of these two expressions is $x^2 + 4x + 4$; hence the first expression contains the factor $(x + 2)^3$; the remaining factor is $3x - 2$.
Thus the roots are $-2, -2, -2, \dfrac{2}{3}$.

§146. From the data, we see that the sum of n terms of the series $1, 3, 5, 7, \ldots$ is equal to the sum of $n - 3$ terms of the series $12, 13, 14, \ldots$;

hence
$$n^2 = \frac{n - 3}{2}\{24 + (n - 4)\};$$

that is, $n^2 - 17n + 60 = 0$; so that $n = 5$ or 12.

§147.

We have
$$x = \frac{1}{3+}\frac{1}{2+}\frac{1}{1+}\frac{1}{x};$$

that is,
$$x = \frac{3x + 2}{10x + 7};$$

hence
$$5x^2 + 2x - 1 = 0,\ \text{and}\ x = \frac{\sqrt{6} - 1}{5}.$$

§148. The given equation may be written
$$(x + a)^3 - 3bc(x + a) + b^3 + c^3 = 0,$$

or $\quad y^3 - 3bcy + b^3 + c^3 = 0,\ \text{where}\ y = x + a.$

By putting $y = s + t$, and proceeding as in Art. 576, we have
$$st = bc,\ \text{and}\ s^3 + t^3 = -(b^3 + c^3);$$

whence $\quad s^3 = -b^3\ \text{and}\ t^3 = -c^3.$

Hence the values of y are $-(b + c),\ -(\omega b + \omega^2 c),\ -(\omega^2 b + \omega c)$.

§149. By Art. 422, it follows that $a^n - a = M(n)$, and $b^n - b = M(n)$; hence $(a^n + b^n) - (a + b) = M(n)$; and therefore by dividing each side by $a + b$, we have $a^{n-1} - a^{n-2}b + a^{n-3}b^2 - \ldots - ab^{n-2} + b^{n-1} - 1 = M(n)$, since $a + b$ is prime to n.

But by Fermat's theorem $a^{n-1} - 1 = M(n)$, and $b^{n-1} - 1 = M(n)$; hence
$$a^{n-2}b - a^{n-3}b^2 + \ldots + ab^{n-2} = M(n) + 1.$$

§150. The generating function
$$= \frac{1 - abx^2}{(1 - ax)^2(1 - bx)^2} = \frac{1}{a - b}\left\{\frac{a}{(1 - ax)^2} - \frac{b}{(1 - bx)^2}\right\}.$$

Therefore $\qquad (a - b)u_n = na^n x^{n-1} - nb^n x^{n-1}.$

The sum of the series $a + 2a^2 x + 3a^3 x^2 + \ldots + na^n x^{n-1}$ is easily found to be
$$\frac{a - na^{n+1}x^n}{1 - ax} + \frac{a^2 x(1 - a^{n-1}x^{n-1})}{(1 - ax)^2}.$$

§151. Here $a^2 + b^2 + c^2 = -2p$, since $a + b + c = 0$, and $bc + ca + ab = p$.

When $x = a$ in the given equation, $y = \dfrac{b^2 + c^2}{a} = \dfrac{a^2 + b^2 + c^2}{a} - a = -\dfrac{2p}{a} - a$

in the transformed equation.

We have therefore to eliminate x between $x^3 + px + q = 0$ and $y = -\dfrac{2p}{x} - x$,

or $x^2 + xy + 2p = 0$.

Eliminating x^3 we have $x^2 y + px - q = 0$.

From the last two equations we obtain
$$\frac{x^2}{-qy - 2p^2} = \frac{x}{2py + q} = \frac{1}{p - y^2};$$
$$\therefore (2py + q)^2 = (y^2 - p)(qy + 2p^2).$$

§152. If $a + b + c = 0$, we have $(a + b + c)^2 = 0$, that is
$$a^2 + b^2 + c^2 = -2(bc + ca + ab); \text{ hence}$$
$$a^4 + b^4 + c^4 + 2b^2 c^2 + 2c^2 a^2 + 2a^2 b^2$$
$$= 4(b^2 c^2 + c^2 a^2 + a^2 b^2) + 8abc(a + b + c);$$

thus $\qquad a^4 + b^4 + c^4 + 2(b^2 c^2 + c^2 a^2 + a^2 b^2)$
$$\therefore (a^2 + b^2 + c^2)^2 = a^4 + b^4 + c^4 + 2(b^2 c^2 + c^2 a^2 + a^2 b^2)$$
$$= 2(a^4 + b^4 + c^4).$$
[Compare $XXXIV.\ b.\ 11.$]

Put $\qquad a = y + z - 2x,\ b = z + x - 2y,\ c = x + y - 2x;$

then $\qquad a^2 + b^2 + c^2 = 6(x^2 + y^2 + z^2 - yz - zx - xy);$

and the required result follows at once.

§153.

(1) Proceeding as in Art. 576, we have
$$y^3 + z^3 = -133, \text{ and } yz = 10;$$

whence we obtain $\qquad y^3 = -125,\ z^3 = -8.$

Thus the real root is $-5 - 2$ or -7; and one of the imaginary roots is
$$-5\omega - 2\omega^2 = 2 - 3\omega = \frac{7 - 3\sqrt{-3}}{2}.$$

(2) The sum of the roots is $a - a + b - b + c = c$; but the sum is also 4;
hence $c = 4$. Removing the factor $x - 4$ corresponding to this root, we
have $x^4 - 10x^2 + 9 = 0$; that is, $(x^2 - 9)(x^2 - 1) = 0$.

§154. Let Q be the quantity of work done by the man in an hour;
P his pay in shillings per hour;
and H the number of hours he works per day;

then by the question $Q \propto \dfrac{P}{\sqrt{H}} = \dfrac{mP}{\sqrt{H}}$, where m is some constant.

Let W represent the whole work; then in the first case he does $\dfrac{W}{54}$ per hour; hence $\dfrac{W}{54} = \dfrac{m \times 1}{\sqrt{9}}$.

Let x be the required number of days; in this case he takes $16x$ hours to do the work; hence $\qquad \dfrac{W}{16x} = \dfrac{m \times 1\frac{1}{2}}{\sqrt{16}}$;

by division, $\qquad \dfrac{16x}{54} = \dfrac{1 \times 4 \times 2}{3 \times 3}$; whence $x = 3$.

§155. From Art. 383, we have $s_n = \dfrac{1}{3}n(n+1)(n+2)$;

and from Art. 386, we have $\sigma_{n-1} = \dfrac{1}{18} - \dfrac{1}{3n(n+1)(n+2)}$.

Hence $\qquad\qquad\qquad \sigma_{n-1} = \dfrac{1}{18} - \dfrac{1}{9s_n}$;

that is, $\qquad\qquad\qquad 18s_n\sigma_{n-1} = s_n - 2$.

§156.

(1) Multiplying the factor $6x - 1$ by 2, the factor $4x - 1$ by 3, and the factor $3x - 1$ by 4, we have
$$(12x - 1)(12x - 2)(12x - 3)(12x - 4) = 120.$$
Putting $12x = y$, we have $(y - 1)(y - 2)(y - 3)(y - 4) = 120.$

or $\qquad\qquad (y^2 - 5y + 4)(y^2 - 5y + 6) = 120;$

that is, $\qquad\qquad (y^2 - 5y)^2 + 10(y^2 - 5y) - 96 = 0;$

whence $\qquad\qquad (y^2 - 5y - 6)(y^2 - 5y + 16) = 0.$

Thus $\qquad\qquad\qquad y = 6, -1, \dfrac{5 \pm \sqrt{-39}}{2}.$

(2) The factors of 585 are $5, 9, 13$; and it will be found that $\dfrac{92}{585} = \dfrac{1}{5} + \dfrac{1}{9} - \dfrac{2}{13}.$

Hence putting $x^2 - 2x = y$, we shall have
$$\dfrac{1}{5} \cdot \dfrac{y - 3}{y - 8} + \dfrac{1}{9} \cdot \dfrac{y - 15}{y - 24} - \dfrac{2}{13} \cdot \dfrac{y - 35}{y - 48} = \dfrac{1}{5} + \dfrac{1}{9} - \dfrac{2}{13};$$

whence $\qquad \dfrac{1}{5} \cdot \dfrac{5}{y - 8} + \dfrac{1}{9} \cdot \dfrac{9}{y - 24} = \dfrac{2}{13} \cdot \dfrac{3}{y - 48};$

that is, $\qquad\qquad \dfrac{1}{y - 8} + \dfrac{1}{y - 24} = \dfrac{2}{y - 48}.$

Thus $\qquad \left(\dfrac{1}{y - 48} - \dfrac{1}{y - 8}\right) + \left(\dfrac{1}{y - 48} - \dfrac{1}{y - 24}\right) = 0;$

or $\qquad\qquad \dfrac{40}{y - 8} + \dfrac{24}{y - 24} = 0,$ whence $y = 18.$

Thus $x^2 - 2x = 18$, and $x = 1 \pm \sqrt{19}.$

§157. Let x be the required number of years; then putting
$$250 \times \left(\dfrac{9}{10}\right)^x = 25, \text{ we have } \left(\dfrac{9}{10}\right)^x = \dfrac{1}{10}.$$

Hence by taking logarithms, $2x \log 3 - 2x = -1$; that is, $x = \dfrac{1}{2 - 2\log 3}.$

§158. The first series
$$= \left(1 - \dfrac{3}{4}\right)^{-\frac{1}{3}} = \left(\dfrac{1}{4}\right)^{-\frac{1}{3}} = 4^{\frac{1}{3}} = 2^{\frac{2}{3}} = \left(\dfrac{1}{2}\right)^{-\frac{2}{3}} = \left(1 - \dfrac{1}{2}\right)^{-\frac{2}{3}} = \text{ the second}$$
series.

§159. We have
$$1 - \dfrac{x}{\alpha} + \dfrac{x(x - \alpha)}{\alpha\beta} = \dfrac{x(x - \alpha)}{\alpha\beta} - \dfrac{x - \alpha}{\alpha} = \dfrac{(x - \alpha)(x - \beta)}{\alpha\beta};$$

and $\qquad \dfrac{(x - \alpha)(x - \beta)}{\alpha\beta} - \dfrac{x(x - \alpha)(x - \beta)}{\alpha\beta\gamma} = -\dfrac{(x - \alpha)(x - \beta)(x - \gamma)}{\alpha\beta\gamma}.$

hence it is easy to see that the value of the first series on the left is

$$\pm \frac{x-\alpha}{\alpha} \cdot \frac{x-\beta}{\beta} \cdot \frac{x-\gamma}{\gamma} \cdot \frac{x-\delta}{\delta} \dots$$

The required result follows from the identity

$$\frac{x-\alpha}{\alpha} \cdot \frac{x-\beta}{\beta} \cdot \frac{x-\gamma}{\gamma} \dots \times \frac{x+\alpha}{\alpha} \cdot \frac{x+\beta}{\beta} \cdot \frac{x+\gamma}{\gamma} \dots$$

$$= \frac{x^2-\alpha^2}{\alpha^2} \cdot \frac{x^2-\beta^2}{\beta^2} \cdot \frac{x^2-\gamma^2}{\gamma^2} \dots$$

§160. We know that $(n-2)(n-1)n(n+1)(n+2)$, or $n(n^4-5n^2+4)$ is a multiple of 120.

Now $n(n^4-5n^2+60n-56)-n(n^4-5n^2+4) = n(60n-60) = 60n(n-1) = M(120)$; whence the result at once follows.

§161. Let x be the time occupied; y the intervals at which they begin work; n the number of men, and m the amount of work done by one man in one hour; then the total work is measured by $24mn$.

The first man works x hours, the second $x-y$, the third $x-2y$, and the last $x-(n-1)y$.

Thus $\qquad xm + (x-y)m + \dots + \{x - (n-1)y\}m = 24mn$;

that is, $\qquad x + (x-y) + (x-2y) + \dots$ to n terms $= 24n$,

or $\qquad\qquad \dfrac{n}{2}($ first term + last term $) = 24n$;

and therefore $\qquad x + \{x - (n-1)y\} = 48.$

But by the question, $\qquad x - (n-1)y = \dfrac{1}{11}x$;

hence $\qquad\qquad x + \dfrac{x}{11} = 48$; that is $x = 44$.

§162.

(1) We have $\dfrac{x}{y^2-3} = \dfrac{y}{x^2-3} = -\dfrac{7}{x^3+y^3} = \dfrac{x-y}{y^2-x^2} = \dfrac{-1}{x+y}$,

supposing that x and y are unequal. If x and y are equal, we have

$$\frac{x}{x^2-3} = -\frac{7}{2x^3}, \text{ or } 2x^4 + 7x^2 - 21 = 0;$$

whence $\qquad\qquad x^2 = \dfrac{-7 \pm \sqrt{217}}{4} = y^2.$

If x and y are unequal, we have

$$x(x+y) = -(y^2-3), \text{ and } 7(x+y) = x^3+y^3.$$

In this case the solutions are obtained from the equations

$$\left. \begin{array}{l} x^2+xy+y^2=3 \\ x+y=0 \end{array} \right\}, \text{ and } \left. \begin{array}{l} x^2+xy+y^2=3 \\ x^2-xy+y^2=7 \end{array} \right\}.$$

(2) We have $\qquad a^2x + b^2y + c^2z = 0$;

also $\qquad a^2y + b^2z + c^2x = x^3 + y^3 + z^3 - 3xyz = a^2z + b^2x + c^2y$;

so that $\qquad x(b^2-c^2) + y(c^2-a^2) + z(a^2-b^2) = 0$;

hence by cross multiplication, we see that x is proportional to

$$b^4 + c^4 - a^2(b^2+c^2).$$

Put $x = k(b^4 + c^4 - a^2b^2 - a^2c^2)$, & c. By subtracting the second of the given equations from the first, we obtain

$$(x-y)(x+y+z) = b^2 - a^2.$$

Substituting for x, y, z in terms of k, we have

$$2k^2(b^2-a^2)(a^2+b^2+c^2)(a^4+b^4+c^4-b^2c^2-c^2a^2-a^2b^2)$$

$$= b^2 - a^2;$$

that is, $\qquad 2k^2(a^6 + b^6 + c^6 - 3a^2b^2c^2) = 1.$

§163.

The coefficient of x^2
$$= a^3(b-c) + b^3(c-a) + c^3(a-b)$$
$$= -(b-c)(c-a)(a-b)(a+b+c).$$

The coefficient of x
$$= -a^3(b^2 - c^2) - b^3(c^2 - a^2) - c^3(a^2 - b^2)$$
$$= (b-c)(c-a)(a-b)(bc + ca + ab)$$

The term independent of x
$$= abc\{a^2(b-c) + b^2(c-a) + c^2(a-b)\}$$
$$= -abc(b-c)(c-a)(a-b).$$

Hence the equation is equivalent to
$$(a+b+c)x^2 - (bc + ca + ab)x + abc = 0.$$

If the roots are equal
$$(bc + ca + ab)^2 - 4abc(a+b+c) = 0;$$

that is, $b^2 c^2 + c^2 a^2 + a^2 b^2 - 2a^2 bc - 2ab^2 c - 2abc^2 = 0;$

or $(bc + ca - ab)^2 = 4a^2 bc;$ hence $bc + ca - ab = \pm 2\sqrt{a^2 bc};$

that is, $\sqrt{bc} \pm \sqrt{ca} = \pm\sqrt{ab};$ or $\dfrac{1}{\sqrt{a}} \pm \dfrac{1}{\sqrt{b}} \pm \dfrac{1}{\sqrt{c}} = 0.$

§164.

(1) The nth term $= n(n+1)(n+3)$
$$= n(n+1)(n+2) + n(n+1)$$

hence $S = \dfrac{1}{4}n(n+1)(n+2)(n+3) + \dfrac{1}{3}n(n+1)(n+2)$
$$= \dfrac{1}{12}n(n+1)(n+2)(3n+13).$$

(2) The nth term $= \dfrac{n^2}{\underline{|n+2}}$
$$= \dfrac{(n+1)(n+2) - 3(n+2) + 4}{\underline{|n+2}}$$
$$= \dfrac{1}{\underline{|n}} - \dfrac{3}{\underline{|n+1}} + \dfrac{4}{\underline{|n+2}}.$$

Hence the sum
$$= \left(\dfrac{1}{\underline{|1}} + \dfrac{1}{\underline{|2}} + \dfrac{1}{\underline{|3}} + \dots\right) - 3\left(\dfrac{1}{\underline{|2}} + \dfrac{1}{\underline{|3}} + \dfrac{1}{\underline{|4}} + \dots\right)$$
$$+ 4\left(\dfrac{1}{\underline{|3}} + \dfrac{1}{\underline{|4}} + \dfrac{1}{\underline{|5}} + \dots\right)$$
$$= (e-1) - 3(e-2) + 4\left(e - 2 - \dfrac{1}{2}\right) = 2e - 5.$$

§165. By Art. 253, we have
$$\dfrac{b+c+d}{3} > (bcd)^{\frac{1}{3}}; \text{ that is } s - a > 3(bcd)^{\frac{1}{3}}.$$

Similarly,
$$s - b > 3(cda)^{\frac{1}{3}}; \; s - c > 3(dab)^{\frac{1}{3}}; \; s - d > (abc)^{\frac{1}{3}}.$$

By multiplying together all these inequalities we have the required result.

§166.

(1) We have $2\sqrt{x + a} = 2\sqrt{y - a} + 5\sqrt{a};$

and $2\sqrt{x - a} = 2\sqrt{y + a} + 3\sqrt{a};$

by squaring and subtracting, we have
$$8a = -8a + 16a + 20\sqrt{a}\sqrt{y-a} - 12\sqrt{a}\sqrt{y+a};$$

whence
$$5\sqrt{y-a} = 3\sqrt{y+a}; \text{ and } y = \frac{17a}{8}.$$

Substituting for y, we have $2\sqrt{x+a} = \left(5 + \dfrac{3}{\sqrt{2}}\right)\sqrt{a}$;

that is,
$$x + a = \frac{59 + 30\sqrt{2}}{8}a; \text{ or } x = \frac{51 + 30\sqrt{2}}{8}a.$$

(2) We have
$$2(yz + zx + xy) = (x+y+z)^2 - (x^2 + y^2 + z^2);$$
whence
$$yz + zx + xy = 3.$$
Again
$$x^2 + y^2 + z^2 - yz - zx - xy = 0;$$
and therefore $x^3 + y^3 + z^3 - 3xyz = 0$, so that $3xyz = 6$, and $xyz = 2$.
Now x, y, z are the roots of the equation
$$t^3 - (x+y+z)t^2 + (yz + zx + xy)t - xyz = 0.$$
Thus x, y, z are the roots of the equation $t^3 - 3t^2 + 3t - 2 = 0$, or $(t-2)(t^2 - t + 1) = 0$;

whence
$$t = 2, \text{ or } \frac{1 \pm \sqrt{-3}}{2}.$$

§167. From these equations we have
$$l\begin{vmatrix} x & y & z \\ z & x & y \\ y & z & x \end{vmatrix} = \begin{vmatrix} 1 & y & z \\ 1 & x & y \\ 1 & z & x \end{vmatrix};$$

that is, $l(x^3 + y^3 + z^3 - 3xyz) = x^2 + y^2 + z^2 - yz - zx - xy$, or $l(x+y+z) = 1$.

Thus
$$l = m = n = \frac{1}{x+y+z};$$

and therefore
$$3k^2 = (x+y+z)^2.$$

§168. It is easily seen that the numerator vanishes for each of the values $a = 0, b = 0, c = 0$. Hence it must be of the form $kabc$ where k is a numerical quantity. Similarly the denominator is of the form $labc$; hence the value of the fraction is some constant quantity m. To find m put $a = b = c = 1$;

then $m = \dfrac{3+1}{3-1} = 2$.

§169. Put $X = x^2 - yz, Y = y^2 - zx, Z = z^2 - xy$; then the given expression is the product of the three factors $X + Y + Z, X + \omega Y + \omega^2 Z, X + \omega^2 Y + \omega Z$.
Now
$$X + Y + Z = x^2 + y^2 + z^2 - yz - zx - xy$$
$$= (x + \omega y + \omega^2 z)(x + \omega^2 y + \omega z).$$
$$X + \omega Y + \omega^2 Z = x^2 - yz + \omega(y^2 - xz) + \omega^2(z^2 - xy)$$
$$= (x + y + z)(x + \omega y + \omega^2 z).$$

Similarly,
$$X + \omega^2 Y + \omega Z = (x + y + z)(x + \omega^2 y + \omega z).$$

Hence the given expression
$$= \{(x+y+z)(x + \omega y + \omega^2 z)(x + \omega^2 + \omega z)\}^2$$
$$= (x^3 + y^3 + z^3 - 3xyz)^2.$$

§170. Since he walks, drives, and rides the same distance in 22 hours, 11 hours, and $8\frac{1}{4}$ hours respectively, his rates of walking, driving and riding must be proportional ro $\dfrac{1}{22}, \dfrac{1}{11}, \dfrac{1}{8\frac{1}{4}}$; that is, to $3, 6, 8$ respectively.

Suppose then that he walks, drives, and rides $3k, 6k, 8k$ miles in the hour; then if x, y, z be the distances AB, BC, CA, we have

$$\frac{x}{3k} + \frac{y}{6k} + \frac{z}{8k} = 15\frac{1}{2}, \ \frac{x}{6k} + \frac{y}{8k} + \frac{z}{3k} = 13, \ \frac{x+y+z}{3k} = 22.$$

Again since he walks, drives, and rides 1 mile in half an hour, we have

$$\frac{1}{3k} + \frac{1}{6k} + \frac{1}{8k} = \frac{5}{2}; \text{ whence } k = \frac{1}{4}.$$

$$\therefore 8x + 4y + 3z = 465, 4x + 3y + 8z = 360 = 360, x + y + z = 82\frac{1}{2};$$

whence $\qquad\qquad\qquad x = 37\frac{1}{2}, y = 30, z = 15.$

§171.

The expression $\quad = n\{(n^6 - 8) - 7n^2(n^2 - 2)\}$

$$= n(n^2 - 2)(n^4 - 5n^2 + 4)$$

$$= n(n^2 - 2)(n - 2)(n - 1)(n + 1)(n + 2);$$

and is therefore divisible by $\underline{5}$ or 120.

If n is a multiple of 7, the given expression is obviously divisible by 7.

If n is not multiple of 7, then $n^6 - 1 = M(7)$, by Fermat's theorem; and therefore $n^6 - 8 = M(7)$. In this case the expression is also divisible by 7; thus it is divisible by 7×120 or 840.

§172.

(1) Substitute $y = 23 - x$ in the first equation; thus

$$\sqrt{x^2 - 12x + 276} + \sqrt{x^2 - 34x + 529} = 33.$$

But

$$(x^2 - 12x + 276) - (x^2 - 34x + 529) = 11(2x - 23) \text{ identically;}$$

hence $\qquad \sqrt{x^2 - 12x + 276} - \sqrt{x^2 - 34x + 529} = \dfrac{2x - 23}{3}.$

By addition, $\qquad 2\sqrt{x^2 - 12x + 276} = \dfrac{2x + 76}{3};$

whence $\qquad\qquad 9(x^2 - 12x + 276) = (x + 38)^2;$

that is, $x^2 - 23x + 130 = 0$, and therefore $x = 13$ or 10.

(2) From the given equations we have

$$\frac{u}{z} = \frac{a}{b}, \frac{y}{x} = \frac{c}{d}, uy = ac, xz = bd.$$

Hence $\qquad\qquad y = \dfrac{cx}{d}, z = \dfrac{bd}{x}, u = \dfrac{ac}{y} = \dfrac{ad}{x}.$

Substitute in the last equation; thus

$$d\left(x - \frac{cx}{d}\right) = x\left(\frac{ad}{x} - \frac{bd}{x}\right) = d(a - b);$$

whence $\qquad\qquad\qquad x = \dfrac{d(a - b)}{d - c}.$

§173. We have

$$(x + y + z + \ldots)\left(\frac{1}{x} + \frac{1}{y} + \frac{1}{z} + \ldots\right) = n + \left(\frac{x}{y} + \frac{y}{x}\right)$$
$$+ \left(\frac{x}{z} + \frac{z}{x}\right) + \ldots;$$

where n is the number of the quantities $x, y, z, \ldots$

On the right side, each of the expressions within brackets is greater than 2; hence the expression on the right is greater than

$n + \left\{2 \times \dfrac{n(n-1)}{2}\right\}$; that is, greater than n^2.

Thus $\qquad (x + y + z + \ldots)\left(\dfrac{1}{x} + \dfrac{1}{y} + \dfrac{1}{z} + \ldots\right) > n^2.$

Put $\qquad x = \dfrac{s-a}{s}, y = \dfrac{s-b}{s}, z = \dfrac{s-c}{s}, \ldots;$

Then $\qquad x+y+z+\ldots = \dfrac{ns-s}{s} = n-1;$

whence on substitution we have the required result.

§174. Suppose that he bought x owt. of cotton, and exchanged each owt. for y gallons of oil, and sold each gallon for z shillings; then he obtained xyz shillings; hence

$$(x+1)(y+1)(z+1) = xyz + 10169;$$
$$(x-1)(y-1)(z-1) = xyz - 9673;$$

and x, y, z are in G.P. so that $xz = y^2$.

From the first two equations,

$$(yz+zx+xy)+(x+y+z) = 10168;$$
$$(yz+zx+xy)-(x+y+z) = 9672;$$

whence $\qquad yz+zx+xy = 9920,$ and $x+y+z = 248.$

But $\qquad yz+zx+xy = yz+y^2+xy = y(x+y+z);$

hence $\qquad 248y = 9920;$ that is, $y = 40.$

Thus $\qquad x+z = 208,$ and $xz = 1600;$

whence $\qquad x = 200, z = 8.$

§175. The expression vanishes when $x = a$, when $x = b$, and when $x = c$; and is therefore divisible by $(x-a)(x-b)(x-c)$.

Again, the expression is of 5 dimensions in x, but the coefficients of both x^5 and x^4 are zero; for the coefficient of $x^5 = -(b-c)-(c-a)-(a-b) = 0$; and the coefficient of x^4

$$= \Sigma\{a(b-c)+4(b-c)(b+c-a)\} = 0.$$

Hence the given expression $= f(a,b,c)(x-a)(x-b)(x-c)$, where $f(a,b,c)$ is a function of a,b,c of three dimensions. Now the given expression vanishes when $b = c, c = a, a = b$; therefore it must be of the form $k(b-c)(c-a)(a-b)(x-a)(x-b)(x-c)$, where k is constant.

By putting $x = 0$, we have

$$\Sigma a(b-c)(b+c-a)^4 = -kabc(b-c)(c-a)(a-b).$$

Finally putting $a = 3, b = 2, c = 1$, we have $12k = 3 \cdot 0^4 - 4 \cdot 2^4 + 4^4 = 192$; whence $k = 16$.

§176. Putting $y = \dfrac{\beta+\gamma}{\alpha}$, we have $y = \dfrac{\alpha+\beta+\gamma}{\alpha} - 1 = \dfrac{p}{\alpha} - 1$; hence if x has the value α, the y has the value $\dfrac{p}{x} - 1$; so that $y = \dfrac{p}{x} - 1$, or $\dfrac{1}{x} = \dfrac{y+1}{p}$.

Substituting in the equation $\dfrac{r}{x^3} - \dfrac{p}{x} + 1 = 0$, we have

$$\frac{r(y+1)^3}{p^3} - y = 0, \text{ or } r(y+1)^3 - p^3 y = 0.$$

§177.

We have $\qquad (a^2+b^2)(c^2+d^2) = a^2c^2 + b^2d^2 + b^2c^2 + a^2d^2$

$$= (ac \pm bd)^2 + (bc \mp ad)^2 \qquad (80)$$

Write A for $ac \pm bd$ and B for $bc \mp ad$; then
$$(a^2+b^2)(c^2+d^2)(e^2+f^2) = (A^2+B^2)(e^2+f^2)$$
$$= (Ae \pm Bf)^2 + (Be \mp Af)^2 \qquad (81)$$

Thus the product of three factors of the given form can be expressed as the sum of two squares, and the same method may be extended to the case of any number of factors. Also we may notice that since there are two pairs of values for A and B and each pair gives two results in (81), we have four pairs of squares whose respective sums are equal to the product of three factors of the given form; and so on for any number of factors.

By the preceding result we have
$$(a^2 + b^2)(c^2 + d^2) = (ac \pm bd)^2 + (bc \mp ad)^2 = A^2 + B^2,$$
$$(e^2 + f^2)(g^2 + h^2) = (eg \pm fh)^2 + (fg \mp eh)^2 = C^2 + D^2, \text{ suppose;}$$
where A, B, C, D have respectively two pairs of values.

Thus
$$(a^2 + b^2)(c^2 + d^2)(e^2 + f^2)(g^2 + h^2) = (A^2 + B^2)(C^2 + D^2)$$
$$= (AC \pm BD)^2$$
$$+ (BC \mp AD)^2$$
$$= p^2 + q^2 \text{ say.}$$

It is clear that each pair of values A, B, with each pair of values of C, D gives us two pairs of values for p, q; thus we have in all eight solutions. One of these, namely that obtained by taking the upper sign throughout, is
$$p = AC + BD = (ac + bd)(eg + fh) + (bc - ad)(fg - eh),$$
$$q = BC - AD = (bc - ad)(eg + fh) - (ac + bd)(fg - eh).$$
[This solution is due to Professor Steggall.]

§178. We have $(x - y)(x^2 + xy + y^2) = 91$; that is $(x - y)(61 + xy) = 91$. Also $(x - y)^2 + 2xy = 61$; hence by putting $x - y = u$, and $xy = v$, we obtain
$$u(61 + v) = 91, \text{ and } u^2 + 2v = 61.$$
Multiply the first equation by 2, and substitute for v; Thus
$$u(183 - u^2) = 182; \text{ or } u^3 - 183u + 182 = 0.$$

Hence $\qquad (u - 1)(u^2 + u - 182) = 0, \text{ or } u = 1, 13, -14;$

and therefore $v = 30, -54, -\dfrac{135}{2}$.

Thus we have $\qquad \left.\begin{array}{l} x - y = 1, \\ xy = 30 \end{array}\right\} ; \quad \left.\begin{array}{l} x - y = 13, \\ xy = -54 \end{array}\right\} ; \quad \left.\begin{array}{l} x - y = -14, \\ xy = -\dfrac{135}{2} \end{array}\right\}.$

§179. The number of ways is equal to the coefficient of x^{2m} in the expansion of $(x^0 + x^1 + x^2 + \ldots + x^m)^4$.

This expression $= \left(\dfrac{1 - x^{m+1}}{1 - x}\right)^4 = (1 - x^{m+1})^4(1 - x)^{-4}.$

Thus we have to find the coefficient of x^{2m} in $(1 - 4x^{m+1})(1 - x)^{-4}$.

The coefficient of x^r in $(1 - x)^{-4}$ is $\dfrac{(r + 1)(r + 2)(r + 3)}{1 \cdot 2 \cdot 3}$;

hence the required coefficient
$$= \frac{1}{6}(2m + 1)(2m + 2)(2m + 3) - \frac{4}{6}m(m + 1)(m + 2)$$
$$= \frac{1}{3}(m + 1)(2m^2 + 4m + 3).$$

§180. We have $\alpha\beta = 1, \alpha + \beta = -p; \gamma\delta = 1, \gamma + \delta = -q;$
hence $\qquad (\alpha - \gamma)(\beta - \gamma) = \alpha\beta - \gamma(\alpha + \beta) + \gamma^2 = \gamma^2 + p\gamma + 1.$

Thus the expression $= (\gamma^2 + p\gamma + 1)(\delta^2 - p\delta + 1)$
$$= \gamma^2\delta^2 + p\gamma\delta(\delta - \gamma) - p^2\gamma\delta$$
$$+ \delta^2 + \gamma^2 - p(\delta - \gamma) + 1$$
$$= 1 + p(\delta - \gamma) - p^2 + \delta^2 + \gamma^2 - p(\delta - \gamma) + 1$$
$$= (\alpha + \delta)^2 - p^2, \text{ since } 2 = 2\gamma\delta,$$
$$= q^2 - p^2.$$

§181. We have $(1 + x)^n = a_0 + a_1x + a_2x^2 + \ldots + a_{m-1}x^{m-1} + \ldots;$
also $\qquad (1 + x)^{-1} = 1 - x + x^2 + \ldots + (-1)^{m-1}x^{m-1} + \ldots;$

hence $(-1)^{m-1} S =$ the coefficient of x^{m-1} in the expansion of $(1+x)^{n-1}$

$$= \frac{(n-1)(n-2)\ldots(\overline{n-1-m-1}+1)}{\underline{|m-1}}.$$

§182. Let a, b, c be the factors of the number; then

$$a^2 + b^2 + c^2 = 2331 \qquad (82)$$

From Art. 431, or by reasoning somewhat in the manner indicated in Art. 432, we see that the number of integers less than the number and prime to it is

$$abc\left(1-\frac{1}{a}\right)\left(1-\frac{1}{b}\right)\left(1-\frac{1}{c}\right);$$

$$\therefore (a-1)(b-1)(c-1) = 7560 \qquad (83)$$

And by Art. 415, $\qquad (a+1)(b+1)(c+1) = 10560 \qquad (84)$

From (83) and (84) by addition and subtraction,

$$abc + a + b + c = 9060 \qquad (85)$$
$$bc + ca + ab + 1 = 1500 \qquad (86)$$

From (82) and (86) $\qquad (a+b+c)^2 = 5329,$

$$\therefore a+b+c = 73;$$

from (4), $\qquad abc = 8987 = 11 \cdot 19 \cdot 43.$

§183.

(1) The roots are $\beta\gamma, \gamma\alpha, \alpha\beta$. Putting $y = \beta\gamma$, we have

$$y = \beta\gamma = \frac{\alpha\beta\gamma}{\alpha} = -\frac{c}{x}; \text{ that is } x = -\frac{c}{y};$$

hence $\qquad -\dfrac{c^3}{y^3} + \dfrac{ac^2}{y^2} - \dfrac{bc}{y} + c = 0.$

(2) We have $\quad 2(x^5 + 1) + x(x^3 + 1) = 12x^2(x+1);$

hence $x = -1$ is a root, and the other roots are given by

$$2(x^4 - x^3 + x^2 - x + 1) + (x^3 - x^2 + x) = 12x^2;$$

or $\qquad 2x^4 - x^3 - 11x^2 - x + 2 = 0.$

Thus $\qquad 2(x^2 + 1)^2 - x(x^2 + 1) - 15x^2 = 0;$

whence $\qquad \{2(x^2+1) + 5x\}\{(x^2+1) - 3x\} = 0.$

§184. The required result at once follows from equating the coefficients of x^n in the expansions of the two series

$$(e^x - e^{-x})^n = 2^n\left(x + \frac{x^3}{\underline{|3}} + \frac{x^5}{\underline{|5}} + \ldots\right)^n;$$

and $\qquad (e^x - e^{-x})^n = e^{nx} - ne^{(n-2)x} + \dfrac{n(n-1)}{\underline{|2}}e^{(n-4)x} + \ldots$

§185. Since $6\sqrt{6} = \sqrt{216}$, it lies between 14 and 15; hence $6\sqrt{6} - 14$ is a proper fraction.

Again $(6\sqrt{6}+14)^{2n+1} - (6\sqrt{6}-14)^{2n+1}$ is an integer, and therefore $(6\sqrt{6}-14)^{2n+1}$ must be equal to the fractional part of N; that is to F; thus $(6\sqrt{6}+14)^{2n+1} = N$, and $(6\sqrt{6}-14)^{2n+1} = F.$

Thus $\qquad NF = (216 - 196)^{2n+1} = 20^{2n+1}.$

§186.

(1) We have $2(yz + zx + xy) = (x+y+z)^2 - x^2 - y^2 - z^2 = 4;$

that is, $\qquad yz + zx + xy = 2.$

Again, $\qquad (x+y+z)^3 = \Sigma x^3 + 3\Sigma x^2 y + 6xyz;$

that is, $\qquad 3\Sigma x^2 y + 6xyz = 8 + 1 = 9.$

But $\qquad \Sigma x^2 y = \Sigma x \Sigma yz - 3xyz = 4 - 3xyz;$

hence $12 - 3xyz = 9$, so that $xyz = 1.$

Now x, y, z are the roots of $t^3 - (x + y + z)t^2 + (yz + zx + xy)t - xyz = 0$; that is, of $t^3 - 2t^2 + 2t - 1 = 0$, or $(t - 1)(t^2 - t + 1) = 0$.

(2) We have $(x - y + z)(x + y - z) = a^2$;
$$(-x + y + z)(x + y - z) = b^2; \quad (x - y + z)(-x + y + z) = c^2.$$
Multiplying together the second and third equations, and dividing by the first we have $(-x + y + z)^2 = \dfrac{b^2 c^2}{a^2}$. Hence

$$-x + y + z = \pm\frac{bc}{a}, \quad x - y + z = \pm\frac{ca}{b}, \quad x + y - z = \pm\frac{ab}{c}.$$

From the given equations we see that these results are to be taken all with the positive or all with the negative sign.

§187. Let x denote the number of Scotch Conservatives, and therefore the number of Welsh Liberals. The number of Scotch Liberals is therefore $60 - x$; hence the Scotch liberal majority is $60 - 2x$, and therefore the number of welsh members is 30. The Irish Liberal majority $= \dfrac{3}{2}(60 - 2x) = 90 - 3x$. We may then represent the number of members by the following table

	Conservatives	Liberals
English	y,	z;
Scotch	x,	$60 - x$;
Welsh	$30 - x$,	x;
Irish	u,	$u + 90 - 3x$.

Thus we have the following equations:
$$z + u - 3x + 150 = y + 15; \text{ that is, } 3x + y - z - u = 135;$$
$$y + u + 30 = 2z + 5; \text{ that is, } y - 2z + u = -25;$$
$$y - z = 2u + 90 - 3x + 10; \text{ that is, } 3x + y - z - 2u = 100;$$
$$y + z + 60 + 30 + 2u + 90 - 3x = 652;$$
$$\text{that is, } -3x + y + z + 2u = 472.$$

From the first and third equations, we have $u = 35$; hence
$$3x + y - z = 170, \quad y - 2z = -60, \quad -3x + y + z = 402.$$
Adding together the first and the last of these equations, we have $2y = 572$ or $y = 286$; hence $z = 173$; and $x = 19$.

§188. It is easy to prove that the expression on the left contains the factor $(b - c)(c - a)(a - b)$; the remaining factor being of three dimensions and symmetrical in a, b, c must be of the form $k\Sigma a^3 + l\Sigma a^2 b + mabc$, where $k, l,$ and m are numerical.

A comparison of the terms involving a^5 shows that $k = 1$.

Again there is no term involving a^4 on the left; while on the right these terms arise from $(b - c)\{-a^2 + a(b + c) - bc\}\{ka^3 + la^2(b + c) + \ldots\}$; hence $k(b + c) - l(b + c) = 0$; whence $l = k = 1$.

To find m, put $a = 2, b = 1, c = -1$ in the identity
$$a^5(c - b) + b^5(a - c) + c^5(b - a) =$$
$$(b - c)(c - a)(a - b)(k\Sigma a^3 + l\Sigma a^2 b + mabc);$$
thus $2^5(-2) + 1^5(3) - 1^5(-1) = (2)(-3)(1)(8k + 4l - 2m)$;
that is $10 = 8k + 4l - 2m$; whence $m = 1$.

§189. Keeping the lowest row unaltered, we see that the determinant
$$= \begin{vmatrix} a^3 - 1 & 3a^2 - 3 & 3a - 3 & 0 \\ a^2 - 1 & a^2 + 2a - 3 & 2a - 2 & 0 \\ a - 1 & 2a - 2 & a - 1 & 0 \\ 1 & 3 & 3 & 1 \end{vmatrix} = \begin{vmatrix} a^3 - 1 & 3a^2 - 3 & 3a - 3 \\ a^2 - 1 & a^2 + 2a - 3 & 2a - 2 \\ a - 1 & 2a - 2 & a - 1 \end{vmatrix}$$

$$= (a-1)^3 \begin{vmatrix} a^2+a+1 & 3a+3 & 3 \\ a+1 & a+3 & 2 \\ 1 & 2 & 1 \end{vmatrix} = (a-1)^3 \begin{vmatrix} a^2+a-2 & 3a-3 & 0 \\ a-1 & a-1 & 0 \\ 1 & 2 & 1 \end{vmatrix}$$

$$= (a-1)^5 \begin{vmatrix} a+2 & 3 & 0 \\ 1 & 1 & 0 \\ 1 & 2 & 1 \end{vmatrix} = (a-1)^5 \begin{vmatrix} a+2 & 3 \\ 1 & 1 \end{vmatrix} = (a-1)^6.$$

§190. We have $\quad \dfrac{a+c}{ac} + \dfrac{a+c-2b}{ac-b(a+c)+b^2} = 0;$

that is, $\quad b^2(a+c) - b\{(a+c)^2 + 2ac\} + 2ac(a+c) = 0;$

or $\qquad\qquad \{b(a+c) - 2ac\}\{b - (a+c)\} = 0;$

hence $\qquad\qquad b = a+c, \text{ or } b(a+c) - 2ac = 0.$

§191.

(1) Denote the roots by $a, a+2, 11-2a$, the sum of these roots being 13. Since the sum of the products of the roots two at a time is 15,

we have $\qquad\qquad a^2 + 2a + (2a+2)(11-2a) = 15;$

that is, $\qquad\qquad 3a^2 - 20a - 7 = 0, \text{ or } (3a+1)(a-7) = 0.$

Again $a(a+2)(11-2a) = -189$; this equation is satisfied by $a = 7$, but not by $a = -\dfrac{1}{3}$. Thus the roots are $7, 9, -3$.

(2) The equation whose roots are $2 \pm \sqrt{-3}$ is $x^2 - 4x + 7 = 0$.

Now $x^4 - 4x^2 + 8x + 35 = (x^2 - 4x + 7)(x^2 + 4x + 5)$; hence the other roots are given by $x^2 + 4x + 5 = 0$.

§192. We have $\qquad a_1 = a - \dfrac{a-b}{3}; \; b_1 = b + \dfrac{a-b}{3};$

thus $\qquad a_1 + b_1 = a+b, \text{ and } a_1 - b_1 = \dfrac{a-b}{3}.$

Similarly, $\qquad a_2 = a_1 - \dfrac{a_1 - b_1}{3} = a - \dfrac{a-b}{3} - \dfrac{a-b}{3^2};$

$$b_2 = b_1 + \dfrac{a_1 - b_1}{3} = b + \dfrac{a-b}{3} + \dfrac{a-b}{3^2};$$

thus $\qquad a_2 + b_2 = a+b; \; a_2 - b_2 = \dfrac{a_1 - b_1}{3} = \dfrac{a-b}{3^2}.$

Again,

$$a_3 = a_2 - \dfrac{a_2 - b_2}{3} = a - \dfrac{a-b}{3} - \dfrac{a-b}{3^2} - \dfrac{a-b}{3^3}; \text{ and so on.}$$

Thus $\qquad a_n = a - (a-b)\left(\dfrac{1}{3} + \dfrac{1}{3^2} + \dfrac{1}{3^3} + \ldots + \dfrac{1}{3^n}\right);$

that is, $\qquad a_n = a - \dfrac{1}{2}(a-b)\left(1 - \dfrac{1}{3^n}\right).$

Similarly $\qquad b_n = b + \dfrac{1}{2}(a-b)\left(1 - \dfrac{1}{3^n}\right).$

When n is infinite,

$$a_n = a - \dfrac{1}{2}(a-b) = \dfrac{1}{2}(a+b);$$

and $\qquad b_n = b + \dfrac{1}{2}(a-b) = \dfrac{1}{2}(a+b).$

§193. By an easy reduction, we see that the left side

$$= w^3(x+y+z) + w^2\{2(x^2+y^2+z^2) + yz+zx+xy\}$$
$$+ w(x^3 + y^3 + z^3 - 2xyz) + xyz(x+y+z)$$
$$= -w^2(x+y+z)^2 + w^2\{2(x^2+y^2+z^2) + yz+zx+xy\}$$
$$+ w(x^3 + y^3 + z^3 - 2xyz) - xyzw$$
$$= w^2(x^2+y^2+z^2 - yz-zx-xy) + w(x^3+y^3+z^3 - 3xyz)$$

$$= w\{-(x+y+z)(x^2+y^2+z^2-yz-zx-xy)$$
$$+ (x^3+y^3+z^3-3xyz)\}$$
$$= 0.$$

§194. Suppose that the expression is not altered by interchanging a and b; then $a + \dfrac{bc-a^2}{a^2+b^2+c^2} = b + \dfrac{ac-b^2}{a^2+b^2+c^2}$; hence $a - b = \dfrac{(a-b)c+(a^2-b^2)}{a^2+b^2+c^2}$;

that is, $1 = \dfrac{a+b+c}{a^2+b^2+c^2}$, since $a-b$ is not zero.

Thus $a - c = \dfrac{a^2-c^2+ab-bc}{a^2+b^2+c^2}$; that is,

$$a + \frac{bc-a^2}{a^2+b^2+c^2} = c + \frac{ab-c^2}{a^2+b^2+c^2};$$

which shows that the expression is unttered by interchanging a and c.

§195. Let the down trains be denoted by T_1, T_2 and the up trains by T_3, T_4; also suppose that they pass each other y hours after 6 o'clock. Then the number of miles which the trains T_1, T_2, T_3, T_4 respectively travel before they pass each other are

$$x_1 y, \quad x_2\left(y-\frac{3}{4}\right), \quad x_3\left(y-\frac{5}{4}\right), \quad x_4\left(y-\frac{5}{2}\right).$$

Also by question

$$x_1 y + x_3\left(y-\frac{5}{4}\right) = m = x_1 y + x_4\left(y-\frac{5}{2}\right);$$

whence

$$(x_1+x_3)y = m + \frac{5}{4}x_3; \text{ and } (x_1+x_4)y = m + \frac{5}{2}x_4.$$

From these equations,

$$4y = \frac{4m+5x_3}{x_1+x_3} = \frac{4m+10x_4}{x_1+x_2}.$$

Again, $\quad x_1 y = x_2\left(y-\dfrac{3}{4}\right)$; whence $4y = \dfrac{3x_2}{x_2-x_1}$.

By equating the three values found for y we obtain the required result.

§196. The left side $\dfrac{\left(1+\dfrac{1}{2}x+\dfrac{3}{8}x^2\right)+\left(1+\dfrac{1}{2}y+\dfrac{3}{8}y^2\right)}{1+\left(1-\dfrac{1}{2}x-\dfrac{1}{8}x^2\right)\left(1-\dfrac{1}{2}y-\dfrac{1}{8}y^2\right)}$

$$= \frac{\left\{1+\dfrac{1}{4}(x+y)+\dfrac{3}{16}(x^2+y^2)\right\}}{1-\dfrac{1}{4}(x+y)-\dfrac{1}{16}(x^2-2xy+y^2)}$$

$$= \left\{1+\frac{1}{4}(x+y)+\frac{3}{16}(x^2+y^2)\right\}$$

$$\left\{1-\frac{1}{4}(x+y)-\frac{1}{16}(x^2-2xy+y^2)\right\}^{-1}$$

$$= \left\{1+\frac{1}{4}(x+y)+\frac{3}{16}(x^2+y^2)\right\}$$

$$\left\{1+\frac{1}{4}(x+y)+\frac{1}{16}(x^2-2xy+y^2)+\frac{1}{16}(x+y)^2\right\}$$

$$= \left\{1+\frac{1}{4}(x+y)+\frac{3}{16}(x^2+y^2)\right\}\left\{1+\frac{1}{4}(x+y)+\frac{1}{8}(x^2+y^2)\right\}$$

$$= 1+\frac{1}{2}(x+y)+\frac{1}{16}\{3(x^2+y^2)+2(x^2+y^2)+(x+y)^2\}$$

$$= 1 + \frac{1}{2}(x+y) + \frac{1}{8}(3x^2 + xy + 3y^2).$$

§197. If S_1 denotes the sum of the series, S_2 the sum of the squares, and P the sum of the products two at a time, we know that $2P = S_1^2 - S_2$; hence $P = 0$ when $S_1^2 = S_2$.

Now
$$S_1 = \frac{n}{2}\{2a - (n-1)b\}.$$

$$S_2 = na^2 - 2ab\{1 + 2 + 3 + \ldots + (n-1)\}$$
$$+ b^2\{1^2 + 2^2 + 3^2 + \ldots + (n-1)^2\}$$
$$= na^2 - n(n-1)ab + \frac{(n-1)n(2n-1)}{6}b^2.$$

Hence if $S_1^2 = S_2$, we have

$$\frac{n}{4}\{2a - (n-1)b\}^2 = a^2 - (n-1)ab + \frac{(n-1)(2n-1)}{6}b^2;$$

$$\therefore (n-1)a^2 - (n-1)^2ab + \frac{1}{12}(n-1)(3n^2 - 7n + 2)b^2 = 0;$$

or
$$a^2 - (n-1)ab + \frac{1}{12}(3n^2 - 7n + 2)b^2 = 0.$$

Hence
$$\frac{2a}{b} = (n-1) \pm \sqrt{(n-1)^2 - \frac{1}{3}(3n^2 - 7n + 2)}$$

$$= (n-1) \pm \sqrt{\frac{1}{2}(n+1)};$$

putting $n+1 = 3m^2$, we have $\dfrac{2a}{b} = 3m^2 \pm m - 2 = (3m \mp 2)(m \pm 1)$.

§198. If $n = 2m$, we may take the terms of the series in the following pairs :
$$\alpha \pm \beta, \ \alpha \pm 3\beta, \ \alpha \pm 5\beta, \ \ldots, \ \alpha \pm (2m-1)\beta.$$

Now
$$\{\alpha + (2r-1)\beta\}^3 + \{\alpha - (2r-1)\beta\}^3 = 2\{\alpha^3 + 3(2r-1)^2\alpha\beta^2\}.$$

Hence
$$S = 2\alpha \sum_{r=1}^{r=m}\{\alpha^2 + 3(2r-1)^2\beta^2\}.$$

But
$$1^2 + 2^2 + 3^3 + \ldots + (2m)^2 = \frac{2m(2m+1)(4m+1)}{6};$$

and
$$1^2 + 2^2 + 3^2 + \ldots + m^2 = \frac{m(m+1)(2m+1)}{6};$$

multiplying the second of these results by 4 and subtracting from the first, we have

$$1^2 + 3^2 + 5^2 + \ldots + (2m-1)^2 = \frac{2m(2m+1)(2m-1)}{6}.$$

$$\therefore S = 2a\{ma^2 + m(2m+1)(2m-1)\beta^2\} = na\{a^2 + (n^2-1)\beta^2\}.$$

§199. This is equivalent to showing that
$$a^6 + b^6 + c^6 > a^2b^2c^2(bc + ca + ab);$$

$$\text{or } \frac{a^6}{b^2c^2} + \frac{b^6}{c^2a^2} + \frac{c^6}{a^2b^2} > bc + ca + ab.$$

Now
$\left(\dfrac{a^3}{bc} - \dfrac{b^3}{ca}\right)^2$ is positive; hence $\dfrac{a^6}{b^2c^2} + \dfrac{b^6}{c^2a^2} > \dfrac{2a^2b^2}{c^2}$.

Thus
$$\frac{a^6}{b^2c^2} + \frac{b^6}{c^2a^2} + \frac{c^6}{a^2b^2} > \frac{a^2b^2}{c^2} + \frac{b^2c^2}{a^2} + \frac{c^2a^2}{b^2}.$$

Again,
$\left(\dfrac{bc}{a} - \dfrac{ca}{b}\right)^2$ is positive; hence $\dfrac{b^2c^2}{a^2} + \dfrac{c^2a^2}{b^2} > 2c^2$.

Thus $\dfrac{b^2c^2}{a^2} + \dfrac{c^2a^2}{c^2} + \dfrac{a^2b^2}{c^2} > a^2 + b^2 + c^2.$

Finally it is well known that $a^2 + b^2 + c^2 > bc + ca + ab$. Hence a fortiori the required result is true.

§200. Let x be the time in hours after which B dismounts; then A has gone ux miles, and B and C have each gone vx. Now B continues to walk $\dfrac{a - vx}{u}$ hours; therefore the whole time occupied is $x + \dfrac{a - vx}{u}$, for his walking pace is the same as A's. When C starts back to meet A they are $(v - u)x$ miles apart; therefore if they meet in p hours we have

$$p(u + v) = (v - u)x; \text{ whence } p = \frac{(v - u)x}{u + v}.$$

Again they meet $(x + p)u$ miles from the starting point, so that the distance remaining is $a - (x + p)u$ miles, and the time occupied in driving this distance is $\dfrac{a - (x + p)u}{v}$ hours.

Now the number of hours after B dismounts

$$= \frac{a - vx}{u} = p + \frac{a - (x + p)u}{v}.$$

From this equation we obtain

$$x(u + v) + up = a, \text{ or } p = \frac{a - x(u + v)}{u}.$$

By equating this to the former value found for p, we obtain $x = \dfrac{a(u + v)}{v^2 + 3uv}$.

Hence the whole time occupied $= x + \dfrac{a - vx}{u} = \dfrac{a}{v}\cdot\dfrac{3v + u}{3u + v}$ hours.

§201. We may represent the city by rectangle whose sides are a and b. Let a, running N. and S., be vertical, and b, running E. and W., be horizontal. Then it is clear that whatever route is chosen the man has to travel a distance equal to a in the vertical direction and a distance equal to b in the distance equal to a in the horizontal direction. Now b is the aggregate of $m - 1$ horizontal distances, and a is the aggregate of $n - 1$ vertical distances, and the $m + n - 2$ potions which make his whole path may occur in any order. Thus the number of ways is equal to the number of permutations of $m + n - 2$ things $m - 1$ of which are of one kind and $n - 1$ are of another kind.

§202. Put u for $\sqrt[4]{x + 27}$, and v for $\sqrt[4]{55 - x}$; then $u^4 + v^4 = 82$, and $u + v = 4$. Raise both sides of the equation to the 4th power; then we have

$$u^4 + v^4 + 4uv(u^2 + v^2) + 6u^2v^2 = 256;$$
$$82 + 4uv(u^2 + 2uv + v^2) - 2u^2v^2 = 256;$$

that is, $\qquad 82 + 64uv - 2u^2v^2 = 256;$

or $\qquad u^2v^2 - 32uv + 87 = 0; \text{ whence } uv = 29, \text{ or } 3.$

Also $\qquad\qquad u + v = 4;$

$$\therefore u = 2 \pm 5\sqrt{-1}, \text{ or } 3, \text{ or } 1;$$
$$\therefore x + 27 = (2 \pm 5\sqrt{-1})^4, \text{ or } 81, \text{ or } 1.$$

§203. If S_{2n} denotes the sum of $2n$ terms of the series

$$ab + (a + x)(b + x) + (a + 2x)(b + 2x) + \ldots;$$
$$S_{2n} = 2nab + x(a + b)(1 + 2 + 3 + \ldots \text{ to } 2n - 1 \text{ terms})$$
$$+ x^2(1^2 + 2^2 + 3^2 + \ldots \text{ to } 2n - 1 \text{ terms})$$
$$= 2nab + n(2n - 1)(a + b)x + \frac{1}{3}n(2n - 1)(4n - 1)x^2.$$

By writing n for $2n$, we have

$$S_n = nab + \frac{1}{2}n(n - 1)(a + b)x + \frac{1}{6}n(n - 1)(2n - 1)x^2.$$

$$\therefore S_{2n} - 2S_n = n^2(a+b)x + n^2(2n-1)x^2 = n^2 x\{a+b+(2n-1)x\}.$$

If l is the last term,

$$l - ab = (a + \overline{2n-1} \cdot x)(b + \overline{2n-1} \cdot x) - ab$$
$$= (2n-1)x\{a+b+(2n-1)x\}.$$

$$\therefore S_{2n} - 2S_n : l - ab = n^2 : 2n-1;$$

which proves the proposition, since $S_{2n} - 2S_n$ or $(S_{2n} - S_n) - S_n$ denotes the excess of the last n terms over the first n terms.

§204.

(1) Let $\dfrac{p_n}{q_n}$ be the nth convergent; then $p_n = 2p_{n-1} - p_{n-2}$, so that the numerators of the successive convergents from a recurring series, whose scale of relation is $1 - 2x + x^2$.

Put $$S = p_1 + p_2 x + p_3 x^2 + \ldots;$$

then, as in Art. 325, we have $S = \dfrac{p_1 + (p_2 - 2p_1)x}{1 - 2x + x^2}.$

But $p_1 = 1$, $p_2 = 2$; hence $S = \dfrac{1}{(1-x)^2}$, and $p_n = n$.

Similarly if $$S' = q_1 + q_2 x + q_3 x^2 + \ldots,$$

we shall find $q_n = n+1$. Thus $\dfrac{p_n}{q_n} = \dfrac{n}{n+1}.$

(2) The scale of relation is $1 - 3x - 4x^2$. With the same notation as in the preceding case, we have

$$p_1 + p_2 x + p_3 x^2 + \ldots = \frac{p_1 + (p_2 - 3p_1)x}{1 - 3x - 4x^2}$$
$$= \frac{4}{1 - 3x - 4x^2} = \frac{4}{5}\left(\frac{4}{1-4x} + \frac{1}{1+x}\right).$$

$$q_1 + q_2 x + q_3 x^2 + \ldots = \frac{q_1 + (q_2 - 3p_1)x}{1 - 3x - 4x^2}$$
$$= \frac{3+4x}{1 - 3x - 4x^2} = \frac{1}{5}\left(\frac{16}{1-4x} - \frac{1}{1+x}\right).$$

Thus $p_n = \dfrac{4}{5}\{4^n + (-1)^{n-1}\}$; and $q_n = \dfrac{1}{5}\{4^{n+1} + (-1)^n\}.$

§205. Put $(a-x)(y-z) = \alpha$, $(a-y)(z-x) = \beta$, $(a-z)(x-y) = \gamma$; then after transposition we have to prove that

$$\alpha^4 + \beta^4 + \gamma^4 - 2\beta^2\gamma^2 - 2\gamma^2\alpha^2 - 2\alpha^2\beta^2$$

is zero. Now this last expression has a factor $\alpha + \beta + \gamma$ which from the above values is evidently equal to zero.

§206. The expression whose value is required can be written in the form

$$\frac{-[(n - m\beta)(n - m\gamma)(n + m\alpha)] + \cdots + \ldots]}{(n - m\alpha)(n - m\beta)(n - m\gamma)}.$$

The numerator

$$= -[n^3 - n^2 m(\alpha - \beta - \gamma) + nm^2(\beta\gamma - \gamma\alpha - \alpha\beta) + m^3\alpha\beta\gamma$$
$$+ \text{two similar expressions}]$$
$$= -[3n^3 - n^2 m\Sigma(\alpha - \beta - \gamma) + nm^2\Sigma(\beta\gamma - \gamma\alpha - \alpha\beta) + 3m^3\alpha\beta\gamma]$$
$$= -3n^3 - n^2 m(\alpha + \beta + \gamma) + nm^2(\beta\alpha + \gamma\alpha + \alpha\beta) - 3m^3\alpha\beta\gamma$$
$$= -3n^3 + nm^2 q + 3rm^3,$$

by the properties of the roots of the equation.

The denominator $= n^3 - n^2 m(\alpha + \beta + \gamma)$
$$+ nm^2(\beta\gamma + \gamma\alpha + \alpha\beta) - m^3\alpha\beta\gamma$$
$$= n^3 + nm^2 q + rm^3.$$

$$\therefore \text{ the required expression} = \frac{3rm^3 + nm^2q - 3n^3}{rm^3 + nm^2q + n^3}.$$

§207. If x is the population at the beginning of the year, then the population at end of the year is $x + \dfrac{x}{33} - \dfrac{x}{46} = \dfrac{1531x}{1518}$; hence if n be the required number of years, $\left(\dfrac{1531}{1518}\right)^n x = 2x$; that is,

$$n(\log 1531 - \log 1518) = \log 2; \text{ or } .0037034n = .3010300, \text{ and } n = 81.$$

§208. We have

$$(1 - x^3)^n = (1 - x)^n(1 + x + x^2)^n = (1 - x)^n(a_0 + a_1x + a_2x^2 + \ldots).$$

Equate the coefficients of x^r; then if r is not a multiple of 3 the coefficient of x^r on the left side is zero, and the required result follows immediately.

If r is a multiple of 3 it is of the form $3m$, and on the left the coefficient of x^{3m} is

$$(-1)^m \frac{n!}{m!(n-m)!}, \text{ or } (-1)^{\frac{r}{3}} \frac{n!}{\frac{r}{3}!\left(n - \frac{r}{3}\right)!}.$$

§209. Denote the number of Poles, Trucks, Greeks, Germans, and Italians by x, y, z, u, v respectively; then we have

$$x = \frac{1}{3}u - 1 = \frac{1}{2}v - 3; \ y + u - z - v = 3;$$

$$z + u = \frac{1}{2}(x + y + z + u + v) - 1;$$

$$z + v = \frac{7}{16}(x + y + z + u + v).$$

From the fourth equation, we have $x + y - z - u + v = 2$; subtracting this from the third equation, we get $2u - 2v - x = 1$.

But from the first two equations, $u = 3x + 3$, $v = 2x + 6$; hence

$$6x + 6 - 4x - 12 - x = 1; \text{ and therefore } x = 7;$$

$$\text{whence } u = 24, v = 20.$$

From the third and fifth equations, we have $y - z = -1$ and

$$z + 20 = \frac{7}{16}(y + z + 51);$$

that is, $9z - 7y = 37$; whence $y = 14, z = 15$.

§210. The nth term $= \dfrac{(n+1)(-x)^{n+1}}{n(n+2)} = \dfrac{1}{2}\left(\dfrac{1}{n} + \dfrac{1}{n+2}\right)(-x)^{n+1};$

$$\therefore 2S = \left(-x^2 + \frac{x^3}{2} - \frac{x^4}{3} + \frac{x^5}{4} - \ldots\right) + \left(-\frac{x^2}{3} + \frac{x^3}{4} - \frac{x^4}{5} + \frac{x^5}{6} - \ldots\right)$$

$$= -x\log(1 + x) - \frac{1}{x}\left\{\log(1 + x) - x + \frac{x^2}{2}\right\};$$

that is, $S = \dfrac{1}{2} - \dfrac{x}{4} - \dfrac{1 + x^2}{2x}\log(1 + x).$

§211. By the Binomial Theorem we have

$$(1 - x)^n = 1 - nx + \frac{n(n-1)}{1.2}x^2 - \frac{n(n-1)(n-2)}{1.2.3}x^3 + \ldots$$

$$+ (-1)^{n-2}\frac{n(n-1)}{1.2}x^{n-2} + (-1)^{n-1}nx^{n-1} + (-1)^n x^n.$$

$$(1 - x)^{-(n+1)} = 1 + (n+1)x + \frac{(n+1)(n+2)}{1.2}x^2$$

$$+ \frac{(n+1)(n+2)(n+3)}{1.2.3}x^3 + \ldots$$

Multiply these two results together, and equate the coefficients of x^{n-1}. Then, if S stand for the left-hand member of the proposed identity, we have $(-1)^{n-1}S =$ the coefficient of x^{n-1} in the expansion of $(1-x)^{-1}$.

$$\therefore (-1)^{n-1}S = 1 = (-1)^{2n},$$

that is,

$$S = (-1)^{n+1}.$$

§212. (1) If we form the successive orders of differences, we obtain

$$
\begin{array}{ccccccc}
6, & 24, & 60, & 120, & 210, & 336, \ldots \\
18, & 36, & 60, & 90, & 126, & \ldots \\
18, & 24, & 30, & 36, & \ldots \\
6, & 6, & 6, & \ldots
\end{array}
$$

$$\therefore u_n = 6 + 18(n-1) + \frac{18(n-1)(n-2)}{\underline{|2}} + \frac{6(n-1)(n-2)(n-3)}{\underline{|3}}$$

$$= n(n+1)(n+2).\text{[See Art.396]}$$

$$\therefore S_n = \frac{1}{4}n(n+1)(n+2)(n+3).$$

(2) Here $u_n = (n+1)^2(-x)^{n-1}$; and therefore the series is recurring and $(1+x)^3$ is the scale of relation. [Art. 398.]

Let $\qquad S = \qquad 4 - 9x + 16x^2 - 25x^3 + 36x^4 - \ldots ;$

then $\qquad 3xS = \qquad 12x - 27x^2 + 48x^3 - 75x^4 + \ldots ;$

$\qquad\qquad 3x^2 S = \qquad 12x^2 - 27x^3 + 48x^4 - \ldots ;$

$\qquad\qquad x^3 S = \qquad 4x^3 - 9x^4 + \ldots .$

By addition, $\qquad (1+x)^3 S = 4 + 3x + x^2.$

(3) Put $x = \frac{1}{2}$; then $S = 1 \cdot 3x + 3 \cdot 5x^2 + 5 \cdot 7x^3 + 7 \cdot 9x^4 + \ldots$; thus $u_n = (2n-1)(2n+1)x^n$. Hence the scale of relation is $(1-x)^3$. [Art.398.]

Proceeding as in (2), we shall find $S = \dfrac{3 + 6x - x^2}{(1-x)^3} = 46.$

§213. Add together the first and third rows, and from the sum subtract twice the second row; also subtract twice the first row from the third row; thus

$$
\begin{vmatrix}
4x & 6x+2 & 8x+1 \\
-3 & -4 & 3 \\
1 & -4 & 0
\end{vmatrix} = 0;
$$

hence $\qquad 4x(12) - (6x+2)(-3) + (8x+1)16 = 0;$

and therefore $194x + 22 = 0$; that is, $x = -\dfrac{11}{97}$.

§214.

(1) This follows by adding together the inequalities

$$a^2 + b^2 c^2 > 2abc; \quad b^2 + c^2 a^2 > 2abc; \quad c^2 + a^2 b^2 > 2abc.$$

(2) The two quantities $a^p - b^p$ and $a^q - b^q$ are both positive, or both negative; hence $(a^p - b^p)(a^q - b^q)$ is positive; that is,

$$a^{p+q} + b^{p+q} > a^p b^q + a^q b^p.$$

Similarly, $\qquad a^{p+q} + c^{p+q} > a^p c^q + a^q c^p.$

and $\qquad b^{p+q} + c^{p+q} > b^p c^q + b^q c^p.$

and so on, the number of inequalities being $\frac{1}{2}n(n-1)$.

By addition,

$$(n-1)(a^{p+q} + b^{p+q} + c^{p+q} + \ldots) > \Sigma a^p b^p;$$

hence $\qquad n(a^{p+q} + b^{p+q} + c^{p+q} + \ldots) > \Sigma a^{p+q} + \Sigma a^p b^q;$

which proves the proposition.

§215. The given equation may be written
$$(y-a)(z-a) = a^2 + \alpha;$$
$$(z-a)(x-a) = a^2 + \beta;$$
$$(x-a)(y-a) = a^2 + \gamma;$$
$$\therefore (x-a)(y-a)(z-a) = \pm\{(a^2+\alpha)(a^2+\beta)(a^2+\gamma)\}^{\frac{1}{2}}.$$
Divide this result in succession by each of the given equations.

§216. The given expression
$$= \{1.2^{n-1} + 2.3^{n-1} + \ldots + (n-2)(n-1)^{n-1}\}$$
$$+(n-1)n^{n-1} + n - 1.$$
Now by Fermat's theorem each of the expressions $2^{n-1}, 3^{n-1}, \ldots (n-1)^{n-1}$ is of the form $1 = M(n)$.
∴ the given expression
$$= \{1 + 2 + 3 + \ldots + (n-2)\} + (n-1) + (n-1)n^{n-1} + M(n)$$
$$= \frac{n(n-1)}{2} + (n-1)n^{n-1} + M(n),$$
which is a multiple of n, since $\dfrac{n(n-1)}{2}$ is integral.

§217. The number of ways of making 30 in 7 shots is the coefficient of x^{30} in $(x^0 + x^2 + x^3 + x^4 + x^5)^7$; for this coefficient arises out of the different ways in which 7 of the indices $0, 2, 3, 4, 5$ combine to make 30.
Now
$$(x^5 + x^4 + x^3 + x^2 + 1)^7 = \{x^4(x+1) + x^3 + x^2 + 1\}^7$$
$$= x^{28}(x+1)^7 + 7x^{24}(x+1)^8(x^3 + x^2 + 1)$$
$$+ 21x^{20}(x+1)^6(x^3 + x^2 + 1)^2 + 35x^{16}(x+1)^4(x^3 + x^2 + 1)^3 + \ldots$$
$$= 21 + 7(1 + 15 + 20) + 21(2 + 5)$$
$$= 21 + 252 + 147 = 420.$$

§218. Denote the complete square by $(x+3k)^2$; then since the coefficient of x^4 in the given expression is 0, the complete cube will be $(x-2k)^3$; thus
$$x^5 - bx^3 + cx^2 + dx - e = (x + 3k)^2(x - 2k)^3$$
$$= x^5 - 15k^2x^3 + 10k^3x^2 + 60k^4x - 72k^5.$$
Hence by equating coefficients, we have $15k^2 = b, 10k^3 = c, 60k^4 = d, 72k^5 = e$;

thus
$$36k^2 = \frac{12b}{5} = \frac{9d}{b} = \frac{5e}{c} = \frac{d^2}{c^2}.$$

§219. There are four cases to consider for the bag contain 3 white, 4 white, 5 white, or 6 white balls, and we consider all these to be equally likely.
$$p_1 = \frac{3}{9} \cdot \frac{2}{8} \cdot \frac{1}{7}; \ p_2 = \frac{4}{10} \cdot \frac{3}{9} \cdot \frac{2}{8}; \ p_3 = \frac{5}{11} \cdot \frac{4}{10} \cdot \frac{3}{9}; \ \frac{6}{12} \cdot \frac{5}{11} \cdot \frac{4}{10};$$
that is,
$$p_1 = \frac{1}{84}, \ p_2 = \frac{1}{30}, \ p_3 = \frac{2}{33}, \ p_4 = \frac{1}{11}.$$
$$\therefore \frac{Q_1}{55} = \frac{Q_2}{154} = \frac{Q_3}{280} = \frac{Q_4}{420} = \frac{1}{909}.$$
The chance of drawing a black ball next
$$= Q_1 \times 1 + Q_2 \times \frac{6}{7} + Q_3 \times \frac{3}{4} + Q_4 \times \frac{2}{3}$$
$$= \frac{55}{909} + \frac{6}{7} \times \frac{154}{909} + \frac{3}{4} \times \frac{280}{909} + \frac{2}{3} \times \frac{420}{909} = \frac{677}{909}.$$

§220. Here $2S = (1^2 + 2^2 + 3^2 + \ldots + n^2)^2 - (1^4 + 2^4 + 3^4 + \ldots + n^4)$.

Now by Art. 405,

$$1^4 + 2^4 + 3^4 + \ldots + n^4 = \frac{n^5}{5} + \frac{1}{2}n^4 + B_1\frac{4}{\lfloor 2}n^3 - B_3\frac{4.3.2}{\lfloor 4}n.$$

$$= \frac{n^5}{5} + \frac{n^4}{2} + \frac{n^3}{3} - \frac{n}{30}$$

$$= \frac{n}{30}(n+1)(2n+1)(3n^2 + 3n - 1).$$

Thus

$$2S = \frac{n^2(n+1)^2(2n+1)^2}{36} - \frac{n(n+1)(2n+1)(3n^2+3n-1)}{30}$$

$$= \frac{1}{180}n(n+1)(2n+1)\{5n(2n^2+3n+1) - 6(3n^2+3n-1)\}$$

$$= \frac{1}{180}n(n+1)(2n+1)(n-1)(2n-1)(5n+6).$$

§221. On reduction, we obtain

$$x^2\{a^2(b-c) + \ldots + \ldots\} - x\{a^2(b^2-c^2) + \ldots + \ldots\} + \{a^2bc(b-c) + \ldots + \ldots\} = 0;$$

if the roots of this equation are equal, we must have

$$\{a^2(b^2-c^2) + \ldots + \ldots\}^2 - 4\{a^2(b-c) + \ldots + \ldots\}\{a^2bc(b-c) + \ldots + \ldots\} = 0;$$

The coefficient of a^4 in the expression on the left

$$= (b^2-c^2)^2 - 4bc(b-c)^2 = (b-c)^2\{(b+c)^2 - 4bc\} = (b-c)^4.$$

The coefficient of $\beta^2\gamma^2$

$$= 2(c^2-a^2)(a^2-b^2) - 4abc(c-a)(a-b) - 4ca(c-a)(a-b)$$

$$= 2(c-a)(a-b)\{(c+a)(a+b) - 2ab - 2ca\} = -2(c-a)^2(a-b)^2.$$

Hence the condition reduces to

$$\alpha^4(b-c)^4 + \beta^4(c-a)^4 + \gamma^4(a-b)^4 - 2\beta^2\gamma^2(c-a)^2(a-b)^2$$

$$- 2\gamma^2\alpha^2(a-b)^2(b-c)^2$$

$$- 2\alpha^2\beta^2(b-c)^2(c-a)^2 = 0.$$

But the expression on the left is now of the form

$$x^4 + y^4 + z^4 - 2y^2z^2 - 2z^2x^2 - 2x^2y^2,$$

the factors of which are

$$-(x+y+z)(-x+y+z)(x-y+z)(x+y-z);$$

and this expression vanishes if $x \pm y \pm z = 0$.

§222. Here $2^{n-1}, (n-2)2^{n-3}, \dfrac{(n-4)(n-3)}{1.2}2^{n-5}, \ldots$ are the coefficients of

$x^{n-1}, x^{n-3},$

$x^{n-5}, \ldots$ in the expansions of $(1-2x)^{-1}, (1-2x)^{-2}, (1-2x)^{-3}, \ldots$ respectively. Hence the sum required is equal to the coefficient of x^{n-1} in the expansion of

$$\frac{1}{1-2x} - \frac{x^2}{(1-2x)^2} + \frac{x^4}{(1-2x)^3} - \ldots$$

and this may be regarded as an infinite series without affecting the result we wish to prove.

But this series is a G.P. whose sum

$$= \frac{1}{1-2x} \div \left(1 + \frac{x^2}{1-2x}\right) = \frac{1}{(1-x)^2}.$$

Therefore by equating coefficients of x^{n-1} we obtain the result stated.

§223.

(1) By addition, we have $(x+y+z)^2 = 225$; that is, $x+y+z = \pm 15$.

Again $x^2 - y^2 - 2z(x-y) = 0$; that is, $(x-y)(x+y-2z) = 0$; whence $x = y$, or $x+y = 2z$.

If $x = y$, we get, by equating the second and third of the given expressions,

$x^2 - 2xz + z^2 = -3$, or $x - z = \pm\sqrt{-3}$. Combining this with $2x + z = \pm 15$, we have

$$x = y = \frac{1}{3}(\pm 15 \pm \sqrt{-3}), z = \frac{1}{3}(\pm 15 \pm 2\sqrt{-3}).$$

If $x + y - 2z = 0$, we have by combination with $x + y + z = \pm 15$, the equations $z = \pm 5, x + y = \pm 10$. Substituting in $y^2 + 2zx = 76$, we have

$$y^2 \pm 10(\pm 10 - y) = 76;$$

that is $y^2 \pm 10y + 24 = 0$; whence $y = \pm 4, \pm 6$.

(2) Put $x = a + h, y = b + k, z = c + l$; then from the first two equations,

$$h + k + l = 0, \quad \frac{h}{a} + \frac{k}{b} + \frac{l}{c} = 0;$$

whence $\qquad \dfrac{h}{a(b - c)} = \dfrac{k}{b(c - a)} = \dfrac{l}{c(a - b)} = \lambda$ say.

From the third equation,

$$ah + bk + cl = bc + ca + ab - a^2 - b^2 - c^2;$$
$$\therefore \lambda\{a^2(b - c) + b^2(c - a) + c^2(a - b)\} = bc + ca + ab - a^2 - b^2 - c^2;$$

or $\qquad \lambda = \dfrac{a^2 + b^2 + c^2 - bc - ca - ab}{(b - c)(c - a)(a - b)}.$

§224. Let the points in one line be denoted by $A_1, A_2, A_3, \ldots$ and those in the other line by $B_1, B_2, \ldots, B_m$; and let A_1, B_1 be towards the same parts. Then from a diagram it will be seen that

$A_2 B_1$ will cut	$m - 1$	lines diverging from	A_1;
$A_3 B_1 \ldots\ldots\ldots$	$2(m - 1)$	$\ldots\ldots\ldots\ldots\ldots\ldots\ldots$	A_1, A_2;
$A_4 B_1 \ldots\ldots\ldots$	$3(m - 1)$	$\ldots\ldots\ldots\ldots\ldots\ldots\ldots$	A_1, A_2, A_3;
$\ldots\ldots \quad \ldots\ldots$	$\ldots\ldots$	$\ldots$	$\ldots$
$A_n B_1 \ldots\ldots\ldots$	$(n - 1)(m - 1)$	$\ldots\ldots\ldots\ldots\ldots\ldots\ldots$	$A_1, A_2, \ldots, A_n$.

Again

$A_2 B_2$ will cut	$m - 2$	lines diverging from	A_1;
$A_3 B_2 \ldots\ldots\ldots$	$2(m - 2)$	$\ldots\ldots\ldots\ldots\ldots\ldots\ldots$	A_1, A_2;
$A_4 B_2 \ldots\ldots\ldots$	$3(m - 2)$	$\ldots\ldots\ldots\ldots\ldots\ldots\ldots$	A_1, A_2, A_3;
$\ldots\ldots \quad \ldots\ldots$	$\ldots\ldots$	$\ldots$	$\ldots$
$A_n B_2 \ldots\ldots\ldots$	$(n - 1)(m - 2)$	$\ldots\ldots\ldots\ldots\ldots\ldots\ldots$	$A_1, A_2, \ldots, A_n$.

And so on, taking all the m points $B_1, B_2 \ldots B_m$ in succession. Finally

$A_2 B_{m-1}$ will cut	1	line from	A_1;
$A_3 B_{m-1} \quad \ldots$	2	lines diverging from	A_1, A_2;
$A_4 B_{m-1} \quad \ldots$	3	$\ldots\ldots\ldots\ldots$	A_1, A_2, A_3;
$\ldots\ldots \quad \ldots\ldots$	$\ldots\ldots$	$\ldots$	$\ldots$
$A_n B_{m-1} \quad \ldots$	$n - 1$	$\ldots\ldots\ldots\ldots$	$A_1, A_2, \ldots, A_n$.

We have now enumerated all the points; for $A_2 B_m$ cuts none of the lines from $A_1, A_3 B_m$ cuts none of the lines from A_1, A_2 and so on.

The number of points we have indicated is clearly equal to

$$\{1 + 2 + 3 + \ldots + (n - 1)\}\{(m - 1) + (m - 2) + \ldots + 1\},$$

which is equal to $\qquad \dfrac{n(n - 1)}{2} \pm \dfrac{m(m - 1)}{2}.$

§225. We have

$$x = (x + x^2 + x^5) + a(x + x^2 + x^5)^2 + b(x + x^2 + x^5)^3 + \ldots;$$

for it is obvious that the coefficient of y is 1.

Equating coefficients of powers of x, we have

$$a + 1 = 0, \text{ or } a = -1;$$
$$b + 2a = 0; \text{ that is, } b = -2a = 2;$$
$$c + 3b + a = 0; \text{ that is, } c = -5;$$

$$d + 4c + 3b + 1 = 0; \text{ that is, } d = 13;$$

whence $\qquad a^2 d - 3abc + 2b^3 = 13 - 30 + 16 = -1.$

§226. Denote the price of a calf, pig, and sheep by $x, x - 1, x - 2$ pounds respectively; and suppose that he spent y pounds over each of the different kinds; then we have the equations

$$\frac{y}{x} + \frac{y}{x-1} + \frac{y}{x-2} = 47, \quad \text{and} \quad \frac{y}{x-1} - \frac{y}{x} = \frac{9}{x-2}.$$

From the second equation $y = \dfrac{9x(x-1)}{x-2}$; substituting in the first equation, we have

$$\frac{9x(x-1)}{x-2} \cdot \frac{3x^2 - 6x + 2}{x(x-1)(x-2)} = 47;$$

that is, $\qquad 27x^2 - 54x + 18 = 47x^2 - 188x + 188;$

whence $\qquad (x - 5)(20x - 34) = 0, \text{ and } x = 5.$

§227. If we put $x = 1$ in the result of Example 2, Art. 447, we at once obtain the desired expression for $\log 2$.

We may also proceed as follows.

If $\qquad \dfrac{1}{a_n} - \dfrac{1}{a_{n+1}} = \dfrac{1}{a_n + x_n}, \quad$ then $x_n = \dfrac{a_n^2}{a_{n+1} - a_n}.$

Thus $\qquad \dfrac{1}{a_1} - \dfrac{1}{a_2} = \dfrac{1}{a_1 + a_2 - a_1} \cdot \dfrac{a_1^2}{\ };$

$$\frac{1}{a_1} - \frac{1}{a_2} + \frac{1}{a_3} = \frac{1}{a_1} - \frac{1}{a_2 + x_2}$$

$$= \frac{1}{a_1 + a_2 + x_2 - a_1} \cdot \frac{a_1^2}{\ } = \frac{1}{a_1 + a_2 - a_1 + a_3 - a_2} \cdot \frac{a_1^2}{\ } \cdot \frac{a_2^2}{\ };$$

$$\frac{1}{a_1} - \frac{1}{a_2} + \frac{1}{a_3} - \frac{1}{a_4} = \frac{1}{a_1 + a_2 - a_1 + a_3 - a_2 + x_3} \cdot \frac{a_1^2}{\ } \cdot \frac{a_2^2}{\ }; \text{ and so on.}$$

By putting $a_1 = 1, a_2 = 2, a_3 = 3, \ldots$, the theorem follows at once,

§228. The number of ways required is equal to the coefficient of x^{240} in the expansion of $(x^0 + x^1 + x^2 + \ldots + x^{100})^6$. This expression is equal to

$$\left(\frac{1 - x^{101}}{1 - x}\right)^6 \text{ or } (1 - x^{101})^6 (1 - x)^{-6};$$

hence the number of ways = the coefficient of x^{240} in $(1 - 6x^{101} + 15x^{202})(1 - x)^{-6}.$

The coefficient of x^r in $(1 - x)^{-6}$ is $\dfrac{\underline{|r + 5}}{\underline{|5}\ \underline{|r}}$; thus the coefficient of x^{240} is obtained from the product of

$(1 - 6x^{101} + 15x^{202})$

$$\left(1 + \ldots + \frac{\underline{|43}}{\underline{|5}\ \underline{|38}} x^{38} + \ldots + \frac{\underline{|144}}{\underline{|5}\ \underline{|139}} x^{139} + \ldots + \frac{\underline{|245}}{\underline{|5}\ \underline{|240}} x^{240}\right).$$

§229. Here $u_n = \dfrac{1.3.5 \ldots (4n - 3)(4n - 5)x^{2n-1}}{2.4.6 \ldots (4n - 4)(4n - 2)};$

hence $\qquad \dfrac{u_n}{u_{n+1}} = \dfrac{4n(4n + 2)}{(4n - 3)(4n - 1)} \cdot \dfrac{1}{x^2}.$

Thus if $x < 1$, the series is convergent; if $x > 1$, divergent.

If $\qquad x = 1, \text{ then } Lim \dfrac{u_n}{u_{n+1}} = 1$

$$Lim\, n\left(\frac{u_n}{u_{n+1}} - 1\right) = Lim \frac{n(24n - 3)}{(4n - 3)(4n - 1)} = \frac{3}{2};$$

hence the series is convergent. [Art. 301.]

[This series is the expansion of the expression in Example 105.]

§230. Let the scale of relation be $1 - px - qx^2$; then

$$288 = 40p + 6q, \quad 40 = 6p + q; \quad \text{whence } p = 12, \quad q = -32;$$

and the scale of relation is $1 - 12x + 32x^2$.

As in Art. 328 we find the generating function

$$= \frac{x - 6x^2}{1 - 12x + 32x^2} = \frac{x}{2}\left\{\frac{1}{1 - 4x} + \frac{1}{1 - 8x}\right\};$$

and the coefficient of x^n is $\frac{1}{2}(4^{n-1} + 8^{n-1})$.

$$\therefore S_n = \frac{1}{2}\{1 + 4 + 4^2 + \ldots + 4^{n-1}\} + \frac{1}{2}\{1 + 8 + 8^2 + \ldots + 8^{n-1}\}$$

$$= \frac{1}{2}\cdot\frac{4^n - 1}{3} + \frac{8^n - 1}{7} = \frac{2^{2n-1}}{3} + \frac{2^{3n-1}}{7} - \frac{5}{21}.$$

$$\therefore S_1 + S_2 + S_3 + \ldots + S_n = \frac{1}{3}\Sigma 2^{2n-1} + \frac{1}{7}\Sigma 2^{3n-1} - \frac{5n}{21}$$

$$= \frac{2}{3^2}(2^{2n} - 1) + \frac{4}{7^2}(2^{3n} - 1) - \frac{5n}{21}.$$

§231. The probability required is the sum of the last two terms in the expansion of $\left(\frac{2}{3} + \frac{1}{3}\right)^5$; and therefore is equal to

$$5\left(\frac{2}{3}\right)\left(\frac{1}{3}\right)^4 + \left(\frac{1}{3}\right)^5, \text{ or } \frac{11}{243}.$$

§232. Subtract the second equation from the first; thus

$$2z(x - y) = a^2 - b^2; \text{ so that } x - y = \frac{a^2 - b^2}{2z}.$$

Substitute in the third equation; thus

$$z^2 + \frac{(a^2 - b^2)^2}{4z^2} = c^2, \text{ or } 4z^4 - 4c^2z^2 + (a^2 - b^2)^2 = 0;$$

$$\therefore (2z^2 - c^2)^2 = (-a^2 + b^2 + c^2)(a^2 - b^2 + c^2);$$

hence

$$4z^2 = 2c^2 \pm 2\sqrt{(-a^2 + b^2 + c^2)(a^2 - b^2 + c^2)}$$

$$= \left\{\pm\sqrt{-a^2 + b^2 + c^2} \pm \sqrt{a^2 - b^2 + c^2}\right\}^2.$$

§233. Let k denote exch of the given equal fractions; then

$$x^2 - xy - xz = ak, \quad y^2 - yz - yx = bk.$$

Subtract the first of these equations from the second, and multiply the result by z; thus we have

$$k(b - a)z = (x - y)(z^2 - xz - yz) = (x - y)ck.$$

$$\therefore cx - cy + (a - b)z = 0.$$

Similarly $\qquad bx + (a - c)y - bz = 0.$

Thus $\qquad x : y : z = a(b + c - a) : b(c + a - b) : c(a + b - c);$

by substituting for x, y, z in $ax + by + cz = 0$, we obtain

$$a^3 + b^3 + c^3 = a^2(b + c) + b^2(c + a) + c^2(a + b).$$

§234. If a is one root, then $-a$ is another root; hence

$$a^3 + pa^2 + qa + r = 0, \text{ and } a^3 - pa^2 + qa - r = 0;$$

from which equations we have $a^3 + qa = 0$, and $pa^2 + r = 0$; thus $a^2 = -q$, and therefore $pq = r$.

§235.

(1) The scale of relation is $(1 - x)^4$. [See Art. 398.]

$$S = \qquad 1 + 8x + 27x^3 + 64x^3 + 125x^4 + \ldots$$

$$-4xS = \qquad -4x - 32x^2 - 108x^3 - 256x^4 - \ldots$$

$$6x^2 S = \qquad 6x^2 + 48x^3 + 162x^4 + \dots$$
$$-4x^3 S = \qquad -4x^3 - 32x^4 - \dots$$
$$x^4 S = \qquad x^4 + \dots$$

On the left hand side, it is easy to prove that
the coefficient of x^n
$$= -4n^3 + 6(n-1)^3 - 4(n-2)^3 + (n-3)^3$$
$$= -(n+1)^3;$$
the coefficient of x^{n+1}
$$= 6n^3 - 4(n-1)^3 + (n-2)^3 = 3n^3 + 6n^2 - 4;$$
the coefficient of x^{n+2}
$$= -4n^3 + (n-1)^3 = -3n^3 - 3n^2 + 3n - 1;$$
the coefficient of x^{n+3}
$$= n^3.$$

(2) We have
$$\frac{5n^2 + 12n + 8}{n^2(n+1)^3(n+2)^3} = \frac{(n^3 + 6n^2 + 12n + 8) - (n^3 + n^2)}{n^2(n+1)^3(n+2)^3}$$
$$= \frac{(n+2)^3 - n^2(n+1)}{.n^2(n+1)^3(n+2)^3} = \frac{1}{n^2(n+1)^3} - \frac{1}{(n+1)^2(n+2)^3};$$

thus $\qquad u_n = v_n - v_{n+1},$ where $v_n = \dfrac{1}{n^2(n+1)^3};$

and therefore $\qquad S_n = \dfrac{1}{1^2 \cdot 2^3} - \dfrac{1}{(n+1)^2(n+2)^3}.$

§236. In the identity
$$(1 + a^3 x^4)(1 + a^5 x^8)(1 + a^9 x^{16}) \dots$$
$$= 1 + A_4 x^4 + A_8 x^8 + A_{12} x^{12} + \dots + A_{4n} x^{4n} + \dots$$
Write $a^{\frac{1}{2}} x^2$ for x; then we get
$$(1 + a^5 x^8)(1 + a^9 x^{16})(1 + a^{17} x^{32}) \dots$$
$$= 1 + A_4 a^2 x^8 + A_8 a^4 x^{16} + \dots + A_{4n} a^{2n} x^{8n} + \dots$$
$$\therefore 1 + A_4 x^4 + A_8 x^8 + \dots$$
$$+ A_{8n} x^{8n} + A_{8n+4} x^{8n+4} + \dots$$
$$= (1 + a^3 x^4)(1 + A_4 a^2 x^8 + A_8 a^4 x^{16} + \dots$$
$$+ A_{4n} a^{2n} x^{8n} + \dots).$$
Equate coefficients of x^{8n}; then $A_{8n} = A_{4n} a^{2n}.$
Again, equate coefficients of x^{8n+4}; then $A_{8n+4} = A_{4n} a^{2n} . a^3 = a^3 A_{8n}.$
Now

$A_4 = a^3;$	$A_8 = a^2 A_4 = a^5;$	$A_{12} = a^3 A_8 = a^8;$
$A_{18} = a^4 A_8 = a^9;$	$A_{20} = a^3 A_{18} = a^{12};$	$A_{24} = a^8 A_{12} = a^{14};$
$A_{28} = a^3 A_{24} = a^{17};$	$A_{32} = a^8 A_{18} = a^{17};$	$A_{38} = a^3 A_{32} = a^{20}.$

Hence the first ten terms are
$$1 + a^3 x^4 + a^5 x^8 + a^8 x^{12} + a^9 x^{18} + a^{12} x^{20} + a^{14} x^{24} + a^{17} x^{28}$$
$$+ a^{17} x^{32} + a^{20} x^{38}.$$

§237. Let x and y miles be the distances from A to B and B to C; and suppose that the man rows u miles per hour, and that the stream flows v miles per hour; then we have
$$\frac{x}{u} + \frac{y}{u+v} = 3, \quad \frac{x}{u} + \frac{y}{u-v} = 3\frac{1}{2}, \quad \frac{x+y}{u+v} = 2\frac{3}{4};$$

while it remains to find $\dfrac{x+y}{u-v}$.

From the above equations we have

$$\frac{y}{u-v} - \frac{y}{u+v} = \frac{1}{2}; \text{ that is, } 4vy = u^2 - v^2;$$

and

$$\frac{x}{u} - \frac{x}{u+v} = \frac{1}{4}; \text{ that is, } 4vx = u(u+v);$$

hence by addition,

$$4v(x+y) = (u+v)(2u-v);$$

therefore

$$\frac{2u-v}{4v} = \frac{x+y}{u+v} = \frac{11}{4}; \text{ so that } u = 6v.$$

Thus

$$\frac{x+y}{u-v} = \frac{x+y}{u+v} \cdot \frac{u+v}{u-v} = \frac{11}{4} \times \frac{7}{5} = 3\frac{17}{20}.$$

§238. Here with the usual notation, we have $p_n = 2p_{n-1} + 3p_{n-2}$; thus the numerators of the successive convergents form a recurring series whose scale of relation is $1 - 2x - 3x^2$.

Put
$$S_p = p_1 + p_2 x + p_3 x^2 + \dots$$

then
$$-2xS_p = \quad -2p_1 x - 2p_2 x^2 - \dots$$
$$-3x^2 S_p = \qquad\qquad -3p_1 x^2 - \dots$$

$$\therefore S_p = \frac{p_1 + (p_2 - 2p_1)x}{1 - 2x - 3x^2} = \frac{3}{(1-3x)(1+x)};$$

that is,
$$S_p = \frac{9}{4(1-3x)} + \frac{3}{4(1+x)};$$

and therefore
$$p_n = \frac{9}{4}.3^{n-1} + \frac{3}{4}(-1)^{n-1}$$
$$= \frac{1}{4}\{3^{n+1} + 3(-1)^{n+1}\}$$

In the same way we may shew that

$$S_q = \frac{9}{4(1-3x)} - \frac{1}{4(1+x)}; \text{ and } q_n = \frac{1}{4}\{3^{n+1} - (-1)^{n+1}\}.$$

§239. The equation cannot have a fractional root, for all the coefficients are integers, and that of x^n is 1; it cannot have an even root, for $f(0)$ or p_n is odd, and hence $f(2m)$ will be odd, since all the terms but the last are even. It cannot have an odd root; for if x is odd,

$$x^n = \text{an odd number} = \text{an even number} + 1.$$

Hence
$$f(x) = \text{an even number} + 1 + p_1 + p_2 + \dots + p_n$$
$$= \text{an even number} + f(1)$$
$$= \text{an odd number},$$

and therefore cannot vanish.

Thus the equation cannot have any commensurable root.

[This solution is due to Professor Steggall.]

§240.

(1) By squaring we obtain

$$ax + \alpha + bx + \beta + 2\sqrt{(ax+\alpha)(bx+\beta)} = cx + \gamma;$$

from this equation by transposing and squaring,

$$4(ax+\alpha)(bx+\beta) = \{(c-a-b)x + (\gamma - \alpha - \beta)\}^2;$$

this equation reduces to a simple equation, if $4ab = (c-a-b)^2$; that is, if $\pm 2\sqrt{ab} = c - a - b$; or $c = a + b \pm 2\sqrt{ab}$; whence $\sqrt{c} = \pm\sqrt{a} \pm \sqrt{b}$.

(2) By transposition,

$$\sqrt{6x^2 - 15x - 7} + \sqrt{4x^2 - 8x - 11} = (2x-3) + \sqrt{2x^2 - 5x + 5} \qquad (87)$$

Now we have identically
$$(6x^2 - 15x - 7) - (4x^2 - 8x - 11) = (2x - 3)^2 - (2x^2 - 5x + 5);$$
hence by division,
$$\sqrt{6x^2 - 15x - 7} - \sqrt{4x^2 - 8x - 11} = (2x - 3) - \sqrt{2x^2 - 5x + 5} \qquad (88)$$
From (87) and (88) by addition, $\sqrt{6x^2 - 15x - 7} = 2x - 3$;

whence $\qquad 2x^2 - 3x - 2 = 0$; so that $x = 2$ or $x = -\dfrac{1}{2}$.

§241. At the first draw he may take 3 red, 3 green, or 2 red and 1 green, or 1 red and 2 green. In finding the chance that at the final draw the three balls are of different colors we may evidently leave out of consideration the first of the above cases.

The chance of each of the other cases $= \dfrac{3 \times {}^3C_2}{{}^6C_3} = \dfrac{9}{20}$.

Then after the 3 blue balls have been dropped into the bag, there are either 2 red, 1 green, 3 blue; or 1 red, 2 green, 3 blue.

In each case the chance of drawing one of each color $= \dfrac{1 \cdot 2 \cdot 3}{{}^6C_3} = \dfrac{6}{20}$.

$\therefore$ the chance of the required event $= 2 \times \dfrac{9}{20} \times \dfrac{6}{20} = \dfrac{27}{100}$. Hence the odds against it are 73 to 27; thus he may lay 72 to 27 or 8 to 3 against it.

§242. Here $f(x) = x^4 - 7x^2 + 4x - 3$, and $f'(x) = 4x^3 - 14x + 4$.

Now S_5 is equal to the coefficient of $\dfrac{1}{x^6}$ in the quotient of $f'(x)$ by $f(x)$.

$$
\begin{array}{r|l}
1 & 4 + 0 - 14 + 4 \\
0 & \quad 0 + 28 - 16 + 12 \\
7 & \quad\quad 0 + 0 + 0 + 0 \\
-4 & \quad\quad\quad 0 + 98 - 56 + 42 \\
3 & \quad\quad\quad\quad 0 - 84 - 48 \ \cdots \\
& \quad\quad\quad\quad\quad 0 + 770 \ \cdots \\
\hline
& 4 + 0 + 14 - 12 + 110 + 140 + \ \cdots \ \cdots
\end{array}
$$

Hence $S_5 = 140$. [Compare Art. 563.]

§243. We have $\qquad a^{q-r} b^{r-p} c^{p-q} = 1$. [*V. a.* Ex. 27.]

Hence $\qquad (q - r) \log a + (r - p) \log b + (p - q) \log c = 0$.

Again, $\qquad (q - r)bc + (r - p)ca + (p - q)ab = 0$. [*V. a.* Ex. 8.]

By cross multiplication we see that $q - r$, $r - p$, $p - q$ are proportional to
$$a(b - c) \log a, b(c - a) \log b, c(a - b) \log c;$$
whence the result is evident, since the sum of $q - r$, $r - p$, $p - q$ is zero.

§244. Denote the numbers by x, y, z, u; then
$$x - y + z + u = 8; \qquad (89)$$
$$x^2 + y^2 - z^2 - u^2 = 36; \qquad (90)$$
$$xy + zu = 42; \qquad (91)$$
$$x^3 - y^3 - z^3 - u^3 = 0 \qquad (92)$$

From (90) and (91) we have $(x - y)^2 - (z + u)^2 = -48$;

dividing this equation (89) we have $(x - y) - (z + u) = -6$;

hence $\qquad x - y = 1, \text{and } z + u = 7$.

Now $\qquad x^3 - y^3 = (x - y)^3 + 3xy(x - y) = 1 + 3xy$;

and $\qquad z^3 + u^3 = (z + u)^3 - 3zu(z + u) = 343 - 21zu$;

hence from (92), we have
$$xy + 7zu = 114.$$
Combining this with (91), we find $xy = 30$, $zu = 12$.

Thus $\qquad x - y = 1$, $xy = 30$; $z + u = 7$, $zu = 12$.

§245.

We have
$$T_{n+2} = aT_{n+1} - bT_n;$$

$$\therefore \frac{1}{b^n}(T_{n+1}^2 - aT_{n+1}T_n + bT_n^2) = \frac{1}{b^n}\{bT_n^2 + T_{n+1}(T_{n+1} - aT_n)\}$$

$$= \frac{1}{b^{n-1}}(T_n^2 - T_{n+1}T_{n-1})$$

$$= \frac{1}{b^{n-1}}\{T_n^2 - T_{n-1}(aT_n - bT_{n-1})\}$$

$$= \frac{1}{b^{n-1}}(T_n^2 - aT_nT_{n-1} + bT_{n-1}^2)$$

$$= \frac{1}{b^{n-2}}(T_{n-1}^2 - aT_{n-1}T_{n-2} + bT_{n-2}^2)$$

$$= \ldots\ldots\ldots\ldots\ldots\ldots\ldots\ldots\ldots\ldots\ldots\ldots$$

$$= T_1^2 - aT_1T_0 + bT_0^2,$$
which is independent of n.

§246.

(1)

We have
$$yz + zx + xy = \frac{xyz}{a} = \frac{d^3}{a};$$

also
$$(x + y + z)^2 = b^2 + 2(yz + zx + xy)$$

$$= b^2 + \frac{2d^3}{a}.$$

Now
$$x^3 + y^3 + z^3 - 3xyz = (x + y + z)(x^2 + y^2 + z^2 - yz - zx - xy);$$

$$\therefore c^3 - 3d^3 = \sqrt{b^2 + \frac{2d^3}{a}} \cdot \left(b^2 - \frac{d^3}{a}\right),$$

or
$$a^3(c^3 - 3d^3)^2 = (ab^2 + 2d^3)(ab^2 - d^3)^2.$$

§247. We shall shew that the roots of the equation
$$x^4 - px^3 + qx^2 - rs = s = 0$$

will be in proportion provided $s = \dfrac{r^2}{p^2}$.

Let a, b, c, d, be the roots, and let $\dfrac{a}{b} = \dfrac{c}{d} = k.$

Now $a + b + c + d = p$; $abc + abd + acd + bcd = r$; $abcd = s$;

or $(b + d)(1 + k) = p$; $bdk(b + d)(1 + k) = r$; $b^2d^2k^2 = s.$

Whence $\dfrac{r}{p} = bdk = \sqrt{s}$; that is, $\dfrac{r^2}{p^2} = s.$

In the case of the equation
$$x^4 - 12x^3 + 47x^2 - 72x + 36 = 0;$$

we have $(b + d)(1 + k) = 12$; $k(b + d)^2 + bd(1 + k^2) = 47$;

$$bdk(b + d)(1 + k) = 72; \quad bdk = 6;$$

therefore $b + d = \dfrac{12}{1 + k}$, and $bd = \dfrac{6}{k}$; by substituting these values in the

second relation we get $47 = k\dfrac{144}{(1 + k)^2} + 6\dfrac{1 + k^2}{k}.$

This equation may be written as follows :
$$47 = 144\frac{k}{(1 + k)^2} + 6\frac{(1 + k)^2}{k} - 12.$$

Put y for $\dfrac{k}{(1+k)^2}$, then $144y^2 - 59y + 6 = 0$,

whence $$y = \frac{3}{16} \text{ or } \frac{2}{9}.$$

Thus we obtain $$k = \frac{1}{3}, 3; \text{ or } \frac{1}{2}, 2.$$

Take $k = 3$, then $b + d = \dfrac{12}{1+k} = 3$, and $bd = \dfrac{6}{k} = 2$; whence $b = 2$, $d = 1$;

therefore $a = 6$, $c = 3$.

Note. The 4 values of k correspond to the ways of stating the proportion between a, b, c, d; namely

$$\frac{a}{b} = \frac{c}{d}, \quad \frac{b}{a} = \frac{d}{c}, \quad \frac{a}{c} = \frac{b}{d}, \quad \frac{c}{a} = \frac{d}{b}.$$

§248. The chance that A, B, C all hit $= \dfrac{4}{5} \cdot \dfrac{3}{4} \cdot \dfrac{2}{3} = \dfrac{2}{5}$.

The chance that A alone misses $= \dfrac{1}{5} \cdot \dfrac{3}{4} \cdot \dfrac{2}{3} = \dfrac{1}{10}$.

The chance that B alone misses $= \dfrac{4}{5} \cdot \dfrac{1}{4} \cdot \dfrac{2}{3} = \dfrac{2}{15}$.

The chance that C alone misses $= \dfrac{4}{5} \cdot \dfrac{3}{4} \cdot \dfrac{1}{3} = \dfrac{1}{5}$.

$\therefore$ the required chance $= \dfrac{2}{5} + \dfrac{1}{10} + \dfrac{2}{15} + \dfrac{1}{5} = \dfrac{5}{6}$.

In the second case we have three hypotheses equally likely, and

$$p_1 = \frac{1}{5} \cdot \frac{3}{4} \cdot \frac{2}{3}; \quad p_2 = \frac{1}{4} \cdot \frac{4}{5} \cdot \frac{2}{3}; \quad p_3 = \frac{1}{3} \cdot \frac{4}{5} \cdot \frac{3}{4};$$

that is, $$p_1 = \frac{1}{10}; \quad p_2 = \frac{2}{15}; \quad p_3 = \frac{1}{5};$$

$$\therefore \frac{Q_1}{3} = \frac{Q_2}{4} = \frac{Q_3}{6} = \frac{Q_1 + Q_2 + Q_3}{13} = \frac{1}{13}.$$

$$\therefore Q_3 = \frac{6}{13}.$$

§249. (1) Forming the successive orders of differences, we have

1	0	-1	0	7	28	79 ...
-1	-1	1	7	21	51 ...	
0	2	6	14	30 ...		
2	4	8	16 ...			

Thus $u_n = an^2 + bn + c + d.2^n$. [See Art. 401.]

The constants may be determined from the equations

$$1 = a + b + c + 2d, \quad 0 = 4a + 2b + c + 4d,$$
$$-1 = 9a + 3b + c + 8d, \quad 0 = 16a + 4b + c + 16d;$$

whence we find $a = -1$, $b = 0$, $c = 0$, $d = 1$.

Thus $u_n = 2^n - n^2$; therefore

$$S_n = 2^{n+1} - 2 - \frac{1}{6}n(n+1)(2n+1).$$

(2) By the method of differences it is easy to whew that the general term of the series $\quad -2 \quad 1 \quad 6 \quad 13$

is $\quad -2 + 3(n-1) + (n-1)(n-2)$, or $n^2 - 3$.

Hence the general term of the series is

$$\frac{(n^2-3)2^n}{n(n+1)(n+2)(n+3)} =$$

$$\frac{\{A(n+1)+B\}2^{n+1}}{(n+1)(n+2)(n+3)} - \frac{(An+B)2^n}{n(n+1)(n+2)} \text{ say;}$$

thus $\qquad n^2 - 3 = 2n\{A(n+1) + B\} - (n+3)(An+B).$

This identity is satisfied is $A = 1$, $B = 1$; hence

$$\frac{(n^2 - 3)2^n}{n(n+1)(n+2)(n+3)} = \frac{(n+2)2^{n+1}}{(n+1)(n+2)(n+3)} - \frac{(n+1)2^n}{n(n+1)(n+2)}$$

Thus $\qquad u_n = \dfrac{2^{n+1}}{(n+1)(n+3)} - \dfrac{2^n}{n(n+2)};$

and therefore $\qquad S = \dfrac{2^{n+1}}{(n+1)(n+3)} - \dfrac{2}{3}.$

(3) The given series $= (1 + x + x^2 + x^3 + x^4 + x^5 + x^6 + \ldots) + (2 + 8x^2 + 32x^4 + 128x^6 + \ldots)$

The sum of the first of these series is $\dfrac{1 - x^{n+1}}{1 - x}$.

If n is even, put $n = 2m$, then the sum of the second of the above series is

$$2\frac{1 - (4x^2)^m}{1 - 4x^2} = \frac{2(1 - 2^n x^n)}{1 - 4x^2}$$

If n is odd, put $n = 2m + 1$; then the series consists of $m+1$ terms and its sum

$$= 2\frac{1 - (4x^2)^{m+1}}{1 - 4x^2} = \frac{2(1 - 2^{n+1}x^{n+1})}{1 - 4x^2}$$

§250.

(1) Multiply the second equation by x and the first equation by y and subtract;

thus $\qquad x^3 - y^3 + z(x^2 - y^2) + z^2(x - y) = 0;$

that is $\qquad (x-y)(x^2 + y^2 + z^2 + yz + zx + xy) = 0.$

Similarly $\qquad (x-z)(x^2 + y^2 + z^2 + yz + zx + xy) = 0.$

Hence $\qquad x = y = z;$ or $\quad x^2 + y^2 + z^2 + yz + zx + xy = 0;$

or two of the quantities x, y, z may be equal, and $x^2 + y^2 + z^2 + yz + zx + xy = 0.$

If $x = y = z$, we have $3x^3 = ax$, so that $x = 0$, or $\dfrac{a}{3}$.

If $x^2 + y^2 + z^2 + yz + zx + xy = 0$, we have from the first equation

$$x^2 + xy + xz + ax = 0; \qquad \text{and therefore} \qquad x + y + z = -a;$$

in this case the solution is indeterminate, for the given equations hold if the relations $x + y + z = -a$, and $x^2 + y^2 + z^2 + yz + zx + xy = 0$ are satisfied. We see moreover that the third case need not be discussed.

(2) We have $\dfrac{\dfrac{a}{x}}{y+z-x} = \dfrac{\dfrac{b}{y}}{z+x-y} = \dfrac{\dfrac{c}{z}}{x+y-z} = \dfrac{\dfrac{b}{y} + \dfrac{c}{z}}{2x} = \dfrac{bz + cy}{2xyz}.$

Hence $\qquad bz + cy = cx + az = ay + bx;$

that is, $\qquad cx - cy + (a-b)z = 0, \quad$ and $\quad bx + (a-c)y - bz = 0;$

whence $\qquad \dfrac{x}{a(-a+b+c)} = \dfrac{y}{b(a-b+c)} = \dfrac{z}{c(a+b-c)}.$

Putting each of the above fractionns equal to k, we have

$$k^2 a(-a+b+c)(a-b+c)(a+b-c) = a.$$

§251.

(1) If $\dfrac{1}{a} + \dfrac{1}{b} + \dfrac{1}{c} = \dfrac{1}{a+b+c}$,

then $\qquad (a+b+c)(bc + ca + ab) - abc = 0;$

that is, $\qquad (b+c)(c+a)(a+b) = 0.$

If $b + c = 0$, then $b = -c$; and therefore $b^n = -c^n$, or $b^n + c^n = 0$; in this case each side of the identity to be proved reduces to $\dfrac{1}{a^n}$.

(2) From the given relation, we have
$$-u^6 + v^6 + 3u^2v^2(u^2 - v^2) = 4uv\{1 - u^4v^4 + 2uv(u^2 - v^2)\};$$
or
$$-(u^2 - v^2)^3 = 4uv\{1 - u^4v^4 + 2uv(u^2 - v^2)\}.$$

$$\therefore (u^2 - v^2)^6 = 16u^2v^2\{(1 - 2u^4v^4 + u^8v^8) \tag{93}$$
$$+ 4u^2v^2(u^2 - v^2)(1 - u^4v^4) + 4uv(u^2 - v^2)(1 - u^4v^4)\} \tag{94}$$

But from the given relation, we have
$$4u^2v^2(u^2 - v^2) + 4uv(1 - u^4v^4) = -(u^6 - v^6) - u^2v^2(u^2 - v^2).$$

Multiply each side of this equation by $u^2 - v^2$ and substitute in (93); thus
$$(u^2 - v^2)^6 = 16u^2v^2\{1 - 2u^4v^4 + u^8v^8 - (u^2 - v^2)(u^6 - v^6)$$
$$- u^2v^2(u^2 - v^2)^2\}$$
$$= 16u^2v^2(1 - u^6 - v^6 + u^8v^8)$$
$$= 16u^2v^2(1 - u^6)(1 - v^6).$$

§252.
(1) Here x, y, z are the roots of $t^3 - 3pt^2 + 3qt - r = 0$.
Let $u = y + z - x$; then $u = (y + z + x) - 2x$; so that we may put
$$u = 3p - 2t, \quad \text{or} \quad 2t = 3p - u;$$
hence
$$(3p - u)^3 - 6p(3p - u)^2 + 12q(3p - u) - 8r = 0,$$
or
$$u^3 - 3pu^2 - (9p^2 - 12q)u + 27p^3 - 36qp + 8r = 0.$$
The product of the roots is $-27p^3 + 36qp - 8r$; which proves the first part of the question.

(2) For the second part we have to find the sum of the cubes of the roots of the equation in u. Denote the roots by u_1, u_2, u_3; then
$$\Sigma u = 3p;$$
$$\Sigma u^2 = (\Sigma u)^2 - 2\Sigma u_1 u_2 = (3p)^2 - 2(-9p^2 + 12q) = 27p^2 - 24q.$$
Again by writing u_1, u_2, u_3 successively for u and adding, we have
$$\Sigma u^3 - 3p\Sigma u^2 - (9p^2 - 12q)\Sigma u + 3(27p^3 - 36pq + 8r) = 0;$$
$$\therefore \Sigma u^3 = (81p^3 - 72pq) + (27p^3 - 36pq) - (81p^3 - 108pq + 24r)$$
$$= 27p^3 - 24r.$$

§253. The coefficient of $x^4 = a^2(b + c)^2 - 4a^2bc = a^2(b - c)^2$.
The coefficient of $y^2z^2 = 2bc(a + b)(a + c) - 4abc(b + c) = 2bc(a - b)(a - c)$.
Let
$$a(b - c) = A^2, \quad b(c - a) = B^2, \quad c(a - b) = C^2,$$
then the given expression
$$= A^4x^4 + B^4y^4 + C^4z^4 - 2B^2C^2y^2z^2 - 2C^2A^2z^2x^2 - 2A^2B^2x^2y^2$$
$$= -(Ax + By + Cz)(-Ax + By + Cz)$$
$$(Ax - By + Cz)(Ax + By - Cz).$$

§254. If x, y, z are not integers, we can find an integer p which will make px, py, pz integral.
The expression $x^{px}y^{py}z^{pz}$ is the product of $px + py + pz$ factors, and the arithmetic mean of these factors is
$$\frac{px^2 + py^2 + pz^2}{px + py + pz} \quad \text{or} \quad \frac{x^2 + y^2 + z^2}{x + y + z}$$
Thus
$$\left(\frac{x^2 + y^2 + z^2}{x + y + z}\right)^{px+py+pz} > x^{px}y^{py}z^{pz}. \quad [Art.253]$$
By taking p^{th} root we get the required result.
For the second part see solution of Ex.6 $XIXb$.

§255. The expansion of $(1 - 4y)^{-\frac{1}{2}}$ is

$$1 + p_1 y + p_2 y^2 + p_3 y^3 + \ldots + p_r y^r + \ldots,$$

where $\qquad p_r = \dfrac{\lfloor 2r}{\lfloor r \ \lfloor r}.$ [See Example 33. $XIVb$]

If we put for y its equivalent $x(1 + x)^{-2}$, we shall have a series whose general term is

$$\frac{\lfloor 2r}{\lfloor r \ \lfloor r} x^r \left\{ 1 - 2rx + \frac{2r(2r + 1)}{1.2} x^2 - \ldots \right\}.$$

In this and all subsequent terms pick out the coefficients of x^n and equate their sum to the coefficient of x^n in $(1 + x)(1 - x)^{-1}$.

Then $\qquad 2 = \displaystyle\sum_{r=1}^{r=n} \frac{\lfloor 2r}{\lfloor r \ \lfloor r} (-1)^{n-r} \cdot \frac{2r.(2r + 1)\ldots(n + r - 1)}{\lfloor n - r}$

$$\therefore 1 = \sum_{r=1}^{r=n} (-1)^{n-r} \frac{\lfloor 2r}{\lfloor r \ \lfloor r} \cdot \frac{r(2r + 1)\ldots(n + r - 1)}{\lfloor n - r}$$

$$= \sum_{r=1}^{r=n} (-1)^{n-r} \frac{\lfloor n + r - 1}{\lfloor r \ \lfloor r - 1 \ \lfloor n - r}.$$

§256.
(1) Substitute $z = -(ax + by)$ in the second and third equations; thus $x(ax + by) = ay + b$, and $y(ax + by) = bx + a$.

From the first of these equations, $y = \dfrac{ax^2 - b}{a - bx}$; so that $ax + by = \dfrac{a^2 x - b^2}{a - bx}$.

By substituting in the second equation, we have

$$(ax^2 - b)(a^2 x - b^2) = (bx + a)(bx - a)^2;$$

whence $(a^3 - b^3)(x^3 - 1) = 0$; and therefore the values of x are $1, \omega, \omega^2$.

The values of y and z are obtained from $y = \dfrac{ax^2 - b}{a - bx}$; $z = -(ax + by)$.

(2) From the second and fourth equations, we have

$$(x + y)^2 - (z - u)^2 = 96; \tag{95}$$

but

$$(x + y) + (z - u) = 12; \text{ and therefore } (x + y) - (z - u) = 8;$$

whence $\qquad x + y = 10, \text{ and } z - u = 2$

Now $\qquad x^3 + y^3 = (x + y)^3 - 3xy(x + y) = 1000 - 30xy;$

and $\qquad z^3 - u^3 = (z - u)^3 + 3zu(z - u) = 8 + 6zu.$

By substitution of these values in the third equation, we obtain

$$992 - 30xy - 6zu = 218, \text{ or } 5xy + zu = 129.$$

From this equation and the fourth of the given equations, we find $xy = 21$, and $zu = 24$.

The solutions are therefore given by

$$\left. \begin{array}{l} x + y = 10, \\ xy = 21. \end{array} \right\} \text{ and } \left. \begin{array}{l} z - u = 2, \\ zu = 24. \end{array} \right\}$$

§257. Put $p = q + x$, where x is very small; then

$$\frac{(n + 1)p + (n - 1)q}{(n - 1)p + (n + 1)q} = \frac{2nq + (n + 1)x}{2nq + (n - 1)x}$$

$$= \left(1 + \frac{n+1}{2nq}x\right)\left(1 + \frac{n-1}{2nq}x\right)^{-1}$$

$$= \left(1 + \frac{n+1}{2nq}x\right)\left(1 - \frac{n-1}{2nq}x\right), \text{ neglecting } x^2,$$

$$= \left(1 + \frac{1}{nq}x\right) = \left(1 + \frac{x}{q}\right)^{\frac{1}{n}} = \left(\frac{p}{q}\right)^{\frac{1}{n}}.$$

Taking in terms as far as x^3, the left side of the given equation

$$= \left(1 + \frac{n+1}{2nq}x\right)\left\{1 - \frac{n-1}{2nq}x + \frac{(n-1)^2}{4n^2q^2}x^2 - \frac{(n-1)^3}{8n^3q^3}x^3 + \ldots\right\};$$

and the right side of the given equation

$$= 1 + \frac{x}{nq} - \frac{n-1}{2n^2q^2}x^2 + \frac{(n-1)(n-2)}{6n^3q^3}x^3 - \ldots$$

In these expressions the difference between the coefficients of x^3

$$= \frac{(n-1)^2 - (n+1)(n-1) + 2(n-1)}{4n^2q^2} = 0.$$

The difference between the coefficients of x^3 is

$$= \frac{6(n+1)(n-1)^2 - 3(n-1)^3 - 8(n-1)(n-2)}{48n^3q^3}$$

$$= \frac{(n-1)(3n^2 - 2n + 7)}{48n^3q^3}$$

Thus the difference is of the order $\dfrac{x^3}{q^3}$, and as $\dfrac{x}{q}$ is a decimal beginning

with $r - 1$ ciphers, $\dfrac{x^3}{q^3}$ will be a decimal beginning with at least $3r - 3$

ciphers.

§258. Denote the prices of a lb. of tea and a lb. of coffee by x and y shillings respectively, and the amounts bought by u and v lbs. respectively; then the amount spent $= ux + vy$ shillings

Hence $\qquad \dfrac{5}{6}ux + \dfrac{4}{5}vy = \dfrac{9}{11}(ux + vy);$

that is, $\qquad \dfrac{ux}{6} = \dfrac{vy}{5}, \text{ or } \dfrac{u}{6y} = \dfrac{v}{5x}.$

Again $\qquad vx + uy = ux + vy + 5;$ so that $(x - y)(v - u) = 5.$

Also $\qquad u + v = 54, \text{ and } 6y - 2x = 5.$

Hence $\qquad \dfrac{v+u}{v-u} = \dfrac{54(x-y)}{5} = \dfrac{27(x-y)}{3y-x};$

but $\qquad \dfrac{v+u}{v-u} = \dfrac{5x+6y}{5x-6y};$

Therefore $\qquad \dfrac{5x+6y}{5x-6y} = \dfrac{25(x-y)}{3y-x};$

or $\qquad 70x^2 - 153xy + 72y^2 = 0;$ whence $(2x - 3y)(35x - 24y) = 0.$

Combining $2x - 3y = 0$ and $6y - 2x = 5$, we have $x = 2\frac{1}{2}, y = 1\frac{2}{3}.$

By hypothesis tea costs more than coffee, and therefore $35x - 24y = 0$ is inadmissible.

§259.

Here $\qquad 2s_n = (1 + 2 + 3 + \ldots + n)^2 - (1^2 + 2^2 + 3^3 + \ldots + n^2)$

$$= \frac{n^2(n+1)^2}{4} - \frac{n(n+1)(2n+1)}{6};$$

$$\therefore s_n = \frac{1}{24}(n-1)n(n+1)(3n+2);$$

$$\therefore s_{n-1} = \frac{(n-2)(n-1)n(3n-1)}{24}.$$

Now
$$\Sigma \frac{s_{n-1}}{\lfloor n} = \frac{1}{24}\Sigma\frac{3n-1}{\lfloor n-3} = \frac{1}{24}\Sigma\left\{\frac{3}{\lfloor n-4} + \frac{8}{\lfloor n-3}\right\}$$

$$= \frac{1}{24}\left\{3\left(1 + \frac{1}{\lfloor 1} + \frac{1}{\lfloor 2} + \frac{1}{\lfloor 3} + \ldots\right)\right.$$

$$\left. +8\left(1 + \frac{1}{\lfloor 1} + \frac{1}{\lfloor 2} + \frac{1}{\lfloor 3} + \ldots\right)\right\}$$

$$= \frac{1}{24}(3e + 8e) = \frac{11}{24}e.$$

§260. If $\frac{1}{k}$ is the value of each ratio, we have

$$pa^2 + 2qab + rb^2 = kP \tag{96}$$

$$pac + q(bc - a^2) - rab = kQ \tag{97}$$

$$pc^2 - 2qac + ra^2 = kR \tag{98}$$

Multiply (96) by a, and (97) by b; then by addition, we have

$$pa(a^2 + bc) + qb(a^2 + bc) = k(aP + bQ);$$

that is, $(pa + qb)(a^2 + bc) = k(aP + bQ).$

Similarly from (97) and (98) we obtain

$$(pc - qa)(a^2 + bc) = k(aQ + bR);$$

$$\therefore \frac{pa + qb}{pc - qa} = \frac{aP + bQ}{aQ + bR};$$

that is, $\dfrac{p}{q} = \dfrac{Pa^2 + 2Qab + Rb^2}{Pac + Q(bc - a^2) - Rab}.$

If we eliminate p instead of r, we find

$$\frac{r}{q} = \frac{Pc^2 - 2Qac + Ra^2}{Pac + Q(bc - a^2) - Rab}.$$

§261. Let α, β, γ denote the roots of the cubic equation

$$x^3 + qx + r = 0.$$

Multiply this equation by x^n, substitute in succession α, β, γ for x, and add; then

$$(\alpha^{n+3} + \beta^{n+3} + \gamma^{n+3}) + q(\alpha^{n+1} + \beta^{n+1}$$
$$+ \gamma^{n+1}) + r(\alpha^n + \beta^n + \gamma^n) = 0;$$

but

$$q = \beta\gamma + \gamma\alpha + \alpha\beta = \frac{1}{2}\left\{(\alpha + \beta + \gamma)^2 - (\alpha^2 + \beta^2 + \gamma^2)\right\}$$

$$= -\frac{1}{2}(\alpha^2 + \beta^2 + \gamma^2);$$

and $r = -\alpha\beta\gamma;$ whence the result at once follows.

§262. The expression $\Sigma(\alpha - \beta)^2(\gamma - \delta)^2$ consists of three separate kinds of terms, and when multiplied out and arranged is easily seen to be

$$2\Sigma\alpha^2\beta^2 - 2\Sigma\alpha\beta\gamma^2 + 12\Sigma\beta\gamma\delta.$$

Now $\Sigma\alpha^2\beta^2 = (\Sigma\alpha\beta)^2 - 2\Sigma\alpha.\,\Sigma\alpha\beta\gamma + 2\Sigma\alpha\beta\gamma\delta;$

and $\Sigma\alpha\beta\gamma^2 = \Sigma\alpha.\,\Sigma\alpha\beta\gamma - 4\Sigma\alpha\beta\gamma\delta.$

Thus the given function becomes

$$2(\Sigma\alpha\beta)^2 - 6\Sigma\alpha.\,\Sigma\alpha\beta\gamma + 24\Sigma\alpha\beta\gamma\delta,$$

or $2q^2 - 6pr + 24s.$

[This solution is due to Professor Steggall.]

§263. Denote the number of turkeys, geese, and ducks by x, y, z respectively; then we have
$$x^2 + y^2 + z^2 = 211, \text{ and } x + y + z = 23.$$
Eliminating z, we obtain
$$x^2 + xy + y^2 - 23(x + y) + 159 = 0;$$
hence
$$2x = -(y - 23) \pm \sqrt{-3y^2 + 46y - 107}.$$
Thus
$$-3y^2 + 46y - 107 = u^2 \text{ say;}$$
that is,
$$3y^2 - 46y + 107 + u^2 = 0 \tag{99}$$
whence
$$3y = 23 \pm \sqrt{208 - 3u^2}$$

Thus $208 - 3u^2 = t^2$; hence u must be less than 9; by trial we find that $u = 2, 6, 8$. On substituting in (99) we have
$$3y^2 - 46y + 107 = -4, \text{ or } -36, -64.$$
The integral values of y found from these equations are $3, 11, 9$.

§264. If $a^{\frac{1}{3}} + b^{\frac{1}{3}} + c^{\frac{1}{3}} = 0$, then $a + b + c = 3a^{\frac{1}{3}}b^{\frac{1}{3}}c^{\frac{1}{3}}$;
hence from the given equation, we have
$$3\left\{(y + z - 8x)(z + x - 8y)(x + y - 8z)\right\}^{\frac{1}{3}} = -6(x + y + z);$$
and therefore
$$(y + z - 8x)(z + x - 8y)(x + y - 8z) = -8(x + y + z)^3.$$
Put $x + y + z = p$, then we have
$$(p - 9x)(p - 9y)(p - 9z) = -8p^3;$$
that is,
$$p^3 - 9p^2(x + y + z) + 81p(yz + zx + xy) - 729xyz = -8p^3;$$
or
$$p^3 - 9p^3 + 81p(yz + zx + xy) - 729xyz = -8p^3;$$
or
$$(x + y + z)(yz + zx + xy) - 9xyz = 0;$$
that is,
$$x^2(y + z) + y^2(z + x) + z^2(x + y) - 6xyz = 0;$$
whence the result at once follows.

§265. We have $\dfrac{a}{x + a} - \dfrac{c}{x + c} = \dfrac{d}{x + d} - \dfrac{b}{x + b}$;
$$\therefore \frac{(a - c)x}{(x + a)(x + c)} = \frac{(d - b)x}{(x + b)(x + d)} \tag{100}$$

Thus $(a + b - c - d)x^2 + 2(ab - cd)x + ab(c + d) - cd(a + b) = 0$ (101)
or $x = 0$.

If the given equation has two equal roots, then either equation (101) has two equal roots, or it has a root equal to zero. In this latter case, the absolute term vanishes, so that
$$ab(c + d) - cd(a + b) = 0, \quad \text{or } \frac{1}{a} + \frac{1}{b} + \frac{1}{c} + \frac{1}{d}.$$
The remaining root is then $-\dfrac{2(ab - cd)}{a + b - c - d}$, which is equal to $-\dfrac{2ab}{a + b}$,
for
$$\frac{ab}{a + b} = \frac{cd}{c + d} = \frac{ab - cd}{a + b - c - d}$$
If equation (101) has a pair of equal roots, then
$$(ab - cd)^2 = \left\{ab(c + d) - cd(a + b)\right\}(a + b - c - d) \tag{102}$$
that is,
$$(a - c)(a - d)(b - c)(b - d) = 0;$$
for equation (102) is satisfied when $a = c$, $a = d$, $b = c$, $b = d$, and is of two dimensions in a, and also of two dimensions in b.

Thus one of the quantities a or b is equal to one of the quantities c or d. Suppose that $a = c$; then
$$\text{each of the equal roots} = -\frac{ab - cd}{a + b - c - d} = -a.$$

Similarly in the other cases.

§266.

(1) By multiplying together the second and third equations, we have
$$yz + zx + xy = a^2 b;$$
Thus $\qquad x + y + z = ab, \; yz + zx + xy = a^2 b, \; xyz = a^3;$
hence x, y, z are the roots of
$$t^3 - abt^2 + a^2 bt - a^3 = 0; \quad \text{or} \quad (t - a)(t^2 + at + a^2 - abt) = 0.$$

(2) From the first and second equations, we have
$$z(ay - bx) = ax - by;$$
hence by substituting in $ax + (bx + c)z = a + b + c$, we have
$$ax + \frac{(bx + c)(ax - by)}{ay - bx} = a + b + c;$$
that is,
$$(a^2 - b^2)xy + x(ac + ab + b^2 + bc) - y(bc + a^2 + ab + ac) = 0;$$
or $\qquad (a - b)xy + x(b + c) - y(a + c) = 0;$
whence
$$y = \frac{(b + c)x}{(a + c) - (a - b)x}.$$
Substituting in the equation $cxy + ax + by = a + b + c$, we have
$$\frac{(b + c)(cx + b)x}{(a + c) - (a - b)x} = a + b + c - ax;$$
or $\qquad (bc + c^2 - a^2 + ab)x^2 + Px - (a + c)(a + b + c) = 0.$

Now the given equations are obviously satisfied by $x = 1, \; y = 1, \; z = 1$; hence the other value of $x = \dfrac{(a + c)(a + b + c)}{a^2 - c^2 - ab - bc} = \dfrac{a + b + c}{a - b - c}.$

§267. Let $f(x) = (x - \alpha)(x - \beta) \ldots (x - \epsilon),$

and let $\phi(x)$ be an expression of degree not above the fourth; then $\phi(x) \div f(x)$ may be resolved into partial fractions. We have, as usual,
$$\frac{\phi(x)}{f(x)} = \frac{\phi(\alpha)}{(x - \alpha)(\alpha - \beta)(\alpha - \gamma)(\alpha - \delta)(\alpha - \epsilon)} + \text{similar terms.}$$
If $x = 0$, we obtain
$$\frac{\phi(0)}{f(0)} = \frac{\phi(\alpha)}{-\alpha(\alpha - \beta)(\alpha - \gamma)(\alpha - \delta)(\alpha - \epsilon)}$$
$$+ \frac{\phi(\beta)}{-\beta(\beta - \alpha)(\beta - \gamma)(\beta - \delta)(\beta - \epsilon)} + \ldots$$
For the given example, we take $\phi(x) = x^4$, so that $\phi(0) = 0$; thus
$$\frac{\alpha^3}{(\alpha - \beta)(\alpha - \gamma)(\alpha - \delta)(\alpha - \epsilon)}$$
$$+ \frac{\beta^3}{(\beta - \alpha)(\beta - \gamma)(\beta - \delta)(\beta - \epsilon)} + \ldots = 0.$$
The more general theorem, which can be proved in the same way, is found by resolving $x\phi(x) \div f(x)$ into partial fractions; where $f(x)$ is of n dimensions in x, and $\phi(x)$ of $n - 2$ dimensions. In this case
$$\frac{\phi(\alpha)}{(\alpha - \beta)(\alpha - \gamma) \ldots} + \frac{\phi(\beta)}{(\beta - \alpha)(\beta - \gamma) \ldots} + \ldots = 0.$$
[This solution is due to Professor Steggall.]

§268. Let x, y, z denote the number of Clergymen, Doctors, and Lawyers respectively; u, v, w their average ages; then
$$ux + vy + wz = 2160;$$

$$\frac{ux + vy + wz}{x + y + z} = 36; \text{ so that } x + y + z = 60.$$

$$ux + vy = 39(x + y); vy + wz = 32\frac{8}{11}(y + z); ux + wz = 36\frac{2}{3}(x + z).$$

From these three equations, we have

$$2(ux + vy + wz) = 75\frac{2}{3}x + 71\frac{8}{11}y + 69\frac{13}{33}z.$$

But $\qquad\qquad ux + vy + wz = 36(x + y + z);$

therefore $\qquad 72x + 72y + 72z = 75\frac{2}{3}x + 71\frac{8}{11}y + 69\frac{13}{33}z.$

or $\qquad\qquad 121x - 9y - 86z = 0.$

The increased average age is $\dfrac{x + 6y + 7z}{x + y + z};$

but this is equal to 5; hence $4x - y - 2z = 0.$

From the last two equations, we have by cross multiplication, $\dfrac{x}{4} =$

$\dfrac{y}{6} = \dfrac{z}{5};$ but $x + y + z = 60;$ hence $x = 16, y = 24, z = 20.$

Again $\qquad\qquad 16u + 24v = 39 \times 40; \text{ that is, } 2u + 3v = 195;$

$$24v + 20w = \frac{360}{11} \times 44; \text{ that is, } 6v + 5w = 360;$$

$$16u + 20w = \frac{110}{3} \times 36; \text{ that is, } 4u + 5v = 330;$$

whence $\qquad\qquad u = 45, v = 35, w = 30.$

§269. Let the two expressions be $ax + by$ and $cx + dy$; then we have the identity $a_0x^4 + 4a_1x^3y + \ldots + a_4y^4 = (ax + by)^4 + (cx + dy)^4;$
hence, equating coefficients,

$$a_0 = a^4 + c^4, \ a_1 = a^3b + c^3d,$$

$$a_2 = a^2b^2 + c^2d^2, \ a_3 = ab^3 + cd^3, \ a_4 = b^4 + d^4.$$

From these equations, we obtain

$$a_0a_2 - a_1^2 = a^2c^2(ad - bc)^2;$$

$$a_1a_3 - a_2^2 = abcd(ad - bc)^2;$$

$$a_2a_4 - a_3^2 = b^2d^2(ad - bc)^2;$$

and therefore the condition required is

$$(a_0a_2 - a_1^2)(a_2a_4 - a_3^2) = (a_1a_3 - a_2^2)^2.$$

We may also proceed as follows:

$$bda_0 + aca_2 = bd(a^4 + c^4) + ac(a^2b^2 + c^2d^2) = (ad + bc)(a^3b + c^3d);$$

that is, $\qquad bda_0 - (ad + bc)a_1 + aca_2 = 0.$

Similarly, $\qquad bda_1 - (ad + bc)a_2 + aca_3 = 0,$

and $\qquad\qquad bda_2 - (ad + bc)a_3 + aca_4 = 0;$

from which, by eliminating $bd, ad + bc, ac,$ we have

$$\begin{vmatrix} a_0 & a_1 & a_2 \\ a_1 & a_2 & a_3 \\ a_2 & a_3 & a_4 \end{vmatrix} = 0.$$

A general theorem, of which the above is a particular case, is proved in Salmon's Higher Algebra, Arts. 168, 171.

[This solution is due to Professor Steggall.]

§270. We have $(y^2 + u^2 + w^2)(z^2 + u^2 + v^2) = b^2c^2 = (vw + uy + uz)^2;$
on reduction we obtain

$$(u^2 - yz)^2 + (wu - vy)^2 + (uv - wz)^2 = 0.$$

Since the roots are real we must have

$$u^2 - yz = 0, \ wu - vy = 0, \ uv - wz = 0.$$

From the other equations we obtain similar results; hence
$$u^2 = yz, \quad v^2 = zx, \quad w^2 = xy;$$
$$vw = ux, \quad wu = vy, \quad uv = wz.$$
Substituting these values in the given equations, we have
$$x(x+y+z) = a^2, \quad u(x+y+z) = bc,$$
$$y(x+y+z) = b^2, \quad v(x+y+z) = ca,$$
$$z(x+y+z) = c^2, \quad w(x+y+z) = ab.$$
Hence $(x+y+z)^2 = a^2 + b^2 + c^2$; and thus
$$x = \pm \frac{a^2}{\lfloor a^2+b^2+c^2}; \quad u = \pm \frac{bc}{\lfloor a^2+b^2+c^2}.$$

§271. We have $n+3$ letters in all, of which three are the vowels a, e, o. The consonants can stand in any of the $n+3$ places so long as their position does not involve an ineligible arrangement of vowels.

The vowels in any word can occur in the following orders:
$$(1)\ aeo;\ (2)\ ora;\ (3)\ aoe;\ (4)\ eoa;\ (5)\ eao\ (6)\ oae.$$

Now (1) and (2) cannot stand at all unless all the vowels come together. therefore there will be $2\lfloor n+1$ words which have the vowels arranged in this order.

Now consider any one of the four remaining cases, such as aoe. Here oe must come together in any word, and a must precede oe. Therefore in considering the number of words possible with this arrangement, we have only to select two places out of $n+2$, and then fill up the remaining n places with consonants. This gives rise to $^{n+2}C_2 \times \lfloor n$ words. It will be found that each of the three remaining cases gives this same number of words. Thus on the whole the number of words is $2\lfloor n+1 + 2(n+2)(n+1)\lfloor n$, which easily reduces to the required form.

§272. We have $x^2 - z^2 = z^2 - y^2$; that is, $(x+z)(x-z) = (z+x)(z-y)$.

This equation is satisfied if
$$k(x+z) = l(z+y), \text{ and } l(x-z) = k(z-y);$$
that is, $\quad kx - ly + (k-l)z = 0$, and $lx + ky - (k+l)z = 0$.

By cross multiplication, we obtain
$$\frac{x}{2kl + l^2 - k^2} = \frac{y}{k^2 + 2lk - l^2} = \frac{z}{k^2 + l^2} = \frac{r}{2} \text{ say.}$$

§273. Here the nth convergent is $\dfrac{2(n-2)}{2n-3}$;

hence $\quad u_n = (2n-3)u_{n-1} + 2(n-2)u_{n-2};$

that is, $\quad u_n - 2(n-1)u_{n-1} = -\{u_{n-1} - 2(n-2)u_{n-2}\};$

$$\cdots\cdots\cdots\cdots\cdots\cdots\cdots\cdots\cdots$$

$$u_3 - 2 \cdot 2u_2 = -(u_2 - 2u_1);$$

whence, by multiplication we obtain
$$u_n - 2(n-1)u_{n-1} = (-1)^{n-2}(u_2 - 2u_1).$$

But $p_1 = 1; p_2 = 1; q_1 = 1; q_2 = 2$; hence
$$p_n - 2(n-1)p_{n-1} = (-1)^{n-1}, \quad q_n - 2(n-1)q_{n-1} = 0.$$

Thus
$$q_n = 2(n-1)q_{n-1} = 2^2(n-1)(n-2)q_{n-2} = \cdots = 2^{n-1}\lfloor n-1.$$

Again
$$\frac{p_n}{\lfloor n-1} - \frac{2p_{n-1}}{\lfloor n-2} = \frac{(-1)^{n-1}}{\lfloor n-1};$$

$$\frac{2p_{n-1}}{\lfloor n-2} - \frac{2^2 p_{n-2}}{\lfloor n-3} = \frac{2(-1)^{n-2}}{\lfloor n-1};$$

$$\cdots\cdots\cdots\cdots\cdots\cdots\cdots\cdots\cdots$$

$$\frac{2^{n-2}p_2}{\underline{|1}} - 2^{n-1}p_1 = \frac{2^{n-2}(-1)}{\underline{|1}}$$

hence by addition, $\dfrac{p_n}{\underline{|n-1}} - 2^{n-1} = -2^{n-2} + \dfrac{2^{n-3}}{\underline{|2}} - \dfrac{2^{n-4}}{\underline{|3}} - \ldots;$

that is, $\dfrac{p_n}{2^{n-1}\underline{|n-1}} = 1 - \dfrac{1}{2} + \dfrac{1}{2^2\underline{|2}} - \dfrac{1}{2^3\underline{|3}} + \ldots;$ and therefore $\dfrac{p_n}{q_n} = e^{\frac{-1}{2}}.$

§274.

(1) We have $\dfrac{n}{(n+1)(n+2)} = -\dfrac{1}{n+1} + \dfrac{2}{n+2}.$

Thus the series $= x^2\left(-\dfrac{1}{2} + \dfrac{2}{3}\right) + x^3\left(-\dfrac{1}{3} + \dfrac{2}{4}\right) + x^4\left(-\dfrac{1}{4} + \dfrac{2}{5}\right) + \ldots$

$$= -\left(\frac{x^2}{2} + \frac{x^3}{3} + \frac{x^4}{4} + \ldots\right) + \frac{2}{x}\left(\frac{x^3}{3} + \frac{x^4}{4} + \frac{x^5}{5} + \ldots\right)$$

$$= \{x + \log(1-x)\} + \frac{2}{x}\left\{-x - \frac{x^2}{2} - \log(1-x)\right\}$$

(2) We have $\dfrac{\underline{|n}}{(a+1)(a+2)\ldots(a+n)}$

$$= \frac{1}{a-1}\left\{\frac{\underline{|n}}{(a+1)(a+2)\ldots(a+n-1)} - \frac{\underline{|n+1}}{(a+1)(a+2)\ldots(a+n)}\right\};$$

and $\dfrac{\underline{|1}}{a+1} = \dfrac{1}{a-1}\left(1 - \dfrac{2}{a+1}\right);$

hence $S = \dfrac{1}{a-1}\left\{1 - \dfrac{\underline{|n+1}}{(a+1)(a+2)\ldots(a+n)}\right\}.$

§275.

(1) Put $2x = u, 3y = v, 4z = w;$ then
$$uvw = -36, \quad (u-1)(v+1)(w-1) = -12,$$
$$(u+1)(v-1)(w+1) = -80.$$

Thus from the second equation
$$uvw + (-vw + wu - uv) + (-u + v - w) - 1 = -12;$$
or $\qquad (-vw + wu - uv) + (-u + v - w) = 23.$

Similarly from the third equation
$$(vw - wu + uv) + (-u + v - w) = -43.$$

From the last two equations, we have
$$vw - wu + uv = -33, \text{ and } -u + v - w = -10$$

that is, $v(-w) + (-w)(-u) + (-u)v = 33,$ and $(-u) + v + (-w) = -10.$

Thus $-u, +v, -w$ are the roots of the equation
$$t^3 + 10t^2 + 33t + 36 = 0,$$
or $\qquad (t+3)(t+3)(t+4) = 0;$

and therefore $-u, v, -w$ are the permutations of the quantities $-3, -3, -4;$ that is, $2x, -3y, 4z$ are the permutations of the quantities $3, 3, 4.$

(2) From the equations
$$3ux - 2vy = 14, \quad vx + uy = 14,$$

we have
$$(3u^2 + 2v^2)x = 14(u + 2v), \text{ and } (3u^2 + 2v^2)y = 14(3u - v).$$

But $\qquad\qquad\qquad 3u^2 + 2v^2 = 14.$

$$\therefore x = u + 2v, \quad y = 3u - v;$$
$$\therefore (u + 2v)(3u - v) = 10uv;$$

that is, $3u^2 - 5uv - 2v^2 = 0$, or $(u - 2v)(3u + v) = 0$.

Taking $u = 2v$, and combining with $3u^2 + 2v^2 = 14$, we have
$$u = \pm 2, \quad v = \pm 1.$$

Similarly from $v = -3u$, we have
$$u = \mp \frac{1}{3}\sqrt{\frac{2}{3}}, \quad v = \pm\sqrt{\frac{2}{3}}.$$

§276. Keeping the first row unaltered, multiply the second, third and fourth rows by a; this is equivalent to multiplying the determinant by a^3. Next multiply the first row of the new determinant by b, c, d and subtract from the new second, third, and fourth rows respectively : thus

$$a^3\Delta = \begin{vmatrix} a^2 + \lambda & ab & ac & ad \\ -b\lambda & a\lambda & 0 & 0 \\ -c\lambda & 0 & a\lambda & 0 \\ -d\lambda & 0 & 0 & a\lambda \end{vmatrix} = a^3\lambda^3 \begin{vmatrix} a^2 + \lambda & b & c & d \\ -b & 1 & 0 & 0 \\ -c & 0 & 1 & 0 \\ -d & 0 & 0 & 1 \end{vmatrix}$$

Thus remaining factor is the last determinant, which reduces to
$$a^2 + b^2 + c^2 + d^2 + \lambda.$$

§277. Here $\Sigma a = -p_1$, $\Sigma ab = p_2$, $\Sigma abc = -p_3$;

and therefore $\Sigma a^2 = p_1^2 - 2p_2$.

Now $(\Sigma a)^3 = \Sigma a^3 + 3\Sigma a^2 b + 6\Sigma abc$.   [*Art.* 522]

Thus $-p_1^3 = \Sigma a^3 + 3\Sigma a^2 b - 6p_3$.

Also $\Sigma a^2 \cdot \Sigma a = \Sigma a^3 + \Sigma a^2 b$;

that is $-p_1(p_1^2 - 2p_2) = \Sigma a^3 + \Sigma a^2 b$;

by eliminating $\Sigma a^2 b$ from the last two equations, we have
$$2\Sigma a^3 + 6p_3 = -3p_1(p_1^2 - 2p_2) + p_1^3;$$

or $\Sigma a^3 = -p_1^3 + 3p_1 p_2 - 3p_3$.

The equation whose roots are $\dfrac{1}{a}; \dfrac{1}{b}, \dfrac{1}{c}, \ldots$ is
$$p_n x^n + p_{n-1}x^{n-1} + p_{n-2}x^{n-2} + \ldots = 0;$$

$$\therefore \frac{1}{a} + \frac{1}{b} + \frac{1}{c} + \ldots = -\frac{p_{n-1}}{p_n};$$

$$\therefore (a^2 + b^2 + c^2 + \ldots)\left(\frac{1}{a} + \frac{1}{b} + \frac{1}{c} + \ldots\right) = -\frac{p_{n-1}}{p_n}(p_1^2 - 2p_2)$$

that is, $-p_1 + \Sigma\dfrac{a^2}{b} = -\dfrac{p_{n-1}}{p_n}(p_1^2 - 2p_2)$

§278. Separate $\dfrac{1 + 2x}{1 - x^3}$ into its partial fractions; thus

$$\frac{1 + 2x}{1 - x^3} = \frac{1}{1 - x} + \frac{x}{1 + x + x^2} = (1 - x)^{-1} + x(1 + x + x^2)^{-1}$$

$$= (1 - x)^{-1} + \frac{x}{1 + x}\left(1 + \frac{x^2}{1 + x^2}\right)^{-1} \tag{103}$$

$$= \left\{1 + x + x^2 + x^3 + \ldots\right\}$$

$$+ \frac{x}{1 + x} - \frac{x^3}{(1 + x)^2} + \frac{x^5}{(1 + x)^3} - \ldots$$

Again $\dfrac{1 + 2x}{1 - x^3} = (1 + 2x)(1 + x^3 + x^6 + x^9 + \ldots)$ \tag{104}

In this last expansion every term is of the form x^{3n} or x^{3n+1}.

If we expand each term of (103) we shall have
$$1 + x + x^2 + \ldots + x^r + \ldots$$

$$+x \left\{1 - x + x^2 - \ldots + (-1)^r x^r + \ldots\right\}$$
$$-x^3 \left\{1 - 2x + 3x^2 - \ldots + (-1)^r (r+1)x^r + \ldots\right\}$$
$$+x^5 \left\{1 - 3x + 6x^2 - \ldots + (-1)^r \frac{(r+1)(r+2)}{1.2} x^r + \ldots\right\}$$

Now equate coefficients of x^{3n+2} in this expansion and (104); thus
$$0 = 1 + (-1)^{3n+1} - (-1)^{3n-1}3n + (-1)^{3n-3}\frac{(3n-2)(3n-1)}{1.2}$$
$$+(-1)^{3n-5}\frac{(3n-4)(3n-3)(3n-2)}{1.2.3} + \ldots;$$
on transposing the first term and dividing every term by
$(-1)^{3n+1}$ we get the required result.

Or thus:
By the Binomial Theorem, we see that
$$1, 3n, \frac{(3n-2)(3n-1)}{1.2}, \frac{(3n-4)(3n-3)(3n-2)}{1.2.3}, \ldots$$
are the coefficients of $x^{3n+1}, x^{3n-3}, \ldots$ in the expansions $(1-x)^{-1}, (1-x)^{-2}, (1-x)^{-4}\ldots$, respectively. Hence the sum required is equal to the coefficient of x^{3n+1} in the expansion of the series
$$\frac{1}{1-x} - \frac{x^2}{(1-x)^2} + \frac{x^4}{(1-x)^3} - \ldots,$$
and although the given expression consists only of a finite number of terms, this series may be considered to extend to infinity.

But this last expression is a G.P. whose sum
$$= \frac{1}{1-x} \div \left(1 + \frac{x^2}{1-x}\right) = \frac{1}{1-x+x^2} = \frac{1+x}{1+x^3}$$
$$= (1+x)(1 - x^3 + x^6 - x^9 + \ldots + (-1)^n x^{3n} + \ldots).$$
Thus the given series $= (-1)^n$.

§279. Let x, y denote the number of shots fired by A and B respectively; and suppose that A killed 1 bird in u shots, and B killed 1 bird in v shots; then $\frac{x}{u}$ and $\frac{y}{v}$ denote the numbers of birds killed.
Hence we have the following equations:
$$x^2 + y^2 = 2880;$$
$$xy = \frac{48xy}{uv}; \text{ that is, } uv = 48;$$
$$\frac{x}{u} + \frac{y}{v} = 10;$$
$$\frac{x}{v} - \frac{y}{u} = 5.$$
From these last two equations, we have
$$\frac{x^2 + y^2}{u} = 10x - 5y; \text{ and therefore } u(2x - y) = 576;$$
$$\frac{x^2 + y^2}{u} = 10y + 5x; \text{ and therefore } v(2y + x) = 576;$$
$$\therefore uv(2x - y)(2y + x) = 576 \times 576;$$
$$\therefore (2x - y)(2y + x) = 12 \times 576.$$
$$\therefore \frac{(2x - y)(2y + x)}{x^2 + y^2} = \frac{12 \times 576}{2880} = \frac{12}{5};$$
that is,
$$2x^2 - 15xy + 22y^2 = 0;$$
or
$$(x - 2y)(2x - 11y) = 0.$$
The equation $2x - 11y = 0$ does not lead to integral values of x and y.

Putting $x = 2y$, we have $x = 48$, $y = 24$.

Thus $\dfrac{48}{u} + \dfrac{24}{v} = 10$, and $uv = 48$; whence $u = 8$, $v = 6$.

§280. By Art. 253, we know that
$$a^3 + b^3 + c^3 > 3abc;$$
$$\therefore 2(a^3 + b^3 + c^3)^2 > 18a^2b^2c^2 \tag{105}$$
and
$$3(a^3 + b^3 + c^3)^2 > 9abc(a^3 + b^3 + c^3) \tag{106}$$
Also
$$(b^3 - c^3)^2 + (c^3 - a^3)^2 + (a^3 - b^3)^2 > 0;$$
that is,
$$a^6 + b^6 + c^6 > b^3c^3 + c^3a^3 + a^3b^3;$$
whence
$$(a^3 + b^3 + c^3)^2 > 3(b^3c^3 + c^3a^3 + a^3b^3);$$
$$\therefore 3(a^3 + b^3 + c^3)^2 > 9(b^3c^3 + c^3a^3 + a^3b^3) \tag{107}$$

From (105), (106), (107) by addition, we have
$$8(a^3 + b^3 + c^3)^2 > 9\left\{2a^2b^2c^2 + abc(a^3 + b^3 + c^3) + b^3c^3 + c^3a^3 \right.$$
$$\left. + a^3b^3\right\};$$
that is,
$$8(a^3 + b^3 + c^3)^2 > 9(a^2 + bc)(b^2 + ca)(c^2 + ab).$$

[See Solution to XXXIV b 28]

§281.

We have
$$u_n = (n + 2)u_{n-1} - 2nu_{n-2}.$$
Therefore
$$u_n - 2u_{n-1} = n(u_{n-1} - 2u_{n-2}).$$
Similarly,
$$u_{n-1} - 2u_{n-2} = (n - 1)(u_{n-2} - 2u_{n-3}),$$

....

$$u_3 - 2u_2 = 3(u_2 - 2u_1);$$

Now
$$p_1 = 2,\ q_1 = 3;\quad p_2 = 8,\ q_2 = 8;$$
hence
$$p_n - 2p_{n-1} = 2\lfloor n,\qquad q_n - 2q_{n-1} = \lfloor n;$$
$$2p_{n-1} - 2^2p_{n-2} = 2^2\lfloor n-1,\qquad 2q_{n-1} - 2^2q_{n-2} = 2\lfloor n-1;$$

....

$$2^{n-2}p_2 - 2^{n-1}p_1 = 2^{n-1}\lfloor 2,\qquad 2^{n-2}q_2 - 2^{n-1}q_1 = 2^{n-2}\lfloor 2;$$
$$2^{n-1}p_1 = 2^n\lfloor 1,\qquad 2^{n-1}q_1 = 2^{n-1}\cdot 3 = 2^{n-1}\lfloor 1 + 2^n;$$

whence by addition,
$$p_n = 2^n\lfloor 1 + 2^{n-1}\lfloor 2 + 2^{n-2}\lfloor 3 + \ldots + 2\lfloor n;$$
$$q_n = 2^n + 2^{n-1}\lfloor 1 + 2^{n-2}\lfloor 2 + \ldots + \lfloor n.$$
and therefore
$$p_n = 2q_n - 2^{n+1};$$
$$\therefore \frac{p_n}{q_n} = 2 - \frac{2^{n+1}}{q_n};\text{ and }q_n = \Sigma_0^n 2^r\lfloor n-r.$$

Now
$$\frac{q_n}{2^n} - 1 = \frac{\lfloor 1}{2} + \frac{\lfloor 2}{2^2} + \frac{\lfloor 3}{2^3} + \frac{\lfloor 4}{2^4} + \ldots.$$

If v_n denote the nth term of this series
$$v_{n-1} = \frac{\lfloor n-1}{2^{n-1}}\text{ and }v_n = \frac{\lfloor n}{2^n};$$
hence $\dfrac{v_n}{v_{n-1}} = \dfrac{n}{2}$, and the series is obviously divergent.

Thus $Lim. \dfrac{2^{n+1}}{q_n} = 0$; and therefore $Lim. \dfrac{p_n}{q_n} = 2.$

§282.

We have
$$\frac{p_{3n+3}}{q_{3n+3}} = \frac{1}{a+} \frac{1}{b+} \frac{1}{c+} \frac{p_{3n}}{q_{3n}}.$$

The first three convergence are

$$\frac{1}{a}, \quad \frac{b}{ab+1}, \quad \frac{bc+1}{abc+a+c};$$

$$\therefore \frac{p_{3n+3}}{q_{3n+3}} = \frac{(bc+1)q_{3n} + bp_{3n}}{(abc+a+c)q_{3n} + (ab+1)p_{3n}};$$

and since the convergence are in their lowest terms,

$$p_{3n+3} = bp_{3n} + (bc+1)q_{3n}.$$

§283. Let a, b, c, d taken in order be the sides of a quadrilateral in which a circle can be inscribed; then $a+c = b+d$. First consider the number of quadrilaterals which can be formed when 1 inch is taken for one of the sides.

When $a = 1$, if $c = 4$, there is only one case, namely $b = 2, d = 3$; if $c = 5$, there is also only one case, namely $b = 2, d = 4$. If $c = 6$, there are two cases, namely $b = 2, d = 5$, or $b = 3, d = 4$; similarly if $c = 7$, there are two cases. If $c = 8$, or $c = 9$, there are three cases in each instance; and it is easy to see that if $c = 2m$, or $c = 2m + 1$ there are $m - 1$ cases in each instance. Hence, when one of the sides is 1 inch, the number of quadrilaterals is

$$2\{1 + 2 + 3 + \ldots + (m-2)\} + (m-1), \text{ or } (m-1)^2 \text{ if } n = 2m;$$

and $\qquad 2\{1 + 2 + 3 + \ldots + (m-2)\}, \text{ or } m(m-1), \text{ if } n = 2m + 1.$

(1) Suppose $n = 2m$.

We have seen that if one of the sides is 1, the number of quadrilaterals is $(m-1)^2$.

If one of the sides is 2, the number of quadrilaterals that can be formed with the lines $2, 3, 4, \ldots, 2m$ is, in virtue of the relation

$$(a-1) + (c-1) = (b-1) + (d-1),$$

the same as the number that can be formed with the lines $1, 2, 3, \ldots, 2m - 1$ when one of th sides is 1, and is therefor equal to $(m-1)(m-2)$.

Similarly, if one of the sides is 3, the number of quadrilaterals that can be formed with the lines $3, 4, 5, \ldots, 2m$ is the same as the number that can be formed with the line $1, 2, 3, \ldots, 2m - 2$ when one of the sides is 1, and is therefore equal to $(m-2)^2$.

If one of the sides is 4, the number of quadrilaterals that can be formed with the lines $4, 5, 6, \ldots, 2m$ is $(m-2)(m-3)$; and so on.

Hence the whole number of quadrilaterals

$$= \Sigma(m-1)^2 + \Sigma(m-1)(m-2)$$

$$= \frac{1}{6}(m-1)m(2m-1) + \frac{1}{3}(m-2)(m-1)m \qquad (108)$$

$$= \frac{1}{6}(m-1)m(4m-5) = \frac{1}{24}n(n-2)(2n-5)$$

(2) Suppose $n = 2m + 1$.

If one of the sides is 1, the number of quadrilaterals that can be formed is $m(m-1)$.

As in (108) it is easy to see that the whole number of quadrilaterals

$$= \Sigma m(m-1) + \Sigma(m-1)^2$$

$$= \frac{1}{3}(m-1)m(m+1) + \frac{1}{6}(m-1)m(2m-1)$$

$$= \frac{1}{6}(m-1)m(4m+1) = \frac{1}{24}(n-3)(n-1)(2n-1) \qquad (109)$$

$$= \frac{1}{24}\{n(n-2)(2n-5) - 3\}$$

The two formulae (108) and (109) are both included in the one expression

$$\frac{1}{48}\{2n(n-2)(2n-5)-3+3(-1)^n\}.$$

[The following alternative solution is due to Professor R.S. Health, D.Sc]

Take two rectangular axes AB, AC, and mark off on them a number of successive equal lengths. Let the points of division on each line (beginning with A) be numbered $1, 2, 3 \ldots n$, and let parallels to the axes be drawn through these points.

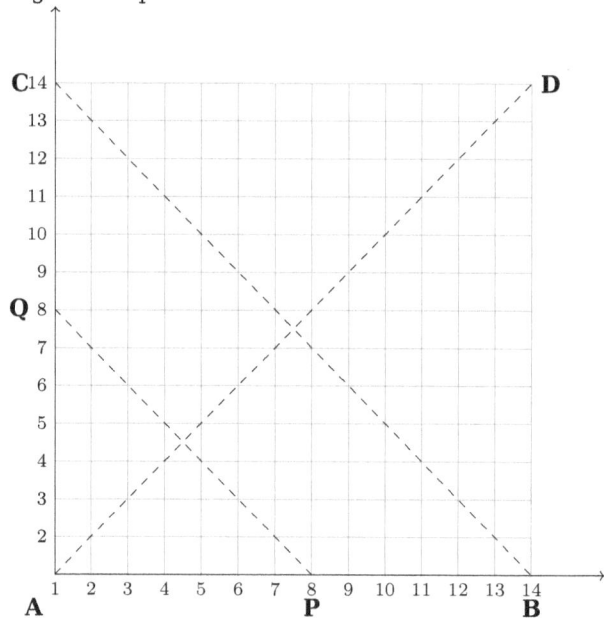

Then any point in the figure represents a combination of two of the numbers $1, 2, 3, \ldots n$. But for our purpose we must exclude the points in the diagonal AD which represents repetitions, $(1,1), (2,2), \ldots$. Now a combination of two points (xy), $(x'y')$ will be suitable if $x+y = x'+y'$; that is, we must select points from the same cross diagonal, such as PQ in the figures, since all such lines are represented by an equation of the form $x + y = constant$. But in any cross diagonal each admissible combination occurs twice over; in PQ for example, we have $(1,8)(8,1)$; $(2,7)(7,2)\ldots$; hence in any cross diagonal points must be chosen from one half of the line only.

(1) Let $n = 2m$. Begin with the central diagonal and proceed towards the point A. This diagonal has m available points, and therefore from these we have $\frac{m(m-1)}{2}$ combinations. Each of the next two diagonals contain $m - 1$ available points, and from each diagonal, we get $\frac{(m-1)(m-2)}{2}$ combinations, and so on. The diagonals $(2,2)$ and $(3,3)$ give no combinations, and therefore the last term of the series is $\frac{2 \cdot 1}{2}$, and this term like the rest occurs twice. Therefore on the whole, remembering that the same combinations also occur above the central

diagonal as we proceed towards D, the number we shall have

$$= \frac{m(m-1)}{2} + 4\left\{\frac{(m-1)(m-2)}{2} + \frac{(m-2)(m-3)}{2} + \ldots + \frac{2.1}{2}\right\}$$

$$= \frac{m(m-1)}{2} + 2\{1.2 + 2.3 + \ldots + (m-2)(m-1)\}$$

$$= \frac{m(m-1)}{2} + \frac{2}{3}(m-2)(m-1)m = \frac{1}{6}m(m-1)(4m-5).$$

(2) Let $n = 2m+1$. Then the diagonals beginning with CB and coming down towards A contain

$$m, \ m, \ m-1, \ m-1, \ \ldots 2, 2$$

available points respectively. Thus the combinationns arising from these diagonals are respectively

$$\frac{m(m-1)}{2}, \ \frac{m(m-1)}{2}, \ \frac{(m-1)(m-2)}{2}, \ \frac{(m-1)(m-2)}{2}, \ \ldots \frac{2 \cdot 1}{2}, \frac{2 \cdot 1}{2}.$$

Also as before, each series of points except the central one occurs again as we pass from CB to D. Thus the whole number of combinations

$$= \frac{m(m-1)}{2} + 2\left\{\frac{m(m-1)}{2} + 2 \cdot \frac{(m-1)(m-2)}{2} + \ldots + 2 \cdot \frac{2 \cdot 1}{2}\right\}$$

$$= \frac{1}{6}(m-1)m(4m+1) \text{ on reduction.}$$

§284. From Ex. 21 of **XXX. b** , we have

$$u_2\phi(n) = \frac{n^3}{3}\left(1 - \frac{1}{a}\right)\left(1 - \frac{1}{b}\right)\left(1 - \frac{1}{c}\right)\ldots$$

$$+ \frac{n}{6}(1-a)(1-b)(1-c)\ldots,$$

and

$$u_3\phi(n) = \frac{n^4}{4}\left(1 - \frac{1}{a}\right)\left(1 - \frac{1}{b}\right)\left(1 - \frac{1}{c}\right)\ldots$$

$$+ \frac{n^2}{4}(1-a)(1-b)(1-c)\ldots,$$

$$\therefore 6nu_2\phi(n) - 4u_3\phi(n) = n^4\left(1 - \frac{1}{a}\right)\left(1 - \frac{1}{b}\right)\left(1 - \frac{1}{c}\right)\ldots = n^3\phi(n);$$

$$\therefore n^3 - 6nu_2 + 4u_3 = 0.$$

§285. Put a, b, c for $y - z, z - x, x - y$ respectively; then $a + b + c = 0$, and we have identically

$$(1 - at)(1 - bt)(1 - ct) = 1 - qt^2 - rt^2,$$

where $q = -(bc + ca + ab)$, and $r = abc$.

Taking logarithms and equating the coefficients of t^n we have

$$\frac{1}{n}(a^n + b^n + c^n) = \text{the coefficient of } t^n \text{ in the series}$$

$$(qt^2 + rt^3) + \frac{1}{2}(qt^2 + rt^3)^2 + \frac{1}{3}(qt^2 + rt^3)^3 + \ldots$$

$=$coefficient of t^n in $t^2(q + rt) + \frac{t^4}{2}(q + rt)^2 + \frac{t^6}{6}(q + rt)^3 + \ldots$.

If $n = 6m \pm 1$, the only terms on the right which need he considered are

$$\frac{t^{4m}}{2m}(q + rt)^{2m} + \frac{t^{4m+2}}{2m+1}(q + rt)^{2m+1} + \frac{t^{4m+4}}{2m+2}(q + rt)^{2m+2} + \ldots$$

$$+ \frac{t^{6m}}{3m}(q + rt)^{3m}.$$

By expanding the binomials in this expression, it is easily seen that the coefficient of every term which contains t^{6m-1} is divisible by qr, and the coefficient of every term which contains t^{6m-1} is divisible by q^2r

Now since $a + b + c = 0$ we have $a^2 + b^2 + c^2 = -2(ab + bc + ca)$;

that is, $(y-z)^2+(z-x)^2+(x-y)^2 = -2(ab+bc+ca)$,
or $x^2+y^2+z^2-yz-zx-xy = -(ab+bc+ca) = q$.

Therefore $(x-y)^n+(y-z)^n+(z-x)^n$ is divisible by $x^2+y^2+z^2-yz-zx-xy$, when n is of the form $6m-1$, and by $(x^2+y^2+z^2-yz-zx-xy)$ when n is of the form $6m+1$.

§286.   For the sake of convenience let us denote the quantities by $a, b, c, d, e, \ldots$; then as in Art.253, we have

$mabcdef\ldots$ to m factors $< a^m+b^m+c^m+d^m+e^m+f^m$
                                $+\ldots$ to m terms;

$macdefg\ldots$ $\ldots\ldots < a^m+c^m+d^m+e^m+f^m+g^m$
                                $+\ldots\ldots\ldots;$

$mabcfgh\ldots$ $\ldots\ldots < a^m+b^m+c^m+f^m+g^m+h^m$
                                $+\ldots\ldots\ldots;$

$\ldots\ldots\ldots\ldots\ldots\ldots\ldots\ldots\ldots\ldots\ldots\ldots\ldots\ldots$

By addition, $mP < \dfrac{\lfloor n-1}{\lfloor n-m \lfloor m-1}\,S$; for the number of times that each
of the terms $a^m, b^m, c^m, \ldots$ will appear in the sum is equal to the number of combinations of (n-1) things taken $(m-1)$ at a time.

§287.   By eliminating x^3, we obtain
$$qx^3 + 3rx + q^2 = 0;$$
that is, $$q(x^2+q) = 3rx;$$
but $$x(x^2+q) = r;$$
$$\therefore \frac{q}{x} = -3x, \quad \text{or } 3x^2+q = 0;$$
and this is the condition that the first equation should have a pair of equal roots.

If each of the equal roots is a, the third root must be $-2a$; hence
$$q = -3a^2 \text{ and } r = -2a^3.$$
Thus the second equation becomes
$$x^3 + 9ax^2 + 15a^2x - 25a^3 = 0,$$
or $$(x-a)(x^2+10ax+25a^2) = 0;$$
and its roots are a, $-5a$, $-5a$.

§288.   If $p+q+r = 0$, then we know that
$$p^4+q^4+r^4-2q^2r^2-2r^2p^2-2p^2q^2 = 0.$$
Hence from the given equation, we have
$$x^4(2a^2-3x^2)^2 + \cdots - 2y^2z^2(2a^2-3y^2)(2a^2-3z^2) + \ldots = 0;$$
or arranging in power of a,
$$4a^4(x^4+y^4+z^4-2y^2z^2-2z^2x^2-2x^2y^2)$$
$$-12a^2(x^6+y^6+z^6-y^4z^2-y^2z^4-z^4x^2-z^2x^4-x^4y^2-x^2y^4) \qquad (110)$$
$$+9(x^8+y^8+z^8-2y^4z^4-2z^4x^4-2x^4y^4) = 0.$$

Denote $x^4+y^4+z^4-2y^2z^2-2z^2x^2-2x^2y^2$ by P;
then $\Sigma x^6 - \Sigma x^4y^2 = (x^2=y^2+z^2)P + 6x^2y^2z^2 = a^2P + 6x^2y^2z^2$;
and $\Sigma x^8 - 2\Sigma y^4z^4 = (x^4+y^4+z^4+2y^2z^2+2z^2x^2+2x^2y^2)P+8x^2y^2z^2(x^2+y^2+z^2)$
$= a^4P + 8a^2x^2y^2z^2$.

Thus (110) becomes
$$= 4a^4P - 12a^2(a^2P + 6x^2y^2z^2) + 9(a^4P + 8a^2x^2y^2z^2) = a^4P.$$

Thus $P = 0$; but
$$P = -(x+y+z)(-x+y+z)(x-y+z)(x+y-z).$$

§289. The equation
$$\frac{x_1}{\theta - b_1} + \frac{x_2}{\theta - b} + \cdots + \frac{x_n}{\theta - b_n} = 1 - \frac{(\theta - a_1)(\theta - a_2)\ldots(\theta - a_n)}{(\theta - b_1)(\theta - b_2)\ldots(\theta - b_n)},$$
when cleared of fractions is of the $(n-1)^{th}$ degree in θ; and in virtue of the given equations it is satisfied by the n values $a_1, a_2, a_3 \ldots a_n$; hence it must be an identity. [Art. 310]

Multiply each side by $\theta - b_1$, and then put $\theta = b_1$; thus
$$x_1 = -\frac{(b_1 - a_1)(b_1 - a_2)\ldots(b_1 - a_n)}{(b_1 - b_2)(b_1 - b_3)\ldots(b_1 - b_n)}.$$

This example is an extension of Art. 586.

§290. As in the example of Art. 498, the determinant of the left-hand side is the square of the determinant
$$\begin{vmatrix} x & y & z \\ y & z & x \\ z & x & y \end{vmatrix};$$
for its constituents are the minors of the several constituents of this determinant.

But the square of this determinant, formed according to the meth-od explained in Art. 498, is the determinant on the right-hand side.

§291. Suppose that A, B, C could do in one day fractions of the work represented by u, v, w respectively, and that they worked for x, y, z days respectively.

Then we have the following equations:
$$ux + vy + wz = 1; \qquad 40(u + v) = 1;$$
$$2vy + 2wz = 1; \qquad (u + v + w)y = 1;$$
$$\frac{2}{3}ux + 4wz = 1; \qquad x - y : x - z = 3 : 5.$$

From the first three equations, we have
$$ux - vy - wz = 0,$$
$$ux + 3vy - 9wz = 0;$$
whence
$$\frac{ux}{3} = \frac{vy}{2} = \frac{wz}{1}.$$

Subtracting the fifth equation from the first, we have
$$u(x - y) + w(z - y) = 0;$$
but
$$\frac{u}{3z} = \frac{w}{x};$$
hence
$$3z(x - y) + x(z - y) = 0,$$
or
$$3yz - 4zx + xy = 0.$$
Again
$$5(x - y) = 3(x - z),$$
or
$$2x = 5y - 3z;$$
also
$$3yz = x(4z - y);$$
hence
$$6yz = (5y - 3z)(4z - y);$$
that is,
$$5y^2 - 17yz + 12z^2 = 0,$$
or
$$(y - z)(5y - 12z) = 0.$$

The root $y - z$ must be rejected because of the equation
$$5(x - y) = 3(x - z).$$
Hence $5y = 12z$, and therefore $2x = 9z$.
thus
$$\frac{x}{45} = \frac{y}{24} = \frac{z}{10};$$
also
$$\frac{ux}{3} = \frac{vy}{2} = \frac{wz}{1},$$
so that
$$15u = 12v = 10w.$$

Substituting in $(u + v)40 = 1$, we find
$$u = \frac{1}{90}, v = \frac{1}{72}, w = \frac{1}{60}.$$
Substituting in $(u+v+w)y = 1$, we obtain $y = 24$; and therefore $x = 45, z = 10$.

§292. Here S_r is the coefficient of a^r in the expansion of
$$(1 + a)(1 + ax)\ldots(1 + ax^{n-1}).$$
Thus
$$(1 + a)(1 + ax)\ldots(1 + ax^{n-1}) = 1 + S_1 a + S_2 a^2 + \ldots + S_r a^r + \ldots.$$
Write ax for a, then
$$(1 + ax)(1 + ax^2)\ldots(1 + ax^n) = 1 + S_1 ax + S_2 a^2 x^2 + \ldots + S_r a^r x^r + \ldots;$$
$$\therefore (1 + ax^n)\{1 + S_1 a + S_2 a^2 + \ldots + S_r a^r + \ldots\}$$
$$= (1 + a)\{1 + S_1 ax + S_2 a^2 x^2 + \ldots + S_r a^r x^r \ldots\}.$$
Equate coefficients of a^{n-r}; thus
$$S_{n-r} + x^n S_{n-r-1} = x^{n-r} S_{n-r} + x^{n-r-1} S_{n-r-1}.$$
$$\therefore (1 - x^{n-r})S_{n-r} = (1 - x^{r+1})x^{n-r-1} S_{n-r-1}$$
Write $r + 1$ for r, then
$$(1 - x^{n-r-1})S_{n-r-1} = (1 - x^{r+2})x^{n-r-2} S_{n-r-2};$$
$$(1 - x^{n-r-2})S_{n-r-2} = (1 - x^{r+3})x^{n-r-3} S_{n-r-3};$$
$$\ldots\ldots\ldots\ldots\ldots = \ldots\ldots\ldots\ldots\ldots$$
$$(1 - x^{r+1})S_{r+1} = (1 - x^{n-r})x^r S_r.$$

Multiply these results together; then, since the product of the binomial factors is the same on each side, we have
$$S_{n-r} = x^r x^{r+1} \ldots x^{n-r-1} S_r$$
$$= S_r x^{r+(r+1)+\ldots+(n-r-1)}$$
$$= S_r x^{\frac{1}{2}(n-1)(n-2r)}.$$

§293. If a, b, c are not integers, we can find an integer m which will make ma, mb, mc integral.

The expression $\left(1 + \dfrac{b - c}{a}\right)^{ma} \left(1 + \dfrac{c - a}{b}\right)^{mb} \left(1 + \dfrac{a - b}{c}\right)^{mc}$ is the product of $ma + mb + mc$ positive factors, since the sum of any two of the quantities a, b, c is grater than the third.

The arithmetic mean of these factors is
$$\frac{(ma + b - c) + (mb + c - a) + (mc + a - b)}{ma + mb + mc}.$$
and is therefore equal to unity.

Hence the above expression is less than $1^{ma+mb+mc}$, or unity. [Art.253]

§294. (1) The given expression
$$= (2b^2 c^2 + 2c^2 a^2 + 2a^2 b^2 - a^4 - b^4 - c^4)(a^2 + b^2 + c^2) - 8a^2 b^2 c^2$$
$$= a^4(b^2 + c^2) + b^4(c^2 + a^2) + c^4(a^2 + b^2) - a^6 - b^6 - c^6 - 2a^2 b^2 c^2$$
$$= (b^2 + c^2 - a^2)(c^2 + a^2 - b^2)(a^2 + b^2 - c^2).$$

(2) We have
$$(x + y + z)^4 + (-x + y + z)^4 = 2\{x^4 + 6x^2(y + z^2) + (y + z)^4\};$$
and $(x - y + z)^4 + (x + y - z)^4 = 2\{x^4 + 6x^2(y - z)^2 + (y - z)^4\}.$
$$\therefore (x + y + z)^4 + (-x + y + z)^4 + (x - y + z)^4 + (x + y - z)^4$$
$$= 4(x^4 + y^4 + z^4 + 6y^2 z^2 + 6z^2 x^2 + 6x^2 y^2).$$
By putting $x = \beta + \gamma$, $y = \gamma + \alpha$, $z = \alpha + \beta$ and dividing throughout by 4, we obtain the required result.

§295. The require sum is equal to the coefficient of x^r in the product of the series

$$1 + x + x^2 + \ldots + x^r + \ldots,$$
$$1 + 2x + 2^2 x^2 + \ldots + 2^r x^r + \ldots,$$
$$1 + 3x + 3^2 x^2 + \ldots + 3^r x^r + \ldots,$$
$$\ldots\ldots\ldots\ldots\ldots\ldots\ldots\ldots\ldots\ldots\ldots$$

and therefore to the coefficient of x^r in the expression

$$\frac{1}{1-x} \cdot \frac{1}{1-2x} \cdot \frac{1}{1-3x} \cdot \frac{1}{1-nx}.$$

Let $\dfrac{1}{(1-x)(1-2x)(1-3x)\ldots(1-nx)} = \dfrac{A}{1-x} + \dfrac{B}{1-2x} + \dfrac{C}{1-3x} + \ldots;$

then by the theory of Partial Fractions, we find $A = \dfrac{(-1)^{n-1}}{(n-1)!}$;

$$B = \frac{(-1)^{n-2} 2^{n-1}}{(n-2)!} = \frac{(-1)^{n-2}(n-1)2^{n-1}}{(n-1)!};$$

$$C = \frac{(-1)^{n-3} 3^{n-1}}{2!(n-3)!} = (-1)^{n-3} \frac{(n-1)(n-2)3^{n-1}}{2!(n-1)!}; \text{ and so on.}$$

Hence the sum required is equal to the coefficient of x^r in

$$\frac{(-1)^{n-1}}{(n-1)!} \left\{ \frac{1}{1-x} - \frac{(n-1)2^{n-1}}{1-2x} + \frac{(n-1)(n-2)3^{n-1}}{2!} \cdot \frac{1}{1-3x} - \ldots \right\};$$

whence the result easily follows.

§296. The given expression is equal to

$$1 - 3n \left\{ 1 - \frac{3n-3}{1 \cdot 2} + \frac{(3n-4)(3n-5)}{1 \cdot 2 \cdot 3} - \ldots \right\}.$$

Now $1, \dfrac{3n-3}{1 \cdot 2}, \dfrac{(3n-5)(3n-4)}{1 \cdot 2 \cdot 3}, \ldots$ are respectively the coefficients of x^{3n-2},

$x^{3n-4}, x^{3n-6}, \ldots$ in the expansion of $(1-x)^{-1}, \dfrac{(1-x)^{-2}}{2}, \dfrac{(1-x)^{-3}}{3}, \ldots$

Thus $1 - \dfrac{3n-3}{1 \cdot 2} + \dfrac{(3n-4)(3n-5)}{1 \cdot 2 \cdot 3} - \ldots$

= the coefficient of x^{3n-2} in $\dfrac{1}{1-x} - \dfrac{x^2}{2(1-x)^2} + \dfrac{x^4}{3(1-x)^3} - \ldots$

This series is the expansion of

$$\frac{1}{x^2} \log\left(1 + \frac{x^2}{1-x}\right), \text{ or } \frac{1}{x^2} \log \frac{1+x^2}{1-x^2}.$$

Therefore the required series

$$= 1 - 3n \left\{ \text{ the coefficient of } x^{3n} \text{ in } \log \frac{1+x^3}{1-x^2} \right\}.$$

Now $\log \dfrac{1+x^3}{1-x^2} = x^3 - \dfrac{x^6}{2} + \dfrac{x^9}{3} - \ldots + (-1)^{p-1} \dfrac{x^{3p}}{p} + \ldots$

$+ \left(x^2 + \dfrac{x^4}{2} + \dfrac{x^6}{3} + \ldots + \dfrac{x^{2p}}{p} + \ldots \right).$

If n is odd, the coefficient of x^{3n} is $(-1)^{n-1}\dfrac{1}{n}$, or $\dfrac{1}{n}$.

if n is even, the coefficient of x^{3n} is $(-1)^{n-1}\dfrac{1}{n} + \dfrac{\frac{2}{n}}{3n}$, or $-\dfrac{1}{3n}$.

Thus the value of the required series is

$$1 - 3n \left(\frac{1}{n} \right), \text{ or } 1 - 3n \left(\frac{1}{3n} \right),$$

according as n is odd or even.

These expressions are equal to $-2, +2$ respectively. Thus in each case the series is equal to $2(-1)^n$.

§297. We have $2a - x = \dfrac{b^2}{u}$, so that $x = 2a - \dfrac{b^2}{u}$;

similarly $u = 2a - \dfrac{b^2}{z}$; $z = 2a - \dfrac{b^2}{y}$; $y = 2a - \dfrac{b^2}{x}$.

Hence $x = 2a - \dfrac{b^2}{2a-} \dfrac{b^2}{2a-} \dfrac{b^2}{2a-} \dfrac{b^2}{x}$.

As in Art. 438, or as in XXXI. a. Ex. 1, we have
$$p_n = a_n p_{n-1} - b_n p_{n-2}, \qquad q_n = a_n q_{n-1} - b_n q_{n-1}.$$
The successive convergence to the continued fraction are
$$\frac{2a}{1}, \; \frac{4a^2 - b^2}{2a}, \; \frac{8a^3 - 4ab^2}{4a^2 - b^2}, \; \frac{16a^4 - 12a^2 b^2 + b^4}{8a^3 - 4ab^2},$$
$$\frac{(16a^4 - 12a^2 b^2 + b^4)x - b^2(8a^3 - 4ab^2)}{(8a^3 - 4ab^2)x - b^2(4a^2 - b^2)}.$$
Equating this last expression to x and simplifying, we have
$$4a(2a^2 - b^2)x^2 - 8a^2(2a^2 - b^2)x + 4ab^2(2a^2 - b^2) = 0;$$
or $\qquad 4a(2a^2 - b^2)(x^2 - 2ax + b^2) = 0.$

Hence unless $2a^2 - b^2 = 0$, we have $x^2 - 2ax + b^2 = 0$. Similar equations hold for y, z, u, and therefore $x = y = z = u$.

If however $2a^2 - b^2 = 0$ the above equation is satisfied; in this case we have
$$x = 2a - \frac{2a^2}{2a-} \frac{2a^2}{2a-} \frac{2a^2}{y} = \frac{-4a^4}{2a^2 y - 4a^3};$$
that is, $x(2a - y) = 2a^2$, which is the remaining equation; hence the given equations are not independent.

§298. From the third equation, $z = -\dfrac{c}{x+y}$; hence substituting in the first two equations, we have
$$ax(x + y) = c(y + 1), \qquad by(x + y) = c(x + 1).$$
From the first of these equations, we find
$$y = \frac{ax^2 - c}{c - ax}, \text{ so that } x + y = \frac{c(x - 1)}{c - ax}.$$
On substitution in the second of the above equations, we obtain
$$b(ax^2 - c)(x - 1) - (x + 1)(c - ax^2) = 0;$$
or $\quad (ab - a^2)x^3 + Px^2 + Qx + (bc - c^2)$ say.

For the discussion of the roots it is immaterial whether a is grater or less than b; let us suppose that a is the grater. There are however two cases to consider, namely when $c > a$, and when $c < a$.

Let us tabulate the signs of the expression
$$b(ax^2 - c)(x - 1) - (x + 1)(c - ax)^2$$
for different values of x.

(1) Suppose $c > a$, so that $\dfrac{c}{a} > \sqrt{\dfrac{c}{a}} > 1$.

When $x = -\infty$, the sign is the same as that of $a^2 - ab$, and therefore is $+$; when $x = -1$, the sign is $+$; when $x = 1$, the sign is $-$;

when $\sqrt{\dfrac{c}{a}}$, the sign is $-$; when $x = \dfrac{c}{a}$, the sign is $+$;

when $x = +\infty$, the sign is $-$.

Hence there are three changes of sign, and therefor three real roots.
[If $a < b$, the expression is negative when $x = -\infty$ and positive when $x = +\infty$, and there are still three changes of sign.]

(2) Suppose $c < a$, so that $1 > \sqrt{\dfrac{c}{a}} > \dfrac{c}{a}$; then

when $x = -\infty$, the sign is $+$;

when $x = -1$, the sign is $-$;

when $x = \dfrac{c}{a}$, the sign is $+$;

when $x = \sqrt{\dfrac{c}{a}}$, the sign is $+$;

when $x = 1$, the sign is $-$;

when $x = +\infty$, the sign is $-$.

Hence as before there are three real roots.

The product of the roots is $\dfrac{bc - c^2}{ab - a^2}$ or $\dfrac{c(b - c)}{a(b - a)}$.

Similarly we can shew that y has three real vaues, and by interchanging a and b, we see that the product of these values is $\dfrac{c(a - c)}{b(a - b)};$ hence the second part of the question follows at once.

Since the values of x and y are real the values of z must be real.

§299. Denote the expression on the left by X; then

$$X = \begin{vmatrix} A & F & E \\ F & B & D \\ E & D & C \end{vmatrix} = \begin{vmatrix} ax - by - cz & bx + ay & az + cx \\ bx + ay & -ax + by - cz & bz + cy \\ az + cx & bz + cy & -ax + by - cz \end{vmatrix}.$$

Multiply both sides by $\begin{vmatrix} 1 & 0 & 0 \\ a & b & c \\ x & y & z \end{vmatrix} = bz - cy.$

Then

$$X(bz - cy) = \begin{vmatrix} ax - by - cz & bx + ay & az + cx \\ x(a^2 + b^2 + c^2) & y(a^2 + b^2 + c^2) & z(a^2 + b^2 + c^2) \\ a(x^2 + y^2 + z^2) & b(x^2 + y^2 + z^2) & c(x^2 + y^2 + z^2) \end{vmatrix}$$

$$= (a^2 + b^2 + c^2)(x^2 + y^2 + z^2) \begin{vmatrix} ax - by - cz & bx + ay & az + cx \\ x & y & z \\ a & b & c \end{vmatrix}.$$

Multiply the second row by $-a$, the third by $-x$ and add to the first; then the last determinant

$$= \begin{vmatrix} -ax - by - cz & 0 & 0 \\ x & y & z \\ a & b & c \end{vmatrix}$$

hence $(bz - cy)X = (a^2 + b^2 + c^2)(x^2 + y^2 + z^2)(ax + by + cz)(bz - cy)$; whence the result follows.

§300. Suppose that at first he walked x miles a day and worked y hours a day, and that the investigation lasted n days.

On the rth day he walked $x + r - 1$ miles and worked $y + r - 1$ hours, and therefore counted $\dfrac{1}{m}(x + r - 1)(y + r - 1)$ words; m being some constant.

Hence $\dfrac{1}{m}\{xy + (x + 1)(y + 1) + (x + 2)(y + 2) + \dots$ to n terms$\} = 232000;$

that is,

$$nxy + (x + y)(1 + 2 + 3 + \dots + \overline{n - 1}) + \{1^2 + 2^2 + 3^2 + \dots + (n - 1)^2)\} = 232000m;$$

or $nxy + \dfrac{n(n - 1)}{2}(x + y) + \dfrac{1}{6}n(n - 1)(2n - 1) = 232000m.$

But $\dfrac{1}{m}xy = 12000$; that is, $xy = 12000m.$

Therefore $\quad n + \dfrac{n(n - 1)}{2}\left(\dfrac{1}{x} + \dfrac{1}{y}\right) + \dfrac{1}{6}n(n - 1)(2n - 1)\dfrac{1}{xy} = \dfrac{116}{6}$ (111)

At the end of half the time he had counted 62000 words; therefore by changing n into $\dfrac{n}{2}$, we have

$$\frac{n}{2} + \frac{n(n-2)}{8}\left(\frac{1}{x}+\frac{1}{y}\right) + \frac{1}{24}n(n-2)(n-1)\frac{1}{xy} = \frac{31}{6} \qquad (112)$$

Multiply this equation by 2 and subtract the result from (111), then

$$\frac{n^2}{4}\left(\frac{1}{x}+\frac{1}{y}\right) + \frac{n^2(n-1)}{4}\frac{1}{xy} = 9.$$

On the last day he counted 72000 words;

therefore $\dfrac{1}{m}(x+n-1)(y+n-1) = 72000$;

that is, $xy + (n-1)xy + (n-1)^2 = 72000m$;

$\therefore\ 1 + (n-1)\left(\dfrac{1}{x}+\dfrac{1}{y}\right) + (n-1)^2\dfrac{1}{xy} = 6$;

or $(n-1)\left(\dfrac{1}{x}+\dfrac{1}{y}\right) + (n-1)^2\dfrac{1}{xy} = 5.$

Thus

$$\frac{n^2}{4}\left\{\frac{1}{x}+\frac{1}{y}+\frac{n-1}{xy}\right\} = 9;\ \text{and}\ (n-1)\left\{\frac{1}{x}+\frac{1}{y}+\frac{n-1}{xy}\right\} = 5;$$

$$\therefore\ \frac{n^2}{4(n-1)} = \frac{9}{5};$$

$\therefore\ 5n^2 - 36n + 36 = 0$, or $(5n-6)(n-6) = 0$.

Substituting $n = 6$, we find from (111) and (112)

$$15\left(\frac{1}{x}+\frac{1}{y}\right) + \frac{55}{xy} = \frac{40}{3};\ 3\left(\frac{1}{x}+\frac{1}{y}\right) + \frac{5}{xy} = \frac{13}{6};$$

whence $\dfrac{1}{x}+\dfrac{1}{y} = \dfrac{7}{12}$, $\dfrac{1}{xy} = \dfrac{1}{12}$;

that is, $x+y = 7$, $xy = 12$; or $x = 3$, $y = 4$.

The End

www.ingramcontent.com/pod-product-compliance
Lightning Source LLC
Chambersburg PA
CBHW021419170526
45164CB00001B/19